Lecture Notes in Computer Science 7826

Commenced Publication in 1973
Founding and Former Series Editors:
Gerhard Goos, Juris Hartmanis, and Jan van Leeuwen

Editorial Board

David Hutchison
  Lancaster University, UK
Takeo Kanade
  Carnegie Mellon University, Pittsburgh, PA, USA
Josef Kittler
  University of Surrey, Guildford, UK
Jon M. Kleinberg
  Cornell University, Ithaca, NY, USA
Alfred Kobsa
  University of California, Irvine, CA, USA
Friedemann Mattern
  ETH Zurich, Switzerland
John C. Mitchell
  Stanford University, CA, USA
Moni Naor
  Weizmann Institute of Science, Rehovot, Israel
Oscar Nierstrasz
  University of Bern, Switzerland
C. Pandu Rangan
  Indian Institute of Technology, Madras, India
Bernhard Steffen
  TU Dortmund University, Germany
Madhu Sudan
  Microsoft Research, Cambridge, MA, USA
Demetri Terzopoulos
  University of California, Los Angeles, CA, USA
Doug Tygar
  University of California, Berkeley, CA, USA
Gerhard Weikum
  Max Planck Institute for Informatics, Saarbruecken, Germany

Weiyi Meng   Ling Feng
Stéphane Bressan   Werner Winiwarter
Wei Song (Eds.)

Database Systems for Advanced Applications

18th International Conference, DASFAA 2013
Wuhan, China, April 22-25, 2013
Proceedings, Part II

Volume Editors

Weiyi Meng
Binghamton University. Department of Computer Science
Binghamton, NY 13902, USA
E-mail: meng@binghamton.edu

Ling Feng
Tsinghua University, Department of Computer Science and Technology
100084 Beijing, China
E-mail: fengling@tsinghua.edu.cn

Stéphane Bressan
National University of Singapore, Department of Computer Science
117417 Singapore
E-mail: steph@nus.edu.sg

Werner Winiwarter
University of Vienna, Research Group Data Analytics and Computing
1090 Vienna, Austria
E-mail: werner.winiwarter@univie.ac.at

Wei Song
Wuhan University, School of Computer Science
430072 Wuhan, China
E-mail: songwei@whu.edu.cn

ISSN 0302-9743 ISSN 1611-3349
ISBN 978-3-642-37449-4 ISBN 978-3-642-37450-0 (eBook)
DOI 10.1007/978-3-642-37450-0
Springer Heidelberg Dordrecht London New York

Library of Congress Control Number: 2013934238
CR Subject Classification (1998): H.2-5, C.2, J.1, J.3

LNCS Sublibrary: SL 3 – Information Systems and Application,
incl. Internet/Web and HCI

© Springer-Verlag Berlin Heidelberg 2013

This work is subject to copyright. All rights are reserved, whether the whole or part of the material is concerned, specifically the rights of translation, reprinting, re-use of illustrations, recitation, broadcasting, reproduction on microfilms or in any other way, and storage in data banks. Duplication of this publication or parts thereof is permitted only under the provisions of the German Copyright Law of September 9, 1965, in its current version, and permission for use must always be obtained from Springer. Violations are liable to prosecution under the German Copyright Law.
The use of general descriptive names, registered names, trademarks, etc. in this publication does not imply, even in the absence of a specific statement, that such names are exempt from the relevant protective laws and regulations and therefore free for general use.

Typesetting: Camera-ready by author, data conversion by Scientific Publishing Services, Chennai, India

Printed on acid-free paper

Springer is part of Springer Science+Business Media (www.springer.com)

Preface

It is our great pleasure to present to you the proceedings of the 18th International Conference on Database Systems for Advanced Applications (DASFAA 2013), which was held in Wuhan, China, in April 2013. DASFAA is a well-established international conference series that provides a forum for technical presentations and discussions among researchers, developers, and users from academia, business, and industry in the general areas of database systems, Web information systems, and their applications.

The call for papers attracted 208 submissions of research papers from 28 countries (based on the affiliation of the first author). After a comprehensive review process, the Program Committee selected 51 regular research papers and 10 short research papers for presentation. The acceptance rate for regular research papers is less than 25%. The conference program also included the presentations of three industrial papers selected by the Industrial Committee chaired by Haixun Wang and Haruo Yokota, and nine demo presentations selected from 19 submissions by the Demo Committee chaired by Hong Gao and Jianliang Xu.

The proceedings also include the extended abstracts of the two invited keynote lectures by internationally known researchers, Katsumi Tanaka (Kyoto University, Japan) and Peter M.G. Apers (University of Twente, The Netherlands), whose topics are on "Can We Predict User Intents from Queries? Intent Discovery for Web Search" and "Data Overload: What Can We Do?", respectively. In addition, an invited paper contributed by the authors of the DASFAA 10-year Best Paper Award winner for the year 2013, Chen Li, Sharad Mehrotra, and Liang Jin, is included. The title of this paper is "Record Linkage: A 10-Year Retrospective." The Tutorial Chairs, Jian Pei and Ge Yu, organized four tutorials given by leading experts on a wide range of topics. The titles and speakers of these tutorials are "Behavior-Driven Social Network Mining and Analysis" by Ee-Peng Lim, Feida Zhu, and Freddy Chua, "Understanding Short Texts" by Haixun Wang, "Managing the Wisdom of Crowds on Social Media Services" by Lei Chen, and "Ranking Multi-valued Objects in a Multi-dimensional Space" by Wenjie Zhang, Ying Zhang, and Xuemin Lin. The Panel Chairs, Aoying Zhou and Jeffrey Xu Yu, organized a stimulating panel on big data research. The panel was chaired by Xiaoyang Sean Wang. This rich and attractive conference program of DASFAA 2013 is published in two volumes of Springer's *Lecture Notes in Computer Science* series.

Beyond the main conference, Bonghee Hong, Xiaofeng Meng, and Lei Chen, who chaired the Workshop Committee, put together three exciting workshops (International DASFAA Workshop on Big Data Management and Analytics, International Workshop on Social Networks and Social Web Mining, and International Workshop on Semantic Computing and Personalization). The workshop papers are included in a separate volume of proceedings also published by Springer in its *Lecture Notes in Computer Science* series.

DASFAA 2013 was primarily sponsored and hosted by Wuhan University of China. It also received sponsorship from the National Natural Science Foundation of China (NSFC), the Database Society of the China Computer Federation (CCF DBS), and the State Key Laboratory of Software Engineering of China (SKLSE). We are grateful to these sponsors for their support and contribution, which were essential in making DASFAA 2013 successful.

The conference would not have been possible without the support and hard work of many colleagues. We would like to express our gratitude to Honorary Conference Chairs, Lizhu Zhou and Yanxiang He, for their valuable advice on all aspects of organizing the conference. Our special thanks also go to the DASFAA Steering Committee for their leadership and encouragement. We are also grateful to the following individuals for their contributions to making the conference a success: the General Co-chairs, Jianzhong Li, Zhiyong Peng and Qing Li, Publicity Co-chairs, Jun Yang, Xiaoyong Du and Satoshi Oyama, Local Arrangements Committee Chair, Tieyun Qian, Finance Co-chair, Howard Leung and Liwei Wang, Web Chair, Liang Hong, Best Paper Committee Co-chairs, Changjie Tang, Hiroyuki Kitagawa and Sang-goo Lee, Registration Chair, Yunwei Peng, Steering Committee Liaison, Rao Kotagiri, APWEB Liaison, Xueming Lin, WAIM Liaison, Guoren Wang, WISE Liaison, Yanchun Zhang, and CCF DBS Liaison, Zhanhuai Li.

Our heartfelt thanks go to all the Program Committee members and external reviewers for reviewing all submitted manuscripts carefully and timely. We also thank all authors for submitting their papers to this conference. Finally, we thank all other individuals and volunteers who helped make the conference program attractive and the conference successful.

April 2013

Weiyi Meng
Ling Feng
Stéphane Bressan
Werner Winiwarter
Wei Song

Organization

Honorary Conference Co-chairs

Lizhu Zhou   Tsinghua University, China
Yanxiang He   Wuhan University, China

Conference General Co-chairs

Jianzhong Li   Harbin Institute of Technology, China
Zhiyong Peng   Wuhan University, China
Qing Li   City University of Hong Kong, China

Program Committee Co-chairs

Weiyi Meng   Binghamton University, USA
Ling Feng   Tsinghua University, China
Stéphane Bressan   National University of Singapore, Singapore

Workshop Co-chairs

Bonghee Hong   Pusan National University, South Korea
Xiaofeng Meng   Renmin University, China
Lei Chen   Hong Kong University of Science and
                             Technology, China

Tutorial Co-chairs

Ge Yu   Northeastern University, China
Jian Pei   Simon Fraser University, Canada

Panel Co-chairs

Aoying Zhou   East China Normal University, China
Jeffery Xu Yu   City University of Hong Kong, China

Demo Co-chairs

Hong Gao   Harbin Institute of Technology, China
Jianliang Xu   Hong Kong Baptist University, China

Industrial Co-chairs

Haixun Wang — Microsoft Research Asia, China
Haruo Yokota — Tokyo Institute of Technology, Japan

Best Paper Committee Co-chairs

Changjie Tang — Sichuan University, China
Hiroyuki Kitagawa — University of Tsukuba, Japan
Sang-goo Lee — Seoul National University, South Korea

Publicity Co-chairs

Jun Yang — Duke University, USA
Xiaoyong Du — Renmin University, China
Satoshi Oyama — Hokkaido University, Japan

Publication Co-chairs

Werner Winiwarter — University of Vienna, Austria
Wei Song — Wuhan University, China

Local Arrangements Chair

Tieyun Qian — Wuhan University, China

Finance Co-chairs

Howard Leung — City University of Hong Kong, China
Liwei Wang — Wuhan University, China

Registration Chair

Yuwei Peng — Wuhan University, China

Web Chair

Liang Hong — Wuhan University, China

Steering Committee Liaison

Rao Kotagiri — University of Melbourne, Australia

WAIM Liaison

Guoren Wang — Northeastern University, China

APWEB Liaison

Xueming Lin   University of New South Wales, Australia

WISE Liaison

Yanchun Zhang   Victoria University, Australia

CCF DBS Liaison

Zhanhuai Li   Northwestern Polytechnical University, China

Program Committees

Research Track

Toshiyuki Amagasa   University of Tsukuba, Japan
Masayoshi Aritsugi   Kumamoto University, Japan
Zhifeng Bao   National University of Singapore, Singapore
Ladjel Bellatreche   Poitiers University, France
Boualem Benatallah   University of New South Wales, Australia
Sourav S Bhowmick   Nanyang Technological University, Singapore
Chee Yong Chan   National University of Singapore, Singapore
Jae Woo Chang   Chonbuk National University, South Korea
Ming-Syan Chen   National Taiwan University, Taiwan
Hong Cheng   Chinese University of Hong Kong, China
James Cheng   Nanyang Technological University, Singapore
Reynold Cheng   University of Hong Kong, China
Byron Choi   Hong Kong Baptist University, China
Yon Dohn Chung   Korea University, South Korea
Gao Cong   Nanyang Technological University, Singapore
Bin Cui   Peking University, China
Alfredo Cuzzocrea   Institute of High Performance Computing and
                           Networking of the Italian National
                               Research Council, Italy
Gill Dobbie   University of Auckland, New Zealand
Eduard C. Dragut   Purdue University, USA
Xiaoyong Du   Renmin University, China
Jianhua Feng   Tsinghua University, China
Jianlin Feng   Sun Yat-Sen University, China
Yunjun Gao   Zhejiang University, China
Wook-Shin Han   Kyung-Pook National University, South Korea
Takahiro Hara   Osaka University, Japan
Bingsheng He   Nanyang Technological University, Singapore
Wynne Hsu   National University of Singapore, Singapore
Haibo Hu   Hong Kong Baptist University, China

Yoshiharu Ishikawa	Nagoya University, Japan
Adam Jatowt	Kyoto University, Japan
Yiping Ke	Institute of High Performance Computing, A*STAR, Singapore
Sang Wook Kim	Hanyang University, South Korea
Young-Kuk Kim	Chungnam National University, South Korea
Hiroyuki Kitagawa	University of Tsukuba, Japan
Hady W. Lauw	Singapore Management University, Singapore
Mong Li Lee	National University of Singapore, Singapore
Sang-goo Lee	Seoul National University, South Korea
Wang-Chien Lee	Pennsylvania State University, USA
Hong-va Leong	Hong Kong Polytechnic University, China
Cuiping Li	Renmin University, China
Guohui Li	Huazhong University of Science and Technology, China
Xiang Li	Nanjing University, China
Xuemin Lin	University of New South Wales, Australia
Jan Lindstrom	IBM Helsinki Lab, Finland
Chengfei Liu	Swinburne University of Technology, Australia
Eric Lo	Hong Kong Polytechnic University, China
Jiaheng Lu	Renmin University, China
Nikos Mamoulis	University of Hong Kong, China
Shicong Meng	IBM Thomas J. Watson Research Center, USA
Xiaofeng Meng	Renmin University, China
Yang-Sae Moon	Kangwon National University, South Korea
Yasuhiko Morimoto	Hiroshima University, Japan
Miyuki Nakano	University of Tokyo, Japan
Vincent T. Y. Ng	Hong Kong Polytechnic University, China
Wilfred Ng	Hong Kong University of Science and Technology, China
Katayama Norio	National Institute of Informatics, Japan
Makoto Onizuka	NTT Cyber Space Laboratories, NTT Corporation, Japan
Sang Hyun Park	Yonsei Universiy, South Korea
Uwe Röhm	University of Sydney, Australia
Ning Ruan	Kent State University, USA
Markus Schneider	University of Florida, USA
Heng Tao Shen	University of Queensland, Australia
Hyoseop Shin	Konkuk University, South Korea
Atsuhiro Takasu	National Institute of Informatics, Japan
Kian-Lee Tan	National University of Singapore, Singapore
Changjie Tang	Sichuan University, China
Jie Tang	Tsinghua University, China
Yong Tang	South China Normal University, China
David Taniar	Monash University, Australia
Vincent S. Tseng	National Cheng Kung University, Taiwan

Vasilis Vassalos	Athens University of Economics and Business, Greece
Guoren Wang	Northeastern University, China
Jianyong Wang	Tsinghua University, China
John Wang	Griffith University, Australia
Wei Wang	University of New South Wales, Australia
Chi-Wing Wong	Hong Kong University of Science and Technology, China
Huayu Wu	Singapore's Institute for Infocomm Research (I2R), Singapore
Xiaokui Xiao	Nanyang Technological University, Singapore
Jianliang Xu	Hong Kong Baptist University, China
Man-Lung Yiu	Hong Kong Polytechnic University, China
Haruo Yokota	Tokyo Institute of Technology, Japan
Jae Soo Yoo	Chungbuk National University, South Korea
Ge Yu	Northeastern University, China
Jeffrey X. Yu	Chinese University of Hong Kong, China
Qi Yu	Rochester Institute of Technology, USA
Zhongfei Zhang	Binghamton University, USA
Rui Zhang	University of Melbourne, Australia
Wenjie Zhang	The University of New South Wales, Australia
Yanchun Zhang	Victoria University, Australia
Baihua Zheng	Singapore Management University, Singapore
Kai Zheng	University of Queensland, Australia
Aoying Zhou	East China Normal University, China
Lei Zou	Peking University, China

Industrial Track

Bin Yao	Shanghai Jiaotong University, China
Chiemi Watanabe	Ochanomizu University, Japan
Jun Miyazaki	Nara Institute of Science and Technology, Japan
Kun-Ta Chuang	National Cheng Kung University, South Korea
Seung-won Hwang	POSTECH, South Korea
Wexing Liang	Dalian University of Technology, China
Ying Yan	Microsoft, USA

Demo Track

Aixin Sun	Nanyang Technological University, Singapore
Chaokun Wang	Tsinghua University, China
Christoph Lofi	National Institute of Informatics, Japan
De-Nian Yang	Institute of Information Science, Academia Sinica, Taiwan

Feida Zhu — Singapore Management University, Singapore
Feifei Li — University of Utah, USA
Guoliang Li — Tsinghua University, China
Ilaria Bartolini — University of Bologna, Italy
Jianliang Xu — Hong Kong Baptist University, China
Jin-ho Kim — Kangwon National University, South Korea
Lipyeow Lim — University of Hawaii at Manoa, USA
Peiquan Jin — USTC, China
Roger Zimmermann — National University of Singapore, Singapore
Shuigeng Zhou — Fudan University, China
Weining Qian — East China Normal University, China
Wen-Chih Peng — National Chiao Tung University, Taiwan
Yaokai Feng — Kyushu University, Japan
Yin Yang — Advanced Digital Sciences Center, Singapore
Zhanhuai Li — Northwestern Polytechnical University, China
Zhaonian Zou — Harbin Institute of Technology, China

External Reviewers

Shafiq Alam
Duck-Ho Bae
Sebastian Bre
Xin Cao
Alvin Chan
Chen Chen
Lisi Chen
Shumo Chu
Xiang Ci
Zhi Dou
Juan Du
Qiong Fang
Wei Feng
Lizhen Fu
Xi Guo
Zhouzhou He
Jin Huang
Min-Hee Jang
Di Jiang
Yexi Jiang
Akimitsu Kanzaki
Romans Kaspeovics
Selma Khouri
Sang-Chul Lee
Sangkeun Lee
Jianxin Li
Lu Li

Sheng Li
Yingming Li
Yong Li
Bangyong Liang
Bo Liu
Cheng Long
Yifei Lu
Lydia M
Youzhong Ma
Silviu Maniu
Jason Meng
Sofian Maabout
Takeshi Misihma
Luyi Mo
Jaeseok Myung
Sungchan Park
Peng Peng
Yun Peng
Yinian Qi
Jianbin Qin
Chuitian Rong
Wei Shen
Hiroaki Shiokawa
Matthew Sladescu
Zhenhua Song
Yifang Sun
Jian Tan

Wei Tan
Ba Quan Truong
Jan Vosecky
Guoping Wang
Liaoruo Wang
Lu Wang
Yousuke Watanabe
Chuan Xiao
Yi Xu
Zhiqiang Xu
Kefeng Xuan
Da Yan
ByoungJu Yang
Xuan Yang
Liang Yao
Jongheum Yeon
Jianhua Yin
Wei Zhang
Xiaojian Zhang
Yutao Zhang
Geng Zhao
Pin Zhao
Xueyi Zhao
Zhou Zhao
Rui Zhou
Xiaoling Zhou
Qijun Zhu

Table of Contents – Part II

Graph Data Management I

Shortest Path Computation over Disk-Resident Large Graphs Based on Extended Bulk Synchronous Parallel Methods 1
  Zhigang Wang, Yu Gu, Roger Zimmermann, and Ge Yu

Fast SimRank Computation over Disk-Resident Graphs 16
  Yinglong Zhang, Cuiping Li, Hong Chen, and Likun Sheng

S-store: An Engine for Large RDF Graph Integrating Spatial Information 31
  Dong Wang, Lei Zou, Yansong Feng, Xuchuan Shen, Jilei Tian, and Dongyan Zhao

Physical Design

Physical Column Organization in In-Memory Column Stores 48
  David Schwalb, Martin Faust, Jens Krueger, and Hasso Plattner

Semantic Data Warehouse Design: From ETL to Deployment à la Carte 64
  Ladjel Bellatreche, Selma Khouri, and Nabila Berkani

A Specific Encryption Solution for Data Warehouses 84
  Ricardo Jorge Santos, Deolinda Rasteiro, Jorge Bernardino, and Marco Vieira

NameNode and DataNode Coupling for a Power-Proportional Hadoop Distributed File System 99
  Hieu Hanh Le, Satoshi Hikida, and Haruo Yokota

Knowledge Management

Mapping Entity-Attribute Web Tables to Web-Scale Knowledge Bases 108
  Xiaolu Zhang, Yueguo Chen, Jinchuan Chen, Xiaoyong Du, and Lei Zou

ServiceBase: A Programming Knowledge-Base for Service Oriented Development 123
  Moshe Chai Barukh and Boualem Benatallah

On Leveraging Crowdsourcing Techniques for Schema Matching
Networks 139
 Nguyen Quoc Viet Hung, Nguyen Thanh Tam, Zoltán Miklós, and
 Karl Aberer

MFSV: A Truthfulness Determination Approach for Fact Statements ... 155
 Teng Wang, Qing Zhu, and Shan Wang

Temporal Data Management

A Mechanism for Stream Program Performance Recovery in Resource
Limited Compute Clusters 164
 Miyuru Dayarathna and Toyotaro Suzumura

Event Relationship Analysis for Temporal Event Search.............. 179
 Yi Cai, Qing Li, Haoran Xie, Tao Wang, and Huaqing Min

Social Networks II

Dynamic Label Propagation in Social Networks..................... 194
 Juan Du, Feida Zhu, and Ee-Peng Lim

MAKM: A MAFIA-Based k-Means Algorithm for Short Text in Social
Networks 210
 Pengfei Ma and Yong Zhang

Detecting User Preference on Microblog........................... 219
 Chen Xu, Minqi Zhou, Feng Chen, and Aoying Zhou

Query Processing II

MUSTBLEND: Blending Visual Multi-Source Twig Query Formulation
and Query Processing in RDBMS 228
 Ba Quan Truong and Sourav S Bhowmick

Efficient SPARQL Query Evaluation via Automatic Data
Partitioning 244
 Tao Yang, Jinchuan Chen, Xiaoyan Wang, Yueguo Chen, and
 Xiaoyong Du

Content Based Retrieval for Lunar Exploration Image Databases....... 259
 Hui-zhong Chen, Ning Jing, Jun Wang, Yong-guang Chen, and
 Luo Chen

Searching Desktop Files Based on Access Logs 267
 Yukun Li, Xiyan Zhao, Yingyuan Xiao, and Xiaoye Wang

An In-Memory/GPGPU Approach to Query Processing for
Aspect-Oriented Data Management 275
    Bernhard Pietsch

Graph Data Management II

On Efficient Graph Substructure Selection 284
    Xiang Zhao, Haichuan Shang, Wenjie Zhang, Xuemin Lin, and
    Weidong Xiao

Parallel Triangle Counting over Large Graphs 301
    Wenan Wang, Yu Gu, Zhigang Wang, and Ge Yu

Data Mining

Document Summarization via Self-Present Sentence Relevance Model... 309
    Xiaodong Li, Shanfeng Zhu, Haoran Xie, and Qing Li

Active Semi-supervised Community Detection Algorithm with Label
Propagation 324
    Mingwei Leng, Yukai Yao, Jianjun Cheng, Weiming Lv, and
    Xiaoyun Chen

Computing the Split Points for Learning Decision Tree in
MapReduce 339
    Mingdong Zhu, Derong Shen, Ge Yu, Yue Kou, and Tiezheng Nie

FP-Rank: An Effective Ranking Approach Based on Frequent Pattern
Analysis 354
    Yuanfeng Song, Kenneth Leung, Qiong Fang, and Wilfred Ng

Applications

A Hybrid Framework for Product Normalization in Online Shopping ... 370
    Li Wang, Rong Zhang, Chaofeng Sha, Xiaofeng He, and Aoying Zhou

Staffing Open Collaborative Projects Based on the Degree of
Acquaintance 385
    Mohammad Y. Allaho, Wang-Chien Lee, and De-Nian Yang

Industrial Papers

Who Will Follow Your Shop? Exploiting Multiple Information Sources
in Finding Followers 401
    Liang Wu, Alvin Chin, Guandong Xu, Liang Du, Xia Wang,
    Kangjian Meng, Yonggang Guo, and Yuanchun Zhou

Performance of Serializable Snapshot Isolation on Multicore Servers 416
  Hyungsoo Jung, Hyuck Han, Alan Fekete, Uwe Röhm, and Heon Y. Yeom

A Hybrid Approach for Relational Similarity Measurement 431
  Zhao Lu and Zhixian Yan

Demo Papers I: Data Mining

Subspace MOA: Subspace Stream Clustering Evaluation Using the MOA Framework 446
  Marwan Hassani, Yunsu Kim, and Thomas Seidl

Symbolic Trajectories in SECONDO: Pattern Matching and Rewriting 450
  Fabio Valdés, Maria Luisa Damiani, and Ralf Hartmut Güting

$ReTweet^p$: Modeling and Predicting Tweets Spread Using an Extended Susceptible-Infected- Susceptible Epidemic Model 454
  Yiping Li, Zhuonan Feng, Hao Wang, Shoubin Kong, and Ling Feng

TwiCube: A Real-Time Twitter Off-Line Community Analysis Tool 458
  Juan Du, Wei Xie, Cheng Li, Feida Zhu, and Ee-Peng Lim

Demo Papers II: Database Applications

TaskCardFinder: An Aviation Enterprise Search Engine for Bilingual MRO Task Cards 463
  Qingwei Liu, Hao Wang, Tangjian Deng, and Ling Feng

EntityManager: An Entity-Based Dirty Data Management System 468
  Hongzhi Wang, Xueli Liu, Jianzhong Li, Xing Tong, Long Yang, and Yakun Li

RelRec: A Graph-Based Triggering Object Relationship Recommender System 472
  Yuwen Dai, Guangyao Li, and Ruoyu Li

IndoorDB: Extending Oracle to Support Indoor Moving Objects Management 476
  Qianyuan Li, Peiquan Jin, Lei Zhao, Shouhong Wan, and Lihua Yue

HITCleaner: A Light-Weight Online Data Cleaning System 481
  Hongzhi Wang, Jianzhong Li, Ran Huo, Li Jia, Lian Jin, Xueying Men, and Hui Xie

Author Index 485

Table of Contents – Part I

Keynote Talks

Data Overload: What Can We Do? 1
  Peter Apers

Can We Predict User Intents from Queries? - Intent Discovery for Web Search - 2
  Katsumi Tanaka

Invited Paper from Recipients of Ten-Year Best Paper Award

Record Linkage: A 10-Year Retrospective 3
  Chen Li, Sharad Mehrotra, and Liang Jin

Social Networks I

Finding Rising Stars in Social Networks 13
  Ali Daud, Rashid Abbasi, and Faqir Muhammad

Expertise Ranking of Users in QA Community 25
  Yuanzhe Cai and Sharma Chakravarthy

Community Expansion in Social Network 41
  Yuanjun Bi, Weili Wu, and Li Wang

Query Processing I

Similarity Joins on Item Set Collections Using Zero-Suppressed Binary Decision Diagrams 56
  Yasuyuki Shirai, Hiroyuki Takashima, Koji Tsuruma, and Satoshi Oyama

Keyword-Matched Data Skyline in Peer-to-Peer Systems 71
  Khaled M. Banafaa, Ruixuan Li, Kunmei Wen, Xiwu Gu, and Yuhua Li

Adaptive Query Scheduling in Key-Value Data Stores 86
  Chen Xu, Mohamed Sharaf, Minqi Zhou, Aoying Zhou, and Xiaofang Zhou

Nearest Neighbor Search

Near-Optimal Partial Linear Scan for Nearest Neighbor Search in High-Dimensional Space 101
    Jiangtao Cui, Zi Huang, Bo Wang, and Yingfan Liu

AVR-Tree: Speeding Up the NN and ANN Queries on Location Data ... 116
    Qianlu Lin, Ying Zhang, Wenjie Zhang, and Xuemin Lin

Top-k Neighborhood Dominating Query 131
    Xike Xie, Hua Lu, Jinchuan Chen, and Shuo Shang

OptRegion: Finding Optimal Region for Bichromatic Reverse Nearest Neighbors 146
    Huaizhong Lin, Fangshu Chen, Yunjun Gao, and Dongming Lu

Index

Generalization-Based Private Indexes for Outsourced Databases 161
    Yi Tang, Fang Liu, and Liqing Huang

Distributed AH-Tree Based Index Technology for Multi-channel Wireless Data Broadcast 176
    Yongtian Yang, Xiaofeng Gao, Xin Lu, Jiaofei Zhong, and Guihai Chen

MVP Index: Towards Efficient Known-Item Search on Large Graphs.... 193
    Ming Zhong, Mengchi Liu, Zhifeng Bao, Xuhui Li, and Tieyun Qian

Indexing Reverse Top-k Queries in Two Dimensions 201
    Sean Chester, Alex Thomo, S. Venkatesh, and Sue Whitesides

Query Analysis

Beyond Click Graph: Topic Modeling for Search Engine Query Log Analysis 209
    Di Jiang, Kenneth Wai-Ting Leung, Wilfred Ng, and Hao Li

Continuous Topically Related Queries Grouping and Its Application on Interest Identification 224
    Pengfei Zhao, Kenneth Wai-Ting Leung, and Dik Lun Lee

Efficient Responsibility Analysis for Query Answers 239
    Biao Qin, Shan Wang, and Xiaoyong Du

Minimizing Explanations for Missing Answers to Queries on Databases 254
    Chuanyu Zong, Xiaochun Yang, Bin Wang, and Jingjing Zhang

XML Data Management

A Compact and Efficient Labeling Scheme for XML Documents 269
  Rung-Ren Lin, Ya-Hui Chang, and Kun-Mao Chao

Querying Semi-structured Data with Mutual Exclusion 284
  Huayu Wu, Ruiming Tang, and Tok Wang Ling

XReason: A Semantic Approach that Reasons with Patterns to Answer
XML Keyword Queries 299
  *Cem Aksoy, Aggeliki Dimitriou, Dimitri Theodoratos, and
  Xiaoying Wu*

History-Offset Implementation Scheme of XML Documents and Its
Evaluations 315
  Tatsuo Tsuji, Keita Amaki, Hiroomi Nishino, and Ken Higuchi

Privacy Protection

On the Complexity of t-Closeness Anonymization and Related
Problems 331
  Hongyu Liang and Hao Yuan

Distributed Anonymization for Multiple Data Providers in a Cloud
System 346
  Xiaofeng Ding, Qing Yu, Jiuyong Li, Jixue Liu, and Hai Jin

Subscription Privacy Protection in Topic-Based Pub/Sub 361
  *Weixiong Rao, Lei Chen, Mingxuan Yuan, Sasu Tarkoma, and
  Hong Mei*

Feel Free to Check-in: Privacy Alert against Hidden Location Inference
Attacks in GeoSNs 377
  Zheng Huo, Xiaofeng Meng, and Rui Zhang

Differentially Private Set-Valued Data Release against Incremental
Updates 392
  Xiaojian Zhang, Xiaofeng Meng, and Rui Chen

Uncertain Data Management

Consistent Query Answering Based on Repairing Inconsistent
Attributes with Nulls 407
  Jie Liu, Dan Ye, Jun Wei, Fei Huang, and Hua Zhong

On Efficient k-Skyband Query Processing over Incomplete Data 424
  *Xiaoye Miao, Yunjun Gao, Lu Chen, Gang Chen, Qing Li, and
  Tao Jiang*

Mining Frequent Patterns from Uncertain Data with MapReduce
for Big Data Analytics... 440
    Carson Kai-Sang Leung and Yaroslav Hayduk

Efficient Probabilistic Reverse k-Nearest Neighbors Query Processing
on Uncertain Data 456
    Jiajia Li, Botao Wang, and Guoren Wang

Efficient Querying of Correlated Uncertain Data with Cached
Results 472
    Jinchuan Chen, Min Zhang, Xike Xie, and Xiaoyong Du

Author Index.. 487

Shortest Path Computation over Disk-Resident Large Graphs Based on Extended Bulk Synchronous Parallel Methods

Zhigang Wang[1], Yu Gu[1], Roger Zimmermann[2], and Ge Yu[1]

[1] Northeastern University, China
wangzhigang_mail@yahoo.cn, {guyu,yuge}@ise.neu.edu.cn
[2] National University of Singapore, Singapore
rogerz@comp.nus.edu.sg

Abstract. The Single Source Shortest Path (SSSP) computation over large graphs has raised significant challenges to the memory capacity and processing efficiency. Utilizing disk-based parallel iterative computing is an economic solution. However, costs of disk I/O and communication affect the performance heavily. This paper proposes a state-transition model for SSSP and then designs two optimization strategies based on it. First, we introduce a tunable hash index to reduce the scale of *wasteful data* loaded from the disk. Second, we propose a new iterative mechanism and design an Across-step Message Pruning (ASMP) policy to deal with the communication bottleneck. The experimental results illustrate that our SSSP computation is 2 times faster than a basic Giraph (a memory-resident parallel framework) implementation. Compared with Hadoop and Hama (disk-resident parallel frameworks), the speedup is 21 to 43.

1 Introduction

The Single Source Shortest Path (SSSP) computation is a classical problem with numerous applications and has been well-studied over the past decades. However, new challenges have been raised by the rapid growth of graph data. For instance, up to March 2012, Facebook has owned about 900 million vertices (i.e., users) and over 100 billion edges. Such large graphs have exceeded the memory capacity of a single machine [1]. Even for memory-resident parallel frameworks [2,3], the data processing capacity of a given cluster is also limited [4]. This problem can be relieved by enlarging the cluster scale, but the consumption will also increase. It is an economic solution if we extend memory-resident parallel frameworks by spilling data on the disk [5]. In this case, how to reduce costs of disk I/O and message communication becomes challenging especially for the iterative computation tasks, such as SSSP.

For *in-memory* algorithms on SSSP, some are difficult to be executed in parallel due to the inherent priority order of relaxation and others perform poorly if data are organized as their sophisticated structures on the disk [6,7]. *External-memory* algorithms with the polynomial I/O complexity have also been proposed [8]. However, the practical performance is unsatisfactory [9] considering the impact of *wasteful data* (load a block of data from the disk but only use a portion). In addition, they are all centralized algorithms and take no account of the communication cost. Recently, G. Malewicz et al. propose a new parallel iterative implementation for SSSP (P-SSSP) and evaluate its performance on Pregel, a memory-resident parallel framework [2] based on the Bulk Synchronous Parallel (BSP) model [10]. Although its outstanding performance is impressive, the runtime will increase rapidly if it is implemented on disk-based frameworks, such as Hama and Hadoop [5,11]. I/O costs incurred by reading *wasteful data* may offset the parallel gains. Furthermore, the large scale of messages will also exacerbate costs of disk-accesses and communication. In this paper, we aim to crack the nut for these two problems of disk-resident P-SSSP over large graphs.

Based on the theoretical and experimental analysis on P-SSSP, we divide iterations into three stages: divergent $\rightarrow$ steady $\rightarrow$ convergent, and then propose a state-transition model. It adopts a *bottom-up* method to evaluate which stage the current iteration belongs to. Afterwards, two optimization policies are designed by analyzing features of the three states.

For divergent and convergent states, the scale of processed data will shade as the iteration progresses, which leads to huge costs of reading *wasteful data*. A tunable hash index is designed to skip *wasteful data* to the utmost extent by adjusting the bucketing granularity dynamically. The time of adjusting depends on the processed data scale instead of inserting or deleting elements, which is different from existing mechanisms [12,13]. In addition, for different adjusting operations (i.e., bucketing granularity), we adopt a Markov chain to estimate their cumulative impacts for iterations and then execute the optimal plan. Another optimization is an Across-step Message Pruning (ASMP) policy. The large scale of messages during the steady state incurs expensive costs of disk I/O and communication. The further analysis shows that a considerable portion of messages are redundant (i.e., the value of a message is not the real shortest distance). By extending BSP, we propose a new iterative mechanism and design the ASMP policy to prune invalid messages which have received. Then a large portion of new redundant messages will not be generated.

Experiments illustrate the runtime of our tunable hash index is 2 times as fast as that of a static one because roughly 80% of *wasteful data* are skipped. The ASMP policy can reduce the message scale by 56% during the peak of communication, which improves the performance by 23%. The overall speedup of our P-SSSP computation compared to a basic implementation of Giraph [3], an open-source clone of Pregel, is a factor of 2. For Hadoop and Hama, the speedup is 21 to 43. In summary, this paper makes the following contributions:

- **State-Transition Model:** We propose a state-transition model which divides iterations of P-SSSP into three states. Then we analyze characteristics of the three states, which is the theoretical basis for optimization policies.
- **Tunable Hash Index:** It can reduce costs of reading *wasteful data* dynamically as the iteration progresses, especially for divergent and convergent states. A Markov chain is used to choose the optimal bucketing granularity.

– **Across-step Message Pruning:** By extending BSP, this policy can prune invalid received messages and avoid the generation of redundant messages. Consequently, the message scale is reduced, especially for the steady state.

The remaining sections are structured as follows. Section 2 reviews the related work. Section 3 gives the state-transition model. Section 4 describes the tunable hash index. Section 5 proposes the Across-step Message Pruning policy. Section 6 presents our performance results. Section 7 concludes and offers an outlook.

2 Related Work

Many algorithms have been proposed for the SSSP computation. However, centralized in-memory algorithms can not process increasingly massive graph data. Advanced parallel algorithms perform poorly if data are spilled on the disk [6,7]. For example, the Δ-stepping algorithm must adjust elements among different buckets frequently [6] and Thorup's method depends on a complex in-memory data structure [7], which is I/O-inefficient. Existing *external-memory* algorithms are dedicated to designing centralized I/O-efficient data structure [8]. Although they have optimized the I/O complexity, the effect is limited for reducing the scale of loaded *wasteful data* because their static mechanisms can not be adjusted dynamically during the computing.

Nowadays, most of existing indexes are in-memory or designed for the *s-t* shortest path [14,15], which is not suitable for SSSP over large graphs. As a pre-computed index, VC-index is proposed to solve the disk-resident SSSP problem [9]. However, this centralized index is still static and requires nearly the same storage space with the initial graph or more. Also, dynamic hash methods for general applications have been proposed, but they are concerned on adjusting the bucketing granularity with changes in the scale of elements [12,13]. While, in our case, the element number (vertices and edges) in a graph is constant.

Implementing iterative computations on parallel frameworks has been a trend. The representative platform is Pregel [2] based on BSP and its open-source implementations, Giraph and Hama [3,5]. Pregel and Giraph are memory-resident, which limits the data processing capacity of a given cluster. This problem also exists for Trinity [16], another well-known distributed graph engine. Although Hama supports disk operations, it ignores the impact of *wasteful data*. For other disk-based platforms based on MapReduce, such as Hadoop, HaLoop and Twister [11,17,18], restricted by HDFS and MapReduce, it is also difficult to design optimization policies to eliminate the impact.

The parallel computing of SSSP can be implemented by a synchronous mechanism, such as BSP [10], or an asynchronous strategy [4]. Compared with the former, although the latter accelerates the spread of messages and improves the speed of convergence, a large scale of redundant messages will be generated, which increases the communication cost. The overall performance of them depends on a graph's density when data are memory-resident [4]. However, for the asynchronous implementation of disk-based SSSP, the frequent update of vertex values will lead to fatal I/O costs, so BSP is more reasonable in this case.

3 State-Transition Model

3.1 Preliminaries

Let $G = (V, E, \omega)$ be a weighted directed graph with $|V|$ vertices and $|E|$ edges, where $\omega : E \to \mathbb{N}^*$ is a weight function. For vertex v, the set of its outgoing neighbors is $adj(v) = \{u|(v,u) \in E\}$. Given a source vertex v_s, $\delta(u)$, the length of a *path* from v_s to u, is defined as $\sum \omega(e)$, $e \in path$. δ is initialized as $+\infty$. The SSSP problem is to find the minimal $\delta(u)$, $\forall u \in V$. By convention, $\delta(u) = +\infty$ if u is unreachable from v_s. We assume that a graph is organized with the adjacency list and each vertex is assigned a unique ID which is numbered consecutively.

The P-SSSP computation proposed by Pregel is composed of a sequence of SuperSteps (i.e., iterations). At the first SuperStep t_1, only v_s sets its $\delta(v_s) = 0$. Then $\forall u \in adj(v_s)$, a message (i.e., candidate shortest distance) is generated: $msg(u) = \langle u, m \rangle$, $m = \delta(v_s) + \omega(v_s, u)$, and sent to u. At t_2, vertex u with a list of $msg(u)$, namely $lmsg(u)$, sets its $\delta(u) = min\{\delta(u), min\{lmsg(u)\}\}$. Here, if $msg_i(u) < msg_j(u)$, that means $m_i < m_j$. If $\delta(u)$ is updated, new messages will be generated and sent to neighbors of u. The remaining iterations will repeat these operations until $\forall v \in V$, its $\delta(v)$ is not be updated. Operations of one SuperStep are executed by several tasks in parallel.

If a large graph exceeds the memory capacity of a given cluster, the topology of the graph is firstly spilled on the disk. Furthermore, the overflowing messages will also be spilled. Data are divided into three parts and respectively stored: message data, vertex data and outgoing edge data. For example, an initial record $\{u, \delta(u), adj(u)\&\omega, lmsg(u)\}$ will be partitioned into three parts: $\{lmsg(u)\}$, $\{u, \delta(u)\}$ and $\{adj(u)\&\omega\}$. $\{u, \delta(u)\}$ and $\{adj(u)\&\omega\}$ are stored in two files respectively but located on the same line, which avoids the cost of re-structuring when sending messages. By this mechanism, the topology of a graph will not be accessed when receiving messages. In addition, we only need to rewrite $\{u, \delta(u)\}$ on the disk and ignore $\{adj(u)\&\omega\}$ when updating δ.

Now, we give two notations used throughout this paper. We define *load ratio* as $LR = |V_l|/|V|$, where $|V_l|$ denotes the scale of vertices loaded from the disk. Another notation is *load efficiency*, which is defined as $LE = |V_p|/|V_l|$, where $|V_p|$ denotes the scale of vertices with received messages $lmsg$.

3.2 Three Stages of Iterations and State-Transition

The iterative process of P-SSSP is a wavefront update from v_s [2]. At the early stage, the message scale of the SuperStep t_i is small because only a few vertices update their δ. However, most of these messages will lead to updating δ at t_{i+1} because a majority of vertices still keep $\delta = +\infty$. Then more new messages will be generated since $|adj(u)| > 1$ generally. Consequently, the message scale will increase continuously. If most of vertices have updated their δ, the speedup of the message scale will decrease. As more and more vertices have found the shortest distance, their δ will be updated no longer. Then the number of messages will reduce until iterations terminate. We have run the P-SSSP implementation on our prototype system over real graphs, S-LJ and USA-RN (described in Section 6). As illustrated in Fig 1(a), the message scale can be simulated as a parabola opening downwards. This curve can also express the trend of processed vertices, since only vertices with received messages will be processed. Furthermore, we divide the process into three stages: divergent → steady → convergent.

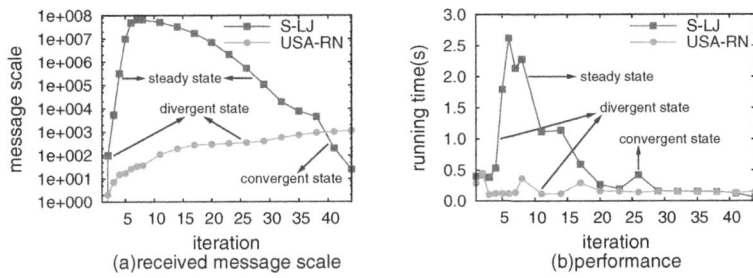

Fig. 1. Three processing stages of P-SSSP

Considering the message scale of divergent and convergent states, we only need to process a small portion of vertices. However, graph data must be loaded in blocks since they are spilled on the disk, which leads to a low LE and incurs considerable costs of reading *wasteful data*. In addition, the running time at the steady state is more than that of other two states obviously (Fig 1(b)). The reason is that massive messages lead to expensive costs of communication and disk-accesses (LR is high because many vertices need to process received messages). To improve the performance, we expect a low LR but a high LE.

We notice that there is no general standard to separate the three states because real graphs have different topology features. For example, S-LJ spreads rapidly at the divergent state. However, it is the opposite for USA-RN, which is a sparser graph. In section 4.2, we will introduce a *bottom-up* method to separate different states according to dynamical statistics.

4 A Tunable Hash Index

4.1 Hash Index Strategy

It is essential to match $\{lmsg(u)\}$ with $\{u, \delta(u)\}$ when updating δ. By a static hash index, we can load messages of one bucket into memory. Then $\{u, \delta(u)\}$ and $\{adj(u)\&\omega\}$ in the same bucket are read from the local disk one by one to complete matching operations. The static hash index can avoid random disk-accesses. Data in buckets without received messages will be skipped, which improves LE, but the effect is limited because $|V_p|$ is changing continuously among the three states. Therefore, we propose a tunable hash index to maximize the scale of skipped graph data by adjusting the bucketing granularity dynamically.

Three parts of data (described in Section 3.1) will be partitioned by the same hash function. Illustrated in Fig 2, index metadatas of buckets are organized as a tree which includes three kinds of nodes: Root Node (T), Message Node (e.g., H^1) and Data Node (e.g., H_1^1). The Message Node is a basic unit for receiving and combining (i.e., only save the minimal $msg(u)$ for vertex u) messages. Initially, every Message Node has one child node, Data Node, which is the basic unit for loading $\{u, \delta(u)\}$ and $\{adj(u)\&\omega\}$. The metadata is a three-tuple $\{R, M, A\}$. R denotes the range of vertex IDs. M is a key-$value$ pair, where key is the number of direct successor nodes and $value = \lceil R.length/key \rceil$. A is a location flag. For a Message Node, it means the location of memory-overflow message files (dir). For a leaf node, it includes the starting offset of $\{u, \delta(u)\}$ and $\{adj(u)\&\omega\}$. For anyone of parallel tasks, we deduce that the number of its Message Nodes depends on B_s, which is the cache size of sending messages. In order to simplify the calculation, we estimate the cache size in the number of messages instead of bytes. B_s is defined by disk I/O costs and communication costs of the real cluster. Limited by the manuscript length, the details are not illustrated here.

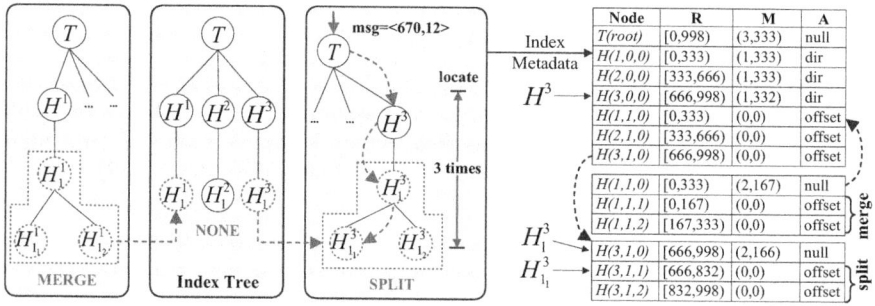

Fig. 2. The tunable hash index

The number of Message Nodes is fixed, but Data Nodes may be split or merged recursively during iterations. The former is only for leaf nodes. If one bucket H_j^i is split into N_j^i child buckets $H_{j_k}^i$, $1 \leq k \leq N_j^i$, that means vertex and outgoing edge data are divided equally in consecutive order. Then, the metadata of H_j^i needs to be updated (e.g., H_1^3). The merging operation is only for a direct predecessor node of leaf nodes. All child nodes will be merged and their parent node becomes a new leaf node (e.g., H_1^1).

$$k = \frac{getID(msg(v)) - \theta}{M.value} + 1 \qquad (1)$$

To skip buckets without messages, we must locate the leaf node every message belongs to. Given a message $msg(v)$, we can locate $H_{j_k}^i$ it belongs to by Formula 1, where θ is the minimal vertex ID in R of H_j^i. The time complexity is $\Omega((h-1) \cdot |E|/N_t)$, where h is the height of the tree and N_t is the number of parallel tasks. In Fig 2, for $\langle 670, 12 \rangle$, we can find the leaf node $H_{1_1}^3$ by locating 3 times.

4.2 Adjust Hash Index Dynamically

Although splitting a leaf node can improve its *load efficiency*, the time of splitting and the number of child nodes are two critical problems.

Algorithm 1. Global Adjusting Type for *Leaf Nodes*
Input : Statistics of the current SuperStep t_i: S; slope of t_{i-1}: K
Output: Global adjusting type of t_{i+1}: AT; slope of t_i: $K^{'}$

1 **Job Master**
2 wait until all tasks report the $vector^k$ and $active^k$
3 $vector \leftarrow \sum_{k=1}^{N_t} vector^k$ /* N_t: the number of parallel tasks */
4 $active \leftarrow \sum_{k=1}^{N_t} active^k$
5 put $active$ into $HistoryQueue$ and estimate $K^{'}$ by the last K_Δ values
6 $AT = \max\{vector(i) | 0 \leq i \leq 2\}$
7 send $\{AT, K^{'}\}$ to each task
8 **Task** k
9 $vector^k \leftarrow \langle 0, 0, 0 \rangle$ /* count the number of every adjusting type */
10 **while** $S^k \neq \phi$ **do**
11 $\quad S_i^k \leftarrow$ remove one from S^k /* S_i^k: the statistics of the ith leaf node */
12 $\quad type = getAdjustType(K, LE_i^k)$ /* type: $Split(0)$, $Merge(1)$, $None(2)$ */
13 $\quad vector^k[type]++$
14 $\quad active^k = active^k + getActive(S_i^k)$ /* the number of processed vertices */
15 send $vector^k$ and $active^k$ to **Job Master**
16 wait until **Job Master** returns $\{AT, K^{'}\}$ and then set $K = K^{'}$

Algorithm 1 is used to obtain a global adjusting type (AT) of the SuperStep t_{i+1}, which solves the first problem. It is also a *bottom-up* method to separate the three states. AT includes $Split$, $Merge$ and $None$. Algorithm 1 runs in a master-slave mode between two consecutive SuperSteps. First, task k judges the expected adjustment type for every leaf node by LE_i^k and K, then records statistics (Steps 10-14). LE_i is *load efficiency* of the ith leaf node at t_i. K is the slope of a fitting curve about $active$'s changing. Second, Job Master sums for all reports (Steps 3-4). $K^{'}$ (i.e., K of t_i) and AT are computed, and then sent to every task (Steps 5-7). Generally, $K_\Delta = 5$ by considering the robustness and veracity. The three states can be separated by AT and $K^{'}$ as follows: the divergent state, $AT \in \{Split, None\} \& K^{'} > 0$; the steady state, $AT \in \{Merge\}$; the convergent state, $AT \in \{Split, None\} \& K^{'} \leq 0$.

In the function $getAdjustType(K, LE_i^k)$, we first try to estimate the effect of $Split$. If it is positive, $type$ is $Split$, else $None$. If $type$ of all child nodes of the same parent node is $None$, we consider merging child nodes. Similarly, if the estimated result is positive, $type$ of them will be changed to $Merge$.

The effect of $Split$ depends on the number of child nodes. We use a Markov chain to find the optimal value, which solves the second problem. For a leaf node

H_j^i which is split into N_j^i child nodes, let $V_{j_k}^i$ be the set of vertices in $H_{j_k}^i$ and $^tV_p^{ij}$ be the set of processed vertices at the SuperStep t, then $^tV_p^{ij} \subseteq \bigcup V_{j_k}^i = V_j^i$, where V_j^i is the vertex set of H_j^i. $^t\Lambda$ denotes the set of child nodes with received messages at t, then $^t\Lambda = \{k | ^tV_p^{ij} \cap V_{j_k}^i \neq \phi, 1 \leq k \leq N_j^i\}$.

Theorem 1. *For H_j^i, let the random variable $X(t)$ be $|^t\Lambda|$ at the SuperStep t, then the stochastic process $\{X(t), t \in T\}$ is a homogeneous Markov chain, where $T = \{0, 1, 2, ..., t_{up}\}$ and t_{up} is an upper bound of the process.*

Proof. In our case, the time set can be viewed as the set of SuperStep counters and the state-space set is $I = \{a_i | 0 \leq a_i \leq N_j^i\}$. In fact, $^t\Lambda$ denotes the distribution of messages among child nodes. At the SuperStep t, vertices send new messages based on their current δ and received messages from t-1. Therefore, $^{t+1}\Lambda$, $^{t+2}\Lambda$, ..., $^{t+n}\Lambda$ only depend on $^t\Lambda$. The transition probability from $^t\Lambda$ to $^{t+1}\Lambda$ is decided by $^t\Lambda$ and δ. So $X(t)$ has the Markov property. Considering I and T are discrete, then $\{X(t), t \in T\}$ is a Markov chain. Furthermore, $P_{xy}(t, t + \Delta t) = P_{xy}(\Delta t)$ in the transition matrix P, so it is also homogeneous.

The original probability can be estimated by a sampling distribution. At t_m, we can get a random sample from $^{t_m}V_p^{ij}$, then the distribution of vertices among N_j^i child buckets can be calculated. Optimistically, we think the probability distribution of going from the state a_x to the state a_y is an arithmetic progression. Its common difference $d = (LE_i) \cdot K$ and the minimal value is $(2y)/x(x+1)$. Then, p_{xy}, the 1-step transition probability, can be calculated. The Δm-step transition probability satisfies the Chapman-Kolmogorov equation. Therefore, $P\{X(t_{m+\Delta m}) = a_y | X(t_m) = a_x\} = p_x(t_m) P_{xy}(\Delta m)$. We can calculate the mathematical expectation about the number of skipped buckets at $t_{m+\Delta m}$:

$$\Phi(N_j^i, t_{m+\Delta m}) = \sum_{x=1}^{N_j^i} \sum_{y=1}^{N_j^i} (N_j^i - x) p_x(t_m) P_{xy}(\Delta m) \qquad (2)$$

Considering the time complexity described in Section 4.1, we can infer the splitting cost $\Psi(N_j^i, \Delta t) = \sum_{k=1}^{\Delta t}(T_{cpu}\Omega(\Delta h|E|/N_t))$, where Δh is the change of height for the index tree after splitting. Specially, $\Delta h = 0$ if $N_j^i = 1$. T_{cpu} is the cost of executing one instruction. $\Delta t = t_{up} - t_m$. If $K < 0$, t_{up} is the max number of iterations defined by the programmer, else $\Delta t = K_\Delta$. The benefit of splitting H_j^i is that data in some child buckets will not be accessed from the disk. Then the saving cost is:

$$\Psi^{'}(N_j^i, \Delta t) = \left(\frac{\mathbb{V}_j^i + \mathbb{E}_j^i}{s_d N_j^i}\right) \sum_{\Delta m=1}^{\Delta t} \Phi(N_j^i, t_{m+\Delta m}) \qquad (3)$$

where $\mathbb{V}_j^i$ is the vertex data scale of H_j^i in bytes, $\mathbb{E}_j^i$ is the outgoing edge data scale and s_d is the speed of disk-accesses. The candidate values of N_j^i are $C = \{\langle n, \rho\rangle | 1 \leq n \leq \varepsilon\}$, where $\rho = \Psi^{'}(n, \Delta t) - \Psi(n, \Delta t)$. ε is a parameter which insures the size of our index will not be more than the given memory capacity.

For *Split*, we find the optimal splitting value γ as follows: first, compute a subset C' of C by choosing the maximal ρ in $\langle n, \rho \rangle$; then, $\forall \langle n, \rho \rangle \in C'$, γ is the minimal n. If $\gamma = 1$, $type = None$, otherwise, $type = Split$. For *Merge*, we view the parent node as a leaf node and assume γ be the real number of its child nodes. Then, if $\rho < 0$, $types$ of its child nodes will be changed to $Merge$.

5 Message Pruning Optimization

In this section, we propose a new iterative pattern, namely EBSP, by extending BSP. EBSP updates δ synchronously but processes messages across-step. By integrating the Across-step Message Pruning (ASMP) policy, the scale of redundant messages can be reduced effectively.

5.1 Analyze Messages of P-SSSP

Definition 1. *Multipath-Vertex*

Given a directed graph, let the collection of vertices be $V_{mul} = \{v | v \in V_i \bigwedge v \in V_j \bigwedge ... \bigwedge v \in V_k, i \neq j \neq k\}$, where V_i is the collection of vertices located i-hop away from the source vertex. Every vertex in V_{mul} is a Multipath-Vertex.

As shown in Fig 3, we assume s is the source vertex, then the 1-hop collection is $V_1 = \{a, b, c, d, e\}$ and the 2-hop collection is $V_2 = \{e, f, g\}$. Obviously, $e \in V_1 \bigcap V_2$, is a Multipath-Vertex. As its successor vertices, f, g, h, i are also Multipath-Vertices. For P-SSSP, the synchronous implementation based on BSP can reduce the number of redundant messages [2]. For example, during the ith SuperStep, vertex u receives two messages $msg_i^{t_1}$ and $msg_i^{t_2}$ at the time of t_1 and t_2, where $t_1 < t_2$. According to the synchronous updating mechanism, if $\delta(u) > msg_i^{t_1} > msg_i^{t_2}$, $msg_i^{t_1}$ is invalid and will be eliminated. Consequently, redundant messages motivated by $msg_i^{t_1}$ will not be generated. However, our in-depth analysis finds that the similar phenomenon will occur again and can not be eliminated by BSP due to the existence of Multipath-Vertices. Considering the following scenario, u receives msg_j at the jth SuperStep, $j = i + 1$. If $msg_i^{t_2} > msg_j$, all messages generated by u at i are still redundant. Furthermore, redundant messages will be spread out continuously along outgoing edges until the max-HOP vertex is affected in the worst case. That incurs extra costs of disk-accesses and communication. In addition, some buckets may not be skipped because they have messages to process, even though the messages are redundant.

Fig 3 illustrates the phenomena in a directed graph. At the 1th SuperStep, $\delta(e)$ is updated to 4, then e sends messages to f and g. However, at the 2th SuperStep, $\delta(e)$ is updated to 2 again (instead of 3). Then, the previous messages are invalid. Consequently, f, g, h, i are processed twice.

5.2 Across-Step Message Pruning

For BSP, δ is updated only by depending on messages from the last SuperStep (in Section 3.1). If we can cumulate more messages before updating δ, then the

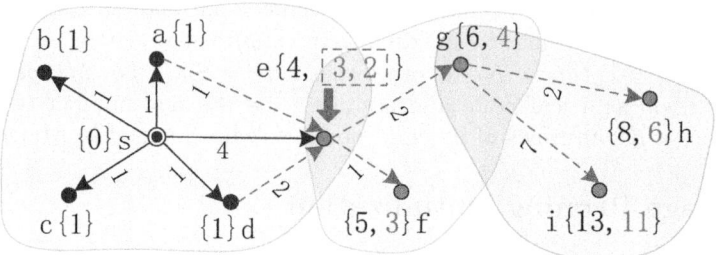

Fig. 3. The analysis of P-SSSP

impact of Multipath-Vertices will be relieved greatly. Consequently, we propose the EBSP model by extending BSP.

Definition 2. *EBSP*

Let M_{i+1} be the set of messages for the SuperStep t_{i+1}. At t_i, it is possible that $M_{i+1}^s \neq \phi$, $M_{i+1}^s \subseteq M_{i+1}$, if messages are sent asynchronously. When processing vertices at t_i, the domain of referenced messages is $M_i \cup M_{i+1}^s$.

EBSP will not affect the correctness of P-SSSP. Based on EBSP, we propose a novel technique called Across-step Message Pruning (ASMP) to relieve the phenomena of disseminating redundant messages. Algorithm 2 introduces the processing of one Message Node H^k. First, we load all received messages from t-1 into memory and put them into M_t^k after combining (in Section 4.1). Then the memory-resident received messages (without combination) of $t+1$ will be used to prune messages in M_t^k (Steps 3-8). A leaf node will be skipped if all of its messages in M_t^k are pruned. By this policy, new messages of $t+1$ will be obtained to optimize the synchronous update mechanism. Instead of combining existing messages, our policy is denoted to avoiding the generation of redundant messages, which is more effective. It can improve the performance of communication and disk-accesses. The scale of redundant messages which are eliminated by ASMP can be estimated by Theorem 2.

Theorem 2. *In Algorithm 2, for one vertex v_r, if $\delta(v_r) > msg_t^k(v_r) > msg_{t+1}^{k_s}(v_r)$, then the maximal number of pruned messages is $\Gamma(v_r)$:*

$$\Gamma(v_r) = \begin{cases} |adj(v_r)|, v_r = v_{maxHOP} \\ |adj(v_r)| + \sum_{\forall v_m \in adj(v_r)} \Gamma(v_m), v_r \neq v_{maxHOP} \end{cases} \quad (4)$$

where v_{maxHOP} is the farthest one among reachable vertices of v_r.

Proof. Normally, if $\delta(v_r) > msg_t^k(v_r)$, $\delta(v_r)$ will be updated and then messages will be sent to $adj(v_r)$. However, in Algorithm 2, $msg_t^k(v_r)$ will be pruned if $msg_t^k(v_r) > msg_{t+1}^{k_s}(v_r)$. Recursively, at the SuperStep $t + 1$, $\forall v_m \in adj(v_r)$, $\delta(v_m)$ will not be updated if $m_{t+1}^k(v_m) > m_{t+2}^{k_s}(v_m)$ or $m_{t+1}^k(v_m) \geq \delta(v_m)$. The pruning effect will not stop until v_{maxHOP} is processed.

Algorithm 2. Across-step Message Pruning

Input : message set for H^k at the SuperStep t and $t+1$: M_t^k, $M_{t+1}^{k_s}$
Output: message set after pruning: $\mathbb{M}_t^k$

1 $V_t^k \leftarrow$ extract vertex IDs from M_t^k
2 $V_{t+1}^{k_s} \leftarrow$ extract vertex IDs from $M_{t+1}^{k_s}$
3 **foreach** $u \in V_t^k \cap V_{t+1}^{k_s}$ **do**
4 $\quad msg_t^k(u) \leftarrow getMsg(M_t^k, u)$
5 $\quad msg_{t+1}^{k_s}(u) \leftarrow \min\{getMsg(M_{t+1}^{k_s}, u)\}$
6 $\quad$ **if** $msg_t^k(u) > msg_{t+1}^{k_s}(u)$ **then**
7 $\quad\quad$ put $msg_t^k(u)$ into the Pruning Set M_p
8 $\mathbb{M}_t^k = M_t^k - M_p$
9 **return** $\mathbb{M}_t^k$

We notice that if $\mathbb{M}_t^k = M_t^k \bigcup M_{t+1}^{k_s}$, δ will also be updated across-step, which is called an Across-step Vertex Updating (ASVU) policy. ASVU can accelerate the spread of messages. Therefore, the iteration will converge in advance compared with ASMP. However, $M_{t+1}^{k_s}$ is only a subset of M_{t+1}^k, so its elements may not be the minimal message value of $t+1$. For example, if $\delta(u) > msg_t^k(u) > msg_{t+1}^{k_s}(u)$, then δ will be updated at t. However, if $msg_{t+1}^{k_s}(u) > msg_{t+1}^k(u)$, $msg_{t+1}^k(u) \in M_{t+1}^k$, messages generated at t are also redundant, which offsets the pruning gains. Specially, compared with ASMP, if $u \in V_{t+1}^{k_s} \land u \notin V_t^k$ and $\delta(u) > msg_{t+1}^{k_s}(u) > msg_{t+1}^k(u)$, extra redundant messages will be generated.

6 Experimental Evaluation

To evaluate our patterns, we have implemented a disk-resident prototype system based on EBSP, namely DiterGraph. Data sets are listed in Table 1. The weight of unweighted graphs is a random positive integer. All optimization policies are evaluated over real graphs [19,20,21]. Then we validate the data processing capacity of DiterGraph over synthetic data sets and compare it with Giraph-0.1, Hama-0.5 and Hadoop-1.1.0. Our cluster is composed of 41 nodes which are connected by gigabit Ethernet to a switch. Every node is equipped with 2 Intel Core i3-2100 CPUs, 2GB available RAM and a Hitachi disk (500GB and 7,200 RPM).

6.1 Evaluation of Tunable Hash Index and Static Hash Index

Fig 4 illustrates the effect of our tunable hash index by comparing it to a static hash index. Their initial bucket number computed based on B_s is equivalent. However, the bucketing granularity of the static hash index will not be adjusted dynamically. In our experiments, we set B_s as 4000 according to the speed of communication and disk-accesses (described in Section 4.1). For USA-RN, we

Table 1. Characteristics of data sets

Data Set	ABBR.	Vertices	Edges	Avg. Degree	Disk Size
Social-LiveJournal1	S-LJ	4,847,571	68,993,773	14.23	0.9GB
Full USA Road Network	USA-RN	23,947,347	5,833,333	0.244	1.2GB
Wikipedia page-to-page	Wiki-PP	5,716,808	130,160,392	22.76	1.5GB
Synthetic Data Sets	Syn-D_x	1-600M	13-8100M	13.5	0.2-114GB

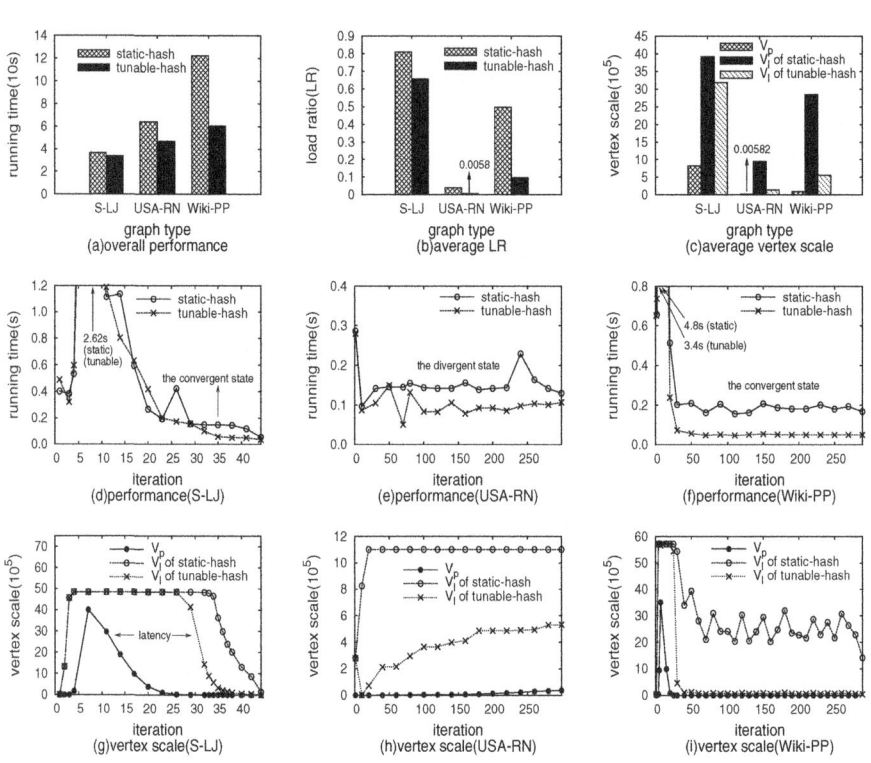

Fig. 4. Tunable hash index vs. static hash index (real data sets, 20 nodes)

only show the statistics of the first 300 iterations. In fact, it requires hundreds of iterations to fully converge because the graph has a huge diameter.

As shown in Fig 4(a), for Wiki-PP, the speedup of our tunable index compared to the static index is roughly a factor of 2. The tunable hash index has reduced its average LR of one iteration by 80.6% (Fig 4(b)), which means a large portion of *wasteful data* have been skipped. Therefore, its average LE ($|V_p|/|V_l|$) is improved by roughly 5 times (Fig 4(c)). The average LR of USA-RN is also reduced by up to 86%, but the overall performance is only improved by 28%, which is less than Wiki-PP. The reason is that USA-RN is a sparse graph, then the essential cost of warm-up (e.g., the initialization overhead of disk operations and communication) occupies a considerable portion of the running time, which affects the overall effect of our index.

For S-LJ, the gain is not as obvious as that of USA-RN and Wiki-PP. By analyzing the performance of every iteration (Fig 4(d)-(f)), we notice that, for USA-RN and Wiki-PP, their resident time of the divergent or convergent state is much longer than that of S-LJ. During these two states, just as illustrated in Fig 4(g)-(i), the scale of *wasteful data* is reduced efficiently by the tunable hash index. For example, P-SSSP over Wiki-PP took 290 iterations to converge. Fig 4(i) shows a large subset of vertices have found the shortest distance within the first 40 iterations. The remaining 250 iterations update less than 3% of δ. Therefore, the cumulate effect of adjustments is tremendous. However, for S-LJ, the number of iterations is only 44. Considering the latency of adjustments (Fig 4(g)), the overall gain is not remarkable.

6.2 Evaluation of ASMP and ASVU

This suit of experiments is used to examine the effect of ASMP and ASVU (described in Section 5.2). They are implemented based on the tunable hash index. As a comparative standard, the baseline method in experiments does not adopt any policies (both ASMP and ASVU) to optimize the message-processing.

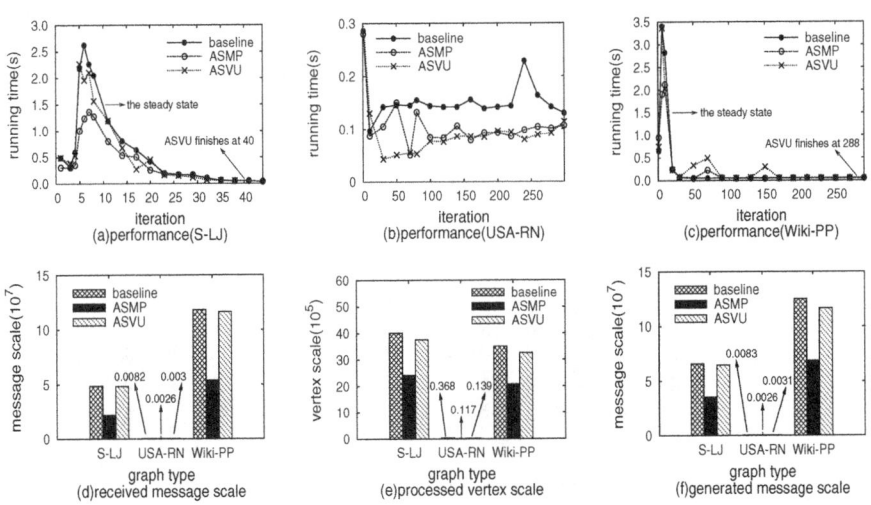

Fig. 5. Analysis on ASMP and ASVU (real data sets, 20 nodes)

As shown in Fig 5(a)-(c), the ASMP policy can optimize the performance of the steady state obviously. Especially for S-LJ, the overall performance can be improved by up to 23% because its resident time of the steady state is relatively longer than that of USA-RN and Wiki-PP. As illustrated in Fig 5(d)-(f), the effect of ASMP is tremendous at the iteration where the received message scale has reached the peak. Exemplified by S-LJ, the number of received messages (M_t) can be reduced by 56%. Then, compared with the baseline method, 45% vertices will be skipped at this iteration, which reduces the cost of disk-accesses (Fig 5(e)). Finally, the scale of new messages also decreases by 46% (Fig 5(f)), which reduces the communication cost. We notice that the iterations of S-LJ and Wiki-PP with ASVU are both completed in advance (Fig 5(a) and (c)) because ASVU can accelerate the spread of messages. However, considering the impact of redundant messages (Fig 5(d)-(f)), the contributions to overall performance of ASVU is not as obvious as that of ASMP. Especially, for S-LJ, the performance of ASMP is 16% faster than that of ASVU.

6.3 Evaluation of Data Processing Capacity and Overall Efficiency

Compared to Giraph, Hama and Hadoop, the P-SSSP implementation on DiterGraph can be executed over large graphs efficiently with limited resources. First, we set the number of nodes as 10. As shown in Figure 6(a), benefitted from our tunable hash index and ASMP, the running time of DiterGraph is two times faster than that of Giraph. Compared with Hadoop and Hama, the speedup is even 21 to 43. We are unable to run P-SSSP on Giraph when the vertex scale is more than 4 million, as the system runs out of memory. Second, we evaluate the scalability of DiterGraph by varying graph sizes and node numbers (Figure 6(b)). Given 40 nodes, when the number of vertices varies from 100 million to 600 million, the increase from 415 seconds to 3262 seconds demonstrates that the running time increases linearly with the graph size. Given the graph size, such as 600 million, the running time decreases from 9998 seconds to 3262 seconds when the number of nodes increases from 10 to 40.

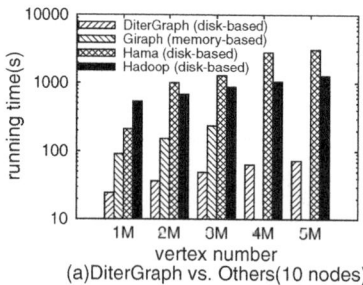
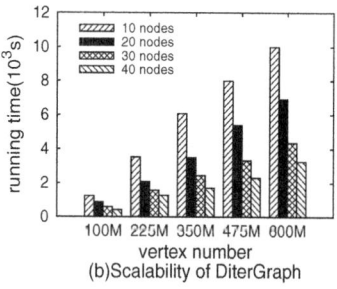

Fig. 6. Data processing capacity and overall efficiency (synthetic data sets)

7 Conclusion and Future Work

In this paper, we propose a novel state-transition model for P-SSSP. Then a tunable hash index is designed to optimize the cost of disk-accesses. By extending BSP, we propose the ASMP policy to reduce the message scale. The extensive experimental studies illustrate that the first policy can optimize the performance during the divergent and convergent states. And the second policy is effective for the steady state. In future work, we will extend our methods for incremental-iterative algorithms, such as the connected components computation, belief propagation and the incremental PageRank computation.

Acknowledgments. This research is supported by the National Natural Science Foundation of China (61272179, 61033007, 61003058), the National Basic Research Program of China (973 Program) under Grant No.2012CB316201, and the Fundamental Research Funds for the Central Universities (N110404006, N100704001).

References

1. Gao, J., Jin, R.M., Zhou, J.S., et al.: Relational Approach for Shortest Path Discovery over Large Graphs. PVLDB 5(4), 358–369 (2012)
2. Malewicz, G., Austern, M.H., Bik, A.J.C., et al.: Pregel: A System for Large-Scale Graph Processing. In: Proc. of SIGMOD, pp. 135–146 (2010)
3. Apache Incubator Giraph, http://incubator.apache.org/giraph/
4. Ewen, S., Tzoumas, K., Kaufmann, M., et al.: Spinning Fast Iterative Data Flows. PVLDB 5(11), 1268–1279 (2012)
5. Apache Hama, http://hama.apache.org/
6. Meyer, U., Sanders, P.: Δ-Stepping: A Parallel Single Source Shortest Path Algorithm. In: Bilardi, G., Pietracaprina, A., Italiano, G.F., Pucci, G. (eds.) ESA 1998. LNCS, vol. 1461, pp. 393–404. Springer, Heidelberg (1998)
7. Thorup, M.: Undirected Single-Source Shortest Paths with Positive Integer Wights in Linear Time. JACM 46(3), 362–394 (1999)
8. Meyer, U., Osipov, V.: Design and Implementation of a Practical I/O-efficient Shortest Paths Algorithm. In: Proc. of ALENEX, pp. 85–96 (2009)
9. Cheng, J., Ke, Y., Chu, S., et al.: Efficient Processing of Distance Queries in Large Graphs: A Vertex Cover Approach. In: Proc. of SIGMOD, pp. 457–468 (2012)
10. Valiant, L.G.: A Bridging Model for Parallel Computation. Communications of the ACM 33(8), 103–111 (1990)
11. Apache Hadoop, http://hadoop.apache.org/
12. Fagin, R., Nievergelt, J., Pippenger, N.: Extendible Hashing - A Fast Access Method for Dynamic Files. TODS 4(3), 315–344 (1979)
13. Litwin, W.: Linear Hashing: A New Tool for File and Table Addressing. In: Proc. of VLDB, pp. 212–223 (1980)
14. Xiao, Y.H., Wu, W.T., Pei, J.: Efficiently Indexing Shortest Paths by Exploiting Symmetry in Graphs. In: Proc. of EDBT, pp. 493–504 (2009)
15. Wei, F.: TEDI: Efficient Shortest Path Query Answering on Graphs. In: Proc. of SIGMOD, pp. 99–110 (2010)
16. Trinity, http://research.microsoft.com/en-us/projects/trinity/
17. Bu, Y., Howe, B., Balazinska, M., et al.: HaLoop: Efficient Iterative Data Processing on Large Clusters. PVLDB 3(1-2), 285–296 (2010)
18. Twister: Iterative MapReduce, http://www.iterativemapreduce.org/
19. SNAP: Network dataset, http://snap.stanford.edu/data/soc-LiveJournal1.html
20. 9th DIMACS, http://www.dis.uniroma1.it/challenge9/download.shtml
21. Using the Wikipedia link dataset, http://haselgrove.id.au/wikipedia.htm

Fast SimRank Computation over Disk-Resident Graphs*

Yinglong Zhang[1,2], Cuiping Li[1], Hong Chen[1], and Likun Sheng[2]

[1] Key Lab of Data Engineering and Knowledge Engineering of Ministry of Education, and Department of Computer Science, Renmin University of China, China
zhang_yinglong@126.com
[2] JiangXi Agricultural University, China

Abstract. There are many real-world applications based on similarity between objects, such as clustering, similarity query processing, information retrieval and recommendation systems. SimRank is a promising measure of similarity based on random surfers model. However, the computational complexity of SimRank is high and several optimization techniques have been proposed. In the paper optimization issue of SimRank computation in disk-resident graphs is our primary focus. First we suggest a new approach to compute SimRank.Then we propose optimization techniques that improve the time cost of the new approach from $O(kN^2D^2)$ to $O(kNL)$, where k is the number of iteration, N is the number of nodes, L is the number of edges, and D is the average degree of nodes. Meanwhile, a threshold sieving method is presented to reduce storage and computational cost. On this basis, an external algorithm computing SimRank in disk-resident graphs is introduced. In the experiments, our algorithm outperforms its opponent whose computation complexity also is $O(kNL)$.

Keywords: SimRank, Random walk, Graph, Similarity.

1 Introduction

The measure of similarity between objects plays significant role in many graph-based applications; examples include recommendation systems, fraud detection, and information retrieval. In contrast to textual content, link structure is a more homogeneous and language independent source of information and it is in general more resistant against spamming [1]. Thus a lot of link-based similarity measures have been proposed.

Among these link-based similarity measures, SimRank [2] is one of promising ones because it was defined based on human intuition: two objects are similar if they are referenced by similar objects[2], and a solid graph theory: random surfers model. Random surfers model is also a theoretical foundation of many other algorithms: PageRank[3], HITS[4], etc.

---
* This work is supported by the Fundamental Research Funds for the Central Universities,and the Research Funds of Renmin University of China(Grant No.12XNH178).

Several algorithms have been proposed on SimRank optimization [1, 5–8]. These optimization algorithms improve efficiency adopting different strategies. To the best of our knowledge, most existing algorithms didn't consider the situation that graph is a large disk-resident graph except [1]. However the research paper [1] approximate SimRank scores using probabilistic approach, thus their result is inherently probabilistic. Can we design a scalable and feasible algorithm to compute SimRank scores when graph is a large disk-resident graph? The issue has inspired our study.

The challenges of computing SimRank scores of a large disk-resident graph are following: first, its time complexity is $O(kN^2D^2)$, where k is the number of iteration, N is number of nodes, and D is the average incoming degree of nodes, and $O(kN^4)$ in the worst case, using original method. Second, the main overhead is I/O when graph cannot be held into main memory and we need to random access data frequently from disk. So we want to design a novel feasible algorithm to reduce I/O.

In this paper, we propose a new approach to compute SimRank scores adopting the random surfer-pairs model which was used to interpret SimRank in the original SimRank proposal [2]. We briefly describe our approach below. From perspective of the model, the SimRank score s(a, b) measures how soon two random surfers are expected to meet at the same node if they started at nodes a and b and randomly walked the graph backwards [2]. Thus SimRank score s(a, b) is summed up all first meeting-time probabilities of two surfers randomly walking from nodes a and b in reversed graph. Our naive and external algorithms are based on that we start from nodes, at which the two surfers can first meet, to compute these first meeting-time probabilities instead of starting from nodes a and b to compute the probabilities.

The main contributions of this paper are the following:

- We propose a new approach to compute SimRank scores.
- Optimization techniques are suggested, which improve time complexity from O (kN^2D^2) to $O(kNL)$, where L is number of edges. A threshold sieving method to accurately estimate SimRank scores is introduced for further improving the efficiency of the approach.
- An external algorithm is designed to compute SimRank scores over disk-resident graphs.
- Experimental results over both synthetic and real data sets show the algorithm is effective and feasible.

The rest of this paper is organized as follows. In the next section SimRank is reviewed, then a new approach is proposed. In sec.3, optimization techniques of the new approach are suggested and a threshold sieving method is introduced. In sec.4, an external algorithm based on the new approach is proposed. In section 5, the experimental results are reported. Section 6 gives the overview of the related works, and concludes paper in section 7.

2 The SimRank Fundamentals

In this section, we first review SimRank technique proposed in [2] (Sect 2.1). Then we propose a new formula directly derived from the random surfer-pairs model which was used to interpret SimRank (Sect 2.2). At last a naive algorithm is given (Sect 2.3).

2.1 SimRank Overview

Given a directed graph $G = (V, E)$ where nodes in V represent objects and edges in E represent relationships between objects. For any $v \in V$, $I(v)$ and $O(v)$ denote in-neighbors and out-neighbors of v, respectively. $I_i(v)$ or $O_j(v)$ is an individual member of $I(v)$, for $1 \leq i \leq |I(v)|$, or of $O(v)$, for $1 \leq j \leq |O(v)|$. The SimRank score of nodes between a and b defined as follows:

$$s(a,b) = \begin{cases} 1, & \text{if } a = b \\ \frac{c \sum_{i}^{|I(a)|} \sum_{j}^{|I(b)|} S(I_i(a), I_j(b))}{|I(a)||I(b)|}, & I(a) \text{ and } I(b) \neq \emptyset \\ 0 & \text{otherwise} \end{cases} \quad (1)$$

where c is constant decay factor and default value of c is 0.6 in the paper.

The naive solution [2] of equation (1) can be reached by iteration to a fixed-point:

$$R_{k+1}(a,b) = \frac{c \sum_{i}^{|I(a)|} \sum_{j}^{|I(a)|} R_k(I_i(a), I_j(b))}{|I(a)||I(b)|}, I(a) \text{ and } I(b) \neq \emptyset \quad (2)$$

,where $R_0(a, b) = 1$(for $a = b$) or $R_0(a, b) = 0$(for $a \neq b$).

The theory of expected-f meeting distance in [2] shows that SimRank score s(a, b) measures how soon two random surfers are expected to meet at the same node if they started at nodes a and b and randomly walked the graph backwards. Based on the theorem, SimRank score between a and b also can be defined:

$$s(a,b) = \sum_{\tau:(a,b) \to (x,x)} P(\tau) c^{l(\tau)} \quad (3)$$

where τ is a tour (paths may have cycles) along which two random suffers walk backwards starting at nodes a and b respectively until they first and only first meet at any node x ; c is a constant decay factor of SimRank.

The tour τ consists of two paths corresponding to the two suffers: $path1=(v_1 , \ldots , v_m , x)$, $path2=(w_1 , \ldots , w_m , x)$; $v_1 = a, w_1 = b$. Obviously the length of $path1$ equals the length of $path2$. Length $l(\tau)$ of tour τ is the length of $path1$ or $path2$. The probability $P(path1)$ of walking on $path1$ is $\prod_{i=1}^{m} \frac{1}{|I(v_i)|}$. For tour τ, the probability $P(\tau)$ of traveling τ is $P(path1)P(path2) = \prod_{i=1}^{m} \frac{1}{|I(v_i)||I(w_j)|}$, or 1 if $l(\tau) = 0$.

In the paper if there is a tour τ along which two random surfers walk backwards starting at two nodes a and b respectively until they first meet at some node x, we say node pairs (a,b) has a meeting-node x correspond to the tour τ of which the length is $l(\tau)$. We also say the node x is the tour τ's meeting-node and the tour τ is a tour corresponding to node pairs (a,b) which we call as start node pair. In the paper we use symbol (a,b) or τ to refer a tour according to context. The start node pair (a,b) and (b,a) are equivalent because similarity scores are symmetric. In the paper, if no specified it means $a \leq b$ for the symbol of (a,b).

2.2 Computing SimRank Based on Random Surfer Model

To compute similarity, using formula (3) we need to obtain all tours for each pair nodes by walking backwards all paths from the two nodes. However it is time consuming to obtain all tours. Now we give a solution of R_k based on formula (3).

Proposition 1. $R_k(a,b)$, the similarity of between a and b on iteration $k+1$, can be computed by:

$$R_{k+1}(a,b) = \sum_{\substack{\tau:(a,b)\to(x,x) \\ l(\tau)\leq k+1}} P(\tau)c^{l(\tau)} \qquad (4)$$

where τ is a tour along which two random suffers walk backwards starting at nodes a and b respectively until they first and only first meet at node x.

The proof is omitted due to the limitation of space[1].

From above proposition, to compute $R_k(a,b)$ we only need to obtain all the corresponding tours of the length equal or less than k and sum up these tours.

2.3 Naïve Algorithm

From Eq.(3) and proposition 1, we know that computing SimRank scores equals obtaining all corresponding tours. How can we efficiently obtain tours?

Observation 1. For tours $\tau:(e,f) \to (x,x)$, τ can be expanded to obtain tours $\tau':(O_i(e),O_j(f)) \to (x,x)$, of which the length is $l(\tau)+1$, by just walking one step from (e,f) to their out neighbor $O(e,f)$. It avoids random walking $l(\tau)+1$ steps starting at $(O_i(e),O_j(f))$ to obtain τ' by only appending path $((O_i(e),O_j(f)),(e,f))$ at the beginning of τ. The length $l(\tau')$ is: $l(\tau)+1$. The probability $P(\tau')$ of traveling τ' is: $P(\tau') = \frac{P(\tau)}{|I(O_i(e))||I(O_j(f))|}$.

Given a tour of which length is k: $\tau:(e,f) \to (x,x)$, in the paper symbol $v_{\tau,k,x}(e,f)$ denotes the value of the tour: $P(\tau)c^{l(\tau)}$; based on observation 1,

$$v_{\tau',k+1,x}(O_i(e),O_j(f)) = c\frac{v_{\tau,k,x}(e,f)}{|I(O_i(e))||I(O_j(f))|} \qquad (5)$$

According to both proposition 1 and observation 1, we have following proposition:

---
[1] Proof can be visited at http://ishare.iask.sina.com.cn/f/34372408.html

Proposition 2. $R_{k+1}(a,b)$, the similarity between a and b on iteration k+1, can be computed by:

$$R_{k+1}(a,b) = R_k(a,b) + \sum_{\substack{\tau:(e,f) \to (x,x) \\ a \in O(e) \wedge b \in O(f)}} c \frac{v_{\tau,k,x}(e,f)}{|I(a)||I(b)|} \quad (6)$$

where c is constant decay factor .

Proof. From observation 1, $\sum_{\substack{\tau:(e,f)\to(x,x) \\ a\in O(e)\wedge b\in O(f)}} c\frac{v_{\tau,k,x}(e,f)}{|I(a)||I(b)|} = \sum_{\substack{\tau:(a,b)\to(x,x) \\ l(\tau)=k+1}} P(\tau)c^{l(\tau)}$.
According to proposition 1, the proposition holds.

Our naïve algorithm based on observation 1, Eq.(5) and proposition 2: for each (x, x) we first walk from the meeting-node to its out neighbors to obtain tours and their values $v_{\tau,1,x}$, then expand tours to obtain other tours and $v_{\tau,2,x}$ just walking one step from current start node pairs to its out neighbors, and so on.

Obviously if a node is a meeting-node, the node has at least two out-neighbors. Our naïve algorithm based on observation 1 and proposition 2. Because the time requirement of naive algorithm is $O(n^3 D^2)$, which is more expensive than that of original method [2], the algorithm is not listed in the paper.

3 Optimization Strategies

Naïve method computes SimRank scores in a depth-first traverse style. We process meeting-nodes one by one: after processing all its tours of which the length is equal less than k for a given meeting-node, we process next one until all meeting-nodes are processed.

However the naïve method is time consuming and not practical. In this section, several optimization techniques are suggested to improve efficiency of the method.

3.1 Breadth-First Computation

Since the depth-first computation is inefficient, we consider computing SimRank scores in a breadth-first traverse manner.

For all meeting-nodes, first we obtain their all tours of which the length is 1. Then, we extend the tours to obtain all tours of which the length is 2 . We continue to extend the tours until the length of tours equal k.

One advantage of using the breadth-first traverse method can improve efficiency. At each iteration for different meeting-node x the corresponding tours are merged by the following formula.

$$v_k(e,f) = \sum_x v_{\tau,k,x}(e,f) \quad (7)$$

The value of the merged tour denoted by $v_k(e,f)$.

And based on E.q.(5),(7) and proposition 2, $R_{k+1}(a,b)$, the similarity between a and b on iteration k+1, can be computed by:

$$R_{k+1}(a,b) = R_k(a,b) + v_{k+1}(a,b) \tag{8}$$

where $v_{k+1}(a,b) = \frac{\sum_{i,j} v_k(I_i(a), I_j(b))}{|I(a)||I(b)|}$.

Another advantage of using the breadth-first traverse is that tours are grouped by first node at each iteration for reducing I/O to external algorithm, the details are discussed in the next section.

3.2 Threshold-Sieved Similarity

Threshold-sieved similarity was first introduced by [5] to filter low and nevertheless non-zero similarity scores because these similarity scores lead to overhead in both storage and computation. However the threshold-sieved similarity in [5] can not apply to our algorithm which adopts different method.

At $k+1$th iteration based value of tours v_k, we compute similarity scores and achieve information of new tours v_{k+1} based on equations(7)(8). After a few iterations, there are many low and nevertheless tours and similarity scores which both lead to heavy overhead in both storage and computation. So we effectively handle desired similarity scores and tours by filtering nevertheless similarity scores and tours.

Given threshold parameter δ, we define threshold-sieved similarity score $R'_k(*,*)$ and tours $v'_k(*,*)$ as follows:

$$R'_0(a,b) = R_0(a,b), R'_k(a,a) = R_k(a,a) = 1 \tag{9}$$

$$v'_0(a,b) = v_0(a,b) = 1, \text{ if } a = b; v'_0(a,b) = v_0(a,b) = 0, \text{ if } a \neq b \tag{10}$$

$$R'_{k+1}(a,b) = R'_k(a,b) + v'_{k+1}(a,b)$$

if either (right-hand side $> \delta$ for k=0 or right-hand side $> (1-c)\delta$ for $k \geq 1$) or $R'_k(a,b) \neq 0$;

$$\tag{11}$$

$$R'_{k+1}(a,b) = 0, \text{ otherwise.} \tag{12}$$

$$v'_{k+1}(a,b) = \frac{c}{|I(a)||I(b)|} \sum_{i,j} v'_k(I_i(a), I_j(b))$$

if right-hand side $> \delta$ for k=0 or right-hand side $> (1-c)\delta$ for $k > 0$

$$\tag{13}$$

$$v'_{k+1}(a,b) = 0, \text{ otherwise.} \tag{14}$$

In definitions (11) to (14), a and b are assumed to be different nodes.

First we give the estimate for threshold-sieved $v'_k(a,b)$ with respective to $v_k(a,b)$:

Proposition 3. *For k=0,1,... the following estimate hold:* $v_k(a,b) - v'_k(a,b) \leq \delta$.

Proof. For k=0, the estimate obviously holds because of $v_0(a,b) = v'_0(a,b)$

For k=1, the difference $v_1(a,b) - v'_1(a,b)$
$= \frac{c}{|I(a)||I(b)|} \sum_{\substack{a \in O(e) \wedge b \in O(f) \\ v_0(e,f) \leq \delta}} v_0(e,f) \leq \delta$, so the proposition holds.

Assume the proposition holds for k ($k > 1$), let us estimate the difference $v_{k+1}(a,b) - v'_{k+1}(a,b)$ for k+1 (two cases):

case 1: if $v'_{k+1}(a,b) = 0$, then from (13) and (14) we have

$$\frac{c}{|I(a)||I(b)|} \sum_{i,j} v'_k(I_i(a), I_j(b)) \leq (1-c)\delta \qquad (15)$$

and

$v_{k+1}(a,b) - v'_{k+1}(a,b) = v_{k+1}(a,b) \leq$ using (15) $\leq v_{k+1}(a,b) + (1-c)\delta - \frac{c}{|I(a)||I(b)|} \sum_{i,j} v'_k(I_i(a), I_j(b)) = (1-c)\delta + \frac{c}{|I(a)||I(b)|} \sum_{i,j} (v_k(I_i(a), I_j(b)) - v'_k(I_i(a), I_j(b))) \leq (1-c)\delta + c\delta = \delta$

case 2: $v'_{k+1}(a,b) \neq 0$, the difference $v_{k+1}(a,b) - v'_{k+1}(a,b)$
$= \frac{c}{|I(a)||I(b)|} \sum_{i,j} (v_k(I_i(a), I_j(b)) - v'_k(I_i(a), I_j(b))) \leq \delta$

thereby showing that indeed the proposition holds for k+1. □

Similar to [5], we also give the following estimate for threshold-sieved similarity scores $R'_k(a,b)$ with respective to conventional similarity scores $R_k(a,b)$:

Proposition 4. *For k=0,1,2,... the following estimate hold:* $R_k(a,b) - R'_k(a,b) \leq \triangle$, *where* $\triangle = k\delta$.

The proof is omitted due to the limitation of space[2].

Proposition 4 states that difference between threshold-sieved $R'_k(a,b)$ and conventional similarity scores $R_k(a,b)$ does not exceed $\triangle$ at worst case. The parameter $\triangle$ is generally chosen to control over the difference by a user. Given $\triangle$, obviously $\delta = \frac{\triangle}{k}$. If $\triangle$ is chosen to be zero, then $\delta = 0$ and $R'_k(a,b) = R_k(a,b)$.

The difference between threshold-sieved and theoretical SimRank scores is same with that of the paper [5] and discussed in details in [5].

4 Tour Algorithm

When the graph is a massive graph and disk-resident, the challenge is how to efficiently achieve the similarity of nodes.

---
[2] Proof can be visited at http://ishare.iask.sina.com.cn/f/34372408.html

Algorithm 1. Tour algorithm
Input:
 edg, c,$\triangle$,K// K is a number of iteration
Output:
 srt // SimRank Result
1: $\delta \leftarrow \frac{\triangle}{K}$
2: read blocks from edg and achieve tours which of length is 1,save merged tours into ct and edg;
3: **for** $k = 2$ to K **do**
4: $par \leftarrow$ **singleStepFromTour**(edg, ct, δ);// 1th stage:get par
5: $ct, srt \leftarrow$ **singleStepFromPartialTour**(ed, par, srt, δ);// 2th stage:get srt
6: **end for**

In the external algorithm called by tour algorithm, we adopt the strategies: breadth-first computation, tours merging, and threshold-sieved similarity.

The tour algorithm will sequentially read and write from three kinds of file: edg, ct and srt.

The disk file edg contains all edges of the graph and read only. Each line of the disk file is a triplet $(tailNode, headNode, p)$ corresponding to an directed edge, where $tailNode$ and $headNode$ are identifies of nodes in the graph, and $p = \frac{1}{|I(headNode)|}$. Edg is sorted by $tailNode$. Since in the algorithm we always need to obtain all out-neighbors of a node, the out-neighbors were clustered into one block by the sorting for reducing I/O.

Algorithm 2. singleStepFromTour
Input:
 input-graph edg,ct, C,δ
Output:
 par file
1: empty par
2: **while** !$edg.eof()$ AND !$ct.eof()$ **do**
3: read data blocks from edg and ct to get partial tours.
4: $\hat{pv}$ are merged into sorted buffer M by Eq. (16)
5: **if** the sorted buffer M is full **then**
6:   M merged with already sorted file par
7: **end if**
8: **end while**

Algorithm 1 is the tour algorithm. In line 1 we achieve tours directly from meeting-nodes, merge tours based on E.q.(7), and store the values into files ct and srt.

At $kth(k \geq 2)$ iteration, the disk file ct contains all tours of which the length exactly equal $k - 1$, and the disk file srt contains similarity of nodes achieved at $k - 1$th iteration. The formats of each line in both files are same: $(n_1 :$

$n_2, v(n_1, n_2), \ldots, n_t, v(n_1, n_t))$, where $v(n, n_i)$ is the value of the tour (n_1, n_i) in ct, $v(n, n_i)$ is the similarity score of (n_1, n_i) in srt, $n_1 \leq n_i$ and $2 \leq i \leq t$. Files are sorted by node n_1.

Algorithm 3. singleStepFromPartialTour
Input:
    edg, par, c, δ
Output:
    ct, srt
1: empty ct
2: **while** $!edg.eof()$ AND $!par.eof()$ **do**
3:    read blocks from edg and par to get new tour.
4:    new tours v_{k+1} are merged into sorted buffer M by Eq. (17)
5:    **if** $!par.eof()$ **then**
6:      **if** the sorted buffer M is full **then**
7:         M merged with already sorted file ct and empty M
8:      **end if**
9:    **end if**
10:   **if** $!par.eof()$ **then**
11:     M and remaining of par be merged with ct and all small values skip due to Eq.(13)(14)
12:   **else**
13:     M be merged with ct and all small values skip due to Eq.(13)(14)
14:   **end if**
15: **end while**
16: **while** $!ct.eof()$ OR $!srt.eof()$ **do**
17:   read blocks from ct and srt, get new score by Eq.(11) (12) and save it into $stemp$
18: **end while**
19: $srt \leftarrow stemp$

At $k(k \geq 2)$ iteration we adopt two-stage strategy (lines 4,5 of algorithm 1) due to the limitation of main memory space:

First stage: we call a method (Algorithm 2) to generate a par file from ct file and edg file. Par contains the partial tours and is sorted by first node of partial node pairs. The values $\hat{pv}_{k+1}$ (its initial value is zero) of partial tours (f, a) can be calculated(a is a out-neighbor of e) by:

$$\hat{pv}_{k+1}(f, a) + = c\frac{v_k(e, f)}{|I(a)|} \qquad (16)$$

Each line of par is $(p_1 : p_2, v(p_1, p_2) \ldots, p_t, v(p_1, p_t))$, where $v(p_1, p_i)$ is the value of partial node pair (p_1, p_i) $(2 \leq i \leq t)$. For each pair nodes (p_1, p_i), the first node is second node of corresponding tours before walking forward the one step, second node of the pair is the out-neighbor which we reach after walking forward the single step. Partial tours are sorted by the first node.

Second stage: the procedure (algorithm 3) is called. In the algorithm 3 we achieve the *ct* file from the *par* file and *edg* file based on following:

$$v_{k+1}(a,b) + = \frac{\hat{pv}_{k+1}(f,a)}{|I(b)|} \tag{17}$$

(b is a out-neighbor of f and the initial value of $v_{k+1}(a,b)$ is zero). Then we obtain new SimRank scores from the new *ct* file and *srt* file based on the equations (11) (12).

4.1 Complexity Analysis

Let us analyze the time requirement of tour algorithm. Let D be average of $|O(a)|$ over all nodes a. At first iteration(line 2 of algorithm 1), we can obtain $D(D-1)$ tours for each meeting-node x. So, at worst, the time requirement is $O(nD^2)$ for all meeting-node x.

At the kth($k > 1$) iteration, we obtain new tours based on current tours by following two stages:

1. Walking forward one step to its out-neighbors from a node which is common first node shared by a group of current tours:$(n_1 : n_2, v(n_1, n_2), \ldots, n_t, v(n_1, n_t))$. At worst, the max size of the group is $n-1$. Time requirement of walking single step from one group of current tours is O(nD). The time cost is $O(n^2D)$ for all groups.(Algorithm 2)

2. Walking forward one step to its out-neighbors from a node which is common first node shared by a group of partial tours. At worst, the max size of the group is also $n-1$. The time cost is $O(n^2D)$ for all groups.(Algorithm 3)

At each iteration, the time requirement to generate sorted file by merging with the sorted buffer at most is $O(nlog_2 n)$.

According to above analysis, the cost to obtain tours is $O(n^2D)$ at each iteration. Because the number of edges is $l = nD$, the total cost of computation SimRank scores is $O(Knl)$, where K is number of iterations.

4.2 I/O Analysis

Tour algorithm sequentially reads(writes) blocks to process data from(to) the files. In contrast, algorithms based on Eq.(2) compute score of (a,b) by random accessing scores of $(I_i(a), I_j(b))$ ($1 \leq i \leq |I(a)|$ and $1 \leq j \leq |I(b)|$) at each iteration because the scores of $(I_i(a), I_j(b))$ can not be clustered in a block and the random accessing causes heavy I/O cost. According to above discussion, tour algorithm is an I/O efficient method comparing algorithms based on Eq.(2).

5 Experimental Studies

In this section, we report our experimental studies on the effectiveness and efficiency of Tour algorithm. We implemented all experiments on a PC with

$i7 - 2620M$ CPU, 8G main memory , running windows 7 operating system. All algorithms are implemented in C++ and the runtime reported in all experiments includes the I/O time.

The first experiment shows feasibility and effectiveness of our tour algorithm. The second experiment shows effectiveness and efficiency of tour algorithm in comparison with excellent algorithms: partial sums and outer summation which are published in [5]. Finally, the third experiment illustrates the feasibility and efficiency of the tour algorithm on the real data. Our real datasets used in experiment are from Stanford Large Network Dataset Collection[3].

5.1 Feasibility and Effectiveness of Tour Algorithm

The time cost of tour algorithm is $O(n^2 D)$. In theory D equals n at worst, in this situation the time cost of our algorithm is $O(n^3)$ and the algorithm is infeasible. So we want to know what the degree value of real graph is in most situation. Stanford Large Network Dataset Collection is a data collection including social networks, web graphs, road networks, internet networks, citation networks, collaboration networks, and communication networks. Table 1 is a statistics on 40 data sets with 10 categories from the Stanford Large Network Dataset Collection. Each rows of table is a data set and its value $\frac{D}{n}$ is largest among datasets of corresponding category. From the table we conclude that the degree $D \ll n$ at worst for real graph in most situation. So our algorithm, of which the time cost is $O(n^2 D)$, is feasible and it is unnecessary to consider the worst case:D equals n.

Table 1. Statistics based on 40 data sets with 10 categories from Stanford Large Network Dataset Collection

Category	Name	Nodes	Edges	Average degree
Social networks	soc-Slashdot0811	77,360	905,468	11.7
Communication networks	email-Enron	36,692	367,662	20
Citation networks	Cit-HepTh	27,770	352,807	12.7
Web graphs	Web-Stanford	281,903	2,312,497	8
Product co-purchasing networks	Amazon0505	410,236	3,356,824	8
Internet peer-to-peer networks	p2p-Gnutella04	10,876	39,994	3.7
Road networks	roadNet-PA	1,088,092	3,083,796	5.7
Signed networks	soc-sign-Slashdot	77,357	516,575	6.7
Location-based online social networks	loc-Brightkite	58,228	214,078	7
Memetracker and Twitter	Twitter7	17,069,982	476,553,560	27.9

Then we compare tour algorithm with original method based on Eq.(2). In this subsection, we set the damping factor $C = 0.8$ and $K = 5$. In this subsection and the next, data sets we used are synthetic data which were produced by scale-free graph generator[4]. Given 3 synthetic data set whose average degrees are all: 3, figure 1 shows that tour algorithm($\Delta = 0$) is better than the original

---

[3] http://snap.stanford.edu/data/index.html
[4] Derek Dreier. Barabasi Graph Generator v1.4. University of California Riverside, Department of Computer Science.

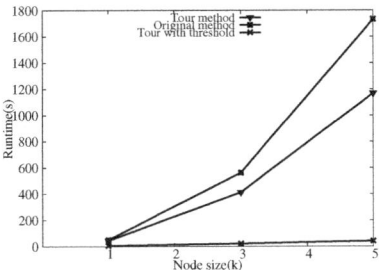

Fig. 1. Tour method VS. original method

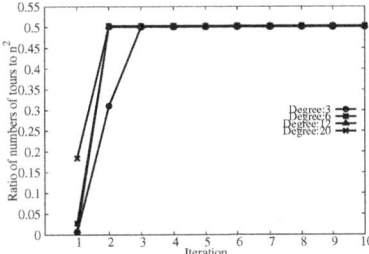

Fig. 2. Ratio of numbers of tours to n^2

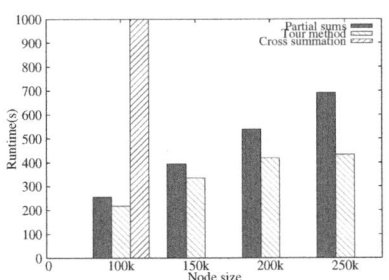

Fig. 3. Tour method VS. partial sums

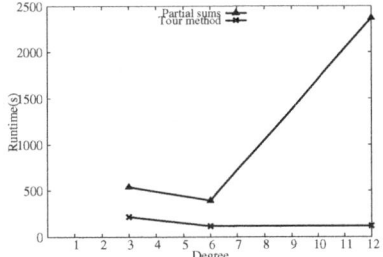

Fig. 4. Tour method VS. partial sums on different node degree

method(threshold tour: $\Delta = 0.01$). The reasons are following: we compute the score of every node pairs (n^2 of these) according to formula (2); Actually there are some node pairs that do not require computing similarity scores, which consume time, because SimRank score of their in-neighbor pairs is zero. However using tour method we obtain tours without these brute forces trying every node pairs at each iteration.

Table 2. Number of tours after each iteration for a graph $n = 5005$ nodes, degree $D = 3$

k	Number of tours			
	$\Delta = 0$		$\Delta = 0.01$	
	Absolute	Relative	Absolute	Relative
1	185,196	$0.007399n^2$	172,131	$0.006877n^2$
2	7,774,172	$0.310594n^2$	771,665	$0.03083n^2$
3	12,522,105	$0.500284n^2$	183,155	$0.007317n^2$
4	12,522,510	$0.5003n^2$	1253	$0.00005n^2$
5	12,522,510	$0.5003n^2$	55	$0.000002n^2$

The computational complexity of tour algorithm depends on number of tours. Figure 2 shows ratio of numbers of tours to n^2 at each iteration on graphs with different degrees and $n = 5000$ nodes without adopting Threshold-sieved technology. Table 2 shows the number of tours at each iteration on the same graph for $\Delta = 0$ and $\Delta = 0.01$ respectively.

5.2 Tour Algorithm *vs* Partial Sums and Outer Summation

In this subsection we compare tour algorithm with partial sums (including selecting essential node pairs) and outer summation. The three algorithms all adopt the threshold-sieved strategy. Partial sums and outer summation are excellent algorithms. Time cost of partial sums is the same as tour algorithm: $O(n^2 D)$. And time cost of outer summation is $O(\frac{n^3}{log_2 n})$. According to the last subsection the degree $D \ll n$ at worst for real graph in most situation so outer summation is infeasible against both partial sums and tour algorithm in practice. The conclusion is validated by our experiment(figure 3). For ease and fairness of comparison, we set the damping factor $C = 0.6, \Delta = 0.05$ and $K = 5$ in this subsection and the next. These parameters are set in accordance with the last experiment of [5]. Figure 3 shows the computation time of the three algorithms over four different size synthetic graphs. The average node degree of the four graphs are all 3. Because the cost time of outer summation is very expensive we run experiment only on the first graph using outer summation. Figure 3 shows tour algorithm is faster than opponents.

In the second experiment, we generate three graphs whose node number all are: 200k. The average degree of the three graphs is 3,6 and 12 respectively. Figure 4 shows result of the experiment: although degree of graphs vary tour algorithm is always better than the partial sums and the tour algorithm is affected less than partial sums by node degree.

Table 3. Computation time of tour algorithm on real graphs

Name of graph	Amazon0505	Amazon0302	web-Stanford	web-go
Number of nodes	410,236	262,111	281,903	875,713
Number of edges	3,356,824	1,234,877	2,312,497	5,105,039
Computation time(seconds)	654	173	2075	995

5.3 Experiments on Real Graphs

In this subsection, four real datasets are used(table 3). Table 3 shows computation time of tour algorithm adopting the threshold-sieved strategy on different real datasets. Two datasets are belong to the category:Product co-purchasing networks, Number of nodes and value $\frac{D}{n}$ of Amazon0505 are the largest among datasets of corresponding category. So the computation time on Amazon0302 is much faster than the time on Amazon0505. The other two datasets are belong to the category: web graphs. Among datasets of corresponding category, value $\frac{D}{n}$ of web-Stanford is the largest and number of nodes of web-go is the largest respectively. Although number of nodes of web-go is larger than that of web-Stanford, the computation time of web-go is faster than that of web-Stanford. The reason is values of tours are small and lots of tours are pruned at each iteration for the web-go graph. In a short, computation time on the different real datasets is accepted and tour algorithm is feasibility and efficiency in the PC environment.

6 Related Works

There are many real-world applications based on similarity between objects, such as clustering, similarity query processing, and information retrieval etc.According to the research [5], similarity measures are outlined two categories: (1) content- or text-based similarity measures, and (2) link-based ones that consider object-to-object relations expressed in terms of links [1, 2, 9–11]. SimRank [2] is one of promising ones among these link-based similarity measures.Several algorithms have been proposed on SimRank optimization [5–7, 1, 8, 12].

Fogaras and Rácz [1] suggested a general framework of SimRank computation based on Monte Carlo method. Their algorithms run in external memory. Their computation is stochastic because their algorithm is based on Monte Carlo method. In comparison, our algorithm is a deterministic solution.

Lizorkin et al. [5] proposed a technique of accuracy estimation and optimization techniques that improve the computational complexity from $O(n^4)$ to $\min(O(nl), O(\frac{n^3}{log_2 n}))$. Three optimization are suggested in their research: partial sums, outer summation and threshold-sieved similarity.

Li et al. [6] presented a approximate SimRank computation algorithms. As discussed in paper [12], the method is not preferable in practice. In [8],Li et al. exploit GPU to accelerate the computation of SimRank on large graphs.

Yu et al. [12] proposed optimization techniques based on matrix method to compute SimRank. The time cost of the technique exactly is $O(Knl)$ for sparse graphs, whereas the cost of our tour algorithm also is $O(Knl)$ in the worst case. For dense graphs they also proposed optimization technique. However, based on our statistics in Table 1, most real datasets are sparse graphs.

In contrast with above mentioned optimization algorithms, Li et al. [7] proposed a Single-Pair SimRank approach to obtain the similarity of a single node-pair. Research[13] extends the similarity join operator to link-based measures. [14] approach SimRank from a top-k querying perspective. Research[15] focus on that the question the most similar k nodes to a given query node on disk-resident Graphs.

7 Conclusions

This paper investigates optimization of SimRank computation for disk-resident graph. First we have proposed a new approach based on the random surfer-pairs model: we start from meeting-node to compute first meeting-time probabilities of two random surfers instead of starting from nodes a and b. Then several optimization techniques have been presented: breadth-first computation and threshold-sieved similarity etc. On this basis, an external algorithm, tour method, has been introduced. At each iteration, tours are grouped by first node to reduce the times of accessing disk data in the tour method. At last, we demonstrate its efficiency and effectiveness on synthetic and real data sets.

Acknowledgments. This work also are supported by the National Natural Science Foundation of China(Grant No.61070056,61033010,61272137,61202114) and National Basic Research Program of China (973 Program)(No. 2012CB316205). This work is partially done when the author visited Sa-Shixuan International Research Centre for Big Data Management and Analytics hosted in Renmin University of China. This Center is partially funded by a Chinese National "111" Project.

References

1. Fogaras, D., Rácz, B.: Practical algorithms and lower bounds for similarity search in massive graphs. IEEE Trans. Knowl. Data Eng. 19(5), 585–598 (2007)
2. Jeh, G., Widom, J.: Simrank: a measure of structural-context similarity. In: KDD, pp. 538–543 (2002)
3. Brin, S., Page, L.: The anatomy of a large-scale hypertextual web search engine. In: Proc.7th International World Wide Web Conference (1988)
4. Kleinberg, J.M.: Authoritative sources in a hyperlinked environment. Journal of the ACM 46(5), 604–632 (1999)
5. Lizorkin, D., Velikhov, P., Grinev, M.N., Turdakov, D.: Accuracy estimate and optimization techniques for simrank computation. VLDB J. 19(1), 45–66 (2010)
6. Li, C., Han, J., He, G., Jin, X., Sun, Y., Yu, Y., Wu, T.: Fast computation of simrank for static and dynamic information networks. In: EDBT, pp. 465–476 (2010)
7. Li, P., Liu, H., Yu, J.X., He, J., Du, X.: Fast single-pair simrank computation. In: SDM, pp. 571–582 (2010)
8. He, G., Feng, H., Li, C., Chen, H.: Parallel simrank computation on large graphs with iterative aggregation. In: KDD, pp. 543–552 (2010)
9. Zhao, P., Han, J., Sun, Y.: P-rank: a comprehensive structural similarity measure over information networks. In: CIKM, pp. 553–562 (2009)
10. Jeh, G., Widom, J.: Scaling personalized web search. In: WWW, pp. 271–279 (2003)
11. Xi, W., Fox, E.A., Fan, W., Zhang, B., Chen, Z., Yan, J., Zhuang, D.: Simfusion: measuring similarity using unified relationship matrix. In: SIGIR, pp. 130–137 (2005)
12. Yu, W., Zhang, W., Lin, X., Zhang, Q., Le, J.: A space and time efficient algorithm for simrank computation. World Wide Web J. 15(3), 327–353 (2012)
13. Sun, L., Cheng, R., Li, X., Cheung, D.W., Han, J.: On link-based similarity join. PVLDB 4(11), 714–725 (2011)
14. Lee, P., Lakshmanan, L.V.S., Yu, J.X.: On top-k structural similarity search. In: ICDE, pp. 774–785 (2012)
15. Sarkar, P., Moore, A.W.: Fast nearest-neighbor search in disk-resident graphs. In: KDD, pp. 513–522 (2010)

S-store: An Engine for Large RDF Graph Integrating Spatial Information

Dong Wang[1], Lei Zou[1,*], Yansong Feng[1], Xuchuan Shen[1], Jilei Tian[2], and Dongyan Zhao[1]

[1] Peking University, Beijing, China
zoulei@pku.edu.cn
[2] Nokia Research Center GEL, Beijing, China
{WangD,zoulei,fengyansong,shenxuchuan,zhaody}@pku.edu.cn,
jilei.tian@nokia.com

Abstract. The semantic web data and the SPARQL query language allow users to write precise queries. However, the lack of spatial information limits the use of the semantic web data on position-oriented query. In this paper, we introduce spatial SPARQL, a variant of SPARQL language, for querying spatial information integrated RDF data. Besides, we design a novel index SS-tree for evaluating the spatial queries. Based on the index, we propose a search algorithm. The experimental results show the effectiveness and the efficiency of our approach.

Keywords: spatial query, RDF graph.

1 Introduction

The Resource Description Framework (RDF)[13] is the W3C's recommendation as the basement of the semantic web. An RDF statement is a triple as ⟨$subject, predicate, object$⟩, which describes a property value of the subject. In real world, a large amount of RDF data are relevant to spatial information. For example, "London locatedIn England" describes a geographic entity London is located in a geographic location England; and the statement "Albert_Einstein hasWonPrize Nobel_Prize" is related to a geographic location of the event, i.e. Sweden.

Recently, researchers have begun to pay attention to the spatial RDF data. In fact, several real-world spatial RDF data sets have already been released, such as YAGO2[1][12], OpenStreetMap[2][10] etc. YAGO2[12] is an RDF data set based on Wikipidea and WordNet. Additionally, YAGO2 integrates GeoNames[3], which is a geographical database that contains more than 10 million geographical names, for expressing the spatial information of the entities and the statements.

---

* Corresponding author.
[1] http://www.mpi-inf.mpg.de/yago-naga/yago/downloads.html
[2] http://planet.openstreetmap.org/
[3] http://www.geonames.org/about.html

Although the traditional spatial databases can manage spatial data efficiently, the "pay-as-you-go" nature of RDF enables spatial RDF data provide more flexible queries. Furthermore, due to the features of Linked Data, spatial RDF data sets are linked to other RDF repositories, which can be queried using both semantic and spatial features. Thus, the spatial information integrated RDF data is more suitable for providing location-based semantic search for users. For example, a user wants to find a physicist who was born in a rectangular area between 59°N 12°E and 69°N 22°E (this area is southern Germany), and won some academic award in a rectangular area between 48.5°N 9.5°E and 49.5°N 10.5°E (it is in Sweden). The query can be represented as a SPARQL-like query in the following. Section 4 gives the formalized definition.

```
SELECT   ?x    WHERE{
         ?x          wasBornIn         ?y     ?l1.
         ?x          hasWonPrize       ?z     ?l2.
         ?y          type_star         city.
         ?z          type              Academic_Prize.}
Filter{IN(?l1,[(59,12),(69,22)] AND IN(?l2,[(48.5,9.5),(49.5,10.5)]}
```

Few SPARQL query engines consider spatial queries, and to the best of our knowledge only two proposals exist in literature. Brodt et al. exploit RDF-3X [14] to build a spatial feature integrated query system [4]. They use GeoRSS GML[8] to express spatial features. The R-tree and RDF-3X indexes are used separately for filtering the entities exploiting the spatial and the RDF semantic features, respectively. Besides, the method only supports the range queries over the spatial entities. YAGO2 demo[4] provides an interface for SPARQL like queries over YAGO2 data. However, the system uses hard-coded spatial predicates on spatial statements. Different from the above approaches, we introduce a hybrid index integrating both the spatial and semantic features, and the range queries and spatial joins are both supported in our solution.

In this paper, we introduce the spatial query over the RDF data, a variant of the SPARQL language for integrating the spatial feature constraint such as the range query and the spatial join. The spatial constraints assert the corresponding entities or events located in an absolute location area or near some entities in the query. For instance, users could search for a football club founded before 1900 nearby London, or a park nearby a specific cinema.

For effectively and efficiently solving the spatial queries, we introduce a tree-style index structure (called SS-tree). The SS-tree index is a hybrid tree-style index integrating the semantic features and the spatial features. Based on SS-tree, we introduce a list of pruning rules that consider both spatial and semantic constraints in the query, and propose a top-down searching algorithm. The tree nodes dissatisfying the signature constraints or the spatial constraints are safely filtered, and the subtrees rooted on the nodes are pruned. We make the following contributions in this paper.

1. We formalize the spatial queries, a variant of SPARQL language, on the RDF data integrating the spatial information. Besides, we introduce two spatial

---

[4] http://www.mpi-inf.mpg.de/yago-naga/yago/demo.html

predicates: the range query predicate and the spatial join predicate. The spatial queries can express both spatial constraints and semantic constraints.
2. We classify the entities into two categories: the spatial entities and the non-spatial entities. Based on these two categories, we build a novel tree-style index integrating the spatial features and semantic features. Additionally, we introduce a list of pruning rules for efficiently exploiting the spatial feature and the semantic feature of the index.
3. We evaluate our approach on a large real-world data set YAGO2, and the result shows our approach outperforms the baseline by several orders of magnitude.

2 Related Work

Many RDF manage systems [1,19,20,14,18,2,5] have been proposed in the past years. RDF-3x[15], Hexstore[18] and gStore[21] are the state-of-the-art RDF manage systems. Since none of the systems takes spatial feature into consideration, all the systems are unsuitable for spatial RDF data management.

Brodt et al.[4] and Virtuoso[7] utilize RDF query engines and spatial index to manage spatial RDF data. [4] uses RDF-3x as the base index, and adds an spatial index for filtering entities before or after RDF-3x join operations. These two approaches only support range query (and spatial join[7]) on entities, and the spatial entities follow the GeoRSS GML[16] model. YAGO2 demo employs six (hard coded) spatial predicates "northOf", "eastOf", "southOf", "westOf", "nearby" and "locatedIn" over statements. Users can construct queries as a list of triple patterns with the spatial predicates. Other spatial queries are not supported. The technical detail and the performance are not reported.

2.1 Spatial Information Integrating in RDF

There are many ways to represent spatial features in RDF data. OpenGIS Simple Features Specification[11] introduces a complex structure for describing complex spatial entities, such as points, lines, polygons, rings etc.. A complex shape can be decomposed into several simpler shapes, where each simple shape has its specific URI. Each spatial entity may be described by a list of statements.

The W3C Geo Vocabulary[3] is a decomposed approach. Each spatial entity is considered as a point with explicit latitude and longitude. Other feature types are not modeled in the W3C Geo Vocabulary.

Geography Markup Language (GML)[8] is an RDF/XML style language, and it can be translated into the RDF format. GeoRSS GML[16] models spatial features as abstract Geometry classes and for subclasses Point, Line, Box and Polygon. Each spatial feature can be translated into a list of RDF statements.

In the above approaches, only spatial entities are modeled in RDF data. YAGO2[12] models spatial features of events, i.e., statements. For example, "Person BornIn Time" must happen in a specific location. Thus, this event has spatial features. Therefore, YAGO2 models statements as SPOTLX tuples $\langle subject,$

⟨predicate, object, time, location, context⟩. In this paper, we only focus on SPOL tuples. The "location" feature is modeled as the W3C Geo Vocabulary format.

3 Preliminary

Since our method is based on our previous work gStore system and VS-tree index[21], to make the paper be self-contained, this section provides a simple overview of gStore and VS-tree. More details can be found at [21]. By encoding each entity and class vertex into a bit string called *signature*, gStore transforms an RDF graph into a data signature graph. Then, a tree-style index VS-tree is proposed over the data signature graph. The VS-tree is an extended S-tree[6]. The nodes on the same level and the corresponding edges constitute a signature graph. A pruning rule for subgraph query over the data signature graph is proposed for executing SPARQL queries.

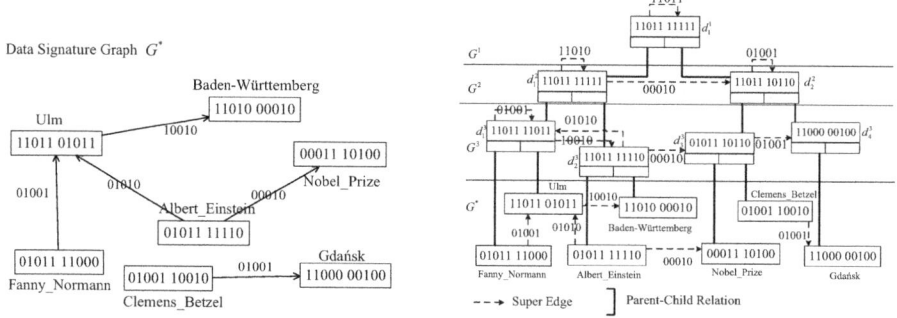

Fig. 1. Signature Graph **Fig. 2.** VS-tree

The signature sig of each subject s depends on all the edges $\{e_1, e_2, \ldots, e_n\}$ adjacent to s. For each e_i, gStore uses a list of hash functions to generate a signature $sig.e_i$, where the front N bits denote the predicate, and the following M bits denote the object. The valid bits depend on the hash code of the corresponding textual information. To determine the valid bits, gStore exploits several hash functions. The signature sig of s follows $sig = sig.e_1|sig.e_2|\ldots|sig.e_n$.

For example, in Figure 4, there's four edges starting from Ulm (#8, #9, #10 and #11). Suppose that we set the first five bits for the predicate and the following five bits for the object, we can get four signatures 1100001000, 1000101010, 1001000010 and 0001100011 corresponding to the four edges. Thus, Ulm can be represented as 1101101011. Figure 3 shows the encoding processing for "Ulm". Figure 1 shows the signature graph of Figure 4. Note that only the entity and class vertices in the RDF graph are encoded.

After the signature graph is generated, the VS-tree is built by inserting nodes into VS-tree sequentially. The corresponding VS-tree is shown in Figure 2.

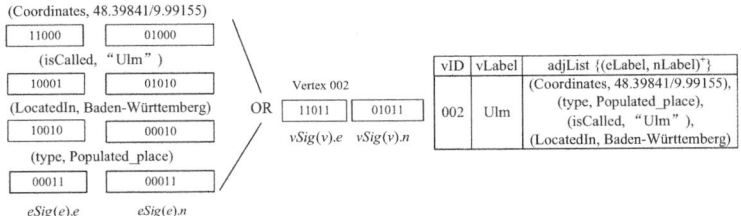

Fig. 3. Encoding Technique

4 Problem Definition

We formally define the spatial RDF and spatial SPARQL query as follow.

Definition 1. *An entity e is called a* spatial entity *if it has an explicit location labeled with the coordinates x and y (for the two-dimensional situation). The other entities are called the* non-spatial entities.

Definition 2. *A statement is a four-tuple $\langle s, p, o, l \rangle$, where s, p, o and l represent for subject, predicate, object and location, respectively. The location feature denotes the location where the statement happens. Note that l can be null. If the l of a statement is not null, the statement is called a* spatial statement. *Otherwise, it's called a non-spatial statement. A collection of statements (including spatial and non-spatial statements) is called a* spatial RDF data set.

Definition 3. *A spatial triple pattern is a four-tuple $\langle s, p, o, l \rangle$, where s, p, o and l represent for subject, predicate, object and location respectively. Each item can be a variable. Note that if l is not a variable, it should be omitted.*

Definition 4. *A spatial query is a list of spatial triple patterns with some spatial filter conditions. If there's no spatial filter condition, the spatial query is reduced to a traditional SPARQL query.*

Figure 4(a) shows a subset of a spatial RDF data set. *Ulm, Baden-Württemberg* and *Gdańsk* are spatial entities, and some statements are spatial statements, such as #1, #2 and #6. Besides, there're a lot of non-spatial entities and non-spatial statements. For example, people have no spatial information, since we can't locate a person on the map. Similarly, the statements like $\langle People\ hasName\ Name \rangle$ are non-spatial statements. In S-store, we use "spatial predicate" to represent the spatial queries. In this stage, we support the range query and the spatial join semantics. In practice, we use $sl(?x)$ for denoting the spatial label of variable $?x$. Besides, $dist(a, b)^5 < r$ denotes the distance between a and b should

---

[5] In this paper, for the ease of the presentation, we adopt the Euclidean distance between two locations. Actually, we can use "the earth's surface distance" to define the distance between two locations based on latitudes and longitudes.

	Subject	Predict	Object	Location(x,y)
#1	Albert_Einstein	BornIn	Ulm	(48.39841,9.99155)
#2	Albert_Einstein	BornOnDate	1879-03-14	(48.39841,9.99155)
#3	Albert_Einstein	type	American_physicists	
#4	Albert_Einstein	type	German_physicists	
#5	Albert_Einstein	type	German_vegetarians	
#6	Albert_Einstein	WonPrize	Nobel_Prize	(59.35,18.0667)
#7	Albert_Einstein	hasName	"Albert Einstein"	
#8	Ulm	Coordinates	48.39841/9.99155	(48.39841,9.99155)
#9	Ulm	isCalled	"Ulm"	(48.39841,9.99155)
#10	Ulm	LocatedIn	Baden-Württemberg	
#11	Ulm	type	Populated_place	(48.39841,9.99155)
#12	Baden-Württemberg	hasCapital	Ulm	
#13	Baden-Württemberg	Coordinates	48.5/9.0	(48.5,9.0)
#14	Baden-Württemberg	LocatedIn	Germany	
#15	Nobel_Prize	type	Academic_awards	
#16	Nobel_Prize	hasName	"Nobel Prize"	
#17	Fanny_Normann	diedIn	Ulm	(48.39841,9.99155)
#18	Fanny_Normann	hasName	"Fanny Normann"	
#19	Fanny_Normann	type	actor	
#20	Clemens_Betzel	diedIn	Gdańsk	(54.35,18.66667)
#21	Clemens_Betzel	diedOnDate	1945-03-27	(54.35,18.66667)
#22	Clemens_Betzel	BornOnDate	1895-06-09	(48.39841,9.99155)
#23	Clemens_Betzel	type	People_from_Ulm	
#24	Gdańsk	Coordinates	54.35/18.66667	(54.35,18.66667)
#25	Gdańsk	Population	455830	

(a)

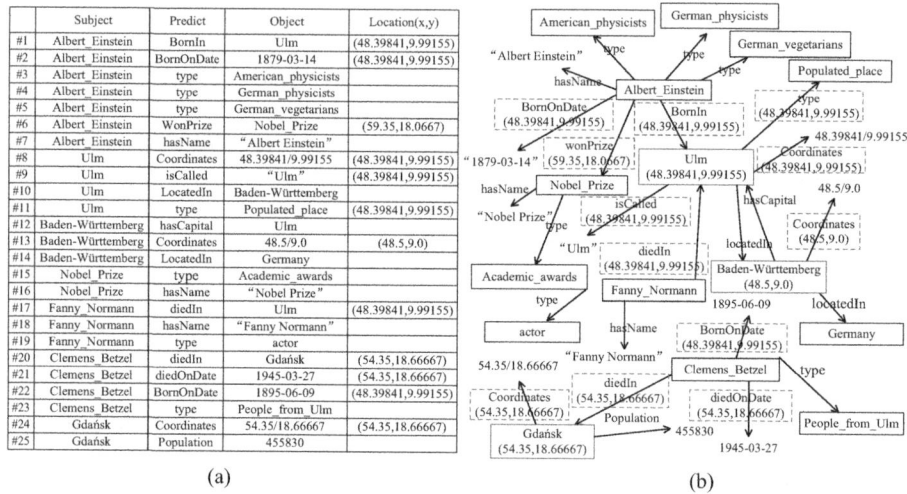

(b)

Fig. 4. Spatial RDF Graph

below the threshold r, where a and b should be a specific location or a variable. If either of a and b is a constant, the query is called a *range query*. If both of a and b are variables, the query is called a *spatial join query*. Note that a spatial query can be a range query and a spatial join query at the same time. The following query examples Q_1 and Q_2 demonstrate the range query and the spatial join query respectively. The former one queries a person who died in a popular place near coordinates (48.39841,9.99155), and the latter one queries two people where the first person died near the place where the second person died.

```
Q1:
Select    ?x    Where
{?x       diedIn       ?y.
?y        type         Populated_place.
} Filter{dist(sl(?y),(48.39841,9.99155))<1}
Q2:
Select    ?x1,?x2      Where
{?x1      diedIn       ?y1    ?l1.
?x2       diedIn       ?y2.
} Filter{dist(sl(?l1),sl(?y2))<1}
```

The spatial RDF data set and the spatial query can be also modeled as graphs (Definitions 5 and 6). The query processing is to find the matches (Definition 7) of a spatial query graph Q in a spatial RDF data graph G. Figure 4(b) shows the graph corresponding to the spatial RDF data set in Figure 4(a), where the spatial entities and the spatial statements are all surrounded by the red rectangles.

Definition 5. *The spatial RDF data graph is denoted as $G = \langle V, E, L_V, L_E, S_V, S_E \rangle$, where*

(1) $V = V_l \cup V_e \cup V_c \cup V_b$ denote all RDF vertexes where V_l, V_e and V_c are the sets of literal vertices, entity vertices, class vertices and blank nodes respectively.

(2) E is the collection of the edges between vertices.

(3) $L_V = \{URI\} \cup \{LiteralValue\}$ is the collection of text label of each vertex, where $v \in \{V_e \cup V_c\} \leftrightarrow label(v) \in \{URI\}$ and $v \in V_l \leftrightarrow label(v) \in \{LiteralValue\}$. For $v \in V_b$, $label(v) = \phi$.

(4) L_E is the collection of edge labels, i.e., all predicates plus null value.

(5) S_V and S_E represent the spatial labels of V and E respectively, where the spatial labels denote where the entity locates (the event happens) in, i.e., the latitude and longitude (only valid for spatial entities and spatial statements).

Definition 6. *The spatial RDF query graph is denoted as $G = \langle V, E, L_V, L_E, SC_V, SC_E \rangle$, where*

(1) $V = V_l \cup V_e \cup V_c \cup V_b \cup V_p$, where V_p denotes the parameter vertices, and V_l, V_e, V_c and V_b are the same as in Definition 5.

(2) E and L_E are the same as in Definition 5.

(3) L_V is the same as in Definition 5. For $v \in V_p$, $label(v) = \phi$.

(4) SC_V and SC_E represent the spatial constraints of V and E respectively, where the spatial constraints can be an absolute area or the relative position for some parameter.

Definition 7. *Consider a spatial RDF graph G and a spatial query graph Q with n vertices $\{v_1, \ldots, v_n\}$. A set of n distinct vertices $\{u_1, \ldots, u_n\}$ in G is said to be a match of Q iff. the following conditions hold:*

1. If $v_i \in \{V_l \cup V_c \cup V_e\}$, $u_i \in \{V_l \cup V_c \cup V_e\}$ and $label(v_i) = label(u_i)$;

2. If $v_i \in V_b$, there is no constraint over u_i;

3. If $v_i \in V_p$, the spatial label $S(u_i)$ must satisfy the spatial constraint $SC(v_i)$;

4. If there is an edge $\overline{v_i v_j}$ from v_i to v_j in Q, there is also an edge $\overline{u_i u_j}$ from u_i to u_j in G. If $\overline{v_i v_j}$ has predicate p and spatial constraint $SC(\overline{v_i v_j})$, $\overline{u_i u_j}$ must have the same predicate p and spatial label $S(\overline{u_i u_j})$ satisfy $SC(\overline{v_i v_j})$.

5 Overview of S-store

S-store employs a hybrid index that integrates both R-tree[9] and VS-tree[21]. Therefore, the pruning strategies of R-tree and gStore are also integrated as the searching strategy for S-store. Our framework consists of the pre-processing, the index construction and the query processing stages.

In the pre-processing stage, we first encode each vertex and edge as a bit string (we call it a *signature*). The encoding technique is shown in Section 3, and more information can be found in [21]. Subsequently, we build the *spatial signature graph $G^\star$*. Figure 5 shows a running example.

In the index construction stage, we construct a tree-style index based on the spatial signature graph for effectively reducing the search space. The index is called *SS-tree*. Figure 7 shows an running example. The nodes on the same level

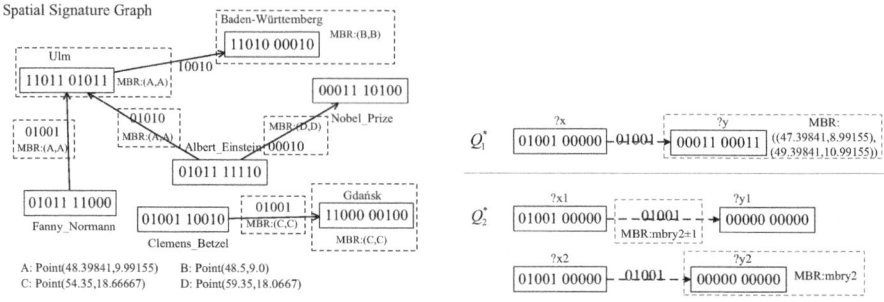

Fig. 5. Spatial Signature Graph **Fig. 6.** Q1 and Q2

of the SS-tree form a spatial signature graph. If there's a match of a query Q in a lower spatial signature graph, there must be a corresponding match in each higher spatial signature graph. Therefore, we need to guarantee that SS-tree is a height-balanced tree.

In the query processing stage, given a query graph Q, we first convert Q into the *spatial signature query graph* $Q^\star$ as in the pre-processing stage. Figure 6 shows the spatial signature query graphs of the Q_1 and the Q_2. Note that, if there is a set of vertices in G matches a query graph Q, there must be a corresponding match in $G^\star$ of $Q^\star$. Subsequently, we implement a top-down searching algorithm over SS-tree to find the matches of $Q^\star$ in $G^\star$. At last, we retrieve the corresponding textual result and return it to the user.

Definition 8. *Given a spatial signature graph $G^\star$ and a spatial signature query graph $Q^\star$ with n signature vertices $\{q_1,\ldots,q_n\}$, a set of distinct signature vertices $\{sig_1,\ldots,sig_n\}$ in $G^\star$ is a match of $Q^\star$ iff. the following conditions hold:*

1. $\forall q_i$, $sig_i.signature \& q_i.signature = q_i.signature$;
2. $\forall q_i$, the spatial label $S(sig_i)$ must satisfy the spatial constraint $SC(q_i)$;
3. If there is an edge $\overline{q_iq_j}$ from q_i to q_j in $Q^\star$, there is also an edge $\overline{sig_isig_j}$ from sig_i to sig_j in $G^\star$, and $\overline{q_iq_j}.signature \& \overline{sig_isig_j}.signature = \overline{q_iq_j}.signature$. If $\overline{q_iq_j}$ has spatial constraint $SC(\overline{q_iq_j})$, $\overline{sig_isig_j}$ must have the spatial label $S(\overline{sig_isig_j})$ satisfy $SC(\overline{q_iq_j})$.

6 Index Construction

In this section, we would introduce our spatial RDF index SS-tree. The index is presented as a tree-style. Generally speaking, we build the SS-tree based on VS-tree in gStore. The difference between S-store and gStore is that S-store can answer spatial queries.

6.1 Spatial Signature Graph Generation

First, we convert a data graph into a spatial signature data graph before building SS-tree. Since the spatial signature data graph can be regarded as a signature graph including spatial features, we generate the signature graph as described in Section 3. Then, for each vertex (v_i) and each edge e_j, we set the $MBR(v_i)$ and the $MBR(e_j)$. The signature and the MBR features of the spatial signature data graph are used to compute the features of the tree nodes on the high level. the unsatisfied tree nodes can be filtered early to save the space and time cost. Due to the space limit, the detailed information is omitted.

6.2 SS-tree Construction

The entities can be separated into two parts based on the spatial features. C_1 is the non-spatial entity set, and C_2 is the spatial entity set. For example, in Figure 4, collection C_1 = {Albert_Einstein, Fanny_Normann, Clemens_Betzel, Nobel_Prize} and collection C_2 = {Ulm, Baden-Württemberg, Gdańsk} respectively.

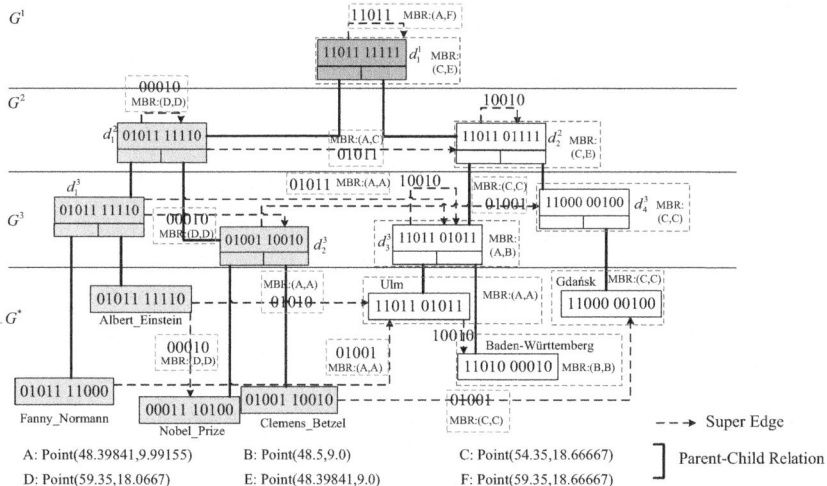

Fig. 7. SS-tree

Based on C_1 and C_2, we can generate two induced spatial signature graphs $G_1^\star$ and $G_2^\star$ from $G^\star$. The induced graph $G_i^\star$ can be composed into $V_i^\star$ and $E_i^\star$, where $V_i^\star$ =the vertices corresponding to C_i, and $E_i^\star = \{\overline{v_k v_l} | v_k \in V_i^\star \wedge v_l \in V_i^\star\}$.

In the following, we use "compute the features" to denote the bottom-up feature constructing process. The process obeys the following rule:
- *SS-tree Rule:* Consider two spatial signature nodes v_1, v_2 and their father nodes n_1, n_2. The following conditions hold:

(1) $n_1.sig \& v_1.sig = v_1.sig$, $n_2.sig \& v_2.sig = v_2.sig$;
(2) $v_1.MBR \in n_1.MBR$, $v_2.MBR \in n_2.MBR$;
(3) If there's an edge $\overline{v_1 v_2}$ between v_1 and v_2, there must be an edge $\overline{n_1 n_2}$ between n_1 and n_2, where $\overline{n_1 n_2}.sig \& \overline{v_1 v_2}.sig = \overline{v_1 v_2}.sig$ and $\overline{v_1 v_2}.MBR \in \overline{n_1 n_2}.MBR$, even if $n_1 = n_2$.

Non-spatial Entities. For collection C_1, we build a VS-tree over $G_1^\star$. The VS-tree constructing method can be found in [21]. After the VS-tree's completed, we compute the MBR features for each edge based on the SS-tree rule (2). T_1 consists of the VS-tree and the spatial features. For example, in Figure 7, the gray nodes and the edges between them compose the sub-SS-tree T_1. The edge $\overline{d_1^3 d_2^3}$ has the spatial feature because the edge between the child node $Albert\_Einstein$ of d_1^3 and the child node $Nobel\_Prize$ of d_2^3 owns a spatial feature.

Spatial Entities. For entity collection C_2, we build a R-tree over $G_2^\star$. Based on the R-tree structure, we first compute the features of the R-tree vertices base on SS-tree rule (1). Then, we add the edges between the upper level vertices of the R-tree and compute the features of the added edges based on the SS-tree rule (3). For example, in Figure 7, the white nodes and the edges between them compose the sub-SS-tree T_2. The signature of the node d_3^3 1101101011 = 1101101011|1101000010, where the former signature belongs to the node "Ulm" and the latter signature belongs to the node Baden-Württemberg, and "Ulm" and Baden-Württemberg are the children of d_3^3.

Combination. Since SS-tree should be a height-balanced tree, we should modify the tree height to balance T_1 and T_2. We employ the following operation called one step growth to increase tree height. Given a tree T, we add a node n as the father of T's root, and n would be the new root of T.

Given trees T_1 and T_2, we grow the lower tree with several steps to ensure the tree heights are equal. And then, we add a node n as new tree T's root, and set the roots of T_1 and T_2 as n's children. The new tree is called T_3.

Based on T_3, we first add all edges $e_{ij} = \{\overline{v_i v_j} | v_i \in G_1^\star \land v_j \in G_2^\star\}$ between $G_1^\star$ and $G_2^\star$, and then add the corresponding edges and compute the features for the edges based on the SS-tree rule (3). For example, the edge between "Albert_Einstein" and "Ulm" is added in this stage, and the corresponding edges $\overline{d_1^3 d_3^3}$ and $\overline{d_1^2 d_2^2}$ are added subsequently.

7 Query Processing

Given a spatial query Q, we first convert the Q to a spatial signature graph $Q^\star$. The converting processing consists of three steps.

(1) Encode the triple patterns as described in Section 6.1.
(2) For each range query predicate, we add the corresponding absolute MBR on the specific variables.
(3) For each spatial join predicate, we add the relevant MBRs on the variables.

The $Q_1^\star$ and $Q_2^\star$ corresponding to Q_1 and Q_2 are shown in Figure 6. The signatures are generated as G to $G^\star$, where the variables contribute no valid bit. The range query predicate of $Q1$ is converted to the absolute MBR binding ?y in $Q_1^\star$, and the spatial join predicate of Q_2 is converted to the relevant MBRs in $Q_2^\star$.

After the corresponding $Q^\star$ is generated, we next search the matches of $Q^\star$ in $G^\star$ exploiting the SS-tree. Consider a spatial signature query graph $Q^\star = \{q_1,\ldots,q_n\}$, we first generate the node candidate set $NodeSet_i$ for each variable q_i, and then verify each candidate in the query candidate set $QSet = \{NodeSet_1 \times \ldots \times NodeSet_n\}$ to generate the matches of $Q^\star$ in $G^\star$. At last, we generate the matches of Q in G based on the matches of $Q^\star$.

7.1 Pruning Rules

For efficiently generating the node candidate set, we have the following five pruning rules. Pruning rule 1 is based on the fact that only spatial entities can be bound by spatial predicates. Pruning rules 2 and 3 are based on that if the distance between v_1 and v_2 is no less than the distance between v_i and v_j where v_i and v_j are the descendant of v_1 and v_2 respectively. Pruning rule 4 is based on that $v.sig \& v_i.sig = v_i.sig$ if v_i is the descendant of v. Pruning rule 5 is based on that if there's no satisfied edge between v_1 and v_2, there's no satisfied edge between v_i and v_j where v_i and v_j are the descendant of v_1 and v_2 respectively. Due to the space limit, we can't state the pruning rules in detail.

Pruning Rule 1. If a variable is bound with a spatial predicate, the subtree T_1 induced by C_1 can be pruned safely.

Pruning Rule 2. Consider a variable v bound with a range query predicate, if there is a tree node n where $v.mbr$ has no intersection with $n.mbr$, the subtree rooted on n can be pruned safely.

Pruning Rule 3. Consider two variables v_i and v_j bound by a spatial join predicate, and $NodeSet_i$ is the candidate set of v_i and $NodeSet_j$ is the candidate set of v_j. Suppose the max distance is set to be $MaxDist$. Let $n_i \in NodeSet_i$, if the distance from MBR of n_i to any node $n_j \in NodeSet_j$ is larger than $MaxDist$, n_i can be safely pruned.

Pruning Rule 4. Consider a variable v, if there is a tree node n where $v.sig \& n.sig! = n.sig$, the subtree rooted on n can be pruned safely.

Pruning Rule 5. Consider two linked variables v_i and v_j with an edge $e = \overline{v_i v_j}$ from v_i to v_j, and $NodeSet_i$ is the candidate set of v_i and $NodeSet_j$ is the candidate set of v_j in the same spatial signature graph. Let $n_i \in NodeSet_i$, if there's no edge from n_i to any node $n_j \in NodeSet_j$, n_i can be safely pruned. What's more, if there's a range predicate on e, the unsatisfied edges are considered nonexistent. The pruning rule is based on the fact that if there's no satisfied edge from n_i to any node $n_j \in NodeSet_j$, there's no satisfied edge from the descendants of n_i to any descendants of the $n_j \in NodeSet_j$.

Algorithm 1 describes the top-down node candidate sets generating process. The use of the pruning rules is shown in Line 9-21.

Algorithm 1. Query Processing
Require: $Q^\star = \langle v_1, \ldots, v_n \rangle$, SS-tree T, root r of T, signature data graph $G^\star$.
Ensure: The node candidate sets $\{NodeSet\}$ of nodes of $Q^\star$ in $G^\star$.
1: Set each $NodeSet_i = r$ //initialize the node candidate set.
2: **while** true **do**
3:   **if** $\forall NodeSet_i \in G^\star$ **then**
4:     return $\{NodeSet\}$ //the sets contains real data points.
5:   **for** all $NodeSet_i$ **do**
6:     $NodeSet_i$ =the children of each node $n_i \in NodeSet_i$
7:     Set $MBR_i = \bigcup\{n | n \in NodeSet_i\}$
8:     **for** all node $n_i \in NodeSet_i$ **do**
9:       **if** $n_i \in T_1 \wedge v_i$ is binding **then**
10:         remove n from $tempNodeSet$ // pruning rule 1.
11:       **if** $v_i.sig \& n_i.sig! = n_i.sig$ **then**
12:         remove n from $tempNodeSet$ //pruning rule 2.
13:       **if** v_i is bound by range query predicates **then**
14:         **if** $intersection(v_i.mbr, n_i.mbr) = \phi$ **then**
15:           remove n from $tempNodeSet$ //pruning rule 3.
16:       **if** $\exists e = \overline{v_i v_j}$ **then**
17:         **if** $n_i.neighbour \cap NodeSet_j = \phi$ **then**
18:           remove n from $tempNodeSet$ //pruning rule 4.
19:       **if** $dist(v_i, v_j) <= l$ **then**
20:         **if** $dist(n_i, MBR_j) > l$ **then**
21:           remove n from $tempNodeSet$ //pruning rule 5.

7.2 Verification

Consider the node candidate set $\{NodeSet\}$, we generate a list of nodes $\langle n_1, \ldots, n_n \rangle$ from each item of $\{NodeSet\}$ respectively, and verify if $\langle n_1, \ldots, n_n \rangle$ forms the connected regions corresponding to the connected regions in $Q^\star$. If $\langle n_1, \ldots, n_n \rangle$ can form, we consider it as a match candidate of $Q^\star$, or we discard it otherwise. The generating process can be realized by employing a BFS algorithm starting from the smallest node candidate sets in each connected region. For example, $Q_2^\star$ has two connected regions. Since $?x_1$ and $?x_2$ have the highest selectivity in each connected region respectively, the $NodeSet_{x_1}$ and $NodeSet_{x_2}$ are selected as the start points. Then, we run BFS from $NodeSet_{x_1}$ and $NodeSet_{x_2}$. If there's an edge $e = \overline{v_k v_l}$ in $Q^\star$, there must be an corresponding edge.

Given a match candidate $Q_c^\star$ of $Q^\star$, we verify if all the spatial constraints are satisfied. The satisfied match candidates are the matches of $Q^\star$. Subsequently, since the encoding technique may bring false positive error, we verify if all edges in Q are satisfied given a match of $Q^\star$. The valid candidates are the matches of Q. Then, the matches of Q is returned to users.

Algorithm 2. $Varification$
Require: node candidates $\{NodeSet\}$, $Q^\star = \langle v_1, \ldots, v_n \rangle$, Q, G.
Ensure: the matches $\{M\}$ of Q.
1: Set the match candidate list of $Q^\star$ $L = \phi$.
2: **for** each connected region $Q_i \subseteq Q^\star$ **do**
3:    Select the $NodeSet_j$ with the smallest size in Q_i.
4:    Set the Q_i's match candidate set $M_c^i = \phi$. //Initialize the match candidate sets.
5:    **for** each node $n_k \in NodeSet_j$ **do**
6:      Run the BFS process from n_k.
7:      **if** $\exists$match candidate m_c^i of Q_i **then**
8:        $M_c^i.add(m_c^i)$. //If all edges are valid, it's a match candidate.
9: Set $M_c^\star = M_c^1 \times \ldots \times M_c^k$. //The match candidates of $Q^\star$.
10: Set $M^\star = \phi$. //The matches of $Q^\star$.
11: **for** each $m_c^\star \in M_c^\star$ **do**
12:    **if** all spatial join predicates are valid on $m_c^\star$ **then**
13:      $M^\star.add(m_c^\star)$.
14: Set $M = \phi$. //The matches of Q.
15: **for** each $m^\star \in M^\star$ **do**
16:    Get the subgraph $m \subseteq G$ corresponding to $m^\star$.
17:    **if** all literal constraints are valid on m **then**
18:      $M.add(m)$.
19: **return** M.

8 Experiments

To the best of our knowledge, only YAGO2 Demo and the system implemented by A. Brodt et al.[4] (DisRDF for short) are available spatial RDF data management system. YAGO2 Demo only accepts range queries over spatial statements based on several hard-coded spatial predicates. DisRDF models the spatial entities with various shapes and only accepts range queries over the spatial entities. Since we support both range query and spatial join semantic over the spatial entities and the spatial statements, the comparisons to other approaches are not applicable. Thus, we focus on the performance and the specific characteristics of our approach.

8.1 Data Set and Setup

Data Set YAGO2 is a real data set based on Wikipedia ,WordNet and GeoNames. The latest version of YAGO2 have more than 10 million entities and 440 million statements. We obtain a spatial RDF data set from YAGO2 by removing some statements that describe the date when another statement is extract or the URL where another statement is extract from. The condensed data set has more than 10 million entities/classes and more than 180 million statements. More than 7 million entities are spatial entities, and more than 90 million statements are spatial statements.

Queries & Setup In order to evaluate our approach, we manually generate 10 sample spatial SPARQL queries that have different features. The sample queries are divided into 5 classes, i.e., A, B, C, D, E. The queries in set A are star queries with the range query predicates over the entities. The queries in set B are the queries with the range query predicates over the entities. The queries in set C are the queries with spatial join predicates over entities. The queries in set D are the queries with range query and spatial join predicates over statements. The queries in set E are combined queries. The queries are given in our technical report [17].

Table 1. The Result Set Size of Queries

	A1	A2	B1	B2	C1	C2	D1	D2	E1	E2
Spatial Queries	3	1177	1	10	18	25	2	23	7	12
SPARQL Queries	10,137,491	8,567	36	50	36	50	36	50	40	50

Table 1 shows the selectivity of each query. In order to show the inefficiency of post-processing method (i.e., finding SPARQL query results by ignoring the spatial constraints and then verifying the candidates by the spatial predicates), we also report the result sizes of all queries discarding the spatial constraints. We run all queries on a PC with an Intel Xeon CPU E5645 running at 2.40 GHz and 16 GB main memory.

Table 2. Statistics of Node Capability

Node Capability	Index Size(MB)	Tree Height	Node Count
30	5,537	6	571,064
50	4,376	5	341,905
100	3,342	4	170,121
150	2,938	4	113,365

Table 3. Statistics of Tree-Construction

Index Style	Node Capability	Index Size(MB)	Tree Height	Node Count
SS-tree	100	3,342	4	170,121
	150	2,938	4	113,365
VS-tree+	100	4,332	3	204,890
	150	3,990	3	138,931

8.2 Evaluating Node Capability

In this subsection, we evaluate whether the different *node capabilities* (i.e., the maximal number of child nodes of each node in the tree index) affect the off-line and the on-line performance. In the evaluation, the node capabilities are set to 30, 50, 100, 150 respectively.

Table 2 shows the storage cost of SS-tree with different node capabilities. Obviously, lower node capability leads to larger node count, higher tree height and larger storage requirement, vice versa. Figure 8 shows the count of node access during search. Clearly, the count of node access depends on the capability of each node. Note that the count of data points involved during search = # of accessed nodes × node capability, which means the operation count on data

S-store: An Engine for Large RDF Graph Integrating Spatial Information   45

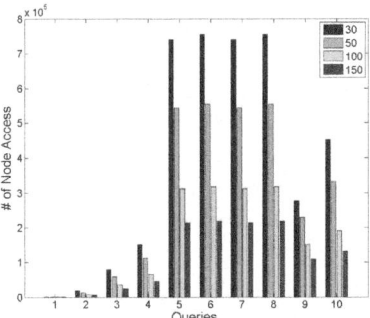 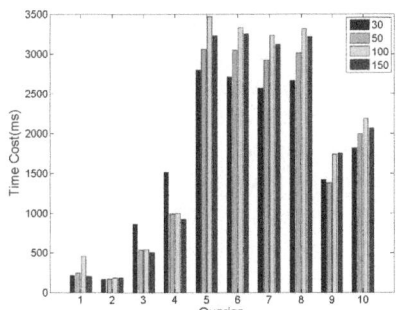

Fig. 8. Node Capability - Nodes Access   **Fig. 9.** Node Capability - Time Cost

points may be lower in the lower node capability situation. Figure 9 reports the query time cost of each query. The query time cost is proportional to the count of data points involved during search in most cases.

8.3 Evaluating Entity Organization

In this subsection, we evaluate whether the different entity organization styles affect the off-line and the on-line performance. We compare the SS-tree and the VS-tree plus spatial features (denoted as VS-tree+). Table 3 shows the results. The SS-tree demands lower storage space than VS-tree+ when the node capabilities are the same. Figure 10 and 11 show the count of nodes accessed and the time cost of each query. As we supposed, SS-tree works better for spatial feature filtering. Since SS-tree organize spatial entities in a R-tree, SS-tree works better on query 1, 2, 3, 4, 9 and 10, where the queries have spatial predicates on nodes.

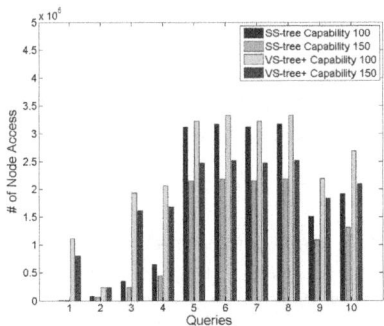 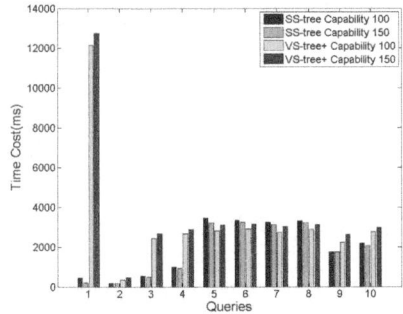

Fig. 10. SS-tree - Nodes Access   **Fig. 11.** VS-tree+ - Time Cost

8.4 Evaluating Performance

For evaluating the efficiency of our approach, we implement a baseline approach based on the method of [4]. The baseline approach adopts the post-processing solution, running SPARQL queries by ignoring the spatial predicates and then refining the candidates by considering the spatial constraints. In this subsection, we make a comparison between S-store and the baseline approach.

In practice, the baseline approach exploits gStore[21] as the RDF management system, and the node capability is set to 150. Besides, the MySQL is used to retrieve the coordinates of the entities and the statements.

The query response times are shown in Table 4, where G-store+ denotes the baseline approach. Since A1 and A2 have many candidate results (see Table 1), the time cost of the baseline is unacceptable. We can't get the results of A1 in reasonable time (more than half an hour), and the time cost for A2 is about 113 seconds. However, our approach (S-store) can answer the query A1 and A2 in 213 and 165 milliseconds, respectively. Although the other queries have just a few candidate results without spatial predicates, S-store still outperforms the baseline approach.

Table 4. The Performance comparison

	Time Cost (ms)									
	A1	A2	B1	B2	C1	C2	D1	D2	E1	E2
S-store	213	165	863	1,518	2,800	2,710	2,571	2,668	1,418	1,816
G-store+	>30min	112,406	5,894	9,555	4,478	4,127	3,624	6,750	5,839	3,779
Speed-up Ratio		99.8%	85.4%	84.1%	37.5%	34.3%	29.1%	60.5%	75.7%	51.9%

9 Conclusions

In this paper, we introduce spatial queries, a variant of SPARQL language, for querying RDF data with spatial features. Spatial queries employ spatial predicates for expressing the range query and the spatial join constraints. Besides, we introduce a novel index called SS-tree for evaluating the spatial queries. Based on SS-tree, we propose several pruning rules and a searching algorithm. The experimental results show the effectiveness and the efficiency of our approach. The spatial queries just cost a few seconds on YAGO2 data set, which has more than 10 million entities and 180 million statements.

Acknowledgments. Dong Wang and Lei Zou were supported by NSFC under Grant No.61003009. Xuchuang Shen and Dongyan Zhao were supported by NSFC under Grant No.61272344 and National High Technology Research and Development Program of China under Grant No. 2012AA011101. Lei Zou's work was partially supported by State Key Laboratory of Software Engineering(SKLSE), Wuhan University, China.

References

1. Abadi, D.J., Marcus, A., Madden, S., Hollenbach, K.: Sw-store: a vertically partitioned dbms for semantic web data management. VLDB J 18(2) (2009)
2. Abadi, D.J., Marcus, A., Madden, S., Hollenbach, K.J.: Scalable semantic web data management using vertical partitioning. In: VLDB (2007)
3. Brickley, D.: Basic geo (wgs84 lat/long) vocabulary. W3C Semantic Web Interest Group (2006), http://www.w3.org/2003/01/geo/
4. Brodt, A., Nicklas, D., Mitschang, B.: Deep integration of spatial query processing into native rdf triple stores. In: Proceedings of the 18th SIGSPATIAL International Conference on Advances in Geographic Information Systems, pp. 33–42. ACM (2010)
5. Broekstra, J., Kampman, A., van Harmelen, F.: Sesame: A generic architecture for storing and querying RDF and RDF schema. In: Horrocks, I., Hendler, J. (eds.) ISWC 2002. LNCS, vol. 2342, pp. 54–68. Springer, Heidelberg (2002)
6. Deppisch, U.: S-tree: A dynamic balanced signature index for office retrieval. In: SIGIR (1986)
7. Erling, O., Mikhailov, I.: RDF support in the virtuoso DBMS. In: Pellegrini, T., Auer, S., Tochtermann, K., Schaffert, S. (eds.) Networked Knowledge - Networked Media. SCI, vol. 221, pp. 7–24. Springer, Heidelberg (2009)
8. Gröger, G., Kolbe, T., Czerwinski, A., Nagel, C.: Opengis city geography markup language (citygml) encoding standard. Open Geospatial Consortium Inc. Reference number of this OGC® project document: OGC (2008)
9. Guttman, A.: R-trees: A dynamic index structure for spatial searching. In: SIGMOD (1984)
10. Haklay, M., Weber, P.: Openstreetmap: User-generated street maps. IEEE Pervasive Computing 7(4), 12–18 (2008)
11. Herring, J.: Opengis® implementation specification for geographic information-simple feature access-part 1: Common architecture. Open Geospatial Consortium, p. 95 (2006)
12. Hoffart, J., Suchanek, F., Berberich, K., Weikum, G.: Yago2: a spatially and temporally enhanced knowledge base from wikipedia. Artificial Intelligence (2012)
13. Klyne, G., Carroll, J., McBride, B.: Resource description framework (rdf): Concepts and abstract syntax. W3C Recommendation 10 (2004)
14. Neumann, T., Weikum, G.: Rdf-3x: a risc-style engine for rdf. PVLDB 1(1) (2008)
15. Neumann, T., Weikum, G.: x-rdf-3x: Fast querying, high update rates, and consistency for rdf databases. PVLDB 1(1) (2010)
16. Singh, R., Turner, A., Maron, M., Doyle, A.: Georss: Geographically encoded objects for rss feeds (2008)
17. Wang, D., Zou, L., Feng, Y., Shen, X., Tian, J., Zhao, D.: S-store: An engine for large rdf graph integrating spatial information (2013), http://www.icst.pku.edu.cn/intro/leizou/TR/2013/TR-DB-ICST-PKU-2013-001.pdf
18. Weiss, C., Karras, P., Bernstein, A.: Hexastore: sextuple indexing for semantic web data management. PVLDB 1(1) (2008)
19. Wilkinson, K.: Jena property table implementation. In: SSWS (2006)
20. Wilkinson, K., Sayers, C., Kuno, H.A., Reynolds, D.: Efficient rdf storage and retrieval in jena2. In: SWDB (2003)
21. Zou, L., Mo, J., Chen, L., Özsu, M., Zhao, D.: gstore: answering sparql queries via subgraph matching. Proceedings of the VLDB Endowment 4(8), 482–493 (2011)

Physical Column Organization in In-Memory Column Stores

David Schwalb, Martin Faust, Jens Krueger, and Hasso Plattner

Hasso Plattner Institute, Potsdam, Germany

Abstract. Cost models are an essential part of database systems, as they are the basis of query performance optimization. Disk based systems are well understood and sophisticated models exist to compare various data structures and to estimate query costs based on disk IO operations. Cost models for in-memory databases shift the focus from disk IOs to main memory accesses and CPU costs. However, modeling memory accesses is fundamentally different and common models do not apply anymore.

In this work, we examine the plan operations scan with equality selection, scan with range selection, positional lookup and insert in in-memory column stores regarding different physical column organizations. We consider uncompressed columns, bit compressed and dictionary encoded columns with sorted and unsorted dictionaries. Furthermore, we discuss tree indices on columns and dictionaries and present a detailed parameter evaluation, considering the number of distinct values, value skewness and value disorder. Finally, we present and evaluate a cost model based on cache misses for estimating the runtime of the discussed plan operations.

1 Introduction

In-memory column stores commence to experience a growing attention by the research community. They are traditionally strong in read intensive scenarios with analytical workloads. A recent trend introduces column stores for the backbone of business applications as a combined solution for transactional and analytical processing. This approach introduces high performance requirements as well for read performance as also for write performance to the systems.

Typically, optimizing read and write performance of data structures results in trade-offs, as e.g. higher compression rates introduce overhead for writing, but increase read performance. The underlying idea of this paper is a database system, which supports different data structures with unique performance characteristics, allowing to switch and choose the used structures at runtime depending on the current, historical or expected future workloads. This paper will not provide a complete description or design of such a system, but focuses on selected data structures for in-memory column stores.

Our contributions are i) a detailed parameter discussion and analysis for the operations scan with equality selection, scan with range selection, lookup and insert on different physical column organizations in in-memory column stores and ii) a cache based cost model for each operation and column organization.

The remainder of the paper is structured as follows. Section 2 gives an overview of related work, followed by a system definition in Section 3. Section 4 introduces the considered plan operators and their implementation, followed by a discussion of parameter influences in Section 5. Then, Section 6 introduces a cache miss based cost model estimating the costs for the discussed plan operations, followed by Section 7 introducing column and dictionary indices and their respective costs. Section 8 closes the paper with concluding remarks.

2 Background and Related Work

This section gives an overview and background of related work regarding in-memory column stores, followed by work concerning cost models for main memory databases and cache effects.

Recent research started questioning the separation of transactional and analytical systems and introduced efforts of uniting both systems again [4,6,8,12,13]. The back-bone of such a system's architecture could be a compressed in-memory column-store, as proposed in [4,12]. Column oriented databases have proven to be advantageous for read intensive scenarios [9,14], especially in combination with an in-memory architecture. Such a system has to handle contradicting requirements for many performance aspects. The question becomes which column oriented data structures are used in combination with light-weight compression techniques, enabling the system to find a balanced trade-off between the contradicting requirements. This paper aims at studying these trade-offs and at analyzing possible data structures.

Relatively little work has been done on researching main memory cost models. This probably is due to the fact, that modeling the performance of queries in main memory is fundamentally different than in disk based systems were IO access is clearly the most expensive part. In in-memory databases, query costs consist of memory and cache access costs on the one hand and CPU costs on the other hand. Manegold and Kersten [10] describe a generic cost model for in-memory database systems, to estimate the execution costs of database queries based on their cache misses. The main idea is to describe and model reoccurring basic patterns of main memory access. More complex patterns are modeled by combining the basic access patterns with a presented algebra. In contrast to the cache-aware cost model from Manegold which focusses on join operators, we compare scan and lookup operators on different physical column layouts.

The influences of the memory hierarchy on application performance has been extensively studied in literature. Various techniques have been proposed to measure costs of cache misses and pipeline stalling. Most approaches are based on handcrafted micro benchmarks exposing the respective parts of the memory hierarchy. Barr, Cox and Rixner [2] study the penalties occurring when missing the translation look-aside buffer (TLB) in systems with radix page tables like the x86-64 system and compare different page table organizations. Due to the page

Table 1. Parameter symbol overview

Description	Unit	Symbol	Description	Unit	Symbol
Value Domain	-	$\mathbb{V}$	Number distinct values	-	d
Dictionary	-	$\mathbb{D}$	Uncompr. Value-Length	bytes	e
Number of rows	-	r	Compr. Value-Length	bits	e_c
Value Disorder	-	u	Query Selectivity	rows	s
Value Skewness	-	k	value-id of v_i	-	$id(v_i)$

table access, the process of translating a virtual to a physical address can induce additional TLB cache misses, depending on the organization of the page table. Babka and Tuma [1] present a collection of experiments investigating detailed parameters and provide a framework measuring performance relevant aspects of the memory architecture of x86-64 systems. The experiments vary from determining the presence and size of caches, the cache line sizes to measure cache miss penalties.

3 System Definition

This section gives a formal definition of the used system, considered physical column organizations and examined parameters. We consider a database consisting of a set of tables $\mathbb{T}$. A table $t \in \mathbb{T}$ consists of a set of attributes $\mathbb{A}_t$. The number of attributes of a table t will be denoted as $|\mathbb{A}_t|$. We assume the value domain $\mathbb{V}$ of each attribute $a \in \mathbb{A}_t$ to be finite and require the existence of a total order ρ over $\mathbb{V}$. In particular, we define e as the value length of attribute a and assume $\mathbb{V}$ to be the set of alphanumeric strings with the length e. An attribute a is a sequence of r values $v \in \mathbb{D}$ with $\mathbb{D} \subseteq \mathbb{V}$, where r is also called number of rows of a and $\mathbb{D}$ also called the dictionary of a.

Table 1 gives an overview of the examined parameters. $\mathbb{D}$ is a set of values $\mathbb{D} = \{v_1, ..., v_n\}$. We define $d := |\mathbb{D}|$ as the number of distinct values of an attribute. In case the dictionary is sorted, we require $\forall_{v_i \in \mathbb{D}} : v_i < v_{i+1}$. In case the dictionary is unsorted, $v_1, ..., v_n$ are in insertion order of the values in attribute a. The position of a value v_i in the dictionary defines its value-id $id(v_i) := i$. For bit-compression, the number of values in $\mathbb{D}$ is limited to 2^b, with b being the number of bits used to encode values in the value vector. We define $e_c := b$ as the compressed value length of a, requiring $e_c \geq \lceil log_2(d) \rceil$ bits. The degree of sortedness of the values in a is described by the measure of disorder denoted by u, based on Knuth's measure of disorder, which describes the minimum amount of elements that need to be removed from a sequence so that the sequence would be sorted [7]. Finally, we define the value skewness k, describing the distribution of values of an attribute, as the exponent characterizing a Zipfian distribution. We chose to model the different distributions by a Zipfian distribution, as the authors in [5] state that the majority of columns analyzed from financial, sales and distribution modules of a enterprise resource planning (ERP) system were following a power-law distribution – a small set of values occurs very often, while the majority of values is rare.

The logical view of a column is a simple collection of values, allowing to append new values, retrieving the value from a position and scanning the complete column with a predicate. How the data is actually stored in memory is not specified. In general, data can be organized in memory in a variety of different ways, e.g. in standard vectors in insertion order, ordered collections or collections with tree indices [13]. In addition to the type of organization of data structures, the used compression techniques are also essential for the resulting performance characteristics. Regarding compression, we will focus on the light weight compression techniques, dictionary encoding and bit compression.

Uncompressed columns store the values as they are inserted in sequential manner, as e.g. used in [8]. In a dictionary encoded column, the actual column contains two containers: the attribute vector and the value dictionary. The attribute vector is a vector storing only references to the actual values of e_c bit, which represent the index of the value in the value dictionary and is also called value-id. For the remainder, we assume $e_c = 32$ bit. The value dictionary may be an unsorted or ordered collection. Usually it is advisable to maintain a tree index structure on top of an unsorted dictionary.

4 Operators

We consider the plan operators scan with equality selection, scan with range selection, positional lookup and inserting new values and discuss their theoretical complexity. These operators were chosen, as we identified them as the most basic operators needed by a database system, assuming a insert only system as proposed in [4,8,12,13]. Additionally, more complex operators can be assembled by combining these basic operators, as e.g. a nested loop join consisting of multiple scans. We differentiate between equality and range selections as they have different performance characteristics due to differences when performing value comparisons introduced by the dictionary encoding.

A **scan with equality selection** sequentially iterates through all values of a column and returns a list of positions where the value in the column equals the searched value. The costs for an equal scan on an uncompressed column are characterized by comparing all r values and by building the result set, resulting in $\mathcal{O}(r \cdot e + s \cdot r)$. On a column with a sorted dictionary, the value-id in the value dictionary of the column for the searched value x is retrieved first by performing a binary search for x in the dictionary. Then, the value-ids of the column are scanned sequentially and each matching value-id is added to the set of results. The costs for an equal scan on a column with a sorted dictionary consist of the binary search cost in the dictionary and comparing each value-id, resulting in $\mathcal{O}(\log d + r \cdot e_c + s \cdot r)$. In contrast to the sorted dictionary case, the search costs for a column with an unsorted dictionary are linear, resulting in a complexity for an equal scan of $\mathcal{O}(d + r \cdot e_c + s \cdot r)$.

A **scan with range selection** sequentially iterates through all values of a column and returns a list of positions where the value in the column is between a *low* and *high* boundary. The implementation of a range scan for an uncompressed column is similar to the equal scan, the comparisons can be performed directly on the values while iterating sequentially through the column. Therefore, the costs are determined by the value length e, the number of rows r and the selectivity s of the scan, resulting in $\mathcal{O}(r \cdot e + s \cdot r)$. The implementation for the range scan on a dictionary encoded column with a sorted dictionary works as follows. First, the value-ids of *low* and *high* are retrieved with a binary search in the dictionary. As the dictionary is sorted $id_{low} < id_{high} \Rightarrow value(id_{low}) < value(id_{high})$ applies. Therefore, the value-ids of the column can be scanned and it can be decided only by comparing with the value-ids of *low* and *high* if the current value-id has to be a part of the result set. The costs are similar to the costs for an equal scan, determined by the binary search costs, the scanning of the column and building the result set, resulting in $\mathcal{O}(\log d + r \cdot e_c + s \cdot r)$. Finally, on an unsorted dictionary, we can not draw any conclusions of the relations between two values based on their value-ids in the dictionary. We iterate sequentially through the value-ids of the column. For each value-id, we perform a lookup retrieving the actual value stored in the dictionary, resulting in a complexity of $\mathcal{O}(r \cdot e_c + r \cdot e + s \cdot r)$.

A **positional lookup** retrieves the value of a given position p from the column. The output is the actual value, as the position is already known. In case of an uncompressed column, the value can be retrieved directly, resulting in a complexity of $\mathcal{O}(e)$. In the case of a dictionary encoded column, the value-id is first retrieved for position p and then a dictionary lookup is performed in order to retrieve the searched value. The costs depend on the compressed and the uncompressed value length, resulting in a complexity of $\mathcal{O}(e_c + e)$.

An **insert operation** appends a new row to a column. As we keep the rows always in insertion order, this can be implemented as a trivial append operation, where we assume that there is enough free and allocated space to store the inserted row. In the case of a dictionary encoded column, we have to check if the value is already in the dictionary. First, a binary search is performed on the dictionary for value v. If v is not found in the dictionary, it is inserted so that the sort order of the dictionary is preserved. In case that v is not inserted at the end of the dictionary a re-encode of the complete column has to be performed, in order to reflect the updated value-ids of the dictionary. After the re-encode or if v was already found in the dictionary, the value-id is appended to the column. The complexity is in $\mathcal{O}(\log d + d + r \cdot e_c + e)$. In case of a column with an unsorted dictionary, we first search for the inserted value in the dictionary by performing a linear search. As the dictionary is not kept in a particular order, the values are always appended to the end of the dictionary. Therefore, no re-encode of the column is necessary. The resulting complexity is $\mathcal{O}(d + e_c + e)$.

5 Parameter Effects

In the previous sections we defined plan operators and discussed their implementations and complexity depending on the parameters defined in Section 3. This section thrives to experimentally verify the theoretical discussion of the

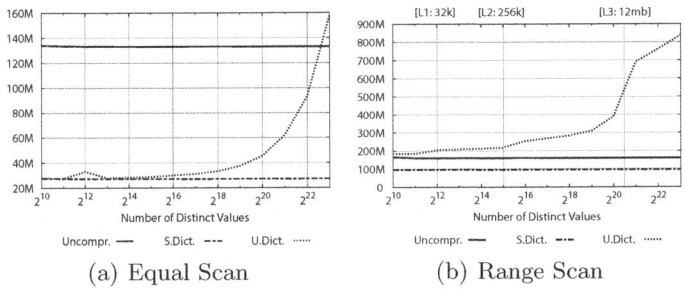

(a) Equal Scan (b) Range Scan

Fig. 1. CPU cycles for (a) equal scan and (b) range scan on one column with number of distinct values d varied from 2^{10} to 2^{23}, $r = 2^{23}$, $u = 2^{23}$, $e = 8$, $k = 0$ and a query selectivity of $s = 2,000$

parameters and their influence on plan operations.[1] Due to space limitations, we only show some detailed experimental results.

Number of Rows. The number of rows r has a linear influence on the performance of scan operations, whereas the time per row stays constant. For positional lookups, the number of rows has no influence on the performance on the lookup operation. When inserting new values into a column, the number of rows r has no influence on the time an actual insert operation takes, regardless if the column is uncompressed or dictionary encoded. However, on dictionary encoded columns with a sorted dictionary, the number of rows has a linear influence on the re-encode operation if necessary.

Number of Distinct Values. We now focus on the number of distinct values d of a column and their influence on scan, insert and lookup operations. When scanning a column with an equal scan, we expect the number of distinct values to influence the dictionary encoded columns, but not the uncompressed column. Figure 1(a) shows the results of an experiment performing an equal scan on a column with 2^{23} rows and d varied from 2^{10} to 2^{23}. We chose a selectivity of 2,000 rows, in order to keep the effect of writing the result set minimal. As expected, the runtime for the scan on the uncompressed column is not affected and we clearly see the linear impact on the column with an unsorted dictionary. In contrast to an equal scan, the implementation of a range scan only differs in the case for an unsorted dictionary. Therefore, the cases for an uncompressed column and a column with a sorted dictionary are the same as discussed for the equal scan operation, as Figure 1(b) shows. In case of an unsorted-dictionary encoded column, Figure 1(b) shows a strong impact of the varied number of distinct values on the runtime. The increase in CPU cycles with increasing distinct values is due to a cache effect. As $u = 2^{23}$, we access the dictionary in a random fashion while iterating over the column. As long as the dictionary is small and fits into

---

[1] All experiments were conducted on an Intel Xeon X5650, with 2x6 cores, hyper-threading, 2.67 GHz and 48 GB main memory. The system had 32 KB L1 data cache (8-way), 256 KB L2 cache (8-way), 12 MB L3 cache (16-way) and a two level TLB with 64 and 512 entries.

the cache, these accesses are relatively cheap. With a growing number of distinct values the dictionary gets too large for the individual cache levels and the number of cache misses per dictionary access increases, resulting in increasing time for the scan operation. Considering a value length of 8 bytes, we can identify jumps slightly before each cache level size of 32KB, 256KB and 12MB.

Value Disorder. When performing an equal scan, the comparisons can be done directly on the value-ids in case of a dictionary encoded column or are done directly on the values in case of an uncompressed column. Therefore, the value disorder does not influence the performance of equal scans. Regarding range scan operations, we see no influence in case of an uncompressed column or a column with a sorted dictionary when varying the disorder of values in a column. In case of a dictionary encoded column with an unsorted dictionary, we se an increase in CPU cycles. In contrast to an equal scan, the range scan operation on an unsorted dictionary has to lookup the actual values in the dictionary in order to compare them. When the value disorder is low, temporal and spatial locality for the dictionary access is high, which results in good cache usage with a high number of cache hits. The greater the disorder gets, the more random the accesses to the dictionary get and the number of cache misses when accessing the dictionary increases, resulting in more CPU cycles for the scan operation. The value disorder has no influence on single positional lookups and inserts.

Value Length. For uncompressed columns, we see an increase for the scan operation with longer values, as expected. However, every 16 bytes we noticed a perfromance jump, due to alignment effects. In case of a dictionary compressed column, we see no significant influence in case of a sorted dictionary based on the value length. When using an unsorted dictionary, the costs for scanning the dictionary are significantly higher and we identify a significant impact of the value length on the total scan costs. In case of an uncompressed column, we see an increase in costs with larger value lengths and the same alignment effect, as the values are compared directly and larger values result in larger costs for comparing the values. Similar to the equal scan, we see no significant impact of the value length for the case of a sorted dictionary but an increase in case of an unsorted dictionary. The costs for a positional lookup do increase linearly with increasing size of values, as the actual values are returned by the lookup operation, resulting in more work for longer values. The costs for inserting new values always increase with larger values, as the costs for writing the values do increase.

Value Skewness. The skewness of values influences the pattern in which the dictionary of a column is accessed when scanning its value-ids and looking them up in the dictionary, the more skew the value distribution, the less cache misses occur. In case of an equal scan, positional lookup or insert we do not have this pattern of scanning the column and accessing the dictionary. Therefore, we do not expect the skewness of values in a column to influence these operations, which is outlined by the experimental result shown in Figure 2(a) for a scan operation with equality selection. In contrast, Figure 2(b) shows the influence of the value skewness on a range scan operation. In case of a dictionary encoded column with an unsorted dictionary, we scan the value-ids of the column sequentially

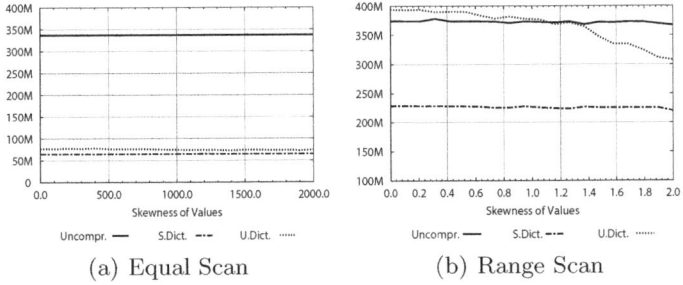

Fig. 2. CPU cycles for (a) equal scan and (b) range scan on one column with value skewness k varied from 0 to 2, $r = 20$ million, $d = 200{,}000$, $u = 20$ million, $e = 8$ and a query selectivity of $s = 2{,}000$

and randomly access the value dictionary (value disorder $u = 20$ million). The more skew the value distribution is, the more likely it gets that a value with a high frequency is accessed and is still in the cache. Therefore the number of cache misses is reduced for skewed value distributions, resulting in a faster range scan operation.

6 Estimating Cache Misses

In the previous section, we found parameters like the influence of the number of distinct values or the value skewness on a scan operator with range selection, that were not inferable based on the theoretical complexity. These influences are based on cache effects, which will be discussed in this section followed by a cost model to predict the number of cache misses for our discussed plan operators. In traditional disk based systems, IO operations are counted and used as the basic unit of measurement. For in-memory database systems IO operations are not of interest and the focus shifts to main memory accesses.

In general, assuming that all data resides in main memory, the total execution time of an algorithm can be separated into the time spent computing T_{CPU} and the time for accessing the data in memory T_{Mem} [10]. Due to increasing processor speeds but stagnating memory speeds, memory access is getting more expensive in relation, as more CPU cycles are wasted while stalling for memory access [3, 10, 11]. In case the considered algorithms are close to be bandwidth bound, T_{Mem} is the dominant factor driving the execution time. Additionally, modeling T_{CPU} requires internal knowledge of the used processor, is very implementation specific and also dependent on the resulting machine code created by the compiler, which makes it hard to model. As our considered operations on the various data structures only perform a small amount of computations while accessing large amounts of data, we assume our algorithms to be bandwidth bound and believe T_{Mem} to be a good estimation of overall costs. The costs for accessing memory can vary heavily due to the underlying memory hierarchy and mechanisms like prefetching and virtual address translation and can be quantified by the number of cache misses on each level in the memory hierarchy.

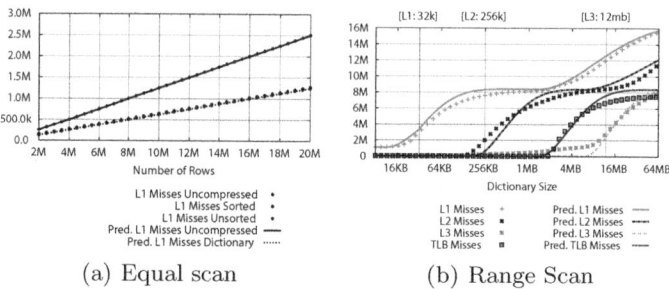

(a) Equal scan (b) Range Scan

Fig. 3. Evaluation of predicted cache misses

We will provide explicit functions to calculate the estimated number of cache misses for each operation and data structure. Some cost functions are based on the work presented in [10], where the authors describe a generic cost model, estimating the execution costs of algorithms based on cache misses by modeling basic access patterns and an algebra to model more complex patterns. We develop own parameterized cost functions estimating the number of cache misses for each operation, specifically designed for the operations and data structures. With the specific parameters for each cache level, the cache misses on that level can be predicted. Furthermore, the total costs can be calculated by multiplying the number of cache misses with the latency of the next level in the hierarchy as proposed in [10]. Measuring the individual cache level latencies requires accurate calibration and is very system specific. As a simpler and more robust estimation, we use the number of cache misses as a direct indicator for the resulting number of cycles, only roughly weighting the different cache levels. The cache level in the hierarchy is indicated by i, whereas the Transaction Look-Aside Buffer (TLB) is treated as an additional level in the hierarchy. The cache line size or block size of a respective level is given by B_i and the size by C_i. The number of cache lines at the level i is denoted by $\#i$. The function $M_i(o, c)$ describes the estimated amount of cache misses for an operation o on a column c. The operations are *escan*, *rscan*, *lookup* and *insert*. The respective physical column organization is given by a subscript indicating A) an uncompressed column with, B) a dictionary encoded column with an unsorted dictionary and C) a dictionary encoded column with a sorted dictionary.

Scan with Equality Selection. An equal scan on uncompressed columns, consists of sequentially iterating over the column, resulting in as many cache misses as the column covers cache lines. In case the column is dictionary encoded with a sorted dictionary, the binary search for the searched value results in log_2 random cache misses, while the sequential scan over the compressed value-ids results in as many cache misses as the compressed column covers cache lines (Equation 1). The number of cache misses for an equal scan on a column with an unsorted dictionary is similar as with a sorted dictionary, but instead of a binary search a linear search is performed, scanning in average half the dictionary (Equation 2). Figure 3(a) shows that the predicted cache misses

$$\mathbf{M_i}(escan_B, c) = \left\lceil \frac{c.\boldsymbol{r} \cdot c.\boldsymbol{e_c}}{B_i} \right\rceil + \log_2(c.\boldsymbol{d} \cdot c.\boldsymbol{e}) \qquad (1)$$

$$\mathbf{M_i}(escan_C, c) = \left\lceil \frac{c.\boldsymbol{r} \cdot c.\boldsymbol{e_c}}{B_i} \right\rceil + \left\lceil \frac{c.\boldsymbol{d} \cdot c.\boldsymbol{e}}{2 \cdot B_i} \right\rceil \qquad (2)$$

follow closely the measured number of misses for the equal scan experiment varying the number of rows.

Scan with Range Selection. In case of an uncompressed column, a scan with range selection iterates sequentially over the uncompressed values, comparing the values with the requested range. Similarly, in case of a sorted dictionary, the searched values are retrieved from the dictionary with a binary search and the value-ids are scanned sequentially for the search value-ids. In both cases, the resulting cache misses are the same as for a scan with equality selection. For unsorted dictionaries, the scan operation sequentially iterates over the column and has to perform a random accesses into the dictionary due to the range selection. Regarding the random access into the dictionary, we assume that every value in the dictionary is accessed at least once. In the best case the access to the dictionary is sequentially utilizing all values in a cache-line. In the worst case, every access to the dictionary may result in a cache miss. The number of cache misses increases with increasing dictionary sizes respective to the cache size and the amount of disorder in the column. Therefore, we model the number of random misses by interpolating between 0 and the number of rows in the column.

In order to smoothly interpolate between two values, we define the following helper functions. I_l is a simple linear interpolation function between y_0 and y_1, whereas t varies from 0 to 1. Furthermore, we define I_d as a decelerating interpolation function.

Based on I_l and I_d, we construct I_c as a cosinus-based interpolation function to smoothly interpolate between two values, as we found this interpolation type to fit well to the cache characteristics. Finally, we introduce I as a helper function modeling a function stepping smoothly from y_0 to y_1 around a location of x_0, whereas ρ indicates the range in which the interpolation and τ the degree of how asymmetric the interpolation is performed. These values might be system specific and can be calibrated as needed.

If the number of covered cache-lines C_i is smaller than the number of available cache-lines $\#_i$, every cache-line is loaded at its first access and remains in the cache. For subsequent accesses, this cache-line is already in the cache and the

$$I_l(y_0, y_1, t) = y_0 + t \cdot (y_1 - y_0) \qquad (3)$$

$$I_d(y_0, y_1, t) = I_l(y_0, y_1, 1 - (1-t)^2) \qquad (4)$$

$$I_c(y_0, y_1, t) = I_l\left(y_0, y_1, \frac{1 - \cos(\pi \cdot I_d\langle 0, 1, t\rangle)}{2}\right) \qquad (5)$$

$$I(x, x_0, y_0, y_1, \rho, \tau) = \begin{cases} y_0 : x < 2^{x_0-\rho} \\ y_1 : x \geq 2^{x_0+\rho*\tau} \\ I_c(y_0, y_1, \frac{\log_2(x)-x_0+\rho}{\rho*(\tau+1)}) : else \end{cases} \quad (6)$$

$$\mathbf{M_i^r}(rscan_C, c) = I(c.\boldsymbol{d}, \log_2(C_i), 0, c.\boldsymbol{r}, \rho, \tau) \quad (7)$$

$$\mathbf{M_i^s}(rscan_C, c) = \max\left(\left\lceil \frac{\mathbf{C}(i,c)}{4096} \right\rceil, \mathbf{C}(i,c) - \frac{\mathbf{M_i^r}}{3}\right) \quad (8)$$

$$\mathbf{M_i^{tlb}}(rscan_C, c) = I(c.\boldsymbol{d}, \log_2(C_{tlb} \cdot 4^i), 0, c.\boldsymbol{r}, \rho, \tau) \quad (9)$$

access does not create an additional cache miss. If $C_i > \#_i$, then already loaded cache-lines may be evicted from cache by loading other cache-lines. Subsequent accesses then have to load the same cache-line again, producing more cache misses. The worst case is that every access to a cache-line has to load the line again, because it was already evicted, resulting in $col.\boldsymbol{r}$ cache misses. Assuming randomly distributed values in a column, the number how often cache-lines are evicted depends on the ratio of the number of cache-lines $\#_i$ and the number of covered cache-lines C_i. With increasing C_i the probability that cache-lines are evicted before they are accessed again increases. Equation 7 outlines the number of random cache misses.

The number of sequential cache misses is calculated in Equation 8 and depends on the success of the prefetcher. In case no or only a few random cache misses occur, the prefetcher has not enough time to load the requested cache lines, resulting in sequential misses. With increasing numbers of random cache misses, the time window for prefetching increases, resulting in less sequential cache misses. Assuming a page size of 4KB, we found $\mathbf{M_i^s}$ to be a good estimation, as a micro benchmark turned out that every three random cache misses when accessing the dictionary leave the prefetcher enough time to load subsequent cache lines. Additionally, we also have to consider extra penalties payed for TLB misses, as outlined by Eqaution 9. In case an address translation misses the TLB and the requested page table entry is not present in the respective cache level another cache miss occurs. In the worst case, this can introduce an additional cache miss for every dictionary lookup.

Finally, the total number of cache misses for a scan operation with range selection on a column with an uncompressed dictionary is given by adding random, sequential and TLB misses. Figure 3(b) shows a comparison of the measured effect of an increasing number of distinct values on a range scan on an uncompressed column with the predictions based on the provided cost functions. The figure shows the number of cache misses for each level and the model correctly predicts the jumps in the number of cache misses.

Lookup. A lookup on an uncompressed column results in as many cache misses as one value covers cache lines on the respective cache level. In case the column is dictionary encoded it makes no difference if the lookup is performed on a column with a sorted or an unsorted dictionary, hence we provide one function $\mathbf{M_i}(lookup_{B/C})$ for both cases.

Insert. The insert operation is the only operation we consider writing to main memory. Although it is not quite accurate, we will treat write access similar as reading from main memory and only consider resulting cache misses. An insert into an uncompressed column is trivial, resulting in as many cache misses as one value covers cache lines on the respective cache level. In case we perform an

$$\mathbf{M_i}(insert_B, c) = \left\lceil \frac{c.e}{B_i} \right\rceil + \left\lceil \frac{c.e_c}{B_i} \right\rceil + \left\lceil \log_2 \left(\frac{c.d \cdot c.e}{B_i} \right) \right\rceil \tag{10}$$

$$\mathbf{M_i^s}(insert_C, c) = \left\lceil \frac{c.e}{B_i} \right\rceil + \left\lceil \frac{c.e_c}{B_i} \right\rceil + \left\lceil \frac{c.d \cdot c.e}{2 \cdot B_i} \right\rceil \tag{11}$$

insert into a column with a sorted dictionary, we first perform a binary search determining if the value is already in the dictionary, before writing the value and value-id, assuming the value was not already in the dictionary (Equation 10). The number of cache misses in the unsorted dictionary case are similar to the sorted dictionary case, although the cache misses for the search depend linearly on the number of distinct values (Equation 11).

7 Index Structures

This section discusses the influence of index structures on top of the evaluated data structures and their influence on the discussed plan operators. First, we extend the unsorted dictionary case by adding a tree structure on top, keeping a sorted order and allowing binary searches. Second, we discuss the influence of inverted indices on columns and extend our model to reflect these changes. As tuples are stored in insertion order, we assume an index to be a separate auxiliary data structure on top of a column, not affecting the placement of values inside the column. Furthermore, we distinguish between column indices and dictionary indices. A column index is built on top of the values of one column, e.g. by creating a tree structure to enable binary search on the physically unsorted values in the column. In contrast, a dictionary index is a B$^+$-Tree built only on the distinct values of a column, enabling binary searching an unsorted dictionary in order to find the position of a given value in the dictionary.

Column and dictionary indices are assumed to be implemented as B$^+$-Tree structures. We denote the fan out of a tree index structure with I_f and the number of nodes needed to store d keys with I_n. The fan out constrains the number n of child nodes of all internal nodes to $I_f/2 \leq n \leq I_f$. I_{B_i} denotes the numbers of cache lines covered per node at cache level i. The number of matching keys for a scan with a range selection is denoted by $q.n_k$ and $q.n_v$ denotes the average number of occurrences of a key in the column.

7.1 Dictionary Index

A dictionary index is defined as a B^+-Tree structure on top of an unsorted dictionary, containing positions referencing to the values in the dictionary. Looking up a record is not affected by a dictionary index as the index can not be leveraged performing the lookup and does not have to be maintained. Also, scans with range selections still need to lookup and compare the actual values as the value-ids of two values still allow no conclusions about which value is larger or smaller.

Regarding equal scans on a column with a dictionary index, we can leverage the dictionary index for retrieving the value-id and perform a binary search. Therefore, the costs for the binary search depend logarithmically on the number of distinct values of the column. When comparing costs for a scan with equality selection for a column using an unsorted dictionary without a dictionary index to a column with a dictionary index, we notice similar costs for the scan operation on columns with few distinct values. However, as the dictionary grows, the costs for linearly scanning the dictionary increase linearly in case of not using a dictionary index and the costs with an index only increase slightly due to the logarithmic cost for the binary search, resulting in better performance when using a dictionary index.

One main cost factor for inserting new values into a column with an unsorted dictionary is the linear search determining if the value is already in the dictionary. This can be accelerated through the dictionary index, although it comes with the costs of maintaining the tree structure. Assuming the new value is not already in the dictionary, the costs for inserting it are writing the new value in the dictionary, writing the compressed value-id, performing the binary search in the index and adding the new value to the index.

Considering the discussed operations, a dictionary encoded column always profits by using a dictionary index. Therefore, we do not provide adapted cost functions for a dictionary index as we do not have to calculate in which cases it is advisable to use. Even insert operations do profit from the index as the dictionary can be searched with logarithmic costs, which outweighs the additional costs of index maintenance.

7.2 Column Index

We assume a column index to be a B^+-Tree structure, similar to the dictionary index described above. However, the index is built on top of the complete column and not only on the distinct values. Therefore, the index does not only store one position, but has to store a list of positions for every value. A column index can be added to any column, regardless of the physical organization of the column. Performing positional lookups does not profit from a column index.

A search with equality selection can be answered entirely by using the column index. Therefore, the costs do not depend on the physical layout of the column and the same algorithm can be used for all column organizations (Equation 12). First, the index is searched for value X by binary searching the tree structure

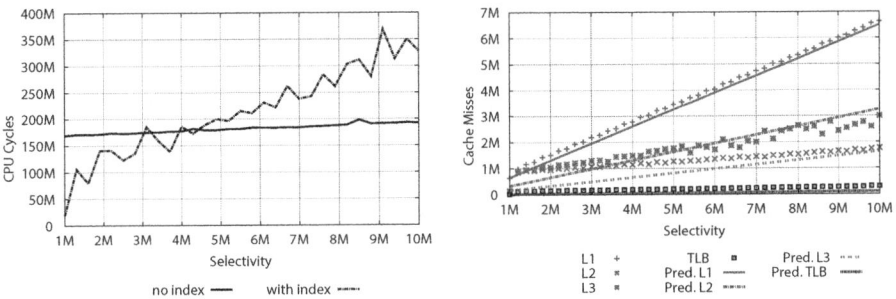

Fig. 4. CPU cycles for (a) scan with a range selection on a column with and without a column index. (b) shows the respective cache misses for the case of using a column index. $r = 10M$, $d = 1M$, $e = 8$, $u = 0$, $k = 0$.

resulting in a list of positions. If the value is not found, an empty list is returned. The resulting list of positions then has to be converted into the output format by adding all positions to the result array. Locating the leaf node for the searched key requires reading $\log_{I_f}(I_n) \cdot I_{B_i}$ cache lines for reading every node from the root node to the searched leaf node, assuming each accessed node lies on a separate cache line. Then, iterating through the list of positions and adding every position to the result array requires to read and write $q.n_v/B_i$ cachelines, assuming the positions are placed sequentially in memory. Searches with range selection can also be answered entirely by using the column index (Equation 13). Assuming the range selection matches any values, we locate the node with the first matching value by performing a binary search on the column index. The number of cache misses for the binary search are $\log_{I_f}(I_n) \cdot I_{B_i}$. Then, we sequentially retrieve the next nodes by following the next pointer of each node until we find a node with a key greater or equal to *high*. Assuming completely filled nodes, this requires reading all nodes containing the $q.n_k$ matching keys, resulting in $q.n_k/I_f$ nodes. For all matching nodes the positions are added to the result array, requiring to read and write $q.n_v/B_i$ cache lines per key. Inserting new values into the physical organization of a column is not affected by a column index. However, the new value has also to be inserted into the column index (Equation 14). The costs incurring for the index maintenance are independent from the physical organization of the column. This requires searching the tree structure for the inserted value, reading $\log_{I_f}(I_n) \cdot I_{B_i}$ cache lines. If the value already exists, the newly inserted position is added to the list of positions of the respective node, otherwise the value is inserted and the tree has to be potentially rebalanced. The costs for rebalancing are in average $\log_{I_f}(I_n) \cdot I_{B_i}$.

Figure 4(a) shows a comparison for a range scan on a column index compared to a column with a sorted dictionary and without an index. The figure shows the resulting CPU cycles for the scan operation with increasing result sizes. For small results the index performs better, but around a selectivity of roughly 4 million the complete scan performs better due to its sequential access pattern.

Figure 4(b) shows the resulting cache misses for the scan operation using the column index and the predictions based on the defined model.

$$\mathbf{M_i}(escan_I) = \log_{I_f}(I_n) \cdot I_{B_i} + 2 \cdot q.n_k \cdot \left\lceil \frac{q.n_v}{B_i} \right\rceil \tag{12}$$

$$\mathbf{M_i}(rscan_I) = \log_{I_f}(I_n) \cdot I_{B_i} + I_{B_i} \cdot \frac{q.n_k}{I_f} + 2q.n_k \cdot \left\lceil \frac{q.n_v}{B_i} \right\rceil \tag{13}$$

$$\mathbf{M_i}(insert_I) = 2 \cdot \log_{I_f}(I_n) \cdot I_{B_i} \tag{14}$$

8 Conclusions

In this paper, we presented a cost model for estimating cache misses for the plan operation equal scan, range scan, positional lookup and insert in a column-oriented in-memory database. We presented a detailed parameter analysis and cost functions predicting cache misses and TLB misses for different column organizations. The number of distinct values has a strong impact on range and equal scans, and renders unsorted dictionaries unusable for columns with a large amount of distinct values and dictionaries larger than available cache sizes. However, if the disorder in the column is low, the penalties payed for range scans are manageable. Additionally, the skewness of values in a column can influence the performance of range scan operators, although the impact is small unless the distribution is extremely skewed. Finally, we presented dictionary and column indices and argued that dictionary encoded columns always profit from using a dictionary index. Uncompressed columns seem to be well suited for classical OLTP workloads with a high number of inserts and mainly single lookups. As the number of scan operations and especially range scans increases, the additional insert expenses pay off, rendering dictionary encoded columns suitable for analytical workloads. Considering mixed workloads, the optimal column organization highly depends on the concrete workload.

References

1. Babka, V., et al.: Investigating Cache Parameters of x86 Family Processors. In: SPEC (2009)
2. Barr, T., et al.: Translation Caching: Skip, Don't Walk (the Page Table). ACM SIGARCH (2010)
3. Drepper, U.: What Every Programmer Should Know About Memory (2007)
4. Grund, M., et al.: HYRISE—A Main Memory Hybrid Storage Engine. VLDB (2010)
5. Hübner, et al: A cost-aware strategy for merging differential stores in column-oriented in-memory DBMS. BIRTE Workshop (2011)
6. Kemper, A., Neumann, T.: HyPer: A hybrid OLTP&OLAP main memory database system based on virtual memory snapshots. In: ICDE (2011)

7. Knuth, D.E.: Art of Computer Programming, vol. 3: Sorting and Searching. Addison-Wesley Professional (1973)
8. Krüger, J., et al.: Fast Updates on Read Optimized Databases Using Multi Core CPUs. VLDB (2011)
9. MacNicol, R., French, B.: Sybase IQ Multiplex — Designed For Analytics. VLDB (2004)
10. Manegold, S.: et al. Generic database cost models for hierarchical memory systems. VLDB (2002)
11. Moore, G.: Cramming more components onto integrated circuits. Electronics 38 (1965)
12. Plattner, H.: A Common Database Approach for OLTP and OLAP Using an In-Memory Column Database. Sigmod (2009)
13. Plattner, H., et al.: In-Memory Data Management: An Inflection Point for Enterprise Applications (2011)
14. Zukowski, M., et al.: MonetDB/X100 - A DBM in The CPU Cache. IEEE Data Eng. Bull. (2005)

Semantic Data Warehouse Design: From ETL to Deployment à la Carte

Ladjel Bellatreche[1], Selma Khouri[1,2], and Nabila Berkani[2]

[1] LIAS/ISAE-ENSMA, France
{bellatreche,selma.khouri}@ensma.fr
[2] National High School for Computer Science (ESI), Algeria
{n_berkani,s_khouri}@esi.dz

Abstract. In last decades, semantic databases ($\mathcal{SDB}$) emerge and become operational databases, since the major vendors provide semantic supports in their products. This is mainly due to the spectacular development of ontologies in several domains like E-commerce, Engineering, Medicine, etc. Contrary to a traditional database, where its tuples are stored in a relational (table) layout, a $\mathcal{SDB}$ stores independently ontology and its instances in one of the three main storage layouts (horizontal, vertical, binary). Based on this situation, $\mathcal{SDB}$ become serious candidates for business intelligence projects built around the Data Warehouse ($\mathcal{DW}$) technology. The important steps of the $\mathcal{DW}$ development life-cycle (user requirement analysis, conceptual design, logical design, ETL, physical design) are usually dealt in isolation way. This is mainly due to the complexity of each phase. Actually, the $\mathcal{DW}$ technology is quite mature for the traditional data sources. As a consequence, leveraging its steps to deal with semantic $\mathcal{DW}$ becomes a necessity. In this paper, we propose a methodology covering the most important steps of life-cycle of semantic $\mathcal{DW}$. Firstly, a mathematical formalization of ontologies, $\mathcal{SDB}$ and semantic $\mathcal{DW}$ is given. User requirements are expressed on the ontological level by the means of the goal oriented paradigm. Secondly, the ETL process is expressed on the ontological level, independently of any implementation constraint. Thirdly, different deployment solutions according to the storage layouts are proposed and implemented using the data access object design patterns. Finally, a prototype validating our proposal using the Lehigh University Benchmark ontology is given.

1 Introduction

Data are the most important asset of organizations since they are manipulated, processed and managed in the organization's daily activity. The best decisions are made when all the relevant data available are taken into consideration. These data are stored in various heterogeneous and distributed sources. To exploit these mine of data, data warehouse ($\mathcal{DW}$) technology showed its efficiency. It aims at materializing data and organizing them in order to facilitate their analysis. A $\mathcal{DW}$ can be seen as a materialized data integration system, where data are viewed in a multidimensional way [3] (i.e., data are organized into *facts* describing subjects of analysis. These data are analyzed according to different *dimensions*). A data integration system offers a global schema representing a unified view of data sources. Formally, it may be defined by a triple: $<\mathcal{G}, \mathcal{S}, \mathcal{M}>$ [18], where $\mathcal{G}$ is the global schema, $\mathcal{S}$ is a set of local schemas that describe the structure of each source participating in the integration process, and $\mathcal{M}$ is a set of assertions relating elements of the global schema $\mathcal{G}$ with elements of the local schemas $\mathcal{S}$. Usually, $\mathcal{G}$ and $\mathcal{S}$ are specified in suitable languages that may allow for the expression of various constraints [8]. In the $\mathcal{DW}$ context, the integration aspect is performed through ETL (Extract-Transform-Load) process, where data are extracted from sources, pre-processed and stored in a target $\mathcal{DW}$ schema [27].

As any information system, the construction of a traditional $\mathcal{DW}$ should pass through a number of phases characterizing its life-cycle: requirements analysis, conceptual design, logical design, ETL process and physical design [11]. Requirement analysis phase identifies which information is relevant to the decisional process by either considering the decision maker needs or the actual availability of data in the operational sources. Conceptual design phase aims at deriving an implementation-independent and expressive conceptual schema. Logical design step takes the conceptual schema and creates a corresponding logical schema on the chosen logical model (relational, multidimensional, hybrid, NoSql models). Usually the relational table layout is advocated in the ROLAP implementation. ETL designs the mappings and the data transformations necessary to load into the logical schema of the $\mathcal{DW}$ the data available at the operational data source level. Physical design addresses all the issues related to the suite of tools chosen for implementation - such as indexing and partitioning. We notice however, that these phases are usually treated in the literature in isolated way. This is due to the difficulty and the complexity of each phase.

Parallel to this, ontologies emerge in several domains like E-commerce, Engineering, Environment, Medicine, etc. The most popular definition of ontology is given by Gruber [12]. An ontology is a formal, explicit specification of a shared conceptualization. Conceptualization refers to an abstract model of some domain knowledge in the world that identifies that domain's relevant concepts [6]. The sharing characteristic reflects that ontology has to be consensual and accepted by a group of experts in a given domain. Based on this definition, we claim that ontologies leverage conceptual models. These nice characteristics have been exploited by academician and industrials to define data integration systems. They contribute in resolving the different syntax and semantics conflicts identified in data sources. Two main architectures of ontology-based data integration system following the $<\mathcal{G}, \mathcal{S}, \mathcal{M}>$ framework are distinguished [3]: (1) in the first architecture, domain ontologies played the role of the global schema ($\mathcal{G}$). A typical example system following this architecture is SIMS [1]. (2) In the second architecture, each source is associated to a local ontology referencing a global ontology (in a priori or a posteriori manners) and mappings are defined between the global and the local ontologies. The MECOTA system [29] is an example of this architecture.

Recently, a couple of studies proposed to store ontologies describing the sense of database instances and those instances in the same repository. Such database is called **Semantic Databases** ($\mathcal{SDB}$). Different $\mathcal{SDB}$s were proposed by both industrial (Oracle, IBM Sor) and academic communities, that differ according to their architectures and their storage layouts: vertical, horizontal, binary (see Section 3). The emergence of $\mathcal{SDB}$s makes these sources candidates for $\mathcal{DW}$ systems. Unfortunately, no complete method considering the particularities of $\mathcal{SDB}$s in $\mathcal{DW}$ design exists. The availability of ontologies may allow defining semantic mappings between schemas of source and the target $\mathcal{DW}$ schema. This leverages the integration process to the ontological level, and frees it from all implementation issues. But, it requires a new step concerning the deployment of the logical schema of the warehouse according the target storage layout.

In this paper, we propose a methodology for designing semantic warehouses, where ontologies are confronted to each step of the life-cycle: requirements analysis, conceptual design, logical design, ETL process, deployment and physical design. The contributions of the paper are:

1. Formalization of a conceptual framework $<\mathcal{G}, \mathcal{S}, \mathcal{M}>$, handling $\mathcal{SDB}$s diversity and definition of a $\mathcal{DW}$ requirements model following a goal-driven approach.
2. Proposition of a complete ontology-based method for designing semantic $\mathcal{DW}$s taking as inputs the framework $<\mathcal{G}, \mathcal{S}, \mathcal{M}>$ and the requirements model.
3. Implementation of a case tool supporting the method and validation of the method through a case study instantiating $<\mathcal{G}, \mathcal{S}, \mathcal{M}>$ framework with Oracle $\mathcal{SDB}$ and LUBM benchmark. To the best of our knowledge, this work is the sole that covers all steps of $\mathcal{DW}$ design using a semantic approach.

The paper is organized as follows. Section 2 presents the related work of $\mathcal{DW}$ design. We focus on ontology-based design methods. Section 3 gives some preliminaries about Semantic Databases. Section 4 describes the design method. Section 5 presents a case study validating our proposal. Section 6 presents the case tool supporting our proposal. Section 7 concludes the paper by summarizing the main results and suggesting future work.

2 Related Work: Towards an Ontological Design

Two main approaches exist for the initial design of $\mathcal{DW}$ [30]: the *supply-driven approach* and the *demand-driven approach*. In the first category, the $\mathcal{DW}$ is designed starting from a detailed analysis of the data sources. User requirements impact on design by allowing the designer to select which parts of source data are relevant for the decision making process. The demand-driven approaches starts from determining the information requirements of $\mathcal{DW}$ users or decision makers. This approach gives user requirements a first role in determining the information contents for analysis. Requirements analysis differs according to the analyzed objects. We distinguish *process-driven*, *user-driven* or *goal-driven* analysis. Process-driven analysis [19,15] analyze requirements by identifying business process of the organization. User-driven analysis [30] identifies requirements of each target user and unifies them in a global model. Goal driven analysis [10] has been frequently used for $\mathcal{DW}$ development. It identifies goals and objectives that guide decisions of the organization at different levels.

As we said before, ontologies played a crucial role in several facets in the process of construction and exploitation of data integration systems: global schema definition, syntactic and semantic conflict resolution, query processing, caching, etc. Similarly, ontologies have been exploited in some steps of $\mathcal{DW}$ design cycle. First in the requirements analysis step, where we proposed in [13] to specify $\mathcal{DW}$ business requirements using an OWL domain ontology covering a set of sources. This projection allows defining the $\mathcal{DW}$ conceptual and then logical model. Ontologies largely contribute in the requirement engineering field to specify, unify, formalize requirements and to reason on them to identify ambiguity, complementary and conflict [23].

Ontologies were timidly used in the ETL step. [4] proposed a method for integrating relational data sources into a $\mathcal{DW}$ schema using an ontology as a global schema playing an intermediary model between the target $\mathcal{DW}$ model and sources schemas. Skoutas et al. [27] automate the ETL process by *constructing* an OWL ontology linking schemas of semi-structured and structured (relational) sources to a target $\mathcal{DW}$ schema.

Other design methods used ontologies for multidimensional modeling tasks. [24] defines the $\mathcal{DW}$ multidimensional model (facts and dimensions) from an OWL ontology by identifying functional dependencies (*Functional Object Properties*) between ontological concepts. A functional property is defined as a property that can have only one (unique) value j for each instance i, i.e. there cannot be two distinct values j_1 and j_2 such that the pairs (i, j_1) and (i, j_2) are both instances of this property. [25] and [22] are two attempts that propose ontological methods combining multidimensional modeling and ETL steps: [25] is based on Skoutas's study [27], and defines an ETL and a design process for analyzing source data stores. It identifies the ETL operations to execute according to a multidimensional model defined. The logical and physical design steps are not considered in this work. [22] considers semantic data provided by the semantic web and annotated by OWL ontologies, from which a $\mathcal{DW}$ model is defined and populated. However, the ETL process in this work is dependent of a specific instance format (triples).

Three observations can be made by analyzing the studied works:

1. We notice that ontologies are used in different design steps separately, proving their usefulness, but no method propose to extend the role of ontologies all along the design steps.
2. The discussed works consider logical schemas of sources as inputs of the $\mathcal{DW}$ system, and make an implicit assumption that the $\mathcal{DW}$ model will be deployed using the same representation (usually using a relational representation). The main contribution of our method compared to these works is that we define the ETL process is fully defined at the ontological level, which

allows the deployment of the $\mathcal{DW}$ model in different platforms according designer recommendations.
3. almost no existing works consider *Semantic Databases* as candidates to build the $\mathcal{DW}$.

3 Preliminaries: Semantic Databases

In this section, we review the main concepts related to semantic databases: ontologies, storage layouts, architectures, etc. to facilitate the presentation of our proposal.

The massive use of ontologies by applications contributes largely of the generation of mountains of data referencing these ontologies. In the first generation of ontological applications, the semantic data are managed in the *main memory* like in *OWLIM* or *Jena*. Due to their growing, scalable solutions were developed. $\mathcal{SDB}$ is one of the most *popular solutions*. It allows the storage of data and their ontology in the same *repository*. Note that one important lesson learned from almost 50 years of database technology is the advantage of data modeling, and the physical and logical data *independence*. Object Management Group (OMG) followed this trend, and defined a design architecture organized in four levels: data of real world, the model, the meta-model and the meta-meta-model.

To ensure the same success of traditional databases, $\mathcal{SDB}$ design has to follow these same levels. Both industrial (like Oracle [31] and IBM SOR [20]) and academic (OntoDB [7]) communities defined $\mathcal{SDB}$ solutions, having different architectures and proposed through an evolving process. Fig. 1 illustrates the evolution of $\mathcal{SDB}$s architectures according to the design levels. With the increasing use of ontologies in different domains, different ontological languages and formalisms have been proposed: RDF, RDFS, OWL, PLIB, KIF (M0 level in Fig. 1). Storing ontologies in a database is made according to a specific storage layout.

The diversity of ontology formalisms proposed gave rise to different storage layouts (M1 level). We distinguish three main *relational* representations [7]: *vertical*, *binary* and *horizontal*. *Vertical* representation stores data in a unique table of three columns (subject, predicate, object) (eg. Oracle). In a *binary* representation, classes and properties are stored in different tables (eg. IBM SOR). *Horizontal* representation translates each class as a table having a column for each property of the class (eg. OntoDB). Other *hybrid* representations are possible.

$\mathcal{SDB}$s can materialize models of different levels, which gave rise to different $\mathcal{SDB}$ architectures. First $\mathcal{SDB}$s (*type I*) proposed a similar architecture as traditional database using two parts: *data part* and the *meta schema part* (catalog). In the data part, ontological instances and also the ontology structure (concepts and properties) are stored (M0 and M1 levels). For example, Oracle [31] $\mathcal{SDB}$ uses vertical representation to store the ontology model and its data in a unique table. A second architecture (*type II*) separates the ontology model from its data that can be stored in different schemas. This architecture is thus composed of three parts [7]: the *meta-schema*, the *ontology model* and the *data schema* (eg.

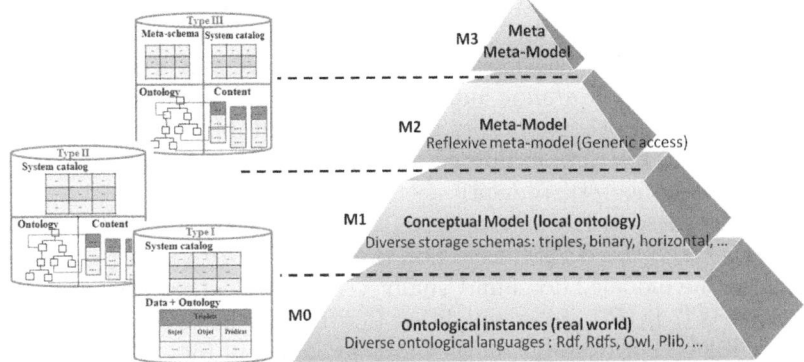

Fig. 1. Evolution of $\mathcal{SDB}$s (ontology models, storage layouts and architectures)

IBM SOR [20] $\mathcal{SDB}$). This architecture outperforms the first one, but the ontology schema is based on the underlying ontology model and is thus static. A third architecture (*type III*) extends the second one by adding a new part which is a reflexive meta-model of the ontology (eg. OntoDB [7]), offering more flexibility to the ontology part (M2 and M3 levels).

4 Our Proposal

Our $\mathcal{DW}$ design follows the mixed approach, where data sources and user requirements have the same role. Another characteristic of our proposal is that it exploits the presence of ontologies. To fulfill our needs, we fix four objectives that we discuss in the next sections:

1. Obj_1: leveraging the integration framework $<\mathcal{G}, \mathcal{S}, \mathcal{M}>$ by an ontology;
2. Obj_2: user requirement have to be expressed by the means of ontology;
3. Obj_3: ETL process has to be defined on ontological level and not on physical or conceptual levels, and
4. Obj_4: The deployment process needs to consider the different storage layouts of semantic $\mathcal{DW}$.

4.1 Obj_1: Integration Framework for SDBs

In this section, we define an integration framework $<\mathcal{G}, \mathcal{S}, \mathcal{M}>$ adapted to $\mathcal{SDB}$ specificities.

The Global Schema $\mathcal{G}$. Schema $\mathcal{G}$ is represented by a Global Ontology (GO). Different languages were defined to describe ontologies. OWL language is the language recommended by W3C consortium for defining ontologies. Description Logics (DLs) [2] present the formalism underlying OWL language. We thus use DLs as a basic formalism for specifying the framework.

In DL, structured knowledge is described using *concepts* denoting unary predicates and *roles* denoting binary predicates. Concepts denote sets of individuals, and roles denote binary relationships between individuals. Two types of concepts and roles are used: *atomic* and *concept descriptions*. Concept descriptions are defined based on other concepts by applying suitable *DL* constructors (eg. intersection, value restriction, limited existential quantification, etc), equipped with a precise set-theoretic semantics.

A knowledge base in DL is composed of two components: the *TBox* (Terminological Box), and the *ABox* (Assertion Box). The *TBox* states the *intentional* knowledge of the modeled domain. Usually, terminological axioms have the form of inclusions: C ⊑ D (R ⊑ S) or equalities: C ≡ D (R ≡ S) (C,D denote concepts, R,S denote roles). For example, in the ontology model of Fig. 2, representing the ontology of LUBM benchmark related to the university domain, the concept *University* can be defined as an *Organization* by specifying the axiom: *University* ⊑ *Organization*. The ABox states the *extensional* knowledge of the domain and defines assertions about individuals. Two types of assertions are possible: concept assertions (Eg. *Student*(Ann)) and role assertions (e.g. *TakeCourse*(Ann, Mathematics)).

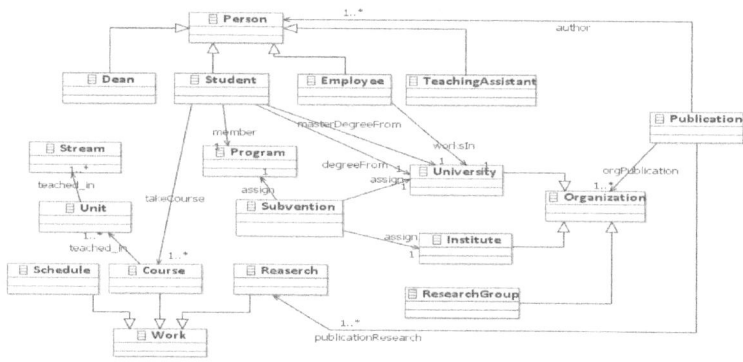

Fig. 2. LUBM global schema

Based on these definitions, the *GO* is formally defined as follows:
$GO :< C, R, Ref(C), formalism >$, such that:

- C: denotes *Concepts* of the model (atomic concepts and concept descriptions).
- R: denotes *Roles* of the model. Roles can be relationships relating concepts to other concepts, or relationships relating concepts to data-values.
- $Ref : C \rightarrow (Operator, Exp(C,R))$. *Ref* is a *function* defining terminological axioms of a DL TBox. *Operators* can be inclusion (⊑) or equality (≡). *Exp(C,R)* is an expression over concepts and roles of *GO* using constructors of DLs such as union, intersection, restriction, etc. (e.g., Ref(Student)→(⊑, Person ⊓ ∀takesCourse(Person, Course))).

- *Formalism* : is the *formalism* followed by the global ontology model like RDF, OWL, etc. Note that the definition of the GO concerns only its TBox, which is usually assumed in $\mathcal{DIS}$.

The Sources S. The set of sources considered are $\mathcal{SDB}$s. Each $\mathcal{SDB}$ is defined by its local ontology (O_i) and its instances part (the ABOX). As, explained previously, the ontology model and its instances can be stored using different storage layouts. $\mathcal{SDB}$s may have different architectures. A $\mathcal{SDB}$ is formally defined as follows $< O_i, I, Pop, SL_{O_i}, SL_I, Ar >$ where:

- O_i: <C, R, Ref, formalism> is the *ontology* model of the $\mathcal{SDB}$.
- I: presents the *instances* (the ABox) of the $\mathcal{SDB}$.
- $Pop : C \rightarrow 2^I$ is a *function* that relates each concept to its instances.
- SL_{O_i}: is the *Storage Layout* of the ontology model (vertical, binary or horizontal).
- SL_I: is the *Storage Layout* of the instances I.
- Ar: is the *architecture* of the $\mathcal{SDB}$.

The Mappings M. Mappings assertions relate a mappable element (MapElmG) of schema G (MapSchemaG) to a mappable element ($MapElmS$) of a source schema (MapSchemaS). These assertions can be defined at the intensional level (TBox) or at the extensional level ($ABox$). Different types of semantic relationships can be defined between mappable elements (Equivalence, Containment or Overlap). Discovering such mappings is related to the domain of schema and ontology matching/alignment, which is out of the scope of this paper. The mapping assertions are formally defined as follows M:< MapSchemaG, MapSchemaS, MapElmG, MapElmS, Interpretation, SemanticRelation >. This formalization is based on [26] meta-model:

- **MapSchemaG** and **MapSchemaS:** present respectively the *mappable schema* of the global and the local ontology.
- **MapElmG** and **MapElmS:** present respectively a *mappable element* of the global and the local ontology schema. This element can be a simple concept, instance or an expression (*Exp*) over the schema.
- **Interpretation:** presents the *Intentional* interpretation or *Extensional* interpretation of the mapping. In our study, the availability of global and local ontologies allows to define intentional mappings.
- **SemanticRelation:** three relationships are possible: *Equivalence, Containment* or *Overlap*. Equivalence states that the connected elements represent the same aspect of the real world. Containment states that the element in one schema represents a more specific aspect of the world than the element in the other schema. Overlap states that some objects described by an element in one schema may also be described by the connected element in the other schema.

4.2 Obj_2: Goal-Oriented Requirements Model

A goal is an objective that the system under consideration should achieve. Identifying goals of users is a crucial task for $\mathcal{DW}$ development. Indeed, the $\mathcal{DW}$ is at the core of a decisional application that needs to analyze the activity of an organization and where goals are important indicators of this activity. After analyzing works of goal-oriented literature, we proposed a Goal model considered as a pivot model since it combines three widespread goal-oriented approaches: KAOS [17], Tropos [28] and iStar [5]. The model is presented in Fig. 3 (right part). Fig. 5 presents a set of goals examples.

Let us take a goal *Goal1:* "*Improve the quality of teaching of courses according to the dean of the university*". The goal model is composed of a main entity *Goal* described by some characteristics (name, context, priority). A goal is issued and achieved by some actors (*dean*). A goal is characterized by two coordinates: (1) a *Result* to analyze (*quality of teaching*) that can be quantified by given formal or semi-formal metrics measuring the satisfaction of the goal (*number of students attending the course*), and (2) some *Criteria* influencing this result (*course*).

Two types of goals are identified: *functional* and *non-functional* goals. A non-functional requirement is defined as an attribute or constraint of the system (such as security, performance, flexibility, etc). Two types of relationships between goals are distinguished (reflexive relations): *AND/OR* relationships decomposing a general goal into sub-goals and *influence* relationships (positive, negative or ambiguous influence). User's goals in our method are used at different stages of the method. First, goals will be used to identify the most relevant data to

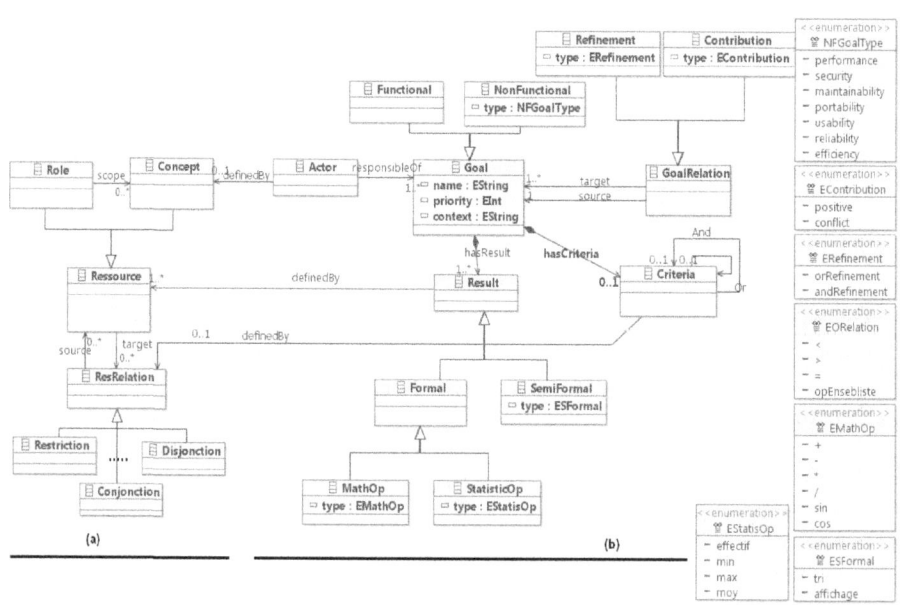

Fig. 3. Requirements model proposed

materialize in the $\mathcal{DW}$. They are also used to identify the multidimensional concepts (facts and dimensions) of the $\mathcal{DW}$ model.

4.3 Design Method

We propose a method for designing a semantic $\mathcal{DW}$ covering the following steps: requirements analysis, conceptual design, logical design, ETL process, deployment and physical design. Fig. 4 illustrates these steps.

Requirements Analysis. This step allows the designer identifying the following: (1) the set of relevant properties used by the target application and (2) the set of treatments it should answer. The first set allows the construction of the *dictionary* containing the relevant concepts required for the application. As the ontology describe all concepts and properties of a given domain, a connection between requirement model and the ontology model is feasible. To do so, we define a connection between coordinates of each goal (*Result* and *Criteria*) and the resources (*concepts* and *roles*) of the *GO* (Fig. 3-left part). This allows the designer to choose the most relevant ontological concepts to express user's goals. Knowing that the *GO* is linked to the data sources, these concepts chosen to express goals inform the designer about the most relevant data to store in the $\mathcal{DW}$ model.

For example, *Goal1* is specified using the following resources of the LUBM ontology (Fig. 2) (*Student*, *Course*, *Work* and *Dean* concepts).

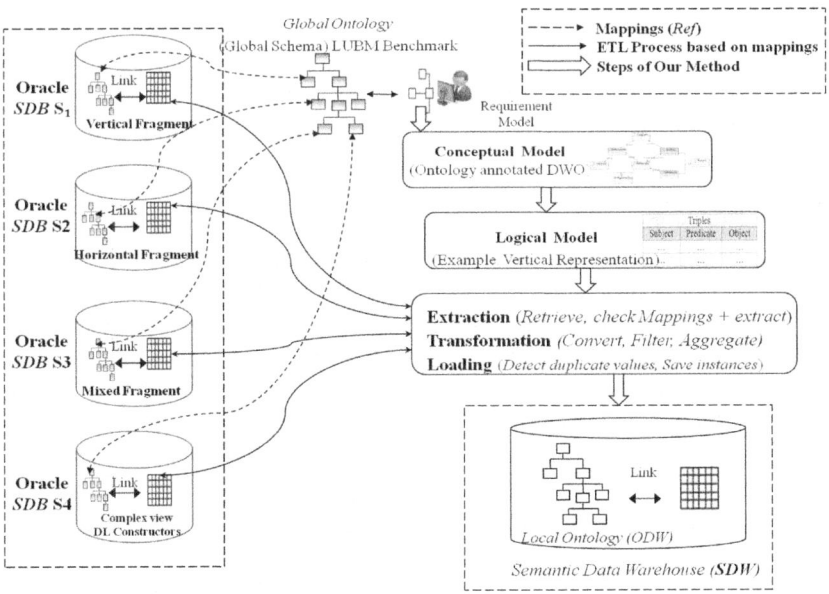

Fig. 4. Design method proposed

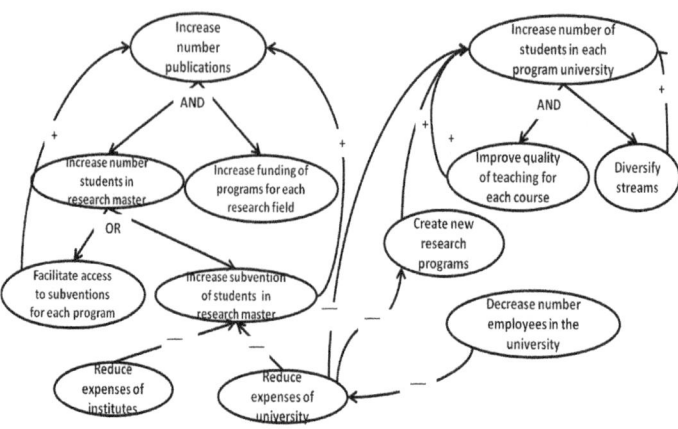

Fig. 5. Goals of university domain

Conceptual Design. A $\mathcal{DW}$ ontology (DWO) (that can be viewed as a conceptual abstraction of the $\mathcal{DW}$) is defined from the global ontology (GO) by extracting all concepts and properties used by user goals. Three scenarios materialize this definition:

1. $DWO = GO$: the GO corresponds exactly to users' requirements,
2. $DWO \subset GO$: the DWO is extracted from the GO,
3. $DWO \supset GO$: the GO does not fulfill all users' requirements.

The designer may extend the DWO by adding new concepts and properties in the case, where the GO does not satisfy all her/his requirements. The concepts belonging to DWO and do not reference any concept of sources are annotated and are set by *null values* in the target warehouse.

DWO is defined, we exploit its automatic reasoning capabilities to correct all inconsistencies. Two usual reasoning mechanisms can first be used: (a) checking the consistency of the ontology (classes and instances) and (b) inferring subsumption relationships. This reasoning is supported by most existing reasoners (*racer*, *Fact++*) and allows the detection of design errors. Another reasoning mechanism is defined in order to propagate influence relationships between goals, as explained in [14]. Influence relationships are used afterwards to explore the multidimensional structure of the $\mathcal{DW}$ model, where we consider 'fact' concepts as central concepts and 'dimensions' as concepts *influencing* them. The multidimensional role of concepts and properties are then discovered and stored as ontological annotations. We propose the algorithm 1 for multidimensional annotations.

Logical Design. The logical $\mathcal{DW}$ model is generated by translating the DWO to a relational model (other data models can be chosen). Several works in the literature proposed methods for translating ontologies described in a given formalism (PLIB, OWL, RDF) to a relational or an object-relational representations [9].

begin
    for *Each goal G* **do**
        Each concept (resp. role) used as a result of G is a fact (resp. measure) candidate;
        Each concept (resp. role) used as a criterion of G is a dimension (resp. dimension attribute) candidate;
        Criteria of goals influencing G are dimension candidates of the measure identified for G;
        Concepts of measures are facts candidates;
        Concepts of dimension attributes are dimension candidates;
        if *fact concept F is linked to a dimension by (1,n) relationship* **then**
        | keep the two classes in the model
        else
        | Reject the dimension class;
        end
        Hierarchies between dimensions are constructed by looking for (1,n) relationships between classes identified as dimensions (for each fact);
    end
    Generalization (is-a) relationships existing in the ontology between facts or between dimensions are added in the model.;
end

Algorithm 1. Multidimensional annotations

Obj_3: **ETL Process.** The goal of the ETL process is to populate the target $\mathcal{DW}$ schema obtained in the previous step, by data of sources. [27] defined ten generic operators typically encountered in an ETL process, which are:

1. EXTRACT (S,C): extracts, from incoming record-sets, the appropriate portion.
2. RETRIEVE(S,C): retrieves instances associated to the class C from the source S.
3. MERGE(S,I): merges instances belonging to the same source.
4. UNION (C,C'): unifies instances whose corresponding classes C and C' belong to different sources S and S'.
5. JOIN (C, C'): joins instances whose corresponding classes C and C' are related by a property.
6. STORE(S,C, I): loads instances I corresponding to the class C in a target data store S.
7. DD(I): detects duplicate values on the incoming record-sets.
8. FILTER(S,C,C'): filters incoming record-sets, allowing only records with values of the element specified by C'.
9. CONVERT(C,C'): converts incoming record-sets from the format of C to the format of C'.
10. AGGREGATE (F, C, C'): aggregates incoming record-sets applying the aggregation function $\mathcal{F}$ (e.g., COUNT, SUM, AVG, MAX) defined in the target data-store.

These operators have to be leveraged to deal with the semantic aspects of sources. Therefore, we propose an Algorithm 2 for populating the DWO schema. The algorithm is based on the generic conceptual ETL operators as presented above. They can then be instantiated according to one of the storage layouts: vertical, binary, horizontal. Each operator will correspond to a defined query.

Based on the framework $<GO,\mathcal{SDB},M>$, the integration process depends on the *semantics of mappings (SemanticRelation)* between GO and local ontologies ($\mathcal{SDB}$), where four semantics mappings are identified: (1) *Equivalent* ($C_{GO} \equiv C_{SDB}$) and (2) *Containment sound* ($C_{GO} \supset C_{SDB}$): where no transformation is needed. Instances are extracted from sources, merged, united or joined then loaded in the target data store. (3) *Containment complete* ($C_{GO} \subset C_{SDB}$): where source instances satisfy only a subset of the constraints required by GO classes, some instances need to be transformed (converted, filtered and aggregated) then merged, unified or joined and finally loaded to the target data store. (4) *Overlap* mappings: where we need to identify the constraints required by GO classes and not applied to the source classes. This case is then treated same as the Containment (Complete) scenario. Algorithm 2 depicts the ETL process based on these four scenarios.

Obj_4: **Deployment and Physical Design.** In a traditional $\mathcal{DW}$, the deployment followed one-to-one rule, where each warehouse table is stored following one storage layout. In a semantic $\mathcal{DW}$, the deployment may followed one-to-many rule (à la carte), where the ontology model and the instances may have several storage layouts independently. Our proposal offers designers the possibility to choose her/his favorite storage (within the DBMS constraints) and then deploy the warehouse accordingly.

The target $\mathcal{DW}$ model defined by our proposal is populated according to a given DBMS. In the next section, a validation of our proposal is given using Oracle DBMS. An Oracle $\mathcal{SDB}$ is used to store the target $\mathcal{DW}$ model and the ontology defining its semantics.

5 Implementation

In order to demonstrate the feasibility of our proposal, we experiment it using Lehigh University BenchMark LUBM[1] (containing 4230 individuals) and **Oracle** $\mathcal{SDB}$. Note that Oracle supports languages RDF and OWL to enable its users to get benefit from a management platform for semantic data. Oracle has defined two subclasses of DLs: *OWLSIF* and *OWLPrime*. We used *OWLPrime*[2] fragment which offers a richer set of DL constructors. The framework $<$GO, $\mathcal{SDB}$s, Mappings$>$ is thus instantiated as follows $<$GO: LUBM Ontology, $\mathcal{SDB}$s: Oracle $\mathcal{SDB}$s, Mappings: defined between local ontologies of sources and LUBM ontology$>$. LUBM Ontology model is used as the global ontology GO, and is presented in Fig. 2.

---

[1] http://www.lehigh.edu/~zhp2/2004/0401/univ-bench.owl
[2] http://www.w3.org/2007/OWL/wiki/OracleOwlPrime

```
begin
    Input:
        DWO: DW Ontology (Schema) and Si: Local Source (SDB)
    Output: DWO populated (schema + instances)
    for Each C : Class of ontology DWO do
        I_{DWO} = φ
        for Each source Si do
            if Cs ≡ C  /* instances in Si satisfy all constraints imposed by DWO*/
            then
                C'= IdentifyClasse (Si, C) /*identify class from Si*/
            else
                if Cs ⊂ C  /*Instances in Si satisfy all constraints imposed by DWO,
                plus additional ones */ then
                    C'= IdentifyClasse (Si, C) /*identify class from Si*/
                else
                    if Cs ⊃ C Or Overlap mappings /* Instances satisfy only a subset of
                    constraints imposed by DWO*/ then
                        if format(C) ≠ format(Cs) then
                            Cconv= CONVERT (C, Cs) /*identify the constraint of format
                            conversion from the source to the target DWO*/
                        end
                        if C represent aggregation constraint then
                            Caggr= AGGREGATE (F, C, Cs) /*identify the constraint of
                            aggregation defined by F*/
                        end
                        if C represents filter constraint then
                            Cfilt= FILTER (Si, C, Cs) /*identify the filter constraint
                            defined in the target DWO*/
                        end
                        C'= ClasseTransformed (Si, C, Cconv, Caggr, Cfilt) /* Associate to the
                        class C' the constraint of conversion, aggregation or filtering
                        defined by Cconv, Caggr and Cfilt*/
                    end
                end
            end
            I_{si}= RETRIEVE (Si, C') /*Retrieve instances of C' and applying constraints
            of conversion, aggregation or filtering if necessary*/
            if more than one instance are identified in the same source then
                I_{DWO}= MERGE (I_{DWO}, I_{si}) /*Merge instances of Si*/
            end
            if classes have the same super class then
                I_{DWO}= UNION (I_{DWO}, I_{si}) /*Unites instances incoming from different
                sources*/
            else
                if classes are related by same property then
                    I_{DWO}= JOIN (I_{DWO}, I_{si}) /* Join incoming instances*/
                end
            end
            if Source contain instances more than needed then
                I_{DWO}= EXTRACT (I_{DWO}, I_{si}) /* Extract appropriate portion of
                instances*/
            end
        end
        STORE(DWO,C, DD(I_{DWO})) /*Detects duplicate values of instances and load
        them in DWO*/
    end
end
```

Algorithm 2. The population of the $\mathcal{DW}$ by the means of ontological ETL operators

The adopted scenario for evaluating our proposal consists in creating 4 Oracle $\mathcal{SDB}$s (S_1, S_2, S_3 and S_4), where each source references the Lehigh University Benchmark (LUBM) ontology. The first three sources are defined using simple mappings between their local ontologies and the LUBM ontology as follows:

1. the first source is defined as a projection on a set of classes of the LUBM ontology. It can be viewed as a vertical fragment of the global ontology (LUMB), where some classes are considered.
2. the second one is defined as a restriction of a set of roles for each LUBM ontology (horizontal fragment).
3. the third one is defined by simultaneously applying the projection and the restriction operations on LUMB ontology (mixed fragment).

Our proposal supports also complex mapping, where a source may be defined as a view over the ontology, i.e., an expression using DL constructors. For instance, our fourth source contains three classes: *Person*, *Student* and *Employee* defined as follows:

1. $S_4.C_1$: Person, Ref (Person) = (Student ∪ Employee) ∩ ∀ member (Person, Organization)
2. $S_4.C_2$: Student, Ref (Student) = Student ∩ ∀takesCourse (Person, Course)
3. $S_4.C_3$: Employee, Ref (Employee) = Person ∩ ∀WorksFor (Person, Organization)

Each source is thus instantiated using our framework as follows S_i: < Oi_{Oracle}, Individuals (triples), Pop is given in tables RDF_link$ and RDF_values$, Vertical, Vertical, type I>. Vertical storage is the relational schema composed of one table of triples (subject, predicate, object). For example: *(Student, type, Class)* for the ontology storage and *(Student#1, type, Student)* and *(Student#1, takeCourse, Course#1)* for the instances storage. And the mapping assertions between global and local ontology of the source are instantiated as follows: < Oi_{Oracle} of each source, GO (LUBM ontology model), Expression over GO, Class of a source S, Intentional interpretation, (Equivalent, Containment or Overlap): *owl:SubClassOf* and *owl:equivalentClass* in OWLPrime>.

We considered requirements presented in Fig. 5. The projection of these requirements on the GO gives the schema of the $\mathcal{DWO}$. For this case study, our $\mathcal{DWO}$ corresponds to scenario 2 ($\mathcal{DWO} \subset GO$). The algorithm 1 is applied to annotate this ontology with multidimensional concepts. The multidimensional schema presented in Fig. 6 is obtained, where three facts are identified (in a dark color) linked to their dimensions.

The $\mathcal{DWO}$ annotated is translated to a relational schema. We also used Oracle $\mathcal{SDB}$s to store the final $\mathcal{DW}$ schema and instances. The $\mathcal{DWO}$ is thus translated into a vertical relational schema (N-Triple file). Oracle $\mathcal{SDB}$ allows to load N-Triple files into a staging table using Oracle's SQL*Loader utility. The ETL algorithm we defined is applied to populate the relational schema. The algorithm uses conceptual ETL operators that must be translated according the vertical representation of Oracle. Oracle offers two ways for querying semantic

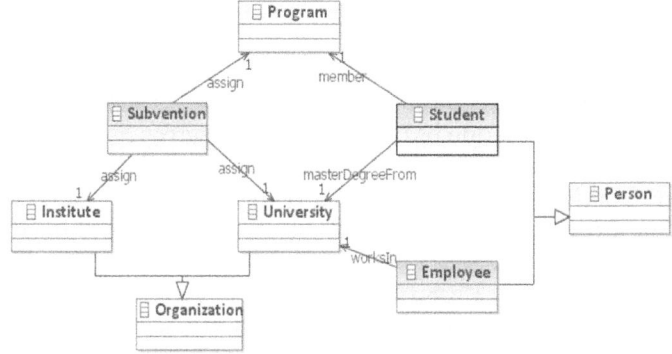

Fig. 6. Multidimensional model generated by our method

data: SQL and SPARQL. We choose SPARQL to express this translation. Due to lack of space, we show the translation of some operators as an example: the namespace of University Ontology of benchmark LUBM: **PREFIX** univ-bench: http://www.lehigh.edu/~zhp2/2004/0401/univ-bench.owl#

EXTRACT operator is translated as follows: *Select ?Instance# Where { ?Instance# rdf:type nameSpace:Class. ?Instance NameSpace:DataProperty value_condition}*

Example 1. Extract students those age = 15 years.
Select ?student Where { ?student rdf:type univ-bench:Student. ?student univ-bench:age 15}

RETRIEVE operator is translated as follows: *Select ?Instances# Where { ?Instances# rdf:type Namespace:Class}*

Example 2. Retrieve instances of the *Student* class.
Select ?InstanceStudent Where { ?InstanceStudent rdf:type univ-bench:Student}

FILTER operator is translated as follows: *Select ?instance ?P where { ?Instance rdf:type namespace:Class ; namespace:P ?P . FILTER (?P > value_condition)}*

Example 3. Filter incoming student instances allowing only those with age is greater than 16 years:
Select ?instanceStudent ?age where {? instanceStudent rdf:type univ-bench:Student ; univ-bench:age ?age . FILTER (?age > 16) }

The result of this integration process is a $\mathcal{DW}$ whose schema is populated by instances selected from Oracle $\mathcal{SDB}$s. For more details, refer to the video available at: http://www.lias-lab.fr/forge/deploymentalacarte/video.html

6 Case Tool

The proposed tool is implemented in Java language and uses OWL API to access ontologies. The tool takes as inputs a set of requirements and a set of $\mathcal{SDB}$s that participates in the construction of the $\mathcal{DW}$. These sources reference a shared ontology formalized in OWL. $\mathcal{SDB}$s are configured by some connection parameters. The first steps (conceptual and logical design) are supported by a model-to-model transformation process. The access to all ontologies is made through the OWL API. Requirements are expressed following our goal oriented model. The DWO is extracted as a module using ProSé plug-in available within Protégé editor, which ensures the logical completeness of the extracted ontology. Fact++ reasoner is invoked to classify the DWO class's taxonomy and to check its consistency. Influence rules defined between goals are implemented using SWRL (Semantic Web Rule Language) language [3], which must be combined with Jess[4] inference engine to execute defined SWRL rules and apply them on the DWO. A parser analyzes the requirements in order to identify the multidimensional aspects of the concepts and the roles by the means of Algorithm 1.

The ETL process is implemented in the tool such that the technical details (the translation of the ETL operators) are hidden to the user. Regarding the deployment of the $\mathcal{DW}$, the tool offers the designer the possibility to choose her/his favorite storage layout and architecture of the target DBMS according her/his requirements. The proposed ETL algorithm is implemented in our tool. Based on the existing mappings between the $\mathcal{SDB}$ schemas and the target $\mathcal{DW}$ schema, the tool allows an automatic extraction of the appropriate data from the $\mathcal{SDB}$ sources, their transformation (filtering, conversion and aggregation) and the computation of the new values in order to obey to the structure of the $\mathcal{DW}$ classes. Then, data are loaded to the appropriate classes of the $\mathcal{DW}$ model.

In order to obtain a generic implementation of the ETL process, we implemented the ETL algorithm using the *Model-View-Controller (MVC) architecture* [16]. We used *Data Access Object (DAO) Design patterns* [21] that implement the access mechanism required to handle the ($\mathcal{SDB}$s). The DAO solution abstracts and encapsulates all access to persistent storage, and hides all implementation details from business components and interface clients. The DAO pattern provides flexible and transparent accesses to different layout storage. Based on the chosen storage layout, the architecture of the $\mathcal{SDB}$s and the target $\mathcal{DW}$ model, the right object DAO is selected.

The demonstration link illustrates the different layers of the architecture which are: **(1) View Layer:** the user interface. **(2) Controller Layer:** represents the events (user actions, changes done on the model and view layers); **(3) Model Layer:** represents DAO layers containing the implementation of the conceptual ETL operators. The tool provides a semantic $\mathcal{DW}$ populated from data of $\mathcal{SDB}$s.

---

[3] http://www.w3.org/Submission/SWRL/
[4] http://www.jessrules.com/

7 Conclusion

Semantic data currently exists everywhere. To facilitate their management, $\mathcal{SDB}$ technology has been proposed. As a consequence, they are candidate for alimenting warehouse projects. After a deep study of $\mathcal{SDB}$s, we first proposed a generic integration framework dedicated to these databases, and a goal model specifying users requirements. We then proposed a method taking these two inputs to design semantic $\mathcal{DW}$s. The method covers the following steps: requirement analysis, conceptual design, logical design, ETL process and physical design. The user holds an important place in the method, and the $\mathcal{DW}$ multidimensional model is essentially defined to achieve its goals. The method is validated through experiments using LUBM benchmark and Oracle $\mathcal{SDB}$s. These experiments show the feasibility of our proposal. The experiment uses . Some mapping assertions are defined between the local ontologies of Oracle $\mathcal{SDB}$s and LUBM ontology schema (considered as the global ontology). The presence of ontologies allows defining semantic mappings between data sources and the target $\mathcal{DW}$ model, *independently of any implementation constraint*. The $\mathcal{DW}$ model can thus be deployed using a given platform chosen by the designer. The method is supported by a case tool implementing all the steps of the method. The tool automates the ETL process where the appropriate data is extracted automatically from the $\mathcal{SDB}$ sources, transformed and cleaned, then loaded to the target $\mathcal{DW}$. The only effort provided by the designer is the translation of the generic conceptual ETL operators according the logical level of the target DBMS.

Currently, we are studying three main issues: (1) the evaluation of our approach by the means of real applications, (2) the consideration of advanced deployment infrastructures (e.g. cloud) and (3) the study of the impact of the evolution of ontologies and user's requirements on our proposal.

References

1. Arens, Y., Hsu, C., Knoblock, C.: Query processing in the SIMS information mediator. In: Readings in Agents, pp. 82–90. Morgan Kaufmann Publishers Inc., San Francisco (1998)
2. Baader, F., Calvanese, D., McGuinness, D., Nardi, D., Patel-Schneider, P. (eds.): The Description Logic Handbook: Theory, Implementation, and Applications. Cambridge University Press (2003)
3. Bellatreche, L., et al.: Contribution of ontology-based data modeling to automatic integration of electronic catalogues within engineering databases. Computers in Industry Journal Elsevier 57(8-9), 711–724 (2006)
4. Calvanese, D., Giacomo, G., Lenzerini, M., Nardi, D., Rosati, R.: Data integration in data warehousing. Int. J. Cooperative Inf. Syst. 10(3), 237–271 (2001)
5. Cares, C., Franch, X., Lopez, L., Marco, J.: Definition and uses of the i* metamodel. In: Proceedings of the 4th International i* Workshop, pp. 20–25 (June 2010)
6. Cruz, I.F., Xiao, H.: The role of ontologies in data integration. Jounal of Engineering Intelligent Systems 13(4), 245–252 (2005)

7. Dehainsala, H., Pierra, G., Bellatreche, L.: OntoDB: An ontology-based database for data intensive applications. In: Kotagiri, R., Radha Krishna, P., Mohania, M., Nantajeewarawat, E. (eds.) DASFAA 2007. LNCS, vol. 4443, pp. 497–508. Springer, Heidelberg (2007)
8. Fagin, R., Kolaitis, P.G., Miller, R.J., Popa, L.: Data exchange: Semantics and query answering. In: ICDT, pp. 207–224 (2003)
9. Gali, A., Chen, C.X., Claypool, K.T., Uceda-Sosa, R.: From ontology to relational databases. In: Wang, S., Tanaka, K., Zhou, S., Ling, T.-W., Guan, J., Yang, D.-q., Grandi, F., Mangina, E.E., Song, I.-Y., Mayr, H.C. (eds.) ER Workshops 2004. LNCS, vol. 3289, pp. 278–289. Springer, Heidelberg (2004)
10. Giorgini, P., Rizzi, S., Garzetti, M.: Goal-oriented requirement analysis for data warehouse design. In: DOLAP 2005, pp. 47–56 (2005)
11. Golfarelli, M.: Data warehouse life-cycle and design. In: Encyclopedia of Database Systems, pp. 658–664. Springer US (2009)
12. Gruber, T.: A translation approach to portable ontology specifications. In Knowledge Acquisition 5(2), 199–220 (1993)
13. Khouri, S., Bellatreche, L.: A methodology and tool for conceptual designing a data warehouse from ontology-based sources. In: DOLAP 2010, pp. 19–24 (2010)
14. Khouri, S., Boukhari, I., Bellatreche, L., Jean, S., Sardet, E., Baron, M.: Ontology-based structured web data warehouses for sustainable interoperability: requirement modeling, design methodology and tool. To appear in Computers in Industry Journal (2012)
15. Kimball, R.: The Data Warehouse Toolkit: Practical Techniques for Building Dimensional Data Warehouses. John Wiley (1996)
16. Krasner, G.E., Pope, S.T.: A cookbook for using the model-view-controller user interface paradigm in smalltalk-80. In: JOOP, pp. 18–22 (August/September 1988)
17. Lamsweerde, A.: Goal-oriented requirements engineering: A guided tour. In: IEEE International Symposium on Requirements Engineering, p. 249 (2001)
18. Lenzerini, M.: Data integration: A theoretical perspective. In: PODS, pp. 233–246 (2002)
19. List, B., Schiefer, J., Tjoa, A.M.: Process-oriented requirement analysis supporting the data warehouse design process a use case driven approach. In: Ibrahim, M., Küng, J., Revell, N. (eds.) DEXA 2000. LNCS, vol. 1873, pp. 593–603. Springer, Heidelberg (2000)
20. Lu, J., Ma, L., Zhang, L., Brunner, J.S., Wang, C., Pan, Y., Yu, Y.: Sor: A practical system for ontology storage, reasoning and search. In: VLDB, pp. 1402–1405 (2007)
21. Matid, D., Butorac, D., Kegalj, H.: Data access architecture in object oriented applications using design patterns. In: IEEE MELECON, May 12-15, pp. 18–22 (2004)
22. Nebot, V., Berlanga, R.: Building data warehouses with semantic web data. Decision Support Systems 52(4), 853–868 (2012)
23. Pires, P.F., Delicato, F.C., Cóbe, R., Batista, T.V., Davis, J.G., Song, J.H.: Integrating ontologies, model driven, and cnl in a multi-viewed approach for requirements engineering. Requirements Engineering 16(2), 133–160 (2011)
24. Romero, O., Abelló, A.: A framework for multidimensional design of data warehouses from ontologies. Data Knowl. Eng. 69(11), 1138–1157 (2010)
25. Romero, O., Simitsis, A., Abelló, A.: *GEM*: Requirement-driven generation of ETL and multidimensional conceptual designs. In: Cuzzocrea, A., Dayal, U. (eds.) DaWaK 2011. LNCS, vol. 6862, pp. 80–95. Springer, Heidelberg (2011)

26. Brockmans, S., Haase, P., Serafini, L., Stuckenschmidt, H.: Formal and conceptual comparison of ontology mapping languages. In: Stuckenschmidt, H., Parent, C., Spaccapietra, S. (eds.) Modular Ontologies. LNCS, vol. 5445, pp. 267–291. Springer, Heidelberg (2009)
27. Skoutas, D., Simitsis, A.: Ontology-based conceptual design of etl processes for both structured and semi-structured data. Int. J. Semantic Web Inf. Syst. 3(4), 1–24 (2007)
28. Susi, A., Perini, A., Mylopoulos, J., Giorgini, P.: The tropos metamodel and its use. Informatica 29, 401–408 (2005)
29. Wache, H., Scholz, T., Stieghahn, H., König-Ries, B.: An integration method for the specification of rule-oriented mediators. In: DANTE, pp. 109–112 (1999)
30. Winter, R., Strauch, B.: A method for demand driven information requirements analysis in data warehousing projects. In: 36th HICSS, p. 231 (2003)
31. Wu, Z., Eadon, G., Das, S., Chong, E., Kolovski, V., Annamalai, M., Srinivasan, J.: Implementing an inference engine for rdfs/owl constructs and user-defined rules in oracle. In: ICDE, pp. 1239–1248 (2008)

A Specific Encryption Solution for Data Warehouses

Ricardo Jorge Santos[1], Deolinda Rasteiro[2], Jorge Bernardino[3], and Marco Vieira[1]

[1] CISUC – FCTUC – University of Coimbra – 3030-290 Coimbra – Portugal
[2] DFM – ISEC – Polytechnic Institute of Coimbra – 3030-190 Coimbra – Portugal
[3] CISUC – ISEC – Polytechnic Institute of Coimbra – 3030-190 Coimbra – Portugal
lionsoftware.ricardo@gmail.com, {dml,jorge}@isec.pt,
mvieira@dei.uc.pt

Abstract. Protecting Data Warehouses (DWs) is critical, because they store the secrets of the business. Although published work state encryption is the best way to assure the confidentiality of sensitive data and maintain high performance, this adds overheads that jeopardize their feasibility in DWs. In this paper, we propose a Specific Encryption Solution tailored for DWs (SES-DW), using a numerical cipher with variable mixes of eXclusive Or (XOR) and modulo operators. Storage overhead is avoided by preserving each encrypted column's datatype, while transparent SQL rewriting is used to avoid I/O and network bandwidth bottlenecks by discarding data roundtrips for encryption and decryption purposes. The experimental evaluation using the TPC-H benchmark and a real-world sales DW with Oracle 11g and Microsoft SQL Server 2008 shows that SES-DW achieves better response time in both inserting and querying, than standard and state-of-the-art encryption algorithms such as AES, 3DES, OPES and Salsa20, while providing considerable security strength.

Keywords: Encryption, Confidentiality, Security, Data Warehousing.

1 Introduction

Data Warehouses (DWs) store extremely sensitive business information. Unauthorized disclosure is therefore, a critical security issue. Although encryption is used to avoid this, it also introduce very high performance overheads, as shown in [16]. Since decision support queries usually access huge amounts of data and substantial response time (usually from minutes to hours) [12], the overhead introduced by using encryption may be unfeasible for DW environments if they are too slow to be considered acceptable in practice [13]. Thus, encryption solutions built for DWs must balance security and performance tradeoff requirements, *i.e.*, they must ensure strong security while keeping database performance acceptable [13, 16].

As the number and complexity of "data-mix" encryption rounds increase, their security strength often improves while performance degrades, and vice-versa. Balancing performance with security in real-world DW scenarios is a complex issue which depends on the requirements and context of each particular environment. Most encryption algorithms are not suitable for DWs, because they have been designed as a general-purpose "one fits all" security solution, introducing a need for specific solutions for DWs capable of producing better security-performance tradeoffs.

Encryption in DBMS can be column-based or tablespace-based. Using tablespace encryption implies losing the ability to directly query data that we do not want or need to encrypt, adding superfluous decryption overheads. Best practice guides such as [14] recommend using column-based encryption for protecting DWs. Thus, we propose a column-based encryption solution and for fairness we compare it with other similar solutions.

In this paper, we propose a lightweight encryption solution for numerical values using only standard SQL operators such as eXclusive OR (XOR) and modulo (MOD, which returns the remainder of a division expression), together with additions and subtractions. We wish to make clear that it is not our aim to propose a solution as strong in security as the state-of-the-art encryption algorithms, but rather a technique that provides a considerable level of overall security strength while introducing small performance overheads, *i.e.*, that presents better security-performance balancing. To evaluate our proposal, we include a security analysis of the cipher and experiments with standard and state-of-the art encryption algorithms such as Order-Preserving Encryption (OPES) [3] and Salsa20 (alias Snuffle) [5, 6], using two leading DBMS.

In summary, our approach has the following main contributions and achievements:

- SES-DW avoids storage space and computational overhead by preserving each encrypted column's original datatype;
- Each column may have its own security strength by defining the number of encryption rounds to execute. This also defines how many encryption keys are used, since each round uses a distinct key (thus, the true key length is the number of rounds multiplied by the length of each round's encryption key). This enables columns which store less sensitive information to be protected with smaller-sized keys and rounds and thus, process faster than more sensitive columns;
- Our solution is used transparently in a similar fashion as the Oracle TDE [11, 14] and requires minimal changes to the existing data structures (just the addition of a new column), and the SES-DW cipher uses only standard SQL operators, which makes it directly executable in any DBMS. This makes our solution portable, low-cost and straightforward to implement and use in any DW;
- Contrarily to solutions that pre-fetch data, by simply rewriting queries we avoid I/O and network bandwidth congestion due to data roundtrips between the database and encryption/decryption mechanism, and consequent response time overhead;
- The experiments show that our technique introduces notably smaller storage space, response and CPU time overheads than other standard and state of the art solutions, for nearly all queries in all tested scenarios, in both inserting and querying data.

The remainder of the paper is organized as follows. In section 2 we present the guidelines and describe our proposal. In Section 3, we discuss its security issues. Section 4 presents experimental evaluations using the TPC-H decision support benchmark and a real-world DW with Oracle 11g and Microsoft SQL Server 2008. Section 5 presents related work and finally, section 6 presents our conclusions and future work.

2 SES-DW: Specific Encryption Solution for Data Warehouses

In this section we point out a set of considerations concerning the use of encryption solutions in DW environments, which guide the requirements that serve as the foundations of our proposal, and then we describe our approach and how it is applied.

2.1 The Foundations of SES-DW

Standard encryption algorithms were conceived for encrypting general-purpose data such as blocks of text, *i.e.*, sets of binary character-values. Standard ciphers (as well as their implementations in the leading DBMS) output text values, while DW data is mostly composed by numerical datatype columns [12]. Most DBMS provide built-in AES and 3DES encryption algorithms and enable their transparent use. However, they require changing each encrypted column's datatype at the core to store the ciphered outputs. To use the encrypted values for querying once decrypted, the textual values must be converted back into numerical format in order to apply arithmetic operations such as sums, averages, etc., adding computational overheads with considerable performance impact. Since working with text values is much more computationally expensive than working with numeric values, standard ciphers are much slower than solutions specifically designed for numerical encryption such as ours, which is specifically designed for numerical values and avoids datatype conversion overheads.

Data in DWs is mostly stored in numerical attributes that usually represent more than 90% of the total storage space [12]. Numerical datatype sizes usually range from 1 to 8 bytes, while standard encryption outputs have lengths of 8 to 32 bytes. Since DWs have a huge amount of rows that typically take up many gigabytes or terabytes of space, even a small increase of any column size required by changing numeric datatypes to textual or binary in order to store encryption outputs introduces very large storage space overheads. This consequently increases the amount of data to process, as well as the required resources, which also degrades database performance. While encrypting text values is mainly not so important for DWs, efficiently encrypting numerical values is critical. In our approach, we preserve the original datatype and length of each encrypted column, to maintain data storage space.

Topologies involving middleware solutions such as [15] typically request all the encrypted data from the database and execute decrypting actions themselves locally. This strangles the database server and/or network with communication costs due to bandwidth consumption and I/O bottlenecks given the data roundtrips between middleware and database, jeopardizing throughput and consequently, response time. Given the typically large amount of data accessed for processing DW queries, previously acquiring all the data from the database for encrypting/decrypting at the middleware is impracticable. Therefore, our approach is motivated by the requirement of using only operators supported by native SQL. This enables using only query rewriting for encrypting and decrypting actions and no external languages or resources need to be instantiated, avoiding data roundtrips and thus, avoiding I/O and network overhead from the critical path when compared to similar middleware solutions.

In what concerns the design of "data mixing" for each of the cipher's rounds, we discard bit shifting and permutations, commonly used by most ciphers, since there is no standard SQL support for these actions. We also discard the use of substitution boxes (*e.g.* AES uses several 1024-byte S-boxes, each of which converts 8-bit inputs to 32-bit outputs). Although complex operations such as the use of S-boxes provide a large amount of data mixing at reasonable speed on several CPUs, thus achieving stronger security strength faster than simple operations, the potential speedup is fairly small and is accompanied by huge slowdowns on other CPUs. It is not obvious that a series of S-box lookups (even with large S-boxes, as in AES, increasing L1 cache pressure on large CPUs and forcing different implementation techniques on small CPUs) is faster than a comparably complex series of integer operations. In contrast, simple operations such as bit additions and XORs are consistently fast, independently from the CPU. Our approach aims to be DBMS platform independent, making it usable in any DW without depending on any programming language or external resource, as well as specific CPU models. Given the requirements described in the former paragraphs, the proposed solution is described in the next subsections.

2.2 The SES-DW Cipher

Considering x the plaintext value to cipher and y the encrypted ciphertext, *NR* the number of rounds, *RowK* a 2^{128} bit encryption key, *Operation*[t] a random binary vector (*i.e.*, each element is 1 or 0), *XorK*[t] and *ModK*[t] as vectors where each element is an encryption subkey with the same bit length as the plaintext x, and $F(t)$ a MOD/XOR mix function (explained further), where t represents each individual encryption round number (*i.e.*, $t = 1...NR$). Figures 1.a and 1.b show the external view of the SES-DW cipher for respectively encrypting and decrypting.

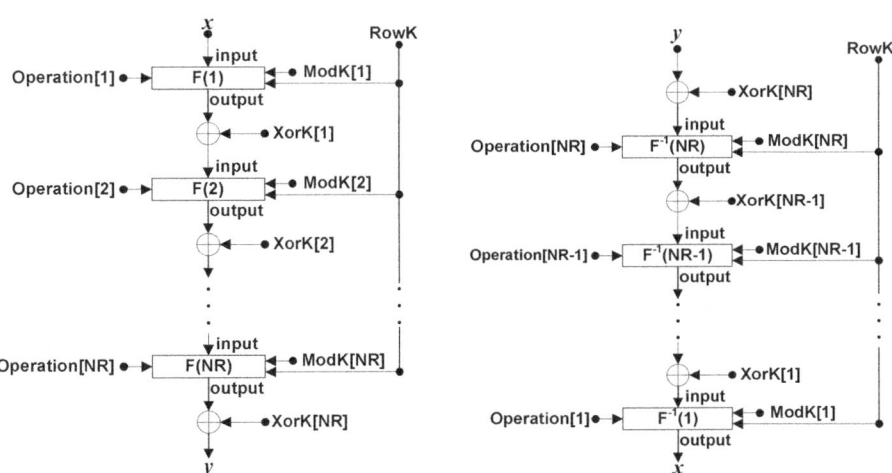

Fig. 1a. The SES-DW encryption cipher   **Fig. 1b.** The SES-DW decryption cipher

As illustrated, we randomly mix MOD with XOR throughout the encryption rounds, given a random distribution of 1 and 0 values of vector *Operation*. In the rounds where *Operation*[*t*] = 0, only XOR is used with the respective *XorK*[*t*]; in rounds where *Operation*[*t*] = 1, we first perform MOD with addition and subtraction using the respective *ModK*[*t*] and *RowK*[*j*], and *TabK*, and afterwards XOR with the respective *XorK*[*t*]. To avoid generating a ciphertext that may overflow the bit length of x it must be assured that the bit length of the term using MOD (*EncryptOutput* + (*RowK*[*j*]) MOD *ModK*[*t*]) - *ModK*[*t*]) is smaller or equal to the bit length of x.

As an example of encryption, consider the encryption of an 8 bit numerical value (x = 126) executing 4 rounds (*NR* = 4), given the following assumptions:

```
Operation = [0, 1, 0, 1]     XorK = [31, 2, 28, 112]
For t=1 (round 1), EncryptOutput = 126 XOR 31 = 97
For t=2 (round 2), EncryptOutput = (97+(15467801 MOD 36)-36) XOR 2 = 64
For t=3 (round 3), EncryptOutput = 64 XOR 28 = 92
For t=4 (round 4), EncryptOutput = ((92+15467801 MOD 19)-19) XOR 112 = 40
```

Thus, Encrypt(126, 4) = 40. In the decryption cipher, shown in Figure 1.b, $F^{-1}(t)$ also represents the reverse MOD/XOR mix function for decryption. Given this, the SES-DW cipher decryption function for decrypting x with *NR* rounds is:

```
FUNCTION Decrypt(x,NR)
    DecryptOutput = x
    FOR t = NR DOWNTO 1 STEP -1
        DecryptOutput = DecryptOutput XOR XorK[t]
        IF Operation[t] = 1 THEN
            DecryptOutput = DecryptOutput - (RowK MOD ModK[t]) + ModK[t]
        END_IF
    END_FOR
RETURN DecryptOutput
```

Considering the encryption example previously shown, we now demonstrate the decryption process for y = 40, given the same *Operation*, *RowK*, *XorK* and *ModK*:

```
For t=4 (round 1), DecryptOutput = (40 XOR 112)-(15467801 MOD 19)+19 = 92
For t=3 (round 2), DecryptOutput = 92 XOR 28 = 64
For t=2 (round 3), DecryptOutput = (64 XOR 2)-(15467801 MOD 36)+36 = 97
For t=1 (round 4), DecryptOutput = 97 XOR 31 = 126
```

Thus, Decrypt(40, 4) = 126, which is the original x plaintext value. Although our cipher only works with numerical values, we maintain the designation of plaintext and ciphertext respectively for the true original input value and ciphered value.

2.3 The SES-DW Functional Architecture

The system's architecture is shown in Figure 2, made up by three entities: 1) the encrypted database and its DBMS; 2) the SES-DW security middleware application; and 3) user/client applications to query the encrypted database. The SES-DW middleware is a broker between the DBMS and the user applications, using the SES-DW encryption and decryption methods and ensuring queried data is securely processed and the proper results are returned to those applications. We assume the DBMS is a trusted server and all communications are made through SSL/TLS secure connections, to protect SQL instructions and returned results between the entities.

A Specific Encryption Solution for Data Warehouses 89

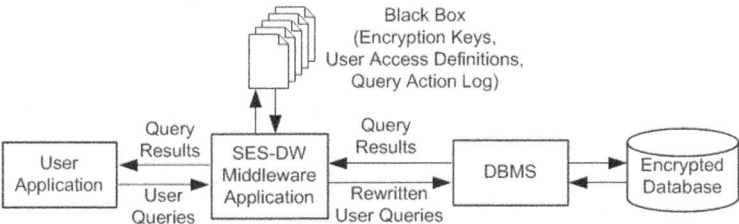

Fig. 2. The SES-DW Data Security Architecture

The Black Box is stored on the database server, created for each encrypted database. This process is similar to an Oracle Wallet, which keeps all encryption keys and definitions for each Oracle Database [14]. However, contrarily to Oracle, where a DBA has free access to the wallet, in our solution only the SES-DW middleware can access the Black Box, *i.e.*, absolutely no user has direct access to its content. In the Black Box, the middleware will store all encryption keys and predefined data access policies for the database. The middleware will also create a history log for saving duplicates of all instructions executed in the database, for auditing and control purposes. All Black Box contents are encrypted using AES with a 256 bit key.

To obtain true results, user actions must go through the security middleware application. Each time a user requests any action, the application will receive and parse the instructions, fetch the encryption keys, rewrite the query, send it to be processed by the DBMS and retrieve the results, and finally send those results back to the application that issued the request. Thus, SES-DW is transparently used, since query rewriting is transparently managed by the middleware. The only change user applications need is to send the query to the middleware, instead of querying the database directly.

To encrypt a database, a DBA requires it through the SES-DW middleware. Entering login and database connection information, the middleware will try to connect to that database. If it succeeds, it creates the Black Box for that database, as explained earlier. Afterwards, the middleware will ask the DBA which tables and columns to encrypt. All the required encryption keys (*RowK*, *XorK*, *ModK*) for each table and column will be generated, encrypted by an AES256 algorithm and stored in the Black Box. Finally, the middleware will encrypt all values in each column marked for encryption. Subsequent database updates must always be done through the middeware, which will apply the cipher to the values and store them directly in the database.

To implement SES-DW encryption in a given table *T*, consider the following: Suppose table *T* with a set of *N* numerical columns $C_i = \{C_1, C_2, ..., C_N\}$ to encrypt and a total set of *M* rows $R_j = \{R_1, R_2, ..., R_M\}$. Each value to encrypt in the table will be identified as a pair (R_j, C_i), where Rj and Ci respectively represent the row and column to which the value refers ($j = \{1..M\}$ and $i = \{1..N\}$). To use the SES-DW cipher, we generate the following encryption keys and requirements:

- An encryption key *TabK*, a 128 bit random generated value, constant for table *T*;
- Vector *RowK*[*j*], with $j = \{1..M\}$, for each row *j* in table T. Each element holds a random 128 bit value;
- Define NR_i with $i = \{1..N\}$, which gives the number of encryption rounds to execute for each column C_i. We define $NR_i = SBL_i/\text{BitLength}(C_i)$, where SBL_i is the desired security bit strength for the *XorK* and *ModK* encryption keys of column C_i

and BitLength(C_i) is the datatype bit length of column C_i (*e.g.* if we want to secure a 16 bit column C_i with a security strength of 256 bits, then the number of encryption rounds would be 256/16 = 16);

- Vectors $XorK_i[t]$ and $ModK_i[t]$, with $t = \{1..NR_i\}$, for each C_i, filled with randomly generated unique values. The bit length of each key is equal to the bit length of each C_i's datatype;
- A vector $Operation_i[t]$, with $t = \{1..NR_i\}$, for each column C_i, filled randomly with 1 and 0 values, so that the count of elements equal to 1 is the same as the count of elements equal to 0 (e.g. $Operation_i$ = [0,1,0,0,1,1,0,1], with NR_i = 8).

Since the number of rows in a DW fact table is often very big, the need to store a $RowK[j]$ encryption key for each row j poses a challenge. If these values were stored in a lookup table separate from table T, a heavy join operation between those tables would be required to decrypt data. Given the typically huge number of rows in fact tables, this must be avoided. For the same reasons, storing $RowK[j]$ in RAM is also impracticable. To avoid table joins, as well as oversized memory consumption, the values of $RowK[j]$ must be stored along with each row j in table T, as an extra column C_{N+1}. This is the only change needed in the DW data structure in order to use SES-DW. To secure the value of $RowK[j]$, it should be XORed with key $TabK$ before being stored. To retrieve the true value of $RowK[j]$ in order to use the SES-DW algorithms, we need to simply calculate (Rj, C_{N+1}) XOR $TabK$.

3 Security Issues

Threat Model. All user instructions are managed by the SES-DW middleware, which transparently rewrites them to query the DBMS and retrieve the results. The users never see the rewritten instructions. For security purposes, the middleware shuts off database historical logs on the DBMS before requesting execution of the rewritten instructions, so they are not stored in the DBMS, since this would disclose the encryption keys. All communications between user applications, the SES-DW middleware and the DBMS are done through encrypted SSL/TLS connections. In what concerns the Black Box, all content is encrypted using the AES 256 algorithm, making it as secure in this aspect as any other similar solution for stored data (*e.g.* Oracle 11g TDE and SQL Server 2008 TDE). The only access to the Black Box content is done by the middleware, which is managed only by the application itself. We assume the DBMS is an untrusted server such as in the Database-As-A-Service (DAAS) model and the "adversary" is someone that manages to bypass network and SES-DW access controls, gaining direct access to the database. We also assume the SES-DW algorithms are public, so the attacker can replicate the encryption and decryption functions, meaning that the goal of the attacker is to obtain the keys in order to break security.

Using Variable Key Lengths and MOD-XOR Mixes. The bit length of the encryption keys $XorK$ and $ModK$ are the same as the bit length of each encrypted column, meaning that an 8 bit sized column datatype will have 8 bit sized encryption keys. It is obvious that using 8 bit keys on their one is not secure at all. However, since all keys are distinct in each round, executing 16 rounds would be roughly equivalent to having a 16*8 = 128 bit key in the encryption process. It is up to the DW security administrator to decide how strongly secure each column should be, which defines how many rounds should be executed, considering the bit length of the column's datatype.

The MOD operator is used in the cipher because it is non-injective, given that for X MOD $Y = Z$, the same output Z, considering Y a constant, can have an undetermined number of possibilities in X as an input that will generate the same value Z (*e.g.* 15 MOD 4=3, 19 MOD 4=3, 23 MOD 4=3, etc). Since MOD operations are non-injective, the encryption rounds using MOD are also non-injective. Given that injectivity is a required property for invertibility, our cipher is thus not directly invertible. It is also true that the same ciphered output values are most likely to come from different original input values. Moreover, randomly using the XOR and MOD operators as the two possible operators for each round also increases the number of possibilities an attacker needs to test in exhaustive searches for the output values of each encryption round, since the attacker does not know the rounds in which MOD is used with XOR and needs to test both hypothesis (XOR and MOD-XOR). Furthermore, if the attacker does not know the security strength chosen for encrypting each column, s/he does not know how many encryption rounds were executed for each ciphered value.

By making the values of $XorK_i$ and $ModK_i$, distinct between columns, we also make encrypted values independent from each other between columns. Even if the attacker breaks security of one column in one table row, the information obtained from discovering the remaining encryption keys is limited. Thus, the attacker cannot infer information enough to break overall security; in order to succeed, s/he must perform recover all the keys for all columns.

Attack Costs. To break security by key search in a given column C_i, the attacker needs to have at least one pair (plaintext, ciphertext) for a row j of C_i, as well as the security bit strength involved, as explained in subsection 2.3, because it will indicate the number of rounds that were executed. In this case, taking that known plaintext, the respective known ciphertext, and the C_{N+1} value (storing $RowK_j$ XOR $TabK$, as explained in subsection 2.3), s/he may then execute an exhaustive key search.

The number of cipher rounds for a column C_i is given by NRi, and β is the bit-length of C_i's datatype. Since half the values of vector *Operation* are zeros and the other half are ones, the probability of occurrences of 1 and 0 is equal, *i.e.*, $Prob(Operation[t]=0) = \frac{1}{2} = Prob(Operation[t]=1)$, where the number of possible values for $Operation[t]$ is 2^{NRi}. Considering β, each $XorK$ and $ModK$ subkey also has a length of β bits and thus, each $XorK$ and $ModK$ subkeys have a search space with 2^{β} possible values. $TabK$ is a 128 bit value, thus with a search space of 2^{128} possible values. Considering the cipher's algorithm and given the probability of {0, 1} values in *Operation*, a XOR is executed in all rounds (NRi), while a MOD is executed before the XOR in half the rounds ($NRi/2$). Given this, the key search space dimension considering the combination of XOR and MOD/XOR rounds is given by $G(x)$:

$$G(x) = \sum_{x=1}^{NRi+\frac{NRi}{2}} F(x) \cdot 2^{(\beta x)+128}$$

$$F(x) = \begin{cases} \binom{NRi-x}{\frac{NRi}{2}-x} & , x = 1 \\ F(x-1) + (-1)^x \binom{NRi-x}{\frac{NRi}{2}-x} & , 2 <= x <= NRi/2 \\ F(x-1) & , NRi/2+1 <= x <= NRi \\ F(x-1) + (-1)^{(x-\frac{NRi}{2})} \binom{x-\frac{NRi}{2}-1}{x-NRi-1} & , NRi+1 <= x <= NRi + NRi/2 - 1 \\ \binom{NRi}{\frac{NRi}{2}} & , x = NRi + NRi/2 \end{cases}$$

Considering Y as the number of attempts to discover the keys, Y is a discrete random variable with support $S = \{1...N\}$, where N represents the search space's dimension. For one attempt, considering a random variable B, it has only two possibilities:

$$B = \begin{cases} 0, & \text{given the attempt is not successful} \\ 1, & \text{given the attempt is successful} \end{cases}$$

Therefore, B follows a Bernoulli distribution with probability $p = Prob(B=1) = 1/N$. Since the number of attempts is limited, given the search space is finite, variable Y also has a finite support $S = \{1...N\}$. The probability of being successful after k attempts is given by: $Prob(Y = k) = Prob(\bar{A} \cap \bar{A} \cap ... \cap \bar{A} \cap A) = \left(1 - \frac{1}{N}\right)^{k-1} \cdot \frac{1}{N}$, $k=1... N$.

Note that the probability of being needed more than m attempts is given by:

$$Prob(Y > m) = \sum_{k=m+1}^{N} Prob(Y = k) = \sum_{k=m+1}^{N} \left(1 - \frac{1}{N}\right)^{k-1} \cdot \frac{1}{N} = (1 - 1/N)^m \cdot \left[\left(1 - \left(1 - \frac{1}{N}\right)^{N-m}\right)\right].$$

The probability of needing n more attempts, given m initial unsuccessful attempts (for $m > 1$ and $n > 1$) is given by $Prob(Y > m+n \mid Y > m) = Prob(Y > m+n) / Prob(Y > m)$, since the event $\{Y > m+n\}$ is contained in $\{Y > m\}$, which means that after having m unsuccessful attempts, being successful after n more attempts only depends on those n additional attempts and not on the initial m attempts, *i.e.*, it does not depend on the past. For the complete search space, the average number of attempts is then given by:

$$\sum_{k=1}^{N} k \cdot Prob(Y = k) = \frac{1}{N} \sum_{k=1}^{N} k \left(1 - \frac{1}{N}\right)^{k-1} = (*).$$

From the series theory it is known that $\sum_{k=0}^{+\infty} x^k = \frac{1}{1-x}$, if $|x|<1$, which is the case in (*) for $\left(1 - \frac{1}{N}\right)$. Thus, $(\sum_{k=1}^{+\infty} x^k)' = \left(\frac{1}{1-x}\right)' \Leftrightarrow \sum_{k=1}^{+\infty} k \cdot x^{k-1} = \frac{1}{(1-x)^2}$, $|x|<1$.

Thus, the average number of attempts for finding the keys is $(*) = \frac{1}{N} \cdot \frac{1}{\left(1-\left(1-\frac{1}{N}\right)\right)^2} =$

N which is equal to the dimension of the key search space (N). Note however, that this is the worst case complexity. It is possible for the attacker to reduce the key search space by chosen plaintext attacks. Since the same *TabK* key is used for encrypting all *RowK*, as explained in previous subsection (C_{N+1}(row j) = *RowK*[j] $\oplus$ *TabK*), the information leakage given by $y_1 \oplus y_2 = (x_1 \oplus TabK) \oplus (x_2 \oplus TabK) \Leftrightarrow y_1 \oplus y_2 = (x_1 \oplus x_2) \oplus (TabK \oplus TabK) \Leftrightarrow y_1 \oplus y_2 = x_1 \oplus x_2$ implies that C_{N+1}(row j) $\oplus$ C_{N+1}(row $j+1$) = *RowK*[j] $\oplus$ *RowK*[$j+1$], reducing the possible search space for *RowK* to 2^{64} instead of 2^{128} in each row. If the attacker manages to use very low *RowK* values, which are most probably smaller than the value of the *ModK* encryption keys (*i.e. RowK<ModK[t]*), then the (*RowK* MOD *ModK[t]*) – *ModK[t]* operation in the cipher will be reduced to *RowK* – *ModK[t]*, thus further reducing complexity. In this case, for example, taking more than one (*plaintext, ciphertext*) pair $y_1 = Encrypt(x_1,2)$ and $y_2 = Encrypt(x_2,2)$ for 2 encryption rounds on the same row, where *Operation*=[0,1]:

$$y_1 \oplus y_2 = (x_1 \oplus XorK[1] + RowK - ModK[2]) \oplus (x_2 \oplus XorK[1] + RowK - ModK[2])$$

Considering that each x_i has a length of β bits, given the encryption key *RowK* has a reduced search space of 2^{64} (as previously mentioned) and each *XorK* and *ModK* have a search space of 2^β, the key search space in this example is given by $2^{2\beta+64}$. Since *XorK[1]* and *ModK[2]* are just half the keys for the 2 round SES-DW, to obtain the remaining *XorK[2]* and *ModK[1]* keys, the search space is incremented by $2^{2\beta}$. Since the number of *XorK* and *ModK* encryption keys is the same as the number of rounds, the generic expression for the reduced key search space in this type of attack is given by $G(x) = 2^{NRi*\beta+64} + 2^{NRi*\beta}$. Note that for an 8 bit value ($\beta = 8$) encrypted by 16 rounds (*NRi* = 16), using 16 *XorK* and *ModK* subkeys with 8 bits each (each total key length for *XorK* and *ModK* is 16*8 = 128 bits), the key search space complexity is $2^{192} + 2^{128} \cong 6,3 \times 10^{57}$, which remains a considerable measure of security strength.

SES-DW Entropy. In information theory, entropy is a measure of randomness or uncertainty. In this context, the term usually refers to Shannon's entropy, which quantifies the randomness of a variable based upon the knowledge of the information contained in its message. The entropy of a discrete variable X with n bits in length is given by the following expression, where $Prob(x_i)$ is the probability of occurrence of each x_i within the probability distribution of all possible integer values $[1…2^n]$:

$$Entropy(X) = -\sum_{i=1}^{2^n}\bigl(Prob(X = x_i).log_2 Prob(X = x_i)\bigr)$$

Since numeric datatype storage sizes are typically 8, 16, 32, 64 or 128 bits, each of our cipher's input/output values (as well as the encryption keys) respectively have a number of 2^8, 2^{16}, 2^{32}, 2^{64}, or 2^{128} possible combinations. While it is computationally fast to obtain the probability distribution in the first case by combining all possible input and encryption key values (with all 8 bit values = $[1...2^8]$) using two cipher rounds (the minimum number of rounds), for the remaining (2^{16}, 2^{32}, 2^{64} and 2^{128}) the task gets exponentially time-expensive. Therefore, after a series of statistical regression experiments using the calculated 8 bit probability distribution for SES-DW, we found that the logarithmic regression ($y = a + b.ln(x)$) generated the most adjusted statistical model for representing the cipher's probability distribution (with $R^2>=0.98$ and a standard error of 0.001). Knowing that the accumulated probability for n bits must be equal to 1, using the logarithmic regression function we must ensure that:

$$\int_1^{2^n} a + b.ln(x)\, dx = 1$$

This expression leads to $Prob(x_i) = \hat{a} + \hat{b}.ln(x_i)$, representing the estimated probability distribution function for n bits SES-DW, where:

$$\hat{a} = \frac{1-n.b.2^n.ln(2)}{2^n-1} + b \quad \wedge \quad \hat{b} = \frac{\bar{X}-\left(2^{n-1}+\frac{1}{2}\right)}{2^{2n-2}-\frac{1}{4}-n.2^{n-1}.\ln(2)}$$

Given $Prob(x)$, the entropy of SES-DW for $n=8$, 16, 32, 64 and 128 bits is shown in Table 1. As seen, the entropy produced for n bits is nearly n, thus meaning the generated ciphertexts are very close to a uniformly random n bit value.

Table 1. Estimated SES-DW entropy values

Number of bits (n)	SES-DW Entropy
8	7,967144
16	15,972308
32	31,979863
64	63,986246
128	127,989741

4 Experimental Evaluation

We used the TPC-H benchmark [17] (1GB and 10GB scale sizes) and a real-world sales DW storing one year of commercial data (taking up 2GB of data). We tested all scenarios using Oracle 11g and Microsoft SQL Server 2008 DBMS, on a Pentium Core2Duo 3GHz CPU with a 1.5TB SATA hard disk and 2GB RAM (512MB of devoted to database memory cache), with Windows 2003 Server. The TPC-H schema has one fact table (*LineItem*), and seven dimension tables. The Sales DW database schema has one fact table (*Sales*) and four dimension tables. In TPC-H setups, four numerical columns of *LineItem* were encrypted (*L_Quantity*, *L_ExtendedPrice*, *L_Tax* and *L_Discount*). In the Sales DW, five numerical columns were encrypted (*S_ShipToCost*, *S_Tax*, *S_Quantity*, *S_Profit*, and *S_SalesAmount*). We compare our solution with the column-based AES128, AES256 and 3DES168 algorithms, and OPES [3] and Salsa20 [5, 6]. OPES and Salsa20 were implemented using C++.

4.1 Analyzing Storage Size and Loading Time

Tables 2 and 3 show the results of data storage size and loading time (in seconds), respectively, for loading the TPC-H 1GB *LineItem* table in each setup. The results in the remaining databases are similar, with absolute values nearly proportional to their database sizes, and due to lack of space and to avoid redundancy are not included. The results shown are an average of six executions for each tested scenario on each DBMS (with standard deviation in Oracle 11g between [2.27, 22.12], and in SQL Server 2008 between [3.19, 20.45]).

Table 2. TPC-H 1GB Lineitem Fact Table Storage Size Overhead

	Oracle TPC-H 1GB Storage Size (Overhead)	SQL Server TPC-H 1GB Storage Size (Overhead)
Standard	772MB	1237MB
AES128/256	1960MB (+1188MB / 154%)	2410MB (+1173MB / 95%)
3DES168	1572MB (+800MB / 104%)	2181MB (+944MB / 76%)
OPES	790MB (+18MB / 2%)	1258MB (+21MB / 2%)
Salsa20	1064MB (+292MB / 38%)	1553MB (+316MB / 26%)
SES-DW	868MB (+96MB / 12%)	1339MB (+102MB / 8%)

Table 3. TPC-H 1GB Lineitem Fact Table Loading Time Overhead

	Oracle TPC-H 1GB Loading Time (Overhead)	SQL Server TPC-H 1GB Loading Time (Overhead)
Standard	253 s	171 s
AES128	608 s (355 s / 141%)	382 s (211 s / 123%)
AES256	636 s (383 s / 152%)	407 s (236 s / 138%)
3DES168	617 s (364 s / 144%)	389 s (218 s / 127%)
OPES	353 s (100 s / 40%)	229 s (58 s / 34%)
Salsa20	419 s (166 s / 66%)	281 s (110 s / 64%)
SES-DW128	279 s (26 s / 10%)	191 s (20 s / 12%)
SES-DW256	294 s (41 s / 16%)	199 s (28 s / 16%)
SES-DW1024	451 s (198 s / 78%)	284 s (113 s / 66%)

As shown, OPES and SES-DW have much smaller storage space overheads (2% to 12%, 18MB to 102MB) than Salsa20 (26% to 38%, 292MB to 316MB), 3DES168 (76% to 104%, 800MB to 944MB) and AES (95% to 154%, 1173MB to 1188MB of overhead). However, in loading time, SES-DW presents the best results by far (10% to 16%, 20 to 41 seconds of overhead). Considering these results, SES-DW is much more efficient, introducing small overheads for similar key sizes. Note that the worst result for SES-DW 1024, which is similar to Salsa20; however, it refers to using 1024 bit encryption keys, far higher than the remaining tested algorithms. Also note that the results for the TPC-H 10GB database are approximately proportional to those of the 1GB database, which means ten times bigger. Since 1GB is actually a very small size for a DW database, it is easy to conclude that the overheads introduced by encryption are extremely significant and may in fact introduce considerable hardware cost.

4.2 Analyzing Database Query Performance

The TPC-H workload included the benchmark queries 1, 3, 6, 7, 8, 10, 12, 14, 15, 17, 19 and 20 (all accessing fact table LineItem). For Sales DW, the workload was a set of 29 queries, all processing the Sales fact table, as a set of usual decision support daily (9 queries), monthly (9 queries) and annual (11 queries) queries. All results are an average from six executions in each scenario (Oracle 11g standard deviations between [0.47, 42.23] and [0.55, 61.34] for 1GB and 10GB TPC-H, respectively, and [0.63, 59.17] for the Sales DW, and SQL Server between [0.56, 49.56] and [0.63, 58.30] for 1GB and 10GB TPC-H, respectively, and [0.47, 66.08] for the Sales DW). Figure 3 shows total workload execution time overhead for each scenario, while Figure 4 shows the same for CPU time overhead. The Standard execution time (execution time of the workload against a non-encrypted database) for each scenario is 492, 5037, and 1766 seconds in Oracle 11g, and 452, 4294, and 1690 seconds in SQL Server 2008, for the 1GB, 10GB TPC-H and Sales DW, respectively.

It can be seen that SES-DW with 128-bit and 256-bit security has the best response and CPU time overheads for all scenarios, followed by Salsa20 and further by AES, while OPES has results leveled between AES and 3DES. Notice that observing the results for the TPC-H database, SES-DW shows better scalability than the remaining ciphers. In fact, SES-DW 1024-bit in the TPC-H 10GB is nearly as fast as Salsa20, the best solution after SES-DW. This means that the relative gains by using SES-DW

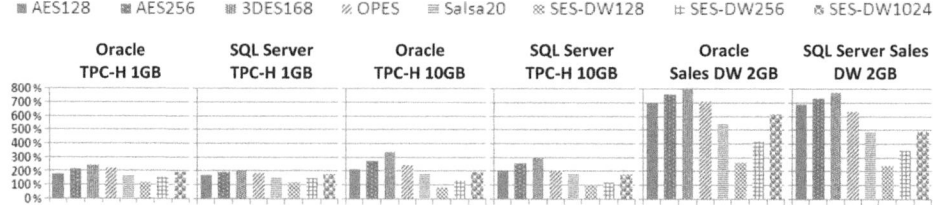

Fig. 3. Total query workload response time overheads (%) for each setup

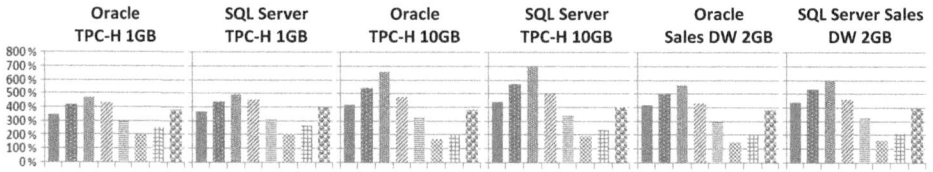

Fig. 4. Total query workload CPU time overheads (%) for each setup

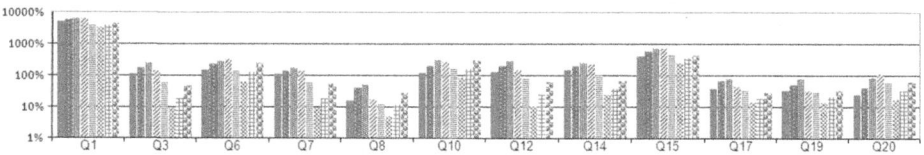

Fig. 5. TPC-H 10GB individual query exec. time overhead p/encrypt. algorithm in Oracle 11g

increases as database size scales up, compared with the remaining ciphers. Notice that being 100% faster in TPC-H 10GB means a saving of 5037 seconds (almost 1,5 hours) in total query workload response time.

Considering these results, since 10GB is actually a small size for a DW database, it is easy to conclude from the overall results that performance overheads introduced by data encryption algorithms in DWs are in fact extremely significant, and even minimum gain in response/CPU time is an important achievement.

The results for individual query execution time in Oracle 11g for TPC-H 10GB scenarios are shown in Figure 5, with a logarithmic scale. These results show that all queries have similar proportional overhead to those of the complete workload. This is also true for all the other scenarios, making it redundant to include all in this section. It can be seen that most queries processed by AES and 3DES have overheads of several orders of magnitude higher than SES-DW.

The number of CPU clock cycles spent on encryption and decryption depends on the algorithm and CPU architecture in which they are executed. As an example, the work in [7] refers that AES [2] with a 128 bit key takes up, on average, 20 clock cycles per encrypted byte on a Pentium IV, for encrypting a 16 byte value, resulting in a total of 20 x 16 = 320 clock cycles. The same algorithm with a 256 bit key takes up an average of 28 clock cycles per encrypted byte, meaning it needs 28*16 = 448 clock cycles for encrypting the same 16 byte value. We measured a speed of 8.53 cycles per byte for SES-DW on a Pentium IV for 128 bits encryption values. This makes SES-DW more than twice as fast as AES 128 on the same CPU model.

5 Related Work

The work in [4] proposes perturbed tables in a DW for preserving privacy that obfuscates data and explain data reconstruction for executing queries. Although providing strong guarantees against privacy breaches, these methods produce errors in data reconstruction, which we avoid. A lightweight database encryption scheme for column-oriented DBMS is proposed in [9], with low decryption overhead. In [3] an Order Preserving Encryption Scheme (OPES) for numeric data is proposed, by flattening and transforming the plain text distribution onto a target distribution, based on value-based buckets. This solution allows any comparison operation to be directly applied on encrypted data. A similar solution for processing queries without decrypting data was proposed by [10], using the database-as-a-service paradigm.

The Data Encryption Standard (DES) [8] is a 64 bit block cipher which uses a 56 bit key. As an enhancement of DES, the Triple DES (3DES) encryption standard was proposed [1]. The 3DES encryption method is similar to the original DES, but it is applied three times to increase the encryption level, using three different 56 bit keys. Thus, the effective key length is 168 bits. The algorithm increases the number of cryptographic operations, making it one of the slowest block cipher methods. The Advanced Encryption Standard (AES) is currently the most used encryption standard [2]. AES provides three key lengths: 128, 192 and 256 bits. It is fast and able to provide stronger encryption, compared to other algorithms such as DES [13]. Brute force attack is the only known effective attack known against it. As we have demonstrated in [16], these ciphers introduce very much performance overhead for DWs.

In the search for more computationally efficient algorithms by exchanging a small number of complex operations such as S-box lookups for longer chains of simpler operations, the Salsa20 (alias Snuffle) family of ciphers [6] was proposed. These ciphers have been well studied and are considered fast high security solutions.

An Enterprise Application Security solution is presented in [15], acting as a wrapper/interface between user applications and the encrypted database server. This solution aims to ensure data integrity and efficient query execution over encrypted databases, by evaluating most queries at the application server and retrieving only the necessary records from the database server.

6 Conclusions and Future Work

We propose an encryption solution specifically designed for enhancing data confidentiality in DWs. This solution is transparent and only require user applications to send their queries to a middleware security broker instead of the DBMS. Only the final processed results are returned to the authorized user applications that requested them. All SQL commands and actions are encrypted and stored in a log by the security broker, which can be audited by any user with administration rights. In the database, the data always stays encrypted, never allowing breaches before queries finish execution. If an attacker bypasses the broker and gains direct access, s/he just sees encrypted "realistic-looking" values. In addition, since data schemas and column-types are preserved and the encrypted data is realistic but not real, our method allows using the database (or "as-is" replicas) for testing purposes and direct querying during application software development, generating realistic but not real results. This also avoids disclosure of the real original data if any attacker bypasses database access control and can retrieve data directly from the database. The proposed solution is independent from DBMS and CPU specific features and requires small computational efforts and can be straightforward and easily implemented in any database. Since it basically works by transparently rewriting user queries, it minimizes efforts in changing user applications and does not jeopardize network and I/O bandwidth. Our technique shows better database performance than standard and state-of-the-art encryption solutions while providing considerable security strength, making it a valid option for balancing performance with security from the DW perspective. As future work, we intend to take advantage of the history log stored in the Black Box in order to manage intrusion detection for attackers that obtain valid database login credentials.

References

1. 3DES, Triple DES, National Bureau of Standards, Nat. Inst. of Standards and Technology (NIST), Fed. Inform. Processing Standards (FIPS) Pub. 800-67, ISO/IEC 18033-3 (2005)
2. AES, Advanced Encryption Standard. NIST, FIPS-197 (2001)
3. Agarwal, R., Kiernan, J., Srikant, R., Xu, Y.: Order-Preserving Encryption for Numeric Data. In: ACM SIG Conf. on Management Of Data, SIGMOD (2004)
4. Agrawal, R., Srikant, R., Thomas, D.: Privacy Preserving OLAP. In: ACM SIG Conf. Management Of Data, SIGMOD (2005)
5. Bernstein, D.J.: Snuffle 2005: The Salsa Encryption Function, http://cr.yp.to/snuffle.html
6. Bernstein, D.J.: The Salsa20 Family of Stream Ciphers. In: Robshaw, M., Billet, O. (eds.) New Stream Cipher Designs. LNCS, vol. 4986, pp. 84–97. Springer, Heidelberg (2008)
7. Bernstein, D.J., Schwabe, P.: New AES software speed records. In: Chowdhury, D.R., Rijmen, V., Das, A. (eds.) INDOCRYPT 2008. LNCS, vol. 5365, pp. 322–336. Springer, Heidelberg (2008)
8. DES, Data Encryption Standard, National Bureau of Standards, Nat. Inst. of Standards and Technology (NIST), Federal Inform. Processing Standards (FIPS) Pub 46 (1977)
9. Ge, T., Zdonik, S.: Fast, Secure Encryption for Indexing in a Column-Oriented DBMS. In: Int. Conf. Data Engineering, ICDE (2007)
10. Hacigumus, H., Iyer, B.R., Li, C., Mehrotra, S.: Executing SQL over Encrypted Data in the Database-Service-Provider Model. ACM C. Management of Data, SIGMOD (2002)
11. Huey, P.: Oracle Database Security Guide 11g. Oracle Corp. (2008)
12. Kimball, R., Ross, M.: The Data Warehouse Toolkit, 2nd edn. Wiley & Sons Inc. (2002)
13. Nadeem, A., Javed, M.Y.: A Performance Comparison of Data Encryption Algorithms. In: IEEE Int. Conf. on Information and Communication Technologies, ICICT (2005)
14. Oracle Corporation, Oracle Advanced Security Transparent Data Encryption Best Practices, Oracle White Paper (July 2010)
15. Radha, V., Kumar, N.H.: EISA - An Enterprise Application Security Solution for Databases. In: Jajodia, S., Mazumdar, C. (eds.) ICISS 2005. LNCS, vol. 3803, pp. 164–176. Springer, Heidelberg (2005)
16. Santos, R.J., Bernardino, J., Vieira, M.: Evaluating the Feasibility Issues of Data Confidentiality Solutions from a Data Warehousing Perspective. In: Cuzzocrea, A., Dayal, U. (eds.) DaWaK 2012. LNCS, vol. 7448, pp. 404–416. Springer, Heidelberg (2012)
17. Transaction Processing Council, The TPC Decision Support Benchmark H, http://www.tpc.org/tpch/default.asp

NameNode and DataNode Coupling for a Power-Proportional Hadoop Distributed File System

Hieu Hanh Le, Satoshi Hikida, and Haruo Yokota

Department of Computer Science, Tokyo Institute of Technology, Japan
{hanhlh,hikida}@de.cs.titech.ac.jp, yokota@cs.titech.ac.jp

Abstract. Current works on power-proportional distributed file systems have not considered the cost of updating data sets that were modified (updated or appended) in a low-power mode, where a subset of nodes were powered off. Effectively reflecting the updated data is vital in making a distributed file system, such as the Hadoop Distributed File System (HDFS), power proportional. This paper presents a novel architecture, a NameNode and DataNode Coupling Hadoop Distributed File System (NDCouplingHDFS), which effectively reflects the updated blocks when the system goes into a high-power mode. This is achieved by coupling the metadata management and data management at each node to efficiently localize the range of blocks maintained by the metadata. Experiments using actual machines show that NDCouplingHDFS is able to significantly reduce the execution time required to move updated blocks by 46% relative to the normal HDFS. Moreover, NDCouplingHDFS is capable of increasing the throughput of the system that is supporting MapReduce by applying an index in metadata management.

Keywords: power-proportionality, HDFS, metadata management.

1 Introduction

Energy-aware commercial off-the-shelf (COTS)-based distributed file systems for cloud applications are increasingly moving toward power-proportional designs, as the configuration of the systems is changeable on demand. Specifically, the system is designed to operate in multiple gears and each gear contains a different number of active nodes. Multi-gear operation is made possible through a number of recent works that focus on power-proportional data placement layouts [1,2]. However, those works have not yet dealt with the reflecting of an updated data set that is modified (or appended) in a low gear mode when several nodes are powered off. In low gear, the currently active nodes should update the modified data instead of the inactive nodes. When the system moves to a high gear, to share the load equally to all active nodes, it is necessary to let the reactivated nodes catch up with the modification of the data set.

In addition to normal operations, the process of reflecting the updated data set increases several costs of metadata management (MDM) and data transference inside the system. Carrying out this process effectively is vital in realizing power proportionality for a distributed file system, such as the Hadoop Distributed File System (HDFS) [3], which is already widely used as a distributed file system for effective big data processing in the cloud. In the current HDFS architecture, reflecting updated files is ineffectively restrained at the NameNode because of access congestion in the metadata information of blocks.

This paper presents a novel architecture called the NameNode and DataNode Coupling HDFS (NDCouplingHDFS), which is designed to effectively reflect updated data in the power-proportional HDFS. NDCouplingHDFS couples MDM and data management to localize the range of blocks maintained by the metadata. Through this idea, the process is effectively distributed to multiple nodes as the load is shared among the nodes and each node can focus on its own work because all the necessary information is located locally.

Moreover, to raise the efficiency of reflecting updated data, it is preferable to eliminate the bottleneck of MDM at the single NameNode in a normal HDFS by using distributed MDM. Taking the locality of the file system into consideration, we suggest two approaches of distributed MDM based on a tree structure, namely static directory partitioning and the B-tree-based index method. In the first approach, we divide the namespace of the system among all the nodes, as each node will maintain a subpart of the directory hierarchy. In the second approach, we apply the parallel index technique, called Fat-Btree [4], which is used in current database management to manage the metadata of the file system. Our main contributions are the following.

- NDCouplingHDFS is proposed to solve the problem of reflecting updated (or appended) data sets when the power-proportional file system shifts from low gear to a higher gear.
- NDCouplingHDFS improves the IO throughput of the metadata operation of the HDFS by implementing distributed MDM with an index technique.
- An empirical experiment to evaluate NDCouplingHDFS is performed on actual machines. The empirical experimental results show that NDCouplingHDFS is able to significantly reduce the execution time to transfer updated blocks by 46% relative to a normal HDFS.

The remainder of this paper is organized as follows. Related work is introduced in Sect. 2. Section 3 describes our proposed system with the architecture and data flow. Section 4 presents a performance evaluation of our proposals. Conclusions and future work are discussed in Sect. 5.

2 Related Work

RABBIT [1] is the first work that aims to provide power proportionality to an HDFS by focusing on read performance. RABBIT uses the equal-work data layout policy using data replication. However, RABBIT does not yet consider the cost of reflecting updated data in low gear. Kim et al. [2] suggest a fractional replication method to achieve a balance between the power consumption and performance of a system. Their work considers the problem of identifying a suitable time to gear down and save power.

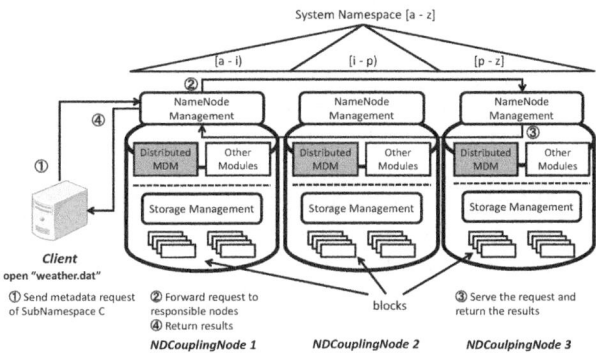

Fig. 1. A NameNode and DataNode Coupling HDFS architecture and data flow

Write Off-loading [5] is motivated by the goal of saving power through spinning down unnecessary disks. It allows write requests on spun-down disks to be temporarily redirected to other active disks in the file system. As a result, this technique lengthens the spin-down durations, thereby achieves additional power saving. Although not aiming to provide power proportionality, the idea could be considered as a solution for multigear file systems dealing with updated data when the system operates in low gear.

In previous work, we have taken into consideration the cost of updated data reflection relating to the size of moving data in a power-proportional HDFS [6]. As the size of moving data is small, the reflection process could be shortened.

3 NDCouplingHDFS

In this part, the assumptions employed in this paper is given. Then, the architecture of our system and two methods for distributed MDM are described. Finally, we present the system's behavior in reflecting updated data.

3.1 Assumptions and Conditions

In our proposal, we employed the following assumptions and conditions.

1. Data layout policy: The scope of this paper is limited to the MDM and the cost of reflecting updated data at power-proportional file systems. In low gear, the data from inactive nodes are replicated at other, active nodes.
2. Replication: When data are replicated at other nodes, their metadata are also replicated at the same node.
3. Failure: We suppose that all nodes in the system operate without failure.

3.2 Architecture and Data Flow of NDCouplingHDFS

The architecture and the data flow of NameNode and DataNode Coupling HDFS (NDCouplingHDFS) are shown in Figure 1. NDCouplingHDFS contains a cluster of NDCouplingNodes. There are two types of modules at each node in NDCouplingHDFS: the NameNode Management (NM) and the Storage Management (SM). The NM includes the new distributed MDM and other unmodified modules (such as Block Placement, Block Mapping) as in a normal HDFS. The important difference from a default HDFS is that the namespace of the file system is divided among all the nodes and the local distributed MDM only manages the metadata for files that are locally located. The SM at NDCouplingNode is the SM at DataNode in a normal HDFS.

Next, the data flow for the client interacting with NDCouplingHDFS is explained using Fig. 1. At first, the client randomly connects to a node to access the file system (open *weather.dat*). At this node, the request is forwarded to the corresponding node that contains the metadata of this file by distributed MDM. Then, the distributed MDM at this node looks for the file's metadata and sends the result back to the client. Finally, based on this result, the client opens connections to the responsible nodes to retrieve or store the file's blocks.

3.3 Distributed Metadata Management

In this part, we describes two approaches of employing distributed MDM to identify the responsible NDCouplingNode that contains the metadata for the accessed files.

Static Directory Partitioning Method. In this paper, we first try the static directory partitioning (SDP) method in distributing the namespace to multiple nodes in the system. Here, subparts of the directory hierarchy are manually assigned to individual nodes. All the nodes in the system have the mapping information about which node is responsible for what subpart of the file system directory. The system can process the request at most one hop to determine the appropriate nodes because the subparts of the hierarchy are treated as independent structures.

Fat-Btree-Based Method. This method applies Fat-Btree to perform distributed MDM. Fat-Btree is an update-conscious parallel B-tree structure that was originally proposed in database management as an indexing technique for efficient data management [4, 7]. Because of the parallel tree structure, the distributed MDM based on Fat-Btree achieves higher performance for search query processing while maintaining good locality tracking of the file system.

Alternative Techniques. To realize good performance with distributed MDM, many recent systems distribute the metadata across multiple nodes utilizing distributed hash table [8,9]. However, distributing metadata by hashing eliminates all hierarchical localities such as the POSIX directory access semantics.

3.4 Updated Data Reflection

Here, we describe the behavior of NDCouplingHDFS in serving the updated-data requests in low gear and reflecting the updated data when the system changes to

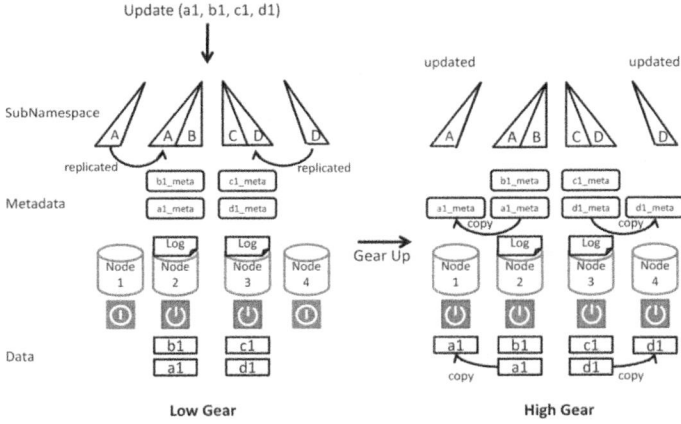

Fig. 2. Operations at updated data reflection processes of NDCouplingHDFS

high gear by reactivating a subset of nodes. In the normal HDFS, basically all the operations are similar however because there is only a single NameNode that is in charge of MDM, all the metadata operations are proccessed at the NameNode. Figure 2 shows an example of a four-node system in which each node maintains a subNamespace of the system. In low gear, Node 1 and Node 4 are inactive, and their maintenance data are consequently replicated at Node 2 and Node 3. During low gear, the part of the new updated data that is maintained by inactive nodes are reflected at predefined active nodes. Information about the data, the temporary node, and the intended node is saved into a Log file. In this example, Node 2 will update the data (here is $a1$) that should be updated by Node 1.

When the system changes to high gear by reactivating nodes (Node 1 and Node 4), the following four-step operations are carried out.

Step 1: Transfer Updated Metadata. The active nodes check the Log files and transfers only the different metadata to the reactivated nodes.

Step 2: Issue Block Transfer Commands. Next, the MDM searches for updated file blocks using the information in Log file. It then issues the block transfer command by filling the block transfer queue of each SM with the block and destination node paired information. After each constant heartbeat, the SM receives a command and transfers the blocks to the destination nodes. There are two considerable approaches for issuing a command. The **sequential issuance method** repeats the above search-and-issue operation for each transferred file, while the **batch issuance method** first looks for all the blocks and their destination nodes and then places them into a queue.

Step 3: Transfer Updated Blocks. When the SM receives the command issued by MDM, it sends the blocks to the destination nodes. However, in the current implementation in this part of the HDFS, for each block, the system has to open a new connection to the destination node. In order to reduce the cost of opening new network connection, we suggest the **batch transfer method** which

Table 1. Characteristics of the configurations used in updated-data reflection experiments

Configuration	NormalHDFS	SSS	SBS	SBB	FBB
Metadata management	Centralized	SDP	SDP	SDP	Fat-Btree
Command issuance	Sequential	Sequential	Batch	Batch	Batch
Block transference	Sequential	Sequential	Sequential	Batch	Batch
Updated metadata transference	-	○	○	○	○

Table 2. Experimental environment

# Gears	2
# nodes Low Gear	8
# nodes High Gear	16
# updated files	16000
file size	1MB

Table 3. Specification of a node

CPU	TM8600 1.0GHz
Memory	DRAM 4GB
NIC	1000 Mb/s
OS	Linux 3.0 64bit
Java	JDK-1.7.0

Table 4. HDFS information and parameters

version	0.20.2
$max.rep\text{-}stream$	100
$heartbeat\_interval$	1

sends all the blocks through just a single connection. The current implementation in the HDFS is called the **sequential transfer method**.

Step 4: Reflect Updated Metadata. The MDM updated the metadata for the newly arrived files as in the default HDFS based on the notifications from SM.

4 Experimental Evaluation

We carried out an empirical experiment with actual machines to verify the effectiveness of NDCouplingHDFS in terms of reducing the cost of updated-data reflection when the system shifts to higher gear. Next, we examined the effectiveness of distributed MDM relating to the scalability of metadata operations.

4.1 Updated-Data Reflection

To verify the effectiveness of each contribution proposed in Sect. 3, we prepared five configurations which are formed from the combinations of distributed MDM, command issuance method and block transference method. Table 1 shows the characteristics of these configurations.

Experimental Environment. We compare the proposed NDCouplingHDFS with the normal HDFS by changing the configuration of the system (Tab. 2). Both systems operate in two gears, a Low Gear and a High Gear with different number of active nodes (eight and 16 nodes). For **NormalHDFS**, there is one further node to be in charge of the NameNode. Because we address MDM in this paper, the number of appended files when the system operates at Low Gear is fixed at 16000 dividing equally to 16 nodes. Here, we use low-power-consuming ASUS Eeebox EB1007 machines, whose specifications are given in Tab. 3. The $max.rep\text{-}stream$, which specifies the maximum number of blocks that can be replicated by a SM at the same time, is set to 100. To efficiently perform the updated data reflection, the communication frequency between NM and SMs is maximized by setting $heartbeat\_interval$ to one (Tab. 4).

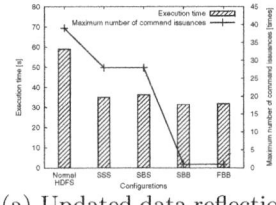 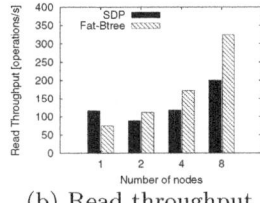 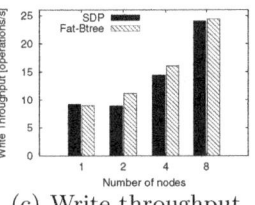

(a) Updated data reflection (b) Read throughput (c) Write throughput

Fig. 3. Experiment results

Experimental Results. Figure 3(a) shows the execution time for reflecting the updated data with different configurations. The left vertical axis shows the execution time from the time that the system begins to change from low gear to high gear until all the just-activated nodes catch up with the most current status of the updated data set. The right vertical axis shows the maximum number of transfer block command issuances, which is the number of times that the SM has to make a connection with the MDM to drain the block transfer queue.

Performance of NDCouplingHDFS. To confirm the NDCouplingHDFS's performance, we focus on the experimental results of **NormalHDFS** and **SSS**, the simplest configuration of NDCouplingHDFS, in Fig. 3(a). We see that NDCouplingHDFS has significantly reduced cost (nearly 41%) in reflecting updated data. In the HDFS, because of the high load at the NameNode with the processing of 8000 files that should be replicated to eight nodes, it requires about 40 connections between the NM and SM to drain the block transfer queue of the SM (about 58 seconds). Meanwhile, the process is distributed to eight nodes in NDCouplingHDFS, hence overall is completed in only about 34 seconds.

Performance of the Command Issuance. From the results of **SSS** and **SBS**, we see that the batch command issuance provided a slightly worse result than did sequential command issuance. The reason is that the SMs in **SBS** wasted several first connections to the NM before it had finished retrieving all 1000 updated files' data. On the other hand, the SM in **SSS** can perform the block replication process immediately from the very first communication.

Performance of the Block Transfer Method. Figure 3(a) shows that **SBB** reduces the execution time of the process to 31 seconds compared with **SBS**. This means that batch block transfer was able to reduce the cost of opening a new network connection for sending blocks. In total, SDP-based NDCouplingHDFS was able to reduce the execution time required for reflecting the updated data by 46% relative to **NormalHDFS**.

Fat-Btree-Based Method. There was little difference between the performance of **FBB** and **SBB**. The cost of the latter is slightly less by 0.5 seconds owing to the lower cost of MDM operations. This is due to the process of transferring incremental metadata, as the Fat-Btree-based method has to transfer more information than SDP because of the complex structure.

4.2 Distributed MDM Performance

In this part, we report the performance evaluation relating to the scalability of metadata operations to confirm the effect of SDP and Fat-Btree-based methods. The configurations of this experiment are shown in Tab. 5.

Table 5. Workload used in distributed MDM performance evaluation experiment

Fat-Btree leaf fanout	16
Data size ($\#files$)	3000
Number of nodes	1, 2, 4, 8
File size	1KB
#write accesses per node	$\frac{\#files}{\#nodes}$
#read accesses per node	$\#files$

Experimental Results. Figure 3(b) and 3(c) show the read and write throughput of two evaluated methods. Here, the operation includes searching/creating for the metadata and reading/writing the physical data of the query file. Figure 3(b) shows that the read performance of the Fat-Btree method significantly scales out. The good balance of the parallel B-tree structure means that the read requests are effectively distributed to all the nodes; hence, the overall throughput increased as the number of nodes increased. In contrast, in the SDP method, the throughput slightly decreased as the number of nodes increased from one to two. The reason is that the cost of opening a new connection to other responsible nodes is much larger than the cost of searching for the responsible metadata. From Fig. 3(c) which describes the overall throughput for write requests, the Fat-Btree method is seen not to provide such a considerable efficiency compare with the SDP method because of the high synchronization cost inside the B-tree structures during an update. Overall, the Fat-Btree is believed more suitable for the read-mostly workloads in MapReduce applications.

5 Conclusion and Future Work

In this paper, we first described the problem of inefficient reflection of updated data in power-proportional distributed file system and then proposed the NDCouplingHDFS architecture, which couples metadata management and data management at each node to solve it. Empirical experiments verified that our solution was able to shorten the execution time required to reflect updated data by 46% relative to the time required by the default HDFS. Moreover, NDCouplingHDFS was able to increase the throughput of the system supporting MapReduce by applying an index in metadata management. In the future, we would like to carry out more experiments with different workloads and a larger scale of nodes. Moreover, we would like to develop a system that integrates NDCouplingHDFS with suitable data placement to provide power proportionality.

Acknowledgements. This work is partly supported by Grants-in-Aid for Scientific Research from Japan Science and Technology Agency (A) (#22240005).

References

1. Amur, H., Cipar, J., Gupta, V., Ganger, G.R., Kozuch, M.A., Schwan, K.: Robust and Flexible Power-proportional Storage. In: Proc. the 1st ACM Symposium on Cloud Computing, SoCC 2010, pp. 217–228 (2010)
2. Kim, J., Rotem, D.: Energy Proportionality for Disk Storage using Replication. In: Proc. the 14th Int'l Conference on Extending Database Technology, pp. 81–92 (2011)
3. Apache Hadoop: HDFS Hadoop Wiki, http://wiki.apache.org/hadoop/HDFS
4. Yokota, H., Kanemasa, Y., Miyazaki, J.: Fat-Btree: An Update Conscious Parallel Directory Structure. In: Proc. the 15th Int'l Conference on Data Engineering, pp. 448–457. IEEE Computer Society (1999)
5. Narayanan, D., Donnelly, A., Rowstron, A.: Write Off-loading: Practical Power Management for Enterprise Storage. In: Proc. 6th USENIX Conference on File and Storage Technologies, pp. 253–267 (2008)
6. Le, H.H., Hikida, S., Yokota, H.: An Evaluation of Power-proportional Data Placement for Hadoop Distributed File Systems. In: Proc. Cloud and Green Computing, pp. 752–759. IEEE Computer Society (2011)
7. Yoshihara, T., Kobayashi, D., Yokota, H.: A Concurrency Control Protocol for Parallel B-tree Structures Without Latch-coupling for Explosively Growing Digital Content. In: Proc. the 11th Int'l Conference on Extending Database Technology: Advances in Database Technology, pp. 133–144. ACM (2008)
8. Rodeh, O., Teperman, A.: zFS-a Scalable Distributed File System using Object Disks. In: Proc. 20th IEEE/11th NASA Goddard Conference on Mass Storage Systems and Technologies (MSST 2003), pp. 207–218. IEEE (2003)
9. Braam, P.: The Lustre Storage Architecture

Mapping Entity-Attribute Web Tables to Web-Scale Knowledge Bases

Xiaolu Zhang, Yueguo Chen, Jinchuan Chen, Xiaoyong Du, and Lei Zou

Renmin University of China
Peking University
{zxlruc2010,chenyueguo,jcchen,duyong}@ruc.edu.cn, zoulei@pku.edu.cn

Abstract. There are many entity-attribute tables on the Web that can be utilized for enriching the entities of knowledge bases (KBs). This requires the schema mapping (matching) between the Web tables and the huge KBs. Existing solutions on schema mapping are inadequate for mapping a Web table and a KB, because of many reasons such as (1) there are many duplicates of entities and their types in a KB; (2) the schema of KB is often implicit, informal, and evolving over time; (3) the KB is typically very large in volume. In this paper, we propose a pure instance-based schema mapping solution to statistically find the effective mapping between a Web table and a KB via the matched data examples. Besides, we propose efficient solutions on finding the matched data examples as well as the overall mapping of a table and a KB. Experiments over real data sets show that our solution is much more accurate than the two baselines of existing solutions. Results also show that our solution is feasible for the mapping of Web tables to large scale KBs.

1 Introduction

The advance of information extraction and data integration techniques has promoted the prosperity of many Web-scale knowledge bases (KBs) such as FreeBase[6], YAGO[21], Linked Data[12]. These KBs typically utilize RDF triples to represent their basic information units. They have been widely used in applications such as semantic search, text understanding and question answering[22,12]. To effectively support these applications, a KB needs have information of a huge number of open domain entities. Many approaches have been tried to enlarge the population of entities in a KB. Although the size of Web-scale KBs keeps growing very fast, the coverage of a single KB is still very limited, compared to the numerous entities in the real world.

The current Web contains billions of tables, among which a huge number of tables (154M found in the Webtables project [8]) contain high-quality relational data. Of these high qualified tables, there are many entity-attribute tables that contain information of some entities of the same type [8,23]. Typically, information of an entity appears in one row with each column representing an attribute of the entity. One typical example of Web tables is the Google Fusion Table [11] where people can publish tabular data as they want. It is possible that some entities in a Web table may have corresponding entries in the KBs, from which we may learn the mapping between the Web table and the KBs. With the mapping, we are able to automatically inject entities of the Web table into the KBs. The problem is therefore a schema mapping problem in which we want to find the mapping between a Web table and a KB.

In the past decades, there have been many approaches[20,15,4,19] proposed for automatic schema mapping, typically focused on finding the mappings of attributes between two schemas. These approaches can be categorized into schema-based approaches[17,16], instance-based approaches[13,14], as well as their combinations [9]. Schema-based approaches use the schema-level information for matching. They cannot be applied in our problem because Web tables often do not have schemas. Even for the KBs, they do not have explicit schemas too. As such, the instance-based approaches[13,14], which consider the data contents for schema mapping, are preferred in our problem. However, there are challenges to apply existing instance-based approaches here because (1) there are many duplicates of entities and their types in a KB; (2) the schema of KB is often implicit, informal, and evolving over time; (3) the KB is typically very large in volume. The matching approaches will be neither efficient nor effective when the whole KB is modeled as a big table of all entities.

To address the above challenges, we propose a novel instance-based schema mapping approach for the integration of Web tables with KBs. In our study, we assume that high qualified entity-attribute Web tables have been extracted. We also assume that literal information of entities has been extracted from the KBs, i.e., the URIs in KBs have been transformed into literal names, so that we can focus on the direct semantic matches of texts in both Web tables and the KBs. Techniques on assigning URIs[12] to the literal information of entities are beyond the scope of the paper. In our approach, the mapping between a Web table and a KB is discovered from the mappings between the tuples of the Web table and the entities of the KB. We propose techniques to efficiently conduct the proposed instance-based schema mapping. The contributions of the paper can be summarized as follows:

- We formalize the table-to-KB schema mapping problem that statistically finds the effective table-to-KB mapping between a Web table and a KB via the matched data examples.
- We propose a technique that is able to tune the number of tuples used for finding the table-to-KB mapping, so that instance-based schema mapping can be conducted in a feasible time without the loss of accuracy too much.
- Extensive experimental results show that the proposed schema mapping solution is very effective for mapping Web tables to KBs, paying a feasible workload for the mapping task.

The rest of the paper is organized as follows. Section 2 introduces some related work. Section 3 states the problem of table-to-KB schema mapping. Then, we discuss efficient solution of schema mapping in Section 4. The experimental study is given in Section 5, followed by the conclusions given in Section 6.

2 Related Work

Schema Mapping. Rahm et al. did a good survey[20] about approaches to automatic schema mapping. The solutions are classified into schema-based[17,16] and instance-based[13,14] based on either only schema information (metadata) or data content is used for finding the mappings between columns of two schemas. Do et al. developed the COMA system[9] for the flexible combination of schema mapping approaches, and then developed a tool called COMA++[3] to cope with schema and ontology matching. Due to the lack of schema information, instance-based approaches are more suitable for the table-to-KB schema mapping problem. Existing instance-based approaches[13,14] typically utilize the statistical information of data contents within a column for matching. However, it faces with a problem that many predicates (columns) of the KBs contain data of the same type and they often have similar distribution of data values. This causes that a column of the Web table often matches with a large number of predicates in the KBs, although most of them are not good mapping results.

Some recent works of schema mapping[1,2,19] use data examples to filter and refine the detected schema mappings. However, they are not designed for finding the schema mapping. As will be shown in our experiments, only a few data examples are often far from enough for effectively finding the mapping between a Web table and a KB, considering that the KB contains a huge number of predicates (columns) and the examples may not have good matches in the KB.

Web Data Integration. A vast amount of structured information is contained in the Web tables [8]. Cafarella et al. proposed the OCTOPUS system[7], which enables users to create new datasets from those tables extracted from the Web. In recent years, Linked Data[12] has been widely accepted as an important way of integrating the massive Web datasets. It allows the interlinkage of datasets through the RDF links created between data items from different data sources. As of Sep. 2011, there have been 31 billion RDF triples in the Linked Data cloud. However, such a way of integrating Web datasets has some problems: (1) Different RDF datasets may contain multiple copies of the same data (entity). Although entity resolution[5] can help to remove some duplicates, the problem will be still serious due to the huge diversity of the Linked Data. (2) Although interlinked, datasets cannot be integrated as one global schema, due to the diversity and flexibility of the huge amount of individual schemas in the Linked Data. (3) Although very large, the coverage of existing Linked Data is still very limited compared to the numerous entities in the real world. Preda et al.[18] proposed an ANGIE system to dynamically and virtually enrich RDF KB by Web services.

3 Table-to-KB Schema Mapping

An entity-attribute table contains information of entities. We define the column containing entity names as the key column of a table. It satisfies some constraints: 1) no duplicated names in the column, i.e., the cardinality of the key column should be equal to the number of rows of the table; 2) its values are not numerical or IDs (it can be easily detected when it contains sequential numbers or IDs of common prefixes). This is reasonable because one of our assumptions is that the applied entity-attribute Web table should contain information of entities distinguished by their names within a column. If a table has more than one candidate key columns (satisfying the above two constraints), only the first one will be picked as the key column. We ignore tables whose entity names are distributed over multiple columns. An example of a Web table and its key column is shown in Fig.1. Note that our solution fails to integrate a Web table with the KB when it does not contain a key column.

1	Grand Illusion	true	Jean Renoir	France	1937
2	Seven Samurai	true	Akira Kurosawa	Japan	1954
3	The Lady Vanishes	true	Alfred Hitchcock	United Kingdom	1938
4	Amarcord	true	Federico Fellini	Italy	1974
5	The 400 Blows	true	François Truffaut	France	1959
6	The Naked Kiss	false	Samuel Fuller	United States	1964
7	Shock Corridor	false	Samuel Fuller	United States	1963

Fig. 1. An example of a Web table and its key column (squared)

Let $c_1, \ldots, c_d$ be the d columns contained by a table T. A tuple t in the table T can then be represented as $t = [t(1), \ldots, t(d)]$. For a KB E, an entity e is denoted as the set of all triples having the same subject. An entity e may have multiple triples for the same predicate. We denote $e(j)$ as the set of objects that the entity e has on the predicate p_j.

Definition 1 (Mapping Unit). *Given a value/cell $t(i)$ of a tuple t, an object set $e(j)$ of an entity e, and a matching threshold $0 < \theta < 1.0$, there is a mapping unit $m(t(i), e(j)) = j$ between $t(i)$ and $e(j)$, if $\exists o \in e(j)$ such that $t(i) = o$ (when c_i is the key column), or $\exists o \in e(j)$ such that $sim(t(i), o) \geq \theta$ (when c_i is not the key column).*

Similarity measures such as Jaccard similarity can be applied to evaluate the similarity of two sets of words. Given a tuple t and an entity e, when there is a mapping unit $m(t(i), e(j))$, we use $m(i, j) = j$ to represent it for simplicity. For each attribute $t(i)$, we use $M(t(i), e)$ to denote the set of all possible mapping units between $t(i)$ and the object sets of e. Note that $M(t(i), e)$ can be an empty set if $t(i)$ is not similar enough to any object of e. For example, in Figure 2, a mapping unit between the cell *1964*, and the predicate *Release date(4)* is 9. The set of mapping units for the cell *Samuel Fuller* and the given entity is $\{3, 5, 6\}$.

Definition 2 (Mapping Vector). *Given a table T of d columns with $c_{i'}$ as its key column, a mapping vector $M = [m_1, \ldots, m_d]$ defines, for each $m_i \neq 0$, a mapping of a column c_i to a predicate p_{m_i}. It must satisfy that $m_{i'} \neq 0$.*

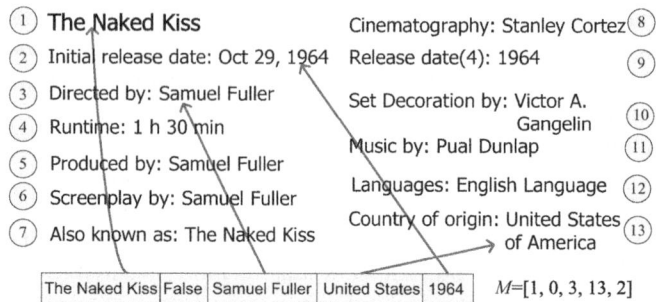

Fig. 2. An example of a mapping vector, with the entity information from Freebase

In the above definition, $m_i = 0$ means that column c_i does not match with any predicate. Given a tuple t and an entity e, for each attribute $t(i)$, we have a mapping unit set $M(t(i), e)$. Then, we are able to generate a set $\mathcal{M}(t, e)$ of mapping vectors by a Cartesian product of the non-empty mapping unit sets of all attributes of t. An example of a mapping vector is shown in Fig.2. A derived mapping vector $M = [m_1, \ldots, m_d]$ satisfies: 1) $m_i \in M(t(i), e)$ if $M(t(i), e) \neq \emptyset$; 2) $m_i = 0$ if $M(t(i), e) = \emptyset$; 3) $m_{i'} \neq 0$. Therefore, $|\mathcal{M}(t, e)| = \prod_{M(t(i),e) \neq \emptyset} |M(t(i), e)|$ if $M(t(i'), e) \neq \emptyset$. $\mathcal{M}(t, e) = \emptyset$ if $M(t(i'), e) = \emptyset$.

Next, we define an important measure of a mapping vector. The confidence of a mapping vector $M = [m_1, \ldots, m_d]$, denoted as $c(M)$, is defined as the number of non-zero mapping units (i.e., $m_i \neq 0$) in M. It describes strength of a mapping vector. For the running example of Fig.2, $c(M) = 4$. The larger the $c(M)$, the better the match between a tuple and an entity (or between a table and a KB). According to this measure, we are able to define significant mapping vectors that we are interested in among all the mapping vectors:

Definition 3 (Significant Mapping Vector). *Given a threshold δ of matched columns, a mapping vector M is a significant mapping vector if $c(M) \geq \delta$.*

A significant mapping vector shows that there are enough (δ) attributes of t that can find their matches in the predicates of e. Given a tuple t, we call those entities in KB that have significant mapping vectors with t (i.e., $\mathcal{M}(t, e) \neq \emptyset$) as the relevant entities of t. Given a table $T' \subseteq T$, we use $\mathcal{M}(T')$ to denote the multiset of all significant mapping vectors of $\mathcal{M}(t, e)$ generated from all combinations of $t \in T'$ and $e \in E$, i.e., $\mathcal{M} = \{M | c(M) \geq \delta, M \in \mathcal{M}(t, e), t \in T', e \in E\}$. As long as $|\mathcal{M}(T')|$ is no less than a user specified size τ, we say that $\mathcal{M}(T')$ is an evidencing multiset of mapping vectors, that can be used for discovering mapping vectors between a table T and the KB E.

Given two mapping vectors $M = [m_1, \ldots, m_d]$ and $M' = [m'_1, \ldots, m'_d]$, we say M' dominates M, denoted as $M' \preceq M$, if $\forall m_i \neq 0$, $m'_i = m_i$. For example, $[1, 0, 3, 13, 2] \preceq [1, 0, 3, 0, 2]$, while $[1, 0, 3, 13, 2] \npreceq [1, 0, 3, 0, 9]$. It is obvious that if $M' \preceq M$ and $M \preceq M'$, we have $M = M'$. With the definition of dominating relationship, we are able to define another interesting measure of a mapping vector, the support of a mapping vector, denoted as $s(M)$. Given a mapping vector M and a multiset $\mathcal{M}(T')$ of mapping vectors detected from a table $T' \subseteq T$ and a KB E, the support of M to the set $\mathcal{M}(T')$, is the number of mapping vectors in $\mathcal{M}(T')$ dominating M. It basically defines the number of evidences (mapping vectors) supporting the mapping vector M.

With the above two measures (the confidence and the support), we are able to define the utility of a mapping vector as:

Definition 4 (Utility Function). *Given an evidencing mapping vector set $\mathcal{M}(T')$, the utility of a mapping vector M is a function of two measures of mapping vectors, $u(M) = c(M)log(s(M))$.*

Note that the utility function may have some other alternatives. We define it as $u(M) = c(M)log(s(M))$ simply because when $c(M)$ is enlarged, the number of mapping vectors it dominates is exponentially enlarged.

With the above definitions, our major problem is that:

Definition 5 (Table-to-KB Schema Mapping). *Given a table T and a KB E, three parameters θ, δ, and τ, find the significant mapping vector (if exists) that having the maximum utility to a multiset $\mathcal{M}(T')$ satisfying either $|\mathcal{M}(T')| \geq \tau$ or $T' = T$.*

Our solution to this problem has two main steps:

1. To find a sub table $T' \subseteq T$ that is able to generate an evidencing multiset of mapping vectors $\mathcal{M}(T')$ such that either $|\mathcal{M}(T')| \geq \tau$ or $T' = T$;
2. To find the significant mapping vector having the maximal utility to $\mathcal{M}(T')$, called as the table-to-KB mapping vector.

As an instance-based schema mapping solution for huge KBs, the efficiency issue is a very important challenge. We will discuss how to address it in details in the following section.

4 An Efficient Solution for Table-to-KB Schema Mapping

4.1 Generating an Evidencing Multiset of Mapping Vectors

The evidencing multiset of mapping vectors are generated from tuple-entity pairs. Obviously, it is not necessary to generate mapping vectors from all possible tuple-entity pairs because most of them do not form any significant mapping vector or even any mapping unit. In our solution, the evidencing multiset of mapping vectors $\mathcal{M}(T')$ is obtained by merging the significant mapping vectors generated from some selected tuples of the table T.

A Baseline Solution. For each tuple t, we want to efficiently get the relevant entities of t. This is achieved by first finding mapping units for the attributes of t. To obtain the mapping units of an attribute $t(i)$, a baseline solution is to utilize the inverted indexes for the objects (attributes) of entities in the KB. Given an attribute $t(i)$ of n distinct words and a matching threshold θ, according to the

Jaccard similarity, the objects that can form mapping units with $t(i)$ must share at least θn common words of $t(i)$. Accordingly, those candidate objects are found by merging the inverted indexes of words in $t(i)$, filtering objects presenting less than θn inverted indexes. After that, a refining process is required by computing the Jaccard similarities between the candidate objects and $t(i)$. Those whose similarities are no less than θ form mapping units of $t(i)$. With the lists of mapping units for different attributes $t(i), i = 1, \ldots, d$, we are able to generate the significant mapping vectors of $\mathcal{M}(t, e)$ by the combination (Cartesian product) of mapping units of the same entity e, in different lists.

The major issue of the baseline solution is the low efficiency. Considering that a table of hundreds of cells, if each cell invokes an above keyword search process, the total time cost for processing a table can be as large as tens to hundreds of seconds, not feasible for practical solutions. As such, we propose a very efficient solution for generating the evidencing mapping vector set $\mathcal{M}(T')$.

Fast Evidencing Set Generation. In entity-attribute Web tables, the number of rows may vary significantly, from tens to thousands. When a table contains a large number of entities, on one hand, many of them may not have relevant entities in KB. It will be better if we can quickly judge whether there are relevant entities of a given tuple, so that we can efficiently prune the tuples without relevant entities by avoiding the expensive search process over them. On the other hand, a large table may contain many tuples having relevant entities in KB. We may not use all of these tuples with relevant entities to find mapping vectors, because the table-to-KB mapping vector is likely to converge when the set $\mathcal{M}(T')$ grows up to certain size. As such, it is beneficial if we find mapping vectors from some selected tuples of the table T.

To efficiently judge whether a tuple contains relevant entities or not, we propose to apply memory-based indexes. To be a relevant entity e of a tuple t, according to the Definition 1 and Definition 2, the entity name of t in the key column must exactly match with an object of e. In other words, we expect that the name of a relevant entity matches with the name of the entity described by t. For this purpose, we are able to create inverted indexes for only the entity names in the KB. Because entity names are often short texts, the inverted indexes of all entity names do not cost too much space. They can be held in the main memory, which leads to an efficient solution to filter irrelevant entities by only their names. Given an entity name of a tuple, we are able to find candidates of its relevant entities based on the indexes. Note that information (predicates and objects) of entities are stored externally, and indexed by their entity IDs. To check whether a candidate relevant entity is a real relevant entity or not, we need load information of the candidate entity from external devices, and compare them with the tuple t based on the Definition 3. Consequently, the major cost of generating the evidencing multiset of mapping vectors $\mathcal{M}(T')$ comes from loading entities from external devices by paying expensive I/Os.

To save the cost of generating $\mathcal{M}(T')$, we need control the number of entities loaded from external devices. On the other hand, to guarantee the accuracy of table-to-KB schema mapping, the size of $\mathcal{M}(T')$ should be significant enough.

This somehow contradicts with the reduction of the number of entities to be loaded. To address this conflict, we propose to load candidate relevant entities of tuples that potentially have less ambiguity, i.e., those tuples having less candidate relevant entities (found by the in-memory indexes) are firstly used for loading candidate relevant entities and generating the significant mapping vectors. Algorithm 1 shows the detailed process of our solution to efficiently generate significant mapping vector set. Note that, the generation process terminates when τ significant mapping vectors have been found.

Algorithm 1. Generating significant mapping vectors

Input: T, a table with c_i as the key column
Input: τ, the maximal number of mapping vectors in $\mathcal{M}(T')$
Output: $\mathcal{M}(T')$

1 set $\mathcal{M}(T')$ as $\emptyset$
2 find candidate relevant entities for all the cells in the key column c_i using in-memory indexes
3 rank all the tuples as a list T', based on the number of the candidate relevant entities they have, in an ascending order
4 **while** $|\mathcal{M}(T')| < \tau$ **do**
5   pick a tuple t from the top of the list T'
6   retrieve all candidate relevant entities of t
7   **foreach** *retrieved candidate* e **do**
8   compute the significant mapping vectors between t and e
9   insert them into $\mathcal{M}(T')$
10   **if** $|\mathcal{M}(T')| \geq \tau$ **then**
11   break
12 return $\mathcal{M}(T')$

4.2 Finding the Table-to-KB Mapping Vector

Given a set of significant mapping vectors $\mathcal{M}(T')$, to find the Table-to-KB mapping vector, a straightforward solution is to enumerate all the significant mapping vectors dominated by each detected significant mapping vector in $\mathcal{M}(T')$. Then, we can compute the utility of each significant mapping vector to the set $\mathcal{M}(T')$, and find the significant mapping vector (if exists) with maximal utility. For a significant mapping vector $M \in \mathcal{M}(T')$, it totally dominates $2^{c(M)} - 2^{\delta} + 1$ significant mapping vectors. As a result, when $c(M)$ is relatively large, there will be a large number of significant mapping vectors to be enumerated, simply for one mapping vector M. The enumeration will be very time-consuming. To solve this problem, we propose a best first search algorithm that is able to incrementally find significant mapping vectors in a bottom-up manner. It avoids the enumeration process by heuristically expanding the mapping vector of maximal utility having been found. To do this, we need record the significant mapping vectors in $\mathcal{M}(T')$ in a way of using binary words. See details in Fig.3.

For each column c_i, one binary word is used for each predicate that the column c_i maps to, according to the significant mapping vectors of $\mathcal{M}(T')$. A binary

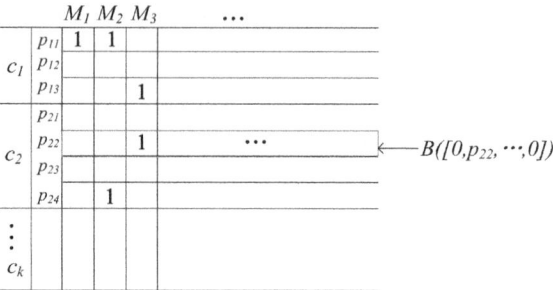

Fig. 3. Binary words for recording significant mapping vectors

word contains $|\mathcal{M}(T')|$ bits, with each bit records whether a significant mapping vector maps to the predicate in the column c_i. In this way, a significant mapping vector can be represented by a column in the bitmap. For an example of Fig.3, $M_1 = [p_{11}, 0, \ldots]$, $M_2 = [p_{11}, p_{24}, \ldots]$ and $M_3 = [p_{13}, p_{22}, \ldots]$. By using the binary words, given a mapping vector M, we are able to efficiently find the significant mapping vectors dominating it in the set $\mathcal{M}(T')$. This can be achieved by the conjunction of all binary words of mapping units of M. For example, to find the significant mapping vectors dominating $M = [p_{11}, p_{24}, 0, \ldots, 0]$, we need conduct a bitwise AND operation over the binary words of c_1-p_{11} (the first row) and c_2-p_{24} (the seventh row). According to the derived conjunctive binary word, we can make sure that M_2 is a significant mapping vector dominating M. With the conjunctive binary word of a mapping vector, the support of a mapping vector $s(M)$ can be efficiently computed by counting the number of ones in the conjunctive binary word. With binary words, we are able to design the Algorithm 2 to efficiently find table-to-KB mapping vector.

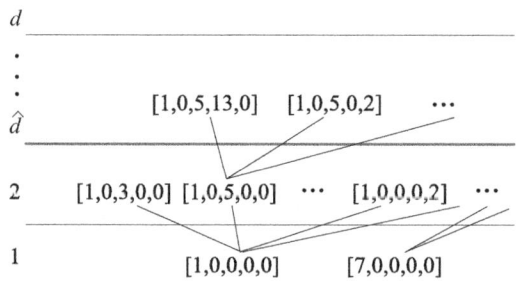

Fig. 4. An example of expanding mapping vectors in Algorithm 2

As shown in Algorithm 2, it maintains a heap H for mapping vectors to be expanded. The algorithm initiates with mapping vectors determined simply by the key column. Each time it picks a mapping vector M of maximal utility from H for expanding. The expansion is conducted only it may potentially contain the mapping vector of maximal utility, determined by the upper bound of the

Algorithm 2. Search table-to-KB mapping vector

Input: $\mathcal{M}(T')$, in the format of binary words
Output: M_{max}, the table-to-KB mapping vector

1 set H as $\emptyset$, be a heap recording all mapping vectors to be expanded
2 generate mapping vectors (with confidence 1) only for the key column, and insert them into H as seeds to be expanded
3 set $u_{max} = 0$, $M_{max} = 0$
4 **while** $H \neq \emptyset$ **do**
5   pop the mapping vector M of maximal utility from H
6   **if** $d \cdot s(M) > u_{max}$ **then**
7   expand M by introducing a new column has not been expanded by M
8   **foreach** *expanded mapping vector M'* **do**
9   **if** $d \cdot s(M') > u_{max}$ **then**
10   push M' into H
11   **if** $u(M') > u_{max}$ and $c(M') \geq \delta$ **then**
12   set u_{max} as $u(M')$
13   set M_{max} as M'
14 return M_{max}

utilities of all mapping vectors expanded from M (i.e., $d \cdot s(M)$). The expansion terminates when no mapping vector exists in H. Finally, M_{max} is returned as the table-to-KB mapping vector if it is a significant mapping vector. Fig.4 illustrates the best first expansion process. Note that, to avoid the repetition of expansion, each mapping vector only expands in columns to the right of the most right column that it has expanded. For example, in Fig.4, the mapping vector $M = [1, 0, 5, 0, 0]$ can only expand its last two columns, instead of the second one.

5 Evaluation

5.1 Experimental Settings

We use the DBPedia 3.7 dataset[1] as the KB in our experiments. The version we used contains 3.64 million entities. The most popular types of entities in DBPedia include persons, places, music albums, films, etc. We extracted literal information from the subjects and objects (by removing the URI prefixes for semantic matching) of the entities in the KB. There are no standard test sets. In our study, the tests are based on twelve Web tables (listed in Table 1) randomly searched from the Google Fusion Table search interface[2]. They contain entities such as persons, movies and songs. Not all the columns are used for schema mapping. We filter those columns that satisfy one of the conditions: 1) IDs; 2) URLs; 3) more than one third of cells are empty cells. We also remove rows which have more than one half of cells are empty.

---

[1] http://dbpedia.org/
[2] http://www.google.com/fusiontables/

Table 1. Web tables used in experiments

Table	Table Names	No. of columns	No. of rows
1	Emily's list of person	4	206
2	Academy award-winning films	4	1152
3	American films of 1993	4	180
4	American superhero animated	6	40
5	Christmas hit singles in the US	4	443
6	museums	4	78
7	new leadership of American PAC	4	232
8	Miss America	5	80
9	Obama for America	4	1456
10	criterions on Netflix	5	520
11	songs	5	757
12	football players	8	54

Two instance-based approaches[20] are applied as baselines. The first one is based on the word frequency (mentioned in [20]) of columns in tables and predicates in the KB. The predicate with the maximal cosine similarity is selected as a mapping predicate of a column. The second one (COMA++[10]) extends the COMA[3] with two instance-based matchers that utilize certain constraints and linguistic approaches. It applies a propagation algorithm to propagate the similarity values of elements to their parents. Our solution is labeled as T2KB. Note that we also implement an entropy based approach [13] that works by computing the "mutual information" between pairs of columns within each schema. However, its performance is so poor that most columns are not correctly mapped. So we do not show its results for comparison. We use Java to implement our algorithms. The experiments are conducted on a server with a 1.8G 24 Core processor, 128GB memory, running 64-bit Linux Redhat kernel.

5.2 Accuracy Comparison

Table 2 shows the precision and recall of the compared methods over 12 tested tables. The precision of an approach is defined as the number of corrected matched columns (manually evaluated) over the number of detected matched columns. The recall is defined as the number of corrected matched columns over the number of columns in the table. From the results, we can see that, the precision and the recall of T2KB are much better than those of WF and COMA in almost all the cases. The accuracy of WF is the worst. This is reasonable because it is hard to find an ideal predicate simply based on the word frequency, due to the large number of candidate predicates in the KB. The accuracy of COMA is much better than that of WF because it is also based on instance-based schema matching. However, compared to the proposed T2KB, COMA is still not accurate enough. Moreover, it is very inefficient. The mapping of a table requires several minutes in average. For the T2KB approach, we can see that its precision is very high, which means that the discovered matched columns have very high accuracy.

Table 2. Comparison of accuracy

Table	WF(Word Frequency) Prec./Recall	COMA Prec./Recall	T2KB Prec./Recall
1	0.25/0.25	0.67/0.50	1.00/0.75
2	0.25/0.25	0.33/0.25	0.33/0.25
3	0.25/0.25	0.75/0.75	1.00/0.75
4	0.17/0.17	0.67/0.67	0.50/0.33
5	0.50/0.50	1.00/1.00	1.00/0.75
6	0.25/0.25	0.75/0.75	1.00/1.00
7	0.50/0.50	0.33/0.25	0.67/0.50
8	0.00/0.00	0.20/0.20	1.00/0.60
9	0.50/0.50	0.33/0.25	1.00/0.75
10	0.20/0.20	0.60/0.60	1.00/0.80
11	0.40/0.40	0.80/0.80	1.00/0.80
12	0.40/0.25	0.60/0.38	0.80/0.50

To further compare the accuracy of the applied schema mapping solutions, we show the mapping results of two examples. The mapping predicate (if exists) of each column is shown. It is bolded if labeled as a corrected match. Table 3 shows the details of the "criterions on Netflix" table, where both the precision and the recall of T2KB outperform those of the other two approaches. For the T2KB approach, the column *Year* is mapped to the predicate *released*, which is actually more concrete than the simple mapping to the predicate *Year*.

Table 3. Accuracy for the table "criterions on Netflix" table (No. 10)

Method	Title	Streaming	Director	Country	Year	Precison	Recall
WF	nextissue	international	mayor	**nationalorigin**	lastrace	0.20	0.20
COMA	name	data	director	name	year	0.60	0.60
T2KB	**name**	/	**director**	**country**	released	1.00	0.80

Table 4 shows an example where all the solutions do not work well. One reason for the bad mapping results is that columns *awards* and *nominations* contain small numbers, which causes them to be easily mapped to a wrong predicate.

5.3 Impacts of Parameters

We test the impacts of three parameters in the T2KB approach, τ, θ and δ. The results are shown in Fig.5. The performance when tuning the parameter τ (by fixing $\theta = 0.2$, $\delta = 3$) is shown in Fig.5(a)-5(b). As can be seen from Fig.5(a), when enlarging τ, both the precision and the recall climb in general. They converge when τ is enlarged up to around 300. This is reasonable because the more mapping vectors used in $\mathcal{M}(T')$, the higher reliability of the derived table-to-KB mapping vector. In the test cases, the average total numbers of

Table 4. Accuracy for the table "Academy award-winning films" (No. 2)

Method	film	year	awards	nominations	Precision	Recall
WF	englishtitle	yearestimate	yushos	count	0.25	0.25
COMA	name	/	title	m	0.33	0.25
T2KB	name	/	gross	gross	0.33	0.25

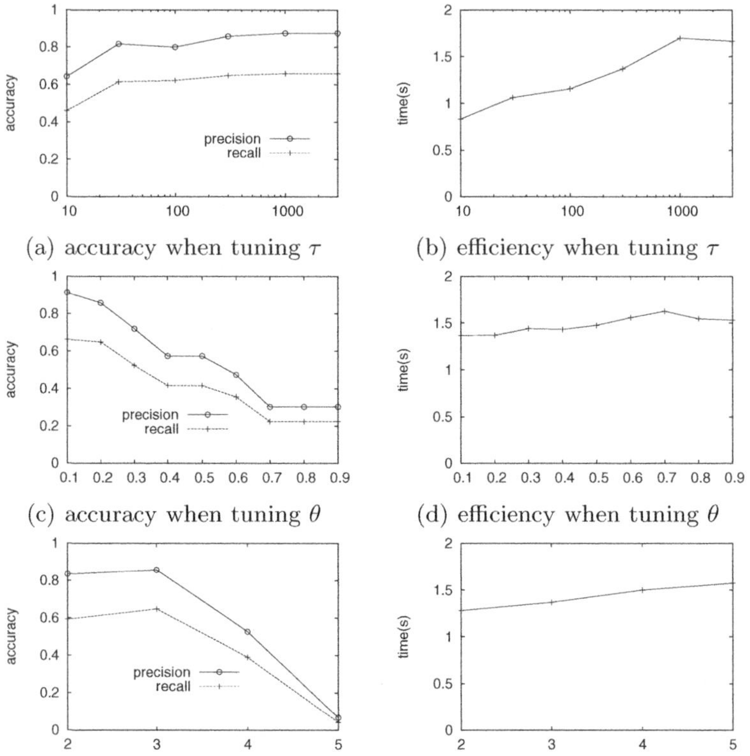

(a) accuracy when tuning τ

(b) efficiency when tuning τ

(c) accuracy when tuning θ

(d) efficiency when tuning θ

(e) accuracy when tuning δ

(f) efficiency when tuning δ

Fig. 5. Impacts of parameters on the performance of the T2KB approach

mapping vectors is around 1000. This explains why the precision and recall converge when τ is larger than around 300.

For the efficiency shown in Fig.5(b), the time cost increases when enlarging τ, simply because more entities need to be loaded and more vectors need to be processed. In general, the T2KB approach can be conducted within a few seconds, which is much more efficient than the COMA approach that typically requires hundreds of seconds for a mapping task. We also test the performance of T2KB when the mentioned baseline solution (in Sec.4.1) is applied for generating significant mapping vectors. It takes 127 seconds in average, far more slower than the proposed solution.

Fig.5(c)-5(d) show the performance of T2KB when θ is tuned from 0.1 to 0.9 (by fixing $\tau = 300$, $\delta = 3$). The enlargement of θ causes the reduction of both precision and recall in general. This is because higher θ causes less mapping vectors in $\mathcal{M}(T')$, leading to a lower accuracy. In other experiments, we set $\theta = 0.2$ by defaults. For the efficiency in Fig.5(d), the time cost slightly increases when enlarging θ, because more entities are need to be loaded for generating enough mapping vectors. There is a drop when θ is enlarged from 0.7 to 0.9, this is because all relevant entities have been loaded for $\theta = 0.7$. When θ is further enlarged, the reduction of the number of mapping vectors in $\mathcal{M}(T')$ saves the cost of generating the table-to-KB mapping vector using Algorithm 2.

For the parameter δ in Fig.5(e)-5(f), we can see that the best accuracy is achieved when $\delta = 3$. This is reasonable because small δ generates too many false positive mapping vectors which may generate some false mapping units in the table-to-KB mapping vector. On the other hand, large δ will cause the number of significant mapping vectors to be very small, which causes the derived table-to-KB mapping vector not reliable. In other experiments, we set $\delta = 3$ by defaults. For efficiency in Fig.5(f), the time cost increases when δ is enlarged in general. The reason is the same as that for the parameter θ in Fig.5(d).

6 Conclusions

We propose a pure instance-based schema mapping solution that finds the table-to-KB schema mapping from the mapping vectors generated from tuples and entities. We show that the proposed solution is more accurate than the two baselines of instance-based approaches for the given table-to-KB schema mapping problem. We also demonstrate that by using the proposed techniques for efficient computing the mapping vector, the proposed instance-based schema mapping solution can be efficiently processed. It is feasible for the schema mapping of large Web tables and huge knowledge bases.

Acknowledgements. This work is supported by the National Science Foundation of China under grant NO.61170010, National Basic Research Program of China (973 Program) No. 2012CB316205 ,and HGJ PROJECT 2010ZX01042-002-002-03.

References

1. Alexe, B., Chiticariu, L., Miller, R.J., Tan, W.C.: Muse: Mapping understanding and design by example. In: ICDE, pp. 10–19 (2008)
2. Alexe, B., ten Cate, B., Kolaitis, P.G., Tan, W.C.: Characterizing schema mappings via data examples. ACM Trans. Database Syst. 36(4), 23 (2011)
3. Aumueller, D., Do, H.H., Massmann, S., Rahm, E.: Schema and ontology matching with coma++. In: SIGMOD Conference, pp. 906–908 (2005)
4. Bellahsene, Z., Bonifati, A., Rahm, E. (eds.): Schema Matching and Mapping. Springer (2011)

5. Benjelloun, O., Garcia-Molina, H., Menestrina, D., Su, Q., Whang, S.E., Widom, J.: Swoosh: a generic approach to entity resolution. VLDB J. 18(1), 255–276 (2009)
6. Bollacker, K.D., Evans, C., Paritosh, P., Sturge, T., Taylor, J.: Freebase: a collaboratively created graph database for structuring human knowledge. In: SIGMOD Conference, pp. 1247–1250 (2008)
7. Cafarella, M.J., Halevy, A.Y., Khoussainova, N.: Data integration for the relational web. PVLDB 2(1), 1090–1101 (2009)
8. Cafarella, M.J., Halevy, A.Y., Wang, D.Z., Wu, E., Zhang, Y.: Webtables: exploring the power of tables on the web. PVLDB 1(1), 538–549 (2008)
9. Do, H.H., Rahm, E.: Coma - a system for flexible combination of schema matching approaches. In: VLDB, pp. 610–621 (2002)
10. Engmann, D., Maßmann, S.: Instance matching with coma++. In: BTW Workshops, pp. 28–37 (2007)
11. Gonzalez, H., Halevy, A.Y., Jensen, C.S., Langen, A., Madhavan, J., Shapley, R., Shen, W.: Google fusion tables: data management, integration and collaboration in the cloud. In: SoCC, pp. 175–180 (2010)
12. Heath, T., Bizer, C.: Linked Data: Evolving the Web into a Global Data Space. Synthesis Lectures on the Semantic Web. Morgan & Claypool Publishers (2011)
13. Kang, J., Naughton, J.F.: On schema matching with opaque column names and data values. In: SIGMOD Conference, pp. 205–216 (2003)
14. Madhavan, J., Bernstein, P.A., Doan, A., Halevy, A.Y.: Corpus-based schema matching. In: ICDE, pp. 57–68 (2005)
15. Madhavan, J., Bernstein, P.A., Rahm, E.: Generic schema matching with cupid. In: VLDB, pp. 49–58 (2001)
16. Melnik, S., Garcia-Molina, H., Rahm, E.: Similarity flooding: A versatile graph matching algorithm and its application to schema matching. In: ICDE, pp. 117–128 (2002)
17. Popa, L., Velegrakis, Y., Miller, R.J., Hernández, M.A., Fagin, R.: Translating web data. In: VLDB, pp. 598–609 (2002)
18. Preda, N., Kasneci, G., Suchanek, F.M., Neumann, T., Yuan, W., Weikum, G.: Active knowledge: dynamically enriching rdf knowledge bases by web services. In: SIGMOD Conference, pp. 399–410 (2010)
19. Qian, L., Cafarella, M.J., Jagadish, H.V.: Sample-driven schema mapping. In: SIGMOD Conference, pp. 73–84 (2012)
20. Rahm, E., Bernstein, P.A.: A survey of approaches to automatic schema matching. VLDB J. 10(4), 334–350 (2001)
21. Suchanek, F.M., Kasneci, G., Weikum, G.: Yago: a core of semantic knowledge. In: WWW, pp. 697–706 (2007)
22. Wu, W., Li, H., Wang, H., Zhu, K.Q.: Probase: a probabilistic taxonomy for text understanding. In: SIGMOD Conference, pp. 481–492 (2012)
23. Yakout, M., Ganjam, K., Chakrabarti, K., Chaudhuri, S.: Infogather: entity augmentation and attribute discovery by holistic matching with web tables. In: SIGMOD Conference, pp. 97–108 (2012)

ServiceBase: A Programming Knowledge-Base for Service Oriented Development

Moshe Chai Barukh and Boualem Benatallah

School of Computer Science & Engineering,
University of New South Wales, Sydney – Australia
{mosheb,boualem}@cse.unsw.edu.au

Abstract. In recent times we have witnessed several advances in modern web-technology that has transformed the Internet into a global deployment and development platform. Such advances include Web 2.0 for large-scale collaboration; Social-computing for increased awareness; as well as Cloud-computing, which have helped virtualized resources over the Internet. As a result, this new computing environment has thus presented developers with ubiquitous access to countless web-services, along with computing resources, data-resources and tools. However, while these web-services enable tremendous automation and re-use opportunities, new productivity challenges have also emerged: The same repetitive, error-prone and time consuming integration work needs to get done each time a developer integrates a new API. To address these challenges we have developed *ServiceBase*, a "programming" knowledge-base, where common service-related low-level logic can be abstracted, organized, incrementally curated and thereby re-used by other application-developers. A framework is also proposed for decomposing and mapping raw service-messages into more common data-constructs, thus making interpreting, manipulating and chaining services further simplified despite their underlying heterogeneity. More so, empowered by this knowledge, we expose a set of APIs to simplify the way web-services can be used in application-development.

Keywords: Service Oriented Architecture, Web-Services, Web 2.0.

1 Introduction

In parallel with cloud computing, we have witnessed several other advances that are transforming the Internet into a global development and deployment platform. These include Web 2.0, Service Oriented Architectures (SOA) and social computing. SOA enables modular and uniform access to heterogeneous and distributed services; Web 2.0 technologies provide a Web-scale sharing infrastructure and platform, while advances in Social-computing are increasing transparency, awareness, and participants' collaboration and productivity. Developers are thus offered with ubiquitous access to a network of logical services along with computing resources, data sources, and tools. In a nutshell, the new computing environment enabled by advances in the above areas, consists of data, computational resources, both virtualized and physical services, and networked devices distributed over the Internet.

This new computing paradigm provides a holistic environment in which users, workers, services, and resources establish on-demand interactions to meet multiple simultaneous goals. More specifically, such Web-services and APIs are widely adopted by programmers to build new applications in various programming languages on top enterprise as well as social media, Internet-of-Things, and crowd and cloud services (from resources to platforms). For instance, a number of added value applications such as Tweetdeck[1] have been built on top the Twitter API. Organizations like Mashery[2] and Apigee[3] are building on these trends to provide platforms for the management of APIs. For instance, ProgrammableWeb now has more than 6,700 APIs in its directory. These services can be combined to build composite applications and higher-level services using service composition techniques [2].

However, while advances in Web service and services composition have enabled tremendous automation and reuse opportunities, new productivity challenges have also emerged. The same repetitive, error-prone and time consuming integration work needs to get done each time a developer integrates a new API, [1,2]. Furthermore, the heterogeneity associated with services also means service-programming has remained a technical and complex task, [3]. For example, the developer would need sound understanding of the different service types and their various access-methods, as well as being able to format input data, or parse and interpret output data in the various different formats that they may be available in, (e.g. XML, JSON, SOAP, Multimedia, HTTP, etc.). In addition to API integration work, programmers may also need to develop additional functionality such as: user management, authentication-signing and access control, tracing, and version management.

In order to address these challenges, we have designed and developed *ServiceBase*, a "programming" knowledge-base, where common service-related low-level logic can be abstracted, organized, incrementally curated and thereby re-used by other application-developers. Architecturally, we have drawn inspiration from *Freebase* and *Wikipedia*, where just as encyclopediatic information is distributed in the form of user-contributed content, similarly, technical knowledge about services could be both populated and shared amongst other developers for the purpose of simplified reuse. More specifically, we offer the following main contributions: (i) We define a unified services representation model to appropriately capture service-knowledge that is organized in our programming-base; (ii) To augment a further level of simplicity, we provide a framework for mapping between native service message formats and more common data-structures. This means that input and/or output messages of services can be decomposed and represented as various types of atomic (string, numeric, binary), or complex (list, tuple) fields, thus making service messages easier to interpret and manipulate; (iii) Empowered by this knowledge stored in the base, we then provide a set of APIs that expose a common-programming interface to developers, thereby simplifying service integration in application development. For example, invoking a service-method could be done in a few simple lines of code, and does not entail the programmer to be aware of and program the low-level details, such

---

[1] http://www.tweetdeck.com
[2] http://www.mashery.com
[3] http://apigee.com

as protocol for access (i.e. REST vs. SOAP), message-formats (i.e. XML vs. JSON), authorization-control (i.e. formulating and signing OAuth request). This is implemented by a service-bus middleware that translates high-level methods to more concrete service calls, by looking-up and then building the necessary information from the knowledge-base. Similarly, we have provided other simple methods for: subscribing to feeds; listening to events; and setting-up callbacks, etc.

2 Unified Web-Services Representation Model

At the most basic level, standardizations such as WSDL have provided an agreed means for describing low-level APIs, and much less formally, the same might be true for the documentation of RESTful Service APIs. However, it is clear most of these have been focused primarily for the purpose of service-description and discovery, where the emphasis on simplifying service-execution is almost neglected, [2,5,7].

In this section, rather than treating services as isolated resources, we describe a unified and hierarchical representation model for the logical organization of services in the knowledge-base.

The type of knowledge captured includes: fundamental service information such as access-protocol, the set of operations and message formats; authorization information in the case of secure-access services; an end-user base that allows users to pre-authorise secure-services, and therefore applications written using ServiceBase, only need to use a single access-token per user, rather than having to manage a set of different token for each secure service per user; message-transformation rules for mapping between raw message-types and relatively more simpler message-field object. A summarized version of the service model has been described in the UML-diagram as illustrated in Figure 1, and elucidated further below:

Services. The class *Service* is an abstract superclass for all specialized service types being implemented. Namely, we support three main types, REST, WSDL and Feed-based services. A *RESTService* extends Service and requires that a *base-endpoint* be specified that makes up the common part of the RESTful call. *WSDLServices* require a *wsdl* source file, whereby the system can then perform runtime parsing in order to simplify the formulation and validation of the service-model. *StreamServices* further extends *RESTServices*, and allows *Feed* sources to be specified. Furthermore, in order to deal with access to secure services using OAuth [8], *AuthorisationInfo* allows for this information to be defined and attached to service-objects. Furthermore, our unique service-model organizes the service into a logical hierarchy, thus enabling children services to inherit the knowledge already registered at the parent. The benefit of this means: not only is it easier to add new services, but it also creates the notion of a service-community (similar to our previous work [4]). In this manner, similar services can be defined with a common-set of operations and share a similar interface, thus simplifying interoperability between services (e.g. replacing one service with another) – we shall give an example of this later in this paper.

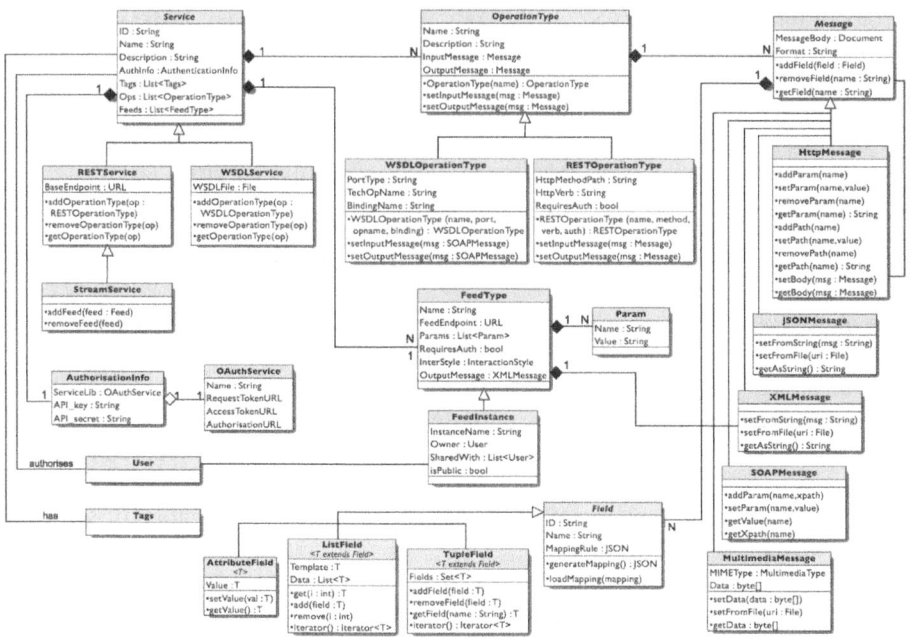

Fig. 1. Summarized view of the Unified Service-Representation Model

OperationType. From a technical standpoint, the notion of a service-operation might take upon different meanings for different service types. For example, in the case of REST, while there are generally four basic operation-types (get, put, post, delete), often what are more relevant to the end-user are the various methods that are available. For example, searching for photos on Flickr might be considered a user-operation, accessed via the endpoint /?method=flickr.photos.search. WSDL services are slightly different in that operations are generally expressed as an RPC call. However, we propose in both cases service-access can be simplified by abstracting the low-level details from the end-user. For example, it would be much more convenient to express a call to Flickr as "/Flickr/getPhotos", and similarly for an operation available in WSDL, such as "/Flights/getBooking", rather than having to specify the more technical low-level details to make the call. In our model, this is therefore supported by the abstract superclass *OperationType*, although since the low-level details of operations may differ between different service representations, we have therefore specialized into various sub-classes.

FeedType. The detection of RSS or ATOM feeds can also be supported and is defined using the *FeedType* class. This specifies the endpoint, the interaction-style used to read feeds, as well as any instance parameters. There are three main interaction-styles supported: *polling* (i.e. periodic pull at a predefined interval); *streaming* (i.e. an open call that allows data to be pushed to the caller); and *publish-subscribe* (i.e. this involves registering to a hub that actively sends data only when new content is available). However, the definition of *FeedType* still remains abstract, in the sense that they may not point to any specific feed-source. For example, a Flickr discussion feed can be defined by the endpoint: http://api.flickr.com/services/feeds/groups_discuss.gne, but also requires a parameter groupID be specified to identify the particular discussion group. Various instances of the feed can now be instantiated, using the *FeedInstance* object, to create several customized instances of this feed without having to re-define the common low-level details. Moreover, as certain feed-sources may require authorisation, the owner can appropriately restrict access to specific feed-instances and not others by sharing only with specific users. Alternatively, it could be defined as public-view.

Messages. The class *Messages* represent the various serialization-types for both incoming and outgoing data that is associated with services. All message-types that we support are specializations of the superclass Message. Messages can be used both at design-time by curators when registering services, as well as at execution-time by developers when interacting with services. An *HttpMessage* encapsulates standard information such as parameters, and payload/body, but also supports parameterized path values, such as "/questions/{q_id}/related". A *MultimediaMessage* allows representing any Internet Media File [9] not already directly supported. For example, Images, PDFs, WordDocuments, etc. While *SOAPMessages* are inherently XML, we provide added support, since we know the schema of the messages. A Parameter of a SOAP message is defined as the triple <name, xpath, value?>, where name is a user-defined name given to the parameter, while xpath is the query used to reach the respective data-field; optionally a pre-defined value can be assigned to this parameter.

Message-Fields. Although all services need to define their native input and output message types, we have also chosen to further decompose any *Message* into a set of user-defined *MessageFields*. There are several benefits for this: (i) Firstly, working with fields means the end-programmer do not need to worry about the low-level logic of working with raw messages, such as formatting and parsing messages. (ii) Secondly, fields provide a unique means for representing similar (yet heterogeneous) services in a common-interface. For example, in the example described later, we show how various database-services could be abstracted to a common set of operations: *put*, *get*, *read* and *delete*, etc. This being despite their underlying heterogeneity (e.g. a JSON versus an XML data-interchange model). (iii) Thirdly, applications written using fields to interact with services means changes in the underlying web-services (i.e. the service-provider modifies their API or message structure or format, etc.), would not require any modifications to the application-code. (Although of course, these changes would need to be made in the mapping-logic, but would only need to be done once, instead of for each and every application that is using the web-service).

We have found three field-types to be appropriate: *AttributeFields* are simple and define a name and value pair; the value-type includes any common primitive type, such as *string*, *integer* and *date*, as well as *binary-array* to handle media-files. We then support two complex types, which may itself contain other atomic or complex fields nested within them. The complex type *ListField* represents an indefinite, ordered list of field-elements of the same type (usually instantiated at 'run-time' to handle an unknown collection of items). While, *TupleField* represents a finite collection of fields, akin to a Struct or Class, which can therefore be of arbitrary type, (but instantiated at 'design-time').

Mapping Rules. However, in order to enable fields, mappings need to be defined, which specifies transformation-rules that map between the defined field and the corresponding raw-message. In some cases, the mapping-rules can be completely generated automatically, such as in the case of WSDL services. However, in the case of REST, which is far more informal and does not prescribe pre-defined schemas, the mappings would need to be specified with some human-assistance. However, the overall benefit is still clear: once a mapping has been registered in the base, others can then reuse it multiple times. Mappings are structured in our system in JSON-format, which to begin with can be generated as a template; it then allows the rules to be entered in order to map to the specific raw-message. The basic template structure for each field-type has been illustrated in Figure 2 below. Although for nested fields, the JSON-template generated would be a corresponding nested structure. Note, the concept of a *nodepath* expression shown for list, has been defined in the work presented at [10], and is only necessary when dealing with raw messages schema that are hierarchically organized XML/JSON data. The value defines the path-to-the-node (i.e. sub-tree) that are to be considered distinct elements of the list.

`"attribute" : {`
`    "value" : "formula"`
`}`
	`"list" : {`
	`    "nodepath" : "formula",`
	`    (nested field)`
	`}`
		`"tuple" : {`
		`    (nested field/s)`
		`}`

(A)	(B)	(C)

Fig. 2. Mapping-rule structures associated with message-field types, where: (A) Attribute-field; (B) List-field; and (C) Collection-field

To specify a rule means to enter specific type of formulae to tell the system how to perform the mapping. For example, for an attribute field, a formula might involve an xpath expression over the native XML message in order to get to the desired node. The formulae would often involve utilising functions; at present we have preloaded our system with the following set of 6 functions:

```
xpath(expr), jsonpath(expr), httpparam(expr),
httppath(expr), httpheader(expr), payload().
```

Although, depending upon which type of message the mapping is loaded upon, the functions defined may behave differently. For instance, if loaded to an *input* message, the function would act to *"write"* to the raw message from the field-values; while if loaded for an *output* message, it would act to *"read"* from the raw message and populate into field-values. Moreover, in a typical service-invocation, mappings would need to be performed both ways: from fields to raw-messages during the input; and vice-versa during the output. Given that we support 5 distinct raw-message types, we have therefore implemented 10 transformation algorithms.

3   ServiceBase System Architecture and Implementation

Figure 3 illustrates the system design and interaction of the main components of the *ServiceBase* system. As mentioned, drawing inspiration from a Web2.0-oriented ecosystem, the service-base acts as a community between *service-curators* (those that primarily add/maintain services in the base), *service-consumers* (mostly application developers integrating services in their implementation), and *end-users* (the final users who ultimately interact with the various service-oriented applications). We begin therefore in this section by describing the main APIs exposed by *ServiceBase*, and in the subsequent sections we introduce a running example, in order to convey the more technical details of the APIs, as it would be used in a real-life scenario.

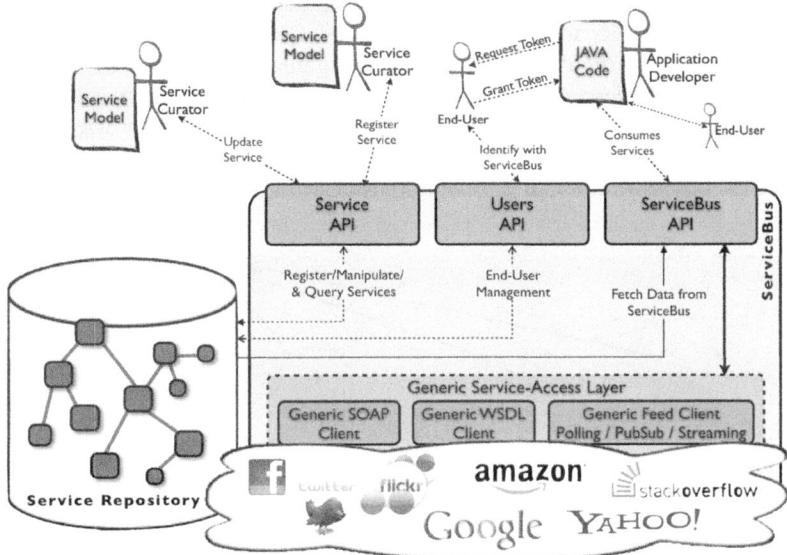

Fig. 3. *ServiceBase* System Architecture

3.1 ServiceBase APIs

The programmatic interface to *ServiceBase* offers the following APIs:

This *ServiceAPI* is primarily used by service-curators in order to register new services into the knowledge-base, but also to: search, explore, update and delete service-definitions that have already been registered.

The *ServiceBusAPI* would primarily be used by application-developers as the main gateway to interact with any of the registered services. In particular, the API provides methods for simplified invocation-calls, feed subscriptions, querying (pull) of feed-events, listening (asynchronous callback push) of events, authorising services, etc.

The *UsersAPI* provide a means for end-users to identify themselves with the service-base. A registered user in the service-base is then able to assign/revoke authorisation privileges to various services. In this manner, application that are written on top of services which require access to secure resources (for example, invoking an operation to get the specified user's collection of Google Docs), can then be further simplified, as the entire logic for handling the secure calls is managed by the service-bus, rather than the application developer. Secure calls can be processed on behalf of a specific user simply by requesting from the user, or having shared, an access-key, (done via OAuth, [8]) which is then passed into the invocation method.

3.2 Service-Modelling Example

As mentioned, services are modeled using the unified services representation model. In order to apply this to a real-world scenario, we consider what the organization of service-entities would look like to model a variety of database services (i.e. database-as-a-service). We also show the process (i.e. the work involved by a typical service-curator) in order to register new services into the base, and the extent of re-use that can be achieved in order to simplify this process. Consider the illustration shown in Figure 4 below – in particular, the hierarchical organization of services entities means adding services as a descendant of a parent service enables inheriting (i.e. re-using) the higher-level knowledge stored. Therefore, at each node: new knowledge could be added, or if inherited, variations or specializations could be made. This organization is the key to enabling incremental growth of the service knowledge base.

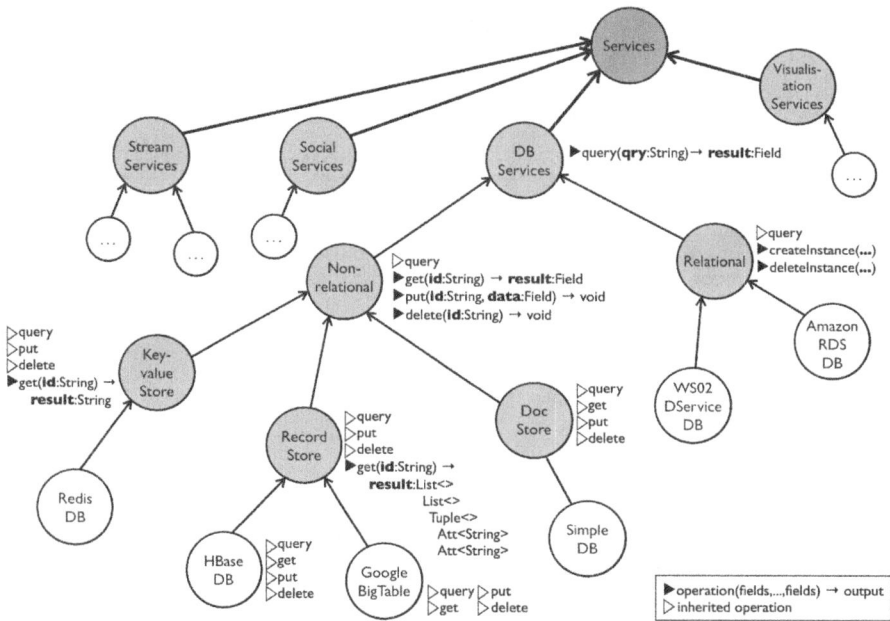

Fig. 4. Organization of DB-Services in *ServiceBase*

In this particular example, we consider DB-services to be split amongst two main types: *relational* and *non-relational* data-stores. It is clear however even at this abstract level, both service types support a *"query"* operation, and therefore this operation-type can be modeled at the uppermost parent level. *Non-relational* database inherits the *DB-service* entity, but defines 3 more operations that could be considered common amongst non-relational stores, similar to the work found at [12]. Namely these are: *"get"*, *"put"* and *"delete"*. As modeling the various operation-types require specifying the input and output message-fields, in some cases we may use the superclass *Field* to support an arbitrary field structure. For example, this is the case for the return type *"result"* of the *"get"* operation (when defined at the high-level). However, non-relational services can be further divided into three main sub-types, those being: *key-value stores*; *record-stores* and *document-stores*. Specializations can thus be applied, for example, the "get" operation of *record-stores*, could be modeled as we have illustrated in Figure 5 shown below.

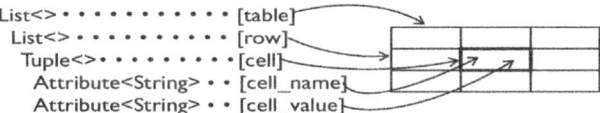

Fig. 5. Illustration of the "get"-operation fields representing a table-record

3.3 Incremental Enrichment of Services in the Knowledge-Base

Given the above model, we may now show how a concrete service, such as *HBase* (a popular non-relational record-stored database-service) could be added. A snippet of the code as shown in the listing below involves defining the new service to inherit the knowledge (i.e. definition of the operation-types, structure of the messages, mapping templates) that has already been defined in the *record-store* entity. This therefore simplifies the process, however it does also require customizations in order to meet the specifics of the concrete service, which cannot otherwise be directly automated.

```
1.  //Retrieves the "RecordStore" service-entity:
2.  Service record_store = ServiceAPI.getServiceByName("RecordStore");
3.
4.  //Defines a new service "HBase" to inherit "RecordStore":
5.  RESTService HBase = new
6.      RESTService
7.          .RESTServiceBuilder("HBase")
8.          .inherit(record_store)
9.          .build();
10.
11. //The operation "get" can be retrieved, since it is inherited:
12. OperationType get = HBase.getOperationType("get");
13.
14. //Customizing the input message to "http" and adding mapping info:
15. HttpMessage msg_in1 = new HttpMessage(get.getInputMessage());
16. msg_in1.getField().loadMapping("in_map1.json");
17. get.setInputMessage(msg_in1);
18.
19. //Customizing the output message to "xml" and adding mapping info:
20. XMLMessage msg_out1 = new XMLMessage(get.getOutputMessage());
21. msg_out1.getField().loadMapping("out_map1.json");
22. get.setOutputMessage(msg_out1);
```

```
1.  //Contents of in_map1.json:
2.  {"Id" : {
3.      "value" : "write(httpparam(id))"
4.  }}
```

```
1.  //Contents of out_map1.json:
2.  {"Table" : {
3.      "nodepath" : "xpath('//Item')",
4.      "Row" : {
5.          "nodepath" : "xpath('//Attribute')",
6.          "Cell" : {
7.              "CellName" : {
8.                  "value" : "read(xpath('//Attribute/Name/text()')"
9.              },
10.             "CellValue" : {
11.                 "value" : "read(xpath('//Attribute/Value/text()')"
12.             }
13.         }
14.     }
15. }}
```

In order to provide additional support, in the mappings shown, we illustrate in bold-text the template that can be generated for the required mapping (i.e. by calling .generateMapping() on the particular field object) – leaving just the specifics to be filled-in. Although, consider now the case where another method could be added to the *record-store* service-entity, for example: getTableList, which takes no message arguments, but returns a list of names, itemizing all the tables defined in the store. The code to add this operation-type is shown in the listing below; and serves to demonstrate how service-entities can grow incrementally. Now, in the case that another concrete record-store service would be added (for example, *Google BigTable*), the new operation-type could easily be inherited.

```
1.   //Retrieves the "RecordStore" service-entity:
2.   Service record_store = ServiceAPI.getServiceByName("RecordStore");
3.
4.   //Define the new operation-type:
5.   OperationType get_tables = new OperationType("GetTableList");
6.
7.   //Define the input message and fields:
8.   Message msg_in = new Message();
9.
10.  //Define the output message and fields:
11.  Message msg_out = new Message();
12.  AttributeField<String> table_names =
         new AttributeField<String>("TableNames");
13.  ListField table_list = new ListField("Tables", table_names);
14.  msg_out.addField(table_list);
15.
16.  //Add input and output messages to operation-type:
17.  get_tables.setInputMessage(msg_in);
18.  get_tables.setOutputMessae(msg_out);
19.
20.  //Add operation-type to service:
21.  record_store.addOperationType(get_tables);
22.
23.  //Update service-entity:
24.  ServiceAPI.updateService(record_store);
```

3.4 Use-Case Scenario

We have shown in the above how service-entities are incrementally enriched. In some cases although it requires some work, once this knowledge has been entered, we can then re-use and utilize for enabling simplified integration to services in application development. We demonstrate this over a simple use-case scenario, which we have implemented: Consider we would like to visualize the contributions of users to a particular Google Document. At present, since the GoogleDoc API only returns the twenty-most recent changes, we are required to log this data ourselves; to do so we utilize *HBase*. This data, which can be stored in a table in *HBase*, can then be queried for analysis in order to formulate the required visualization.

Based on our evaluation, implementing this using traditional means (i.e. without using *ServiceBase*) required a total of approximately 326 lines-of-code, and 3 dependency libraries. Whereas as demonstrated below, a solution using *ServiceBase* could be implemented in less than 77 lines-of-code, with no additional libraries.

The implementation we have devised could be divided into two parts: (i) A deamon process that monitors changes on the particular GDoc, and logs this data into the database-service; (ii) The request-reply function that when called queries this database and returns an appropriate visualization.

ServiceBase: A Programming Knowledge-Base for Service Oriented Development

We first use ServiceBase to create a *FeedInstance* of the generic Google Document `DocChanges` [15] *FeedType* , and then create a subscription to this, as shown below:

```
1. FeedInstance activity_feed =
          new FeedInstance(googleDocs.getFeedTypeByName("ActivityFeed"));
2. activity_feed.setInstanceName("MoshesGDocActivity");
3. activity_feed.setPathParam("user_id", "moshe…@gmail.com");
4. activity_feed.setPublic(false);

5. //Register the feed-instance on the service-bus:
6. ServiceAPI.registerFeedInstance(access_key, activity_feed);

7. //Create a subscription to this event:
8. String subscription_id =
       ServiceBus.subscribe("/GoogleDocs/ActivityFeed/MoshesGDocActivity");
```

Upon doing so, an event-callback could then be written that acts on the event that a new change has occurred. If so, an entry is added to the database to log this change. The code below thus represents the deamon process that could be inserted:

```
1.    public class EventHandlers {
2.      @EventCallback(tag="my_handler_id")
3.      public void MyHandler(Field gdoc_activity, String sub_id){

4.         //Get the data needed from activity field object:
5.         gdoc_activity = (TupleField) gdoc_activity;
6.         String doc_id = gdoc_activity.getField("gdoc_id").getValue();
7.         String author = gdoc_activity.getField("author").getValue();
8.         String desc = gdoc_activity.getField("description").getValue();

9.         //Get DB-service object:
10.        Service HBase = ServiceAPI.getServiceByName("HBase");
11.        ListField row = (ListField)HBase.getOperationTypeByName("put")
12.                               .getInputMessage().getField();

13.        //Formulate input Message-Fields:
14.        TupleField gdoc_id = new TupleField("Cell");
15.        cell.addField(new AttributeField<String>("CellName","gdoc_id"));
16.        cell.addField(new AttributeField<String>("CellValue",doc_id));
17.        row.add(gdoc_id);

18.        TupleField author_username = new TupleField("Cell");
19.        cell.addField(new AttributeField<String>("CellName","usrname"));
20.        cell.addField(new AttributeField<String>("CellValue",author));
21.        row.add(author_username);

22.        TupleField description = new TupleField("Cell");
23.        cell.addField(new AttributeField<String>("CellName", "desc"));
24.        cell.addField(new AttributeField<String>("CellValue", desc));
25.        row.add(description);

26.        //Invoke DB-service to add data:
27.        ServiceBus.invoke(HBase, "put", row);
28.      }
29.    }

30. ServiceBus.addEventListner(access_key, subscription_id,
                          new EventHandlers(), "my_handler_id");
```

Finally, in order to produce the required visualization, we implement the following function that queries the database-service for analysis. (In this example, we simply assume any change counts as 1-point towards the scores that calculates contributions). We use Google Chart API [11] for creating a visualization of data.

```
1.    public String getGraph(String access_key, String doc_id){

2.       //Get Service-entity from bus:
3.       RESTService HBase = ServiceAPI.getServiceByName("HBase");
4.       RESTOperationType getRecord =
                HBase.getOperationTypeByName("GetRecord");
5.       AttributeField<String> id =
                new AttributeField<String>("id", doc_id);

6.       //Invoke service to query DB for log-data:
7.       Field changes =
                ServiceBus.invoke(access_key, HBase, getRecord, id);

8.       //Calculate contributions:
9.       HashMap<String,Integer> user_score =
                new HashMap<String,Integer>();

10.      for(Field rows : (ListField<Field>) changes){
11.         for(Field cell : (ListField<Field>) rows){
12.            cell = (TupleField) cell;
13.            if(cell.get("name").compareTo("username")==0){
14.               String username = ((TupleField)cell).get("value");
15.               if(user_score.get(username)!=null){
16.                  int curr_score = user_score.get(username).parseInt();
17.                  user_score.put(username, curr_score+1);
18.               }
19.               else
20.                  user_score.put(username, 1);
21.            }
22.         }
23.      }

24.      //Use Google Graph API for getting a visualisation URL:
25.      Service googleGraph = ServiceAPI.getServiceByName("GoogleGraph");
26.      ListField pie_data = (ListField) googleGraph
                              .getOperationType("createPieChart)
                              .getInputMessage()
                              .getField();

27.      for(String username : user_score.keySet()){
28.         TupleField chartdata = new TupleField("ChartData");
29.         chartdata.addField(
30.            new AttributeField("label", username);
31.         chartdata.addField(
32.            new AttributeField("value", user_score.get(username));
33.         pie_data.add(chartdata);
34.      }
35.      String url =
                ServiceBus.invoke(null, googleGraph, "createPieChart", pie_data);

36.      return url;
37.   }
```

4 Evaluation

We have evaluated the overall effectiveness of our proposed approach (i.e. to simplify access and integration of web-services in application development), by adopting the above scenario in a user-study. The factors used to measure effectiveness were: (i) The total number of lines-of-code excluding white-space and comments; (ii) Number of extra dependencies needed; and (iii) Time taken to complete task. The study was conducted on a total of five participants, all of which possessed an average to moderately-high level of software development expertise. In order to further balance the evaluation, three participants were asked to attempt the implementation using traditional techniques first, and then secondly using *ServiceBase*; whereas the other two participants were asked to do this in reverse. The results of our study are shown in the graphs as illustrated at Figure 6 below.

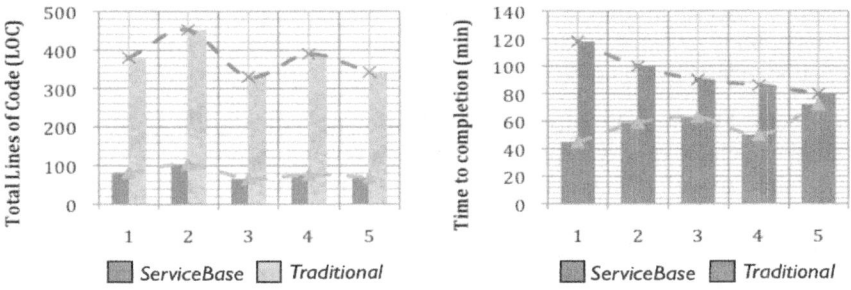

Fig. 6. Evaluation Results for GDocs Contribution Calculator use-case

As an overall analysis, it is clear that across all participants, the number of lines of code and time taken to complete the task is significantly reduced when using *ServiceBase* than in comparison to the traditional development approaches. In general as well, while the implementations using *ServiceBase* did not require any additional libraries, the traditional approaches in contrast required on average at least two to three additional libraries. In light of these results, this evaluation study successfully demonstrates the anticipated benefit of our proposed approach.

5 Related Work

Web-Service Types, Modeling Technique and Concerns. There are clearly two widely accepted representation approaches for services, namely SOAP and REST [2,5,7]. Nonetheless, while both strive to achieve the same underlying goal, there has in fact been much debate about whether "REST has replaced SOAP services!" [6], or questions posed relating to "which one is better?" [16]. While the conclusions of these debates are largely beyond the scope of our exploration, it is clear that RESTful service has by far outweighed SOAP service offerings. In fact, at the time of writing, there has been a reported 500 SOAP services in contrast to over 2,800 RESTful services. The clear reasons for this is due to the fact that RESTful services are by far easier to understand and provides better support for modern web-technology. For example, whereas SOAP enforces XML, RESTful services support a more human-friendly JSON, that also enables increased support for embedding JavaScript and Ruby. SOAP services on the other hand have mainly focused on enterprise resulting in a more verbose architecture. Particularly WSDL guided by the increasingly family of WS-* standards. However, it is precisely the lack of standards surrounding REST that has polarized the community for or against REST being the next generation of web-services technology, [17].

Web-Services Repositories, Access Techniques and Concerns. Ultimately, the value of service-models is assessed by its usefulness, such as: whether services can be stored and explored; and whether the model enables a degree of automated support to utilize them in application development. In the SOAP community, while standards such as UDDI were proposed to act as a global-repository, it seemed the idea soon failed where the emphasis has shifted to simply relying on web-based engines in order to locate services, similar to what is done for RESTful services. For example, *ProgrammableWeb* list thousands of APIs, however clearly not much of the meta-information available would be useful to support or simplify service-execution.

Towards an Abstract Architecture for Uniform Presentation of Resources. To address these challenges, we have thus been motivated to propose a unified service representation model, which is an essential component in order to provide a common interface for interacting with services. This means the heterogeneity of services can be masked by more high-level operations that automate the concrete set of instructions behind the scenes. From an architectural perspective, there are in fact several works that share the same motivation, although for other more specific domains.

For example, in the case of data-storage services, BStore [18] is a framework that allows developers to separate their web application code from the underlying user data-storages. The architecture consists of three components: file-systems, which could be considered as data-storage APIs (or services in our model); the file-system manager acting as the middleware (or service-bus in our model); and applications that require access to the underlying user-data. A common-interface is then proposed for both loading storage-services as well as for applications to access this data.

Another example is SOS [12], which also defines a common-interface to interact with non-relational databases. Similar to the concept of the unified-model, they provide a meta-model approach to map specific interfaces of various systems to a common one. However, since the work mainly deals with data-storage services, the common set of operations is relatively simple. Also relevant is that the work deals with providing a common model for run-time data. In this manner further similarity can be drawn to the message-fields and mapping component of our system, where interestingly they too identify three main constructs for modeling heterogeneous data, which they refer to as String, Collection and Object.

In the case of Feed-based services, the work at [15] presents an architecture for consumers of feeds to organize the services that they are using, share them, or use it to build tools which would implement a decentralized system for publishing, consuming and managing feed-subscription. We identify the middleware in their framework to be the feed-subscription manager (FSM), which decouples consumers from the underlying feed-services. In this case, the common representation model for feeds is expressed via an Atom feed. Common operations to services are then expressed via AtomPub in order to interact with the various underlying feed-sources.

However, in all cases mentioned above, while they share similar concepts of architecture, their applicability is still only limited to a particular domain.

6 Conclusions

Although the Internet continues to flourish with a growing number of APIs, there still lie significant challenges in integrating services in everyday application development. Motivated by this need, we proposed in this paper a platform for simplified access to web-services. In order to achieve this, we first addressed the heterogeneity of various service representation types by proposing a unified service model and mapping framework. Inspired from the Web2.0 paradigm, we design a programming knowledge-base, such that common service-knowledge can be abstracted, organized, incrementally curated and thereby re-used by other developers. Empowered by this, we have implemented a set of APIs that offers a common programming interface for significantly simplified access to service, accordingly, we have conducted an evaluation to verify the overall effectiveness of our proposed work.

References

1. Yu, J., Benatallah, B., Casati, F., Daniel, F.: Understanding Mashup Development. IEEE Internet Computing 12(5), 44–52 (2008)
2. Pautasso, C., Zimmermann, O., Leymann, F.: Restful Web Services vs. "Big" Web Services: Making the Right Architectural Decision. In: 17th International Conference on World Wide Web, pp. 805–814. ACM (2008)
3. Voida, A., Harmon, E., Al-Ani, B.: Homebrew Databases: Complexities of Everyday Information Management in non-profit Organizations. In: Conference on Human Factors in Computing Systems (CHI). ACM Press, Vancouver (2011)
4. Benatallah, B., Dumas, M., Sheng, Q.Z.: Facilitating the rapid development and scalable orchestration of composite web services. Distributed and Parallel Databases 17(1), 5–37 (2005)
5. Alonso, G., Casati, F., Kuno, H., Machiraju, V.: Web services: Concepts, Architectures, and Application, 354 Pages. Springer (2004) ISBN: 978-3-540-44008-6
6. How REST replaced SOAP on the Web, http://www.infoq.com/articles/rest-soap
7. Geambasu, R., Cheung, C., Moshchuk, A., Gribble, S., Levy, H.M.: Organizing and sharing distributed personal web-service data. In: 17th International Conference on World Wide Web, pp. 755–764. ACM Press (2008)
8. OAuth, http://oauth.net/
9. Wikipedia: 'Internet Media Type', http://en.wikipedia.org/wiki/Internet_media_type
10. Kwok, W.: Bidirectional transformation between relational data and XML document with semantic preservation and incremental maintenance. PhD Thesis, University of Hong Kong
11. Google Image Chart API, https://developers.google.com/chart/image/

12. Atzeni, P., Bugiotti, F., Rossi, L.: SOS (Save Our Systems): A uniform programming interface for non-relational systems. In: 15th International Conference on Electronic Conference (EDBT), Berlin, Germany (2012)
13. OrientDB Graph-Document NoSQL DBMS, http://www.orientdb.org/index.htm
14. Amazon Simple DB, http://aws.amazon.com/simpledb/
15. Wilde, E., Liu, Y.: Feed Subscription Management. University of California, Berkley School of Information Report 2011-042 (2011)
16. REST and SOAP: When Should I Use Each? http://www.infoq.com/articles/rest-soap-when-to-use
17. Duggan, D.: Service Oriented Architecture: Entities, Services, and Resources. Wiley-IEEE Computer Society, NJ (2012)
18. Chandra, R., Gupta, P., Zeldovich, N.: Separating Web Applications from User Data Storage with BStore. In: WebApps (2010)

On Leveraging Crowdsourcing Techniques for Schema Matching Networks

Nguyen Quoc Viet Hung[1], Nguyen Thanh Tam[1], Zoltán Miklós[2], and Karl Aberer[1]

[1] École Polytechnique Fédérale de Lausanne
{quocviethung.nguyen,tam.nguyenthanh,karl.aberer}@epfl.ch
[2] Université de Rennes 1
zoltan.miklos@univ-rennes1.fr

Abstract. As the number of publicly-available datasets are likely to grow, the demand of establishing the links between these datasets is also getting higher and higher. For creating such links we need to match their schemas. Moreover, for using these datasets in meaningful ways, one often needs to match not only two, but several schemas. This matching process establishes a (potentially large) set of attribute correspondences between multiple schemas that constitute a *schema matching network*. Various commercial and academic schema matching tools have been developed to support this task. However, as the matching is inherently uncertain, the heuristic techniques adopted by these tools give rise to results that are not completely correct. Thus, in practice, a post-matching human expert effort is needed to obtain a correct set of attribute correspondences.

Addressing this problem, our paper demonstrates how to leverage crowdsourcing techniques to validate the generated correspondences. We design validation questions with contextual information that can effectively guide the crowd workers. We analyze how to reduce overall human effort needed for this validation task. Through theoretical and empirical results, we show that by harnessing natural constraints defined on top of the schema matching network, one can significantly reduce the necessary human work.

1 Introduction

There are more and more services on the internet that enable users to upload and share structured data, including Google Fusion Tables [13], Tableausoftware[1], Factual[2]. These services primarily offer easy visualization of the uploaded data as well as tools to embed the visualisation to blogs or Web pages. As the number of publicly available datasets grows rapidly and they are often fragmented into different sources, it is essential to create the interlinks between these datasets [7]. For example, in Google Fusion Tables, the coffee consumption data are distributed among different tables in that each table represents for a specific region [13]. In order to extract generic information for all regions, we need to aggregate and mine across multiple tables. This raises the challenges for interconnecting table schemas to achieve an integrated view of data.

---
[1] http://www.tableausoftware.com/public
[2] http://www.factual.com/

One of the major challenges in interconnecting the datasets is to establish the connections between attributes of individual schemas that describe the datasets. The process of establishing correspondences between the attributes of two database schemas has been extensively researched, and there is a large body of work on heuristic matching techniques[4, 22]. Beside the research literature, numerous commercial and academic tools, called schema matchers, have been developed. Even though these matchers achieve impressive performance on some datasets, they cannot be expected to yield a completely correct result since they rely on heuristic techniques. In practice, data integration tasks often include a post-matching phase, in which correspondences are reviewed and validated by human experts.

Given our application context, the large number of schemas and (possible) connections between them, the validation task would require an extreme effort. In this paper we demonstrate the use of crowdsourcing techniques for schema matching validation. Specifically, we study a setting in which the two schemas to be matched do not exist in isolation but participate in a larger matching network and connect to several other schemas at the same time. Beside interconnecting structured data on the Internet, there are a number of application scenarios in which such model can be applied, for example schema matching in large enterprises [18, 24] or service mashups [9].

Crowdsourcing techniques have been successfully applied for several data management problems, for example in CrowdSearch [26] or CrowdScreen [19]. McCann et al. [17], have already applied crowdsourcing methods for schema matching. In their work, they focused on matching a pair of schemas, but their methods are not directly applicable for the matching network that is our main interest. Leveraging network information, we define natural constraints that not only effectively guide the crowd workers but also significantly reduce the necessary human efforts.

Our contributions can be summarized as follows.

- We analyze the schema matching problem in networks whose schemas are matched against each other. On top of such networks, we exploit the relations between correspondences to define the matching network constraints.

- We design questions presented to the crowd workers in a systematic way. In our design, we focus on providing contextual information for the questions, especially the transitivity relations between correspondences. The aim of this contextual information is to reduce question ambiguity such that workers can answer more rapidly and accurately.

- We design an aggregate mechanism to combine the answers from multiple crowd workers. In particular, we study how to aggregate answers in the presence of matching network constraints. Our theoretical and empirical results show that by harnessing the network constraints, the worker effort can be lowered considerably.

The rest of the paper is structured as follows. The next section gives an overview of our framework. In Section 3, we describe how to design the questions that should be presented to crowd workers. In Section 4, we formulate the problem of aggregating the answers obtained from multiple workers. Section 5 clarifies our aggregate methods that exploit the presence of matching network constraints. Section 6 presents experimental results. Section 7 summarizes related work, while Section 8 concludes the paper.

2 Overview

Schema matching network is a network of schemas, together with the pairwise attribute correspondences between the attributes of the corresponding schemas. In our setting we suggest that these schema matching networks shall be constructed in the following two-step incremental process: (1) generate pairwise schema matchings using existing tools such as COMA [10] and AMC [20], (2) validate the generated matching candidates by crowd workers (i.e. decide whether the generated correspondence is valid or not). After the first step, the schema matching network is constructed and defined as a tuple (S, C), where S is a set of schemas and C is a set of correspondences generated by matching tools.

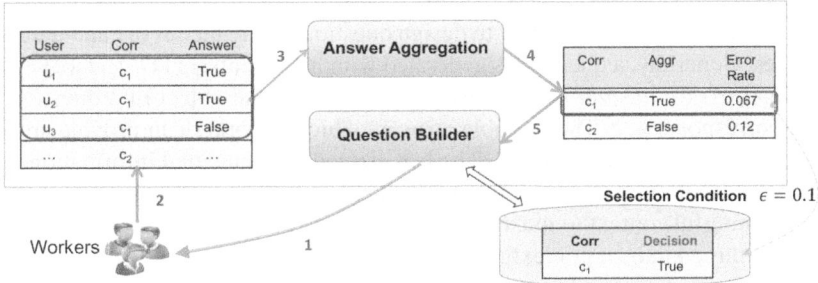

Fig. 1. Architecture of the crowdsourcing framework

For realizing the second second step of validating the correspondences, we propose the framework depicted in Figure 1. The input to our framework is a set of correspondences C. These correspondences are fetched to *Question Builder* component to generate questions presented to crowd workers. A worker's answer is the validation of worker u_i on a particular correspondence $c_j \in C$, denoted as a tuple $\langle u_i, c_j, a \rangle$, where a is the answer of worker u_i on correspondence c_j. Domain values of a are $\{true, false\}$, where $true/false$ indicates c_j is approved/disapproved.

In general, the answers from crowd workers might be incorrect. There are several reasons for this, such as the workers might misunderstand their tasks, they may accidentally make errors, or they simply do not know the answers. To cope with the problem of possibly incorrect answers, we need aggegation mechanisms, realized in the *Answer Aggregation* component. We adopt probabilistic aggregation techniques. We estimate the quality of the aggregated value by comparing the answers from different workers. The aggregated result of a correspondence is a tuple $\langle a^*, e \rangle$, where a^* is the aggregated value, e is the error rate of aggregation. If the error rate e is greater than a pre-defined threshold ϵ, we continue to fetch c into *Question Builder* to ask workers for more answers. Otherwise, we make the decision a^* for the given correspondence. This process is repeated until the halting condition is satisfied. In our framework, the halting condition is that all correspondences are decided.

In our setting, it is reasonable to assume that there is an objective ground truth, i.e., there exists a single definitive matching result that is external to human judgment. However, this truth is hidden and no worker knows it completely. Therefore, we leverage the wisdom of the crowd in order to approximate the hidden ground truth (with the help of our aggregation techniques). However, approximating the ground truth with limited budget raises several challenges: (1) *How to design the questions for effective answers?* (2) *How to make aggregation decision based on the answers from workers?* (3) *How to reduce the number of questions with a given quality requirement?* In the following sections, we will address these challenges.

3 Question Design

In this section, we demonstrate how to design questions using the set of candidate correspondences. Generally, a question is generated with 3 elements: (1) *Object*, (2) *Possible answers* and (3) *Contextual information*. In our system, the object of a question is an attribute correspondence. The possible answers which a worker can provide are either *true* (approve) or *false* (disapprove). The last element is contextual information, which plays a very important role in helping workers answer the question more easily. It provides a meaningful context to make the question more understandable. In our work, we h have used three kinds of contextual information:

- **All alternative targets:** We show a full list of candidate targets generated by matching tools. By examining all possible targets together, workers have can better judge whether the given correspondence is correct or not as opposed to evaluating a single value correspondence. Figure 2(A) gives an example of this design.
- **Transitive closure:** We do not only display all alternatives, but also the transitive closure of correspondences. The goal of displaying the transitive closure is to provide a context that shall help workers to resolve the ambiguity, when otherwise these alternatives are hard to distinguish. For example, in Figure 2(B), workers might not be able to decide which one of two attributes CRM.BirthDate and CRM.Name corresponds to the attribute MDM.BirthName. Thanks to the transitive closure MDM.BirthName → CRM.Name → SRM.BirthName, workers can confidently confirm the correctness of the match between CRM.Name and MDM.BirthName.
- **Transitive violation:** In contrast to transitive closure, this design supports a worker to identify incorrect correspondences. Besides all alternatives, the contextual information contains a circle of correspondences that connects two different attributes of the same schema. For instance, in Figure 2(C), workers might find it difficult to choose the right target among CRM.BirthDate, CRM.Name for MDM.BirthName. The transitive violation CRM.Name → SRM.BirthName → MDM.BirthName → CRM.BirthDate is the evidence that helps worker to reject the match between MDM.BirthName and CRM.BirthDate.

Comparing to the question generating and posting strategy presented in [17], our question design is more general. In our approach, both the pairwise information (i.e., data

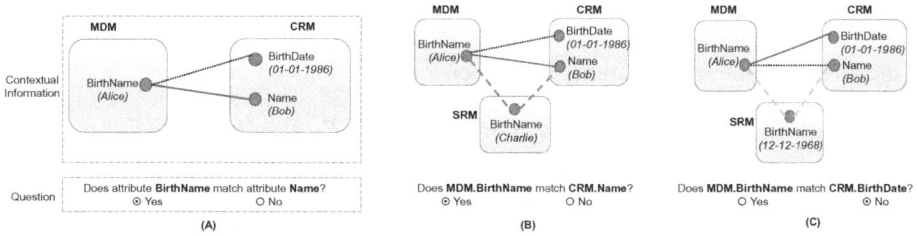

Fig. 2. Question designs with 3 different contextual information: (A) All alternative targets, (B) Transitive closure, (C) Transitive violation

value and all alternatives) and the network-level contextual information (i.e., transitive closure and transitive violation) are displayed to help the workers to answer the question more effectively. To evaluate the effectiveness of the question design, we conducted some experiments in section 6. It turned out that the contextual information proposed as above is critical. Having the contextual information at hand, the workers were able to answer the questions faster and more accurately. Subsequently, the total cost could be substantially reduced since the payment for each task can be decreased [2].

4 Aggregating User Input

In this section we explain our aggregation techniques. After posting questions to crowd workers (as explained in Section 3), for each correspondence $c \in C$, we collect a set of answers π_c (from different workers) in which each element could be *true*(approve) or *false*(disapprove). The goal of aggregation is to obtain the aggregated value a_c as well as estimate the probability that a_c is incorrect. This probability is also called the error rate of the aggregation e_c.

In order to compute the aggregated value a_c and error rate e_c, we first derive the probability of possible aggregations $Pr(X_c)$. In that, X_c is a random variable of aggregated values of c and domain values of X_c is $\{true, false\}$. This value refers to the ground truth, however that is hidden from us, thus we try to estimate this probability with the help of aggregation methods. There are several techniques proposed in the literature to compute this probability such as majority voting [2] and expectation maximization (EM) [8]. While majority voting aggregates each correspondence independently, the EM method aggregates all correspondences simultaneously. More precisely, the input of majority voting is the worker answers π_c for a particular correspondence c, whereas the input of EM is the worker answers $\pi = \bigcup_{c \in C} \pi_c$ for all correspondences.

In this paper, we use EM as the main aggregation method to compute the probability $Pr(X_c)$. The EM method differs from majority voting in considering the quality of workers, which is estimated by comparing the answers of each worker against other workers answers. More precisely, the EM method uses maximum likelihood estimation to infer the aggregated value of each correspondence and measure the quality of that value. The reason behind this choice is that the EM model is quite effective for labeling tasks and robust to noisy workers [23].

After deriving the probability $Pr(X_c)$ for each correspondence $c \in C$, we will compute the aggregation decision $\langle a_c, e_c \rangle = g_\pi(c)$, where a_c is the aggregated value and e_c is the error rate. The aggregation of this decision is formulated as follows:

$$g_\pi(c) = \begin{cases} \langle true, 1 - Pr(X_c = true) \rangle & \text{If } Pr(X_c = true) \geq 0.5 \\ \langle false, 1 - Pr(X_c = false) \rangle & \text{Otherwise} \end{cases} \quad (1)$$

In equation 1, the error rate is the probability of making wrong decision. In order to reduce error rate, we need to reduce the uncertainty of X_c (i.e., entropy value $H(X_c)$). If the entropy $H(X_c)$ is closed to 0, the error rate is closed to 0. For the experiments described in section 6, in order to achieve lower error rate, we need to ask more questions. However, with given requirements of low error rate, the monetary cost is limited and needs to be reduced. In next section, we will leverage the constraints to solve this problem.

5 Leveraging Constraints to Reduce User Efforts

For experiments described in section 6, we found that to achieve lower error rate, more answers are needed. This is, in fact, the trade-off between the cost and the accuracy[26]. The higher curve of Figure 3 depicts empirically a general case of this trade-off.

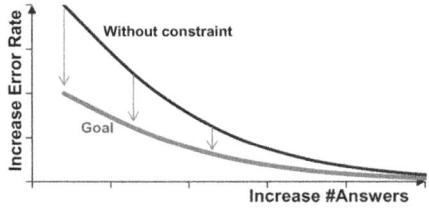

Fig. 3. Optimization goal

We want to go beyond this trade-off by lowering this curve as much as possible. When the curve is lower, with the same error rate, the number of answers is smaller. In other words, with the same number of answers, the error rate is smaller. To achieve this goal, we leverage the network consistency constraints to adjust the error rate with the same number of answers. In this section, we will show how to exploit these constraints.

5.1 Aggregating with Constraints

In section 4, we already formulate the answer aggregation. Now we leverage constraints to adjust the error rate of the aggregation decision. More precisely, we show that by using constraints, it requires fewer answers to obtain aggregated result with the same error rate. In other words, given the same answer set on a certain correspondence, the error rate of aggregation with constraint is lower than the one without constraint. We consider very natural constraints that we assume to hold; in other words we assume that these are hard constraints.

Given the aggregation $g_\pi(c)$ of a correspondence c, we compute the justified aggregation $g_\pi^\gamma(c)$ when taking into account the constraint γ. The aggregation $g_\pi^\gamma(c)$ is obtained similarly to equation 1, except that the probability $Pr(X_c)$ is replaced by the conditional probability $Pr(X_c|\gamma)$ when the constraint γ holds. Formally,

$$g_\pi^\gamma(c) = \begin{cases} \langle true, 1 - Pr(X_c = true|\gamma)\rangle & \text{If } Pr(X_c = true|\gamma) \geq 0.5 \\ \langle false, 1 - Pr(X_c = false|\gamma)\rangle & \text{Otherwise} \end{cases} \quad (2)$$

In the following, we describe how to compute $Pr(X_c|\gamma)$ with 1-1 constraint and circle constraint. Then, we show why the affect of constraints can reduce error rate. We leave the investigation of other types of constraints as an interesting future work.

5.2 Aggregating with 1-1 Constraint

Our approach underlies the intuition illustrated in Figure 4(A), depicting two correspondences c_1 and c_2 with the same source attribute. After receiving the answer set from workers and applying probabilistic model (section 4), we obtained the probability $Pr(X_{c_1} = true) = 0.8$ and $Pr(X_{c_2} = false) = 0.5$. When considering c_2 independently, it is hard to conclude c_2 being approved or disapproved. However, when taking into account c_1 and 1-1 constraint, c_2 tends to be disapproved since c_1 and c_2 cannot be approved simultaneously. Indeed, following probability theory, the conditional probability $Pr(X_{c_2} = false|\gamma_{1-1}) \approx 0.83 > Pr(X_{c_2} = false)$.

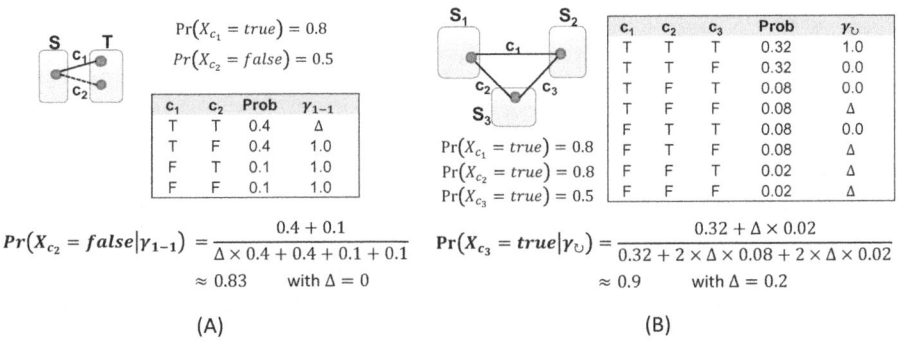

(A) (B)

Fig. 4. Compute conditional probability with (A) 1-1 constraint and (B) circle constraint

In what follows, we will formulate 1-1 constraint in terms of probability and then show how to compute the conditional probability $Pr(X_c|\gamma_{1-1})$.

Formulating 1-1 Constraint. Given a matching between two schemas, let us have a set of correspondences $\{c_0, c_1, \ldots, c_k\}$ that share a common source attribute. With respect to 1-1 constraint definition, there is at most only one c_i is approved (i.e., $X_{c_i} = true$). However there are some exceptions where this constraint does not hold. For instance, the attribute *name* might be matched with *firstname* and *lastname*. But these cases only happen with low probability. In order to capture this observation, we formulate 1-1 constraint as follows:

$$Pr(\gamma_{1-1}|X_{c_0}, X_{c_1}, \ldots, X_{c_k}) = \begin{cases} 1 & \text{If } m \leq 1 \\ \Delta \in [0,1] & \text{If } m > 1 \end{cases} \quad (3)$$

where m is the number of X_{c_i} assigned as *true*. When $\Delta = 0$, there is no constraint exception. In general, Δ is close to 0. The approximated value of Δ can be obtained through statistical model [6].

Computing Conditional Probability. Given the same set of correspondence $\{c_0, c_1, \ldots, c_k\}$ above, let denote p_i as $Pr(X_{c_i} = true)$ for short. Without loss of generality, we consider c_0 be the favourite correspondence whose probability p_0 is obtained from the worker answers. Using the Bayesian theorem and equation 3, the conditional probability of correspondence c_0 with 1-1 constraint γ_{1-1} is computed as:

$$Pr(X_{c_0} = true|\gamma_{1-1}) = \frac{Pr(\gamma_{1-1}|X_{c_0} = true) \times Pr(X_{c_0} = true)}{Pr(\gamma_{1-1})} = \frac{(x + \Delta(1-x)) \times p_0}{y + \Delta(1-y)} \quad (4)$$

where
$$x = \prod_{i=1}^{k} (1 - p_i)$$
$$y = \prod_{i=0}^{k} (1 - p_i) + \sum_{i=0}^{k} [p_i \prod_{j=0, j\neq i}^{k} (1 - p_j)]$$

x can be interpreted as the probability of the case where all other correspondences except c being disapproved. y can be interpreted as the probability of the case where all correspondences being disapproved or only one of them being disaproved. The precise derivation of equation 4 is put in the Appendix.

5.3 Aggregating with Circle Constraint

Figure 4(B) depicts an example of circle constraint for three correspondences c_1, c_2, c_3. After receiving the answer set from workers and applying probabilistic model (section 4), we obtained the probability $Pr(X_{c_1} = true) = Pr(X_{c_2} = true) = 0.8$ and $Pr(X_{c_3} = true) = 0.5$. When considering c_3 independently, it is hard to conclude c_3 being *true* or *false*. However, when taking into account c_1, c_2 under the 1-1 constraint, c_3 tends to be *true* since the circle created by c_1, c_2, c_3 shows an interoperability. Therefore, following probability theory, the conditional probability $Pr(X_{c_3} = true|\gamma_{1-1}) \approx 0.9 > Pr(X_{c_3} = true)$.

In the following we will formulate circle constraint in terms of probability and then show how to compute the conditional probability $Pr(X_c|\gamma_\cup)$.

Formulating Circle Constraint. Following the notion of cyclic mappings in [6], we formulate the conditional probability of a circle as follows:

$$Pr(\gamma_\cup|X_{c_0}, X_{c_1}, \ldots, X_{c_k}) = \begin{cases} 1 & \text{If } m = k+1 \\ 0 & \text{If } m = k \\ \Delta & \text{If } m < k \end{cases} \quad (5)$$

Where m is the number of X_{c_i} assigned as *true* and Δ is the probability of compensating errors along the circle (i.e., two or more incorrect assignment resulting in a correct reformation).

Computing Conditional Probability. Given a closed circle along $c_0, c_1, \ldots, c_k$, let denote the constraint on this circle as $\gamma_\cup$ and p_i as $Pr(X_{c_i} = true)$ for short. Without loss of generality, we consider c_0 to be the favorite correspondence whose probability p_0 is obtained by the answers of workers in the crowdsourcing process. Following the Bayesian theorem and equation 5, the conditional probability of correspondence c_0 with circle constraint is computed as:

$$Pr(X_{c_0} = true | \gamma_\cup) = \frac{Pr(\gamma_\cup | X_{c_0} = true) \times Pr(X_{c_0} = true)}{Pr(\gamma_\cup)} = \frac{(\prod_{i=1}^{k}(p_i) + \Delta(1-x)) \times p_o}{\prod_{i=0}^{k}(p_i) + \Delta(1-y)} \quad (6)$$

where
$$x = \prod_{i=1}^{k}(p_i) + \sum_{i=1}^{k}[(1-p_i)\prod_{j=1, j\neq i}^{k} p_j]$$
$$y = \prod_{i=0}^{k}(p_i) + \sum_{i=0}^{k}[(1-p_i)\prod_{j=0, j\neq i}^{k} p_j]$$

x can be interpreted as the probability of the case where only one correspondence among $c_1, \ldots, c_k$ except c_0 is disapproved. y can be interpreted as the probability of the case where only one correspondence among $c_0, c_1, \ldots, c_k$ is disapproved. The detail derivation of equation 6 is put in the Appendix.

5.4 Aggregating with Multiple Constraints

In general settings, we could have a finite set of constraints $\Gamma = \{\gamma_1, \ldots, \gamma_n\}$. Let denote the aggregation with a constraint $\gamma_i \in \Gamma$ is $g_\pi^{\gamma_i}(c) = \langle a_c^i, e_c^i \rangle$, whereas the aggregation without any constraint is simply written as $g_\pi(c) = \langle a_c, e_c \rangle$. Since the constraints are different, not only could the aggregated value a_c^i be different ($a_c^i \neq a_c^j$) but also the error rate e_c^i could be different ($e_c^i \neq e_c^j$). In order to reach a single decision, the challenge then becomes how to define the multiple-constraint aggregation $g_\pi^\Gamma(c)$ as a combination of single-constraint aggregations $g_\pi^{\gamma_i}(c)$.

Since the role of constraints is to support reducing the error rate and the aggregation $g_\pi(c)$ is the base decision, we compute the multiple-constraint aggregation $g_\pi^\Gamma(c) = \langle a_c, e_c^\Gamma \rangle$, where $e^\Gamma = min(\{e_c^i | a_c^i = a_c\} \cup e_c)$. Therefore, the error rate of final aggregated value is reduced by harnessing constraints. For the experiments in real datasets described in the next section, we will show that this aggregation reduces a half of worker efforts while preserving the quality of aggregated results.

6 Experiments

The main goal of the following evaluation is to analyze the use of crowdsourcing techniques for schema matching network. To verify the effectiveness of our approach, three experiments are performed: (i) effects of contextual information on reducing question ambiguity, (ii) relationship between the error rate and the matching accuracy, and (iii) effects of the constraints on worker effort. We proceed to report the results on the real datasets using both real workers and simulated workers.

6.1 Experimental Settings

Datasets. We have used 3 real-world datasets: Google Fusion Tables, UniversityApp-Form, and WebForm. They are publicly available on our website [3]. In the experiments, the topology of schema matching network is a complete graph (i.e. all graph nodes are interconnected with all other nodes). In that, the candidate correspondences are generated by COMA [10] matcher.

Worker Simulation. In our simulation, we assume that the ground truth is known in advance (i.e. the ground truth is known for the experimenter, but not for the (simulated) crowd worker). Each simulated worker is associated with a pre-defined reliability r that is the probability of his answer being correct against the ground truth.

6.2 Effects of Contextual Information

In this experiment, we select 25 correct correspondences (i.e., exist in ground truth) and 25 incorrect correspondences (i.e., not exists in ground truth). For each correspondence, we ask 30 workers (Bachelor students) with three different contextual information: (a) all alternatives, (b) transitive closure, (c) transitive violation. Then, we collect the worker answers for each correspondence.

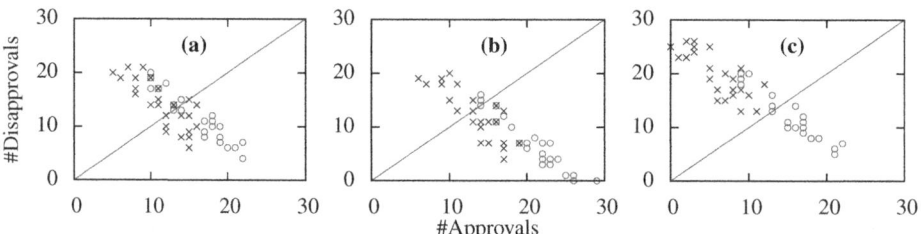

Fig. 5. Effects of contextual information. (a) all alternatives, (b) transitive closure, (c) transitive violation

Figure 5 presents the result of this experiment. The worker answers of each case are presented by a collection of 'x' and 'o' points in the plots. In that, 'o' points indicate correspondences that exist in ground truth, whereas 'x' points indicate correspondences that do not exist in ground truth. For a specific point, X-value and Y-value are the number of workers approving and disapproving the associated correspondences, respectively. Therefore, we expect that the 'o' points are placed at the right-bottom of the coordinate plane, while the 'x' points stay at the left-top of the coordinate plane.

Comparing Figure 5(b) with Figure 5(a), the 'o' points tend to move down to the bottom-right of the baseline (# 'approve' answers increases and # 'disapprove' answers decreases). Whereas, the movement of the 'x' points is not intensive. This can be interpreted that presenting the transitive closure context help workers to give feedback more exactly but also make them misjudge the incorrect correspondences.

---

[3] http://lsirwww.epfl.ch/schema_matching

In order to study the effects of transitive violation, we compare Figure 5(c) with Figure 5(a). Intuitively, the 'x' points move distinctly toward the top-left of the baseline, while the position of 'o' points keeps stable. This observation shows that transitive violations help workers identify the incorrect correspondences, in contrast to the effect of transitive satisfactions mentioned above.

Since in real settings the ground truth is not known before-hand, we cannot choose appropriate design type for each question. Following the principle of maximum entropy, in order not to favour any of the design types, we design each question in type (b) and (c) with probability of 0.5. In case the given correspondence does not involve in any transitive satisfaction and violation, we design its question in type (a).

6.3 Relationship between Error Rate and Matching Accuracy

In order to assess the matching accuracy, we borrow the *precision* metric from information retrieval, which is the ratio of correspondences existing in ground truth among all correspondences whose aggregated value is *true*. However, the ground truth is not known in general. Therefore, we use an indirect metric—error rate—to estimate the matching quality. We expect that the lower error rate, the higher quality of matching results.

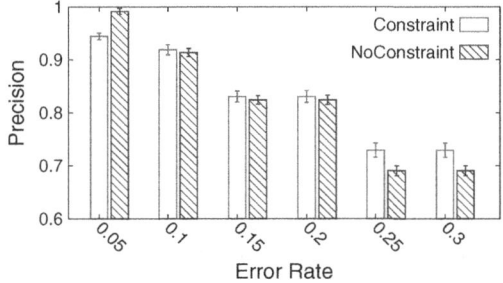

Fig. 6. Relationship between error rate and precision

The following empirical results aim to validate this hypothesis. We conduct the experiment with a population of 100 simulated workers and their reliability scores are generated according to normal distribution $\mathcal{N}(0.7, 0.04)$. Figure 6 depicts the relationship of the error rate and precision. In that, we vary error threshold ϵ from 0.05 to 0.3, meaning that the questions are posted to workers until the error rate of aggregated value is less than the given threshold ϵ. The precision is plotted as a function of ϵ. We aggregate the worker answers by two strategies: without constraint and with constraint. Here we consider both 1-1 constraint and circle constraint as hard constraints, thus $\Delta = 0$.

The key observation is that when the error rate is decreased, the precision approaches to 1. Reversely, when the error rate is increased, the precision is reduced but greater than $1 - \epsilon$. Another interesting finding is that when the error rate is decreased, the value distribution of precision in case of with and without constraint is identical. This indicates our method of updating the error rate is relevant.

In summary, the error rate is a good indicator of the quality of aggregated results. In terms of precision, the quality value is always around $1 - \epsilon$. In other words, the error threshold ϵ can be used to control the real matching quality.

6.4 Effects of the Constraints

In this experiment set, we will study the effects of constraints on the expected cost in real datasets. In Section 5, we already seen the benefit of using constraints in reducing error rate. Therefore, with given requirement of low error, the constraints help to reduce the number of questions (i.e., the expected cost) that need to ask workers. More precisely, given an error threshold ($\epsilon = 0.15, 0.1, 0.05$), we iteratively post questions to workers and aggregate the worker answers until the error rate is less than ϵ. We use simulated workers with reliability r varying from 0.6 to 0.8. Similar to the above experiment, we set $\Delta = 0$. The results are presented in Figure 7.

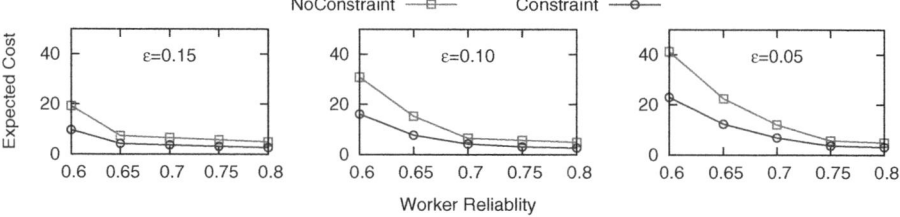

Fig. 7. User Efforts: effects of constraints

A significant observation in the results is that for all values of error threshold and worker reliability, the expected cost of the aggregation with constraints is definitely smaller (approximately a half) than the case without constraints. For example, with worker reliability is $r = 0.6$ and error threshold $\epsilon = 0.1$, the expected number of questions is reduced from 31 (without constraints) to 16 (with constraints). This concludes the fact that the constraints help to reduce the error rate, and subsequently reduce the expected cost.

Another key finding in Figure 7 is that, for both cases (using vs. not using constraints in the aggregation), the expected cost increases significantly as the value for error threshold ϵ decreases. For example, it requires about 20 questions (without constraints) or 10 questions (with constraints) to satisfy error threshold $\epsilon = 0.15$. Whereas, it takes about 40 questions (without constraints) or 20 questions (with constraints) to satisfy error threshold $\epsilon = 0.05$. This result supports the fact that to reduce error rate, we need to ask more questions.

7 Related Work

We now review salient work in schema matching and crowdsourcing areas that are related to our research.

Schema Matching. Database schema matching is an active research field. The developments of this area have been summarized in two surveys [4, 22]. Existing works on schema matching focused mainly on improving quality parameters of matchers, such as precision or recall of the generated matchings. Recently, however, ones started to realize that the extent to what precision and recall can be improved may be limited for general-purpose matching algorithms. Instead of designing new algorithms, there has been a shift towards matching combination and tuning methods. These works include YAM [11], systematic matching ensemble selection [12] or automatic tuning of the matcher parameters [15].

While there is a large body of works on schema matching, the post-matching reconciliation process (that is central to our work) has received little attention in the literature. Recently, there are some works [14, 17, 21] using pay-as-you-go integration method that establishes the initial matching and then incrementally improves matching quality. While the systems in [14, 21] rely on one user only, the framework in [17] relies on multiple users.

Schema Matching Network. The idea of exploiting the presence of a large set of schemas to improve the matchings has been studied before. Holistic matching [25] attempted to exploit statistical co-occurrences of attributes in different schemas and use them to derive complex correspondence. Whereas, corpus-based matching [16] attempted to use a 'corpus' of schemas to augment the evidences that improve exist matchings and exploit constraints between attributes by applying statistical techniques. Network level constraints, in particular the circle constraints, were originally considered in [1, 6] in which they study the establishment of semantic interoperability in a large-scale P2P network. In this paper, we study contextual information and integrity constraints (e.g., 1-1 and circle constraints) on top of the schema matching network.

Crowdsourcing. In recent years, crowdsourcing has become a promising methodology to overcome human-intensive computational tasks. Its benefits vary from unlimited labour resources of user community to cost-effective business models. The book [2] summarized problems and challenges in crowdsourcing as well as promising research directions for the future. A wide range of crowdsourcing platforms, which allows users to work together in a large-scale online community, have been developed such as Amazon Mechanical Turk and CloudCrowd.

On top of these platforms, there are also many crowdsourcing applications that have been built for specific domains. For example, in [26], the crowdsourcing is employed to validate the search results of automated image search on mobile devices. In [3], the authors leveraged the user CAPTCHAs inputs in web forms to recognize difficult words that cannot solved precisely by optical character recognition (OCR) programs.

Regarding the utilization of constraints, there are some previous works such as [5, 27]. In [27], the constraints were used to define the tasks for collaborative planning systems whereas in [5], the constraints were used to check worker quality by quantifying the consistency of worker answers. In our work, the constraints are used to adjust the error rate for reducing worker efforts.

8 Conclusions and Future Work

Using shared datasets in meaningful ways frequently requires interconnecting several sources, i.e., one needs to construct the attribute correspondences between the concerned schemas. The schema matching problem has, in this setting, a completely new aspect: there are more than two schemas to be matched and the schemas participate in a larger *schema matching network*. This network can provide contextual information to the particular matching tasks.

We have presented a crowdsourcing platform that is able to support schema matching tasks. The platform takes the candidate correspondences that are generated by pairwise schema matching and generates questions for crowd workers. The structure of the matching network can be exploited in many ways. First, as this is a contextual information about the particular matching problem, it can be used to generate questions that guide the crowd workers and help them to answer the questions more accurately. Second, natural constraints about the attribute correspondences at the level of the network enable to reduce the necessary efforts, as we demonstrated this through our experiments.

Our work opens up several future research directions. First, one can extend our notion of schema matching network and consider representing more general integrity constraints (e.g., functional dependencies or domain-specific constraints). Second, one can devise more applications which could be transformed into the schema matching network. While our work focuses on schema matching, our techniques, especially the constraint-based aggregation method, can be applied to other tasks such as entity resolution, business process matching, or Web service discovery.

Acknowledgment. This research has received funding from the NisB project - European Union's Seventh Framework Programme (grant agreement number 256955) and the PlanetData project - Network of Excellence (grant agreement number 257641).

References

[1] Aberer, K., Cudré-Mauroux, P., Hauswirth, M.: Start making sense: The Chatty Web approach for global semantic agreements. JWS, 89–114 (2003)
[2] von Ahn, L.: Human computation. In: DAC, pp. 418–419 (2009)
[3] von Ahn, L., Maurer, B., McMillen, C., Abraham, D., Blum, M.: Recaptcha: Human-based character recognition via web security measures. Science, 1465–1468 (2008)
[4] Bernstein, P.A., Madhavan, J., Rahm, E.: Generic Schema Matching, Ten Years Later. PVLDB, 695–701 (2011)
[5] Chen, K.T., Wu, C.C., Chang, Y.C., Lei, C.L.: A crowdsourceable qoe evaluation framework for multimedia content. In: MM, pp. 491–500 (2009)
[6] Cudré-Mauroux, P., Aberer, K., Feher, A.: Probabilistic message passing in peer data management systems. In: ICDE, p. 41 (2006)
[7] Das Sarma, A., Fang, L., Gupta, N., Halevy, A., Lee, H., Wu, F., Xin, R., Yu, C.: Finding related tables. In: SIGMOD, pp. 817–828 (2012)
[8] Dawid, A.P., Skene, A.M.: Maximum likelihood estimation of observer error-rates using the EM algorithm. J. R. Stat. Soc., 20–28 (1979)
[9] Di Lorenzo, G., Hacid, H., Paik, H.: y., Benatallah, B.: Data integration in mashups. In: SIGMOD, pp. 59–66 (2009)

10. Do, H., Rahm, E.: COMA: a system for flexible combination of schema matching approaches. In: PVLDB, pp. 610–621 (2002)
11. Duchateau, F., Coletta, R., Bellahsene, Z., Miller, R.J.: (Not) yet another matcher. In: CIKM. pp. 1537–1540 (2009)
12. Gal, A., Sagi, T.: Tuning the ensemble selection process of schema matchers. JIS, 845–859 (2010)
13. Gonzalez, H., Halevy, A.Y., Jensen, C.S., Langen, A., Madhavan, J., Shapley, R., Shen, W., Goldberg-Kidon, J.: Google fusion tables: web-centered data management and collaboration. In: SIGMOD, pp. 1061–1066 (2010)
14. Jeffery, S.R., Franklin, M.J., Halevy, A.Y.: Pay-as-you-go user feedback for dataspace systems. In: SIGMOD, pp. 847–860 (2008)
15. Lee, Y., Sayyadian, M., Doan, A., Rosenthal, A.S.: eTuner: tuning schema matching software using synthetic scenarios. JVLDB 16, 97–122 (2007)
16. Madhavan, J., Bernstein, P.A., Doan, A., Halevy, A.: Corpus-based schema matching. In: ICDE, pp. 57–68 (2005)
17. McCann, R., Shen, W.: Matching schemas in online communities: A web 2.0 approach. In: ICDE, pp. 110–119 (2008)
18. Nguyen, H., Fuxman, A., Paparizos, S., Freire, J., Agrawal, R.: Synthesizing products for online catalogs. PVLDB, 409–418 (2011)
19. Parameswaran, A.G., Garcia-Molina, H., Park, H., Polyzotis, N., Ramesh, A., Widom, J.: Crowdscreen: algorithms for filtering data with humans. In: SIGMOD, pp. 361–372 (2012)
20. Peukert, E., Eberius, J., Rahm, E.: AMC - A framework for modelling and comparing matching systems as matching processes. In: ICDE, pp. 1304–1307 (2011)
21. Qi, Y., Candan, K.S., Sapino, M.L.: Ficsr: feedback-based inconsistency resolution and query processing on misaligned data sources. In: SIGMOD, pp. 151–162 (2007)
22. Rahm, E., Bernstein, P.A.: A Survey of Approaches to Automatic Schema Matching. JVLDB, 334–350 (2001)
23. Sheng, V.S., Provost, F.: Get Another Label? Improving Data Quality and Data Mining Using Multiple, Noisy Labelers. In: SIGKDD, pp. 614–622 (2008)
24. Smith, K.P., Morse, M., Mork, P., Li, M., Rosenthal, A., Allen, D., Seligman, L., Wolf, C.: The role of schema matching in large enterprises. In: CIDR (2009)
25. Su, W., Wang, J., Lochovsky, F.: Holistic schema matching for web query interfaces. In: Ioannidis, Y., Scholl, M.H., Schmidt, J.W., Matthes, F., Hatzopoulos, M., Böhm, K., Kemper, A., Grust, T., Böhm, C. (eds.) EDBT 2006. LNCS, vol. 3896, pp. 77–94. Springer, Heidelberg (2006)
26. Yan, T., Kumar, V.: CrowdSearch: exploiting crowds for accurate real-time image search on mobile phones. In: MobiSys, pp. 77–90 (2010)
27. Zhang, H., Law, E., Miller, R., Gajos, K., Parkes, D., Horvitz, E.: Human computation tasks with global constraints. In: CHI, pp. 217–226 (2012)

Appendix

Compute Conditional Probability $Pr(X_{c_0}|\gamma_{1-1})$: According to Bayes theorem, $Pr(X_{c_0}|\gamma_{1-1}) = \frac{Pr(\gamma_{1-1}|X_{c_0}) \times Pr(X_{c_0})}{Pr(\gamma_{1-1})}$. Now we need to compute $Pr(\gamma_{1-1})$ and $Pr(\gamma_{1-1}|X_{c_0})$. Let denote $p_i = Pr(X_{c_i} = true)$, for short. In order to compute $Pr(\gamma_{1-1})$, we do following steps: (1) express $Pr(\gamma_{1-1})$ as the sum from the full joint of $\gamma_{1-1}, c_0, c_1, \ldots, c_k$, (2) express the joint as a product of conditionals. Formally, we have:

$$Pr(\gamma_{1-1}) = \sum_{c_0,c_1,\ldots,c_k} Pr(\gamma_{1-1}, X_{c_0}, X_{c_1}, \ldots, X_{c_k})$$
$$= \sum Pr(\gamma_{1-1}|X_{c_0}, X_{c_1}, \ldots, X_{c_k}) \times Pr(X_{c_0}, X_{c_1}, \ldots, Xc_k)$$
$$= 1 \times Pr(X_{c_0}, X_{c_1}, \ldots, X_{c_k}|m(X_{c_0}, X_{c_1}, \ldots, X_{c_k}) \le 1)$$
$$+ \Delta \times Pr(X_{c_0}, X_{c_1}, \ldots, X_{c_k}|m(X_{c_0}, X_{c_1}, \ldots, X_{c_k}) > 1)$$
$$= y + \Delta \times (1 - y)$$

where m is function counting the number of X_{c_i} assigned as *true*
$$y = \prod_{i=0}^{n}(1 - p_i) + \sum_{i=0}^{n}[p_i \prod_{j=0, j \ne i}^{n}(1 - p_j)]$$

Similar to computing $Pr(\gamma_{1-1})$, we also express $Pr(\gamma_{1-1}|X_{c_0})$ as the sum from the full joint of $\gamma_{1-1}, c_1, \ldots, c_k$ and then express the joint as a product of conditionals. After these steps, we have $Pr(\gamma_{1-1}|X_{c_0} = true) = x + \Delta \times (1 - x)$, where $x = \prod_{i=1}^{k}(1 - p_i)$. After having $Pr(\gamma_{1-1})$ and $Pr(\gamma_{1-1}|X_{c_0})$, we can compute $Pr(X_{c_0}|\gamma_{1-1})$ as in equation 4.

Compute Conditional Probability $Pr(X_{c_0}|\gamma_\cup)$: According to Bayes theorem, $Pr(X_{c_0}|\gamma_\cup) = \frac{Pr(\gamma_\cup|X_{c_0}) \times Pr(X_{c_0})}{Pr(\gamma_\cup)}$. In order to compute $Pr(\gamma_\cup|X_{c_0})$ and $Pr(\gamma_\cup)$, we also express $Pr(\gamma_\cup|X_{c_0})$ as the sum from the full joint of $\gamma_{1-1}, c_0, c_1, \ldots, c_k$ and then express the joint as a product of conditionals. After some transformations, we can obtain equation 6.

MFSV: A Truthfulness Determination Approach for Fact Statements[*]

Teng Wang[1,2], Qing Zhu[1,2], and Shan Wang[1,2]

[1] Key Laboratory of the Ministry of Education for Data Engineering and Knowledge Engineering, Renmin University of China, Beijing, China
[2] School of Information, Renmin University of China, Beijing, China
{wangteng,zq,swang}@ruc.edu.cn

Abstract. How to determine the truthfulness of a piece of information becomes an increasingly urgent need for users. In this paper, we propose a method called *MFSV*, to determine the truthfulness of *fact statements*. We first calculate the similarity between a piece of related information and the target fact statement and capture the credibility ranking of the related information through combining *importance ranking* and *popularity ranking*. Based on these, contributions of a piece of related information to the truthfulness determination is derived. Then we propose two methods to determine the truthfulness of the target fact statement. At last, we run comprehensive experiments to show *MFSV*'s availability and high accuracy.

Keywords: Fact statements, credibility ranking, similarity, truthfulness.

1 Introduction

Untruthful information spreads on the Web, which may mislead other users and have a negative impact on user experience. It is required to determine the truthfulness of a piece of information. Information is mainly loaded by sentences. The sentences state *facts*, rather than *opinions*, are called fact statements [10]. In this paper, we mainly focus on positive fact statement. Fact statements, which state correct objective facts, are trustful fact statements, others are called untruthful fact statements. Before determining the truthfulness of a fact statement, a user should specify some part(s) of the fact statement he/she is not sure about. The part(s) is/are called the doubt unit(s) of the fact statement[10]. If the doubt unit(s) is/are specified, the fact statement can be regarded as an answer to a question. If there is only one correct answer to the question, the fact statement to the question is an unique-answer fact statement; otherwise, it is a multi-answer one. In [9][10], the trustful fact statement is picked out from the target fact statement and the alternative fact statements. There are three limitations in these studies: (i) The doubt unit(s) must be specified, otherwise the alternative fact statements

---

[*] This work is partly supported by the Important National Science & Technology Specific Projects of China (Grant No.2010ZX01042-001-002), the National Natural Science Foundation of China (Grant No.61070053), the Graduates Science Foundation of Renmin University of China(Grant No.12XNH177).

can't be found. (ii) If too much information is included by the doubt unit(s), it is very hard to find proper alternative fact statements. (iii) These methods can not be used for multi-answer fact statements, since only one fact statement is considered trustful.

The contributions and the rest of the paper are organized as follows. Section 2 briefly summarize the related works. Section 3 describes how to determine the truthfulness of a fact statement using $MFSV$. Experiments and analysis are shown in Section 4. At last, we conclude this paper.

2 Related Works

Some researchers hold the opinion that credible sources are very likely to present trustful information. They analyze the features (e.g., page keywords, page title, page style) of credible web pages[1][2][3]. Exploiting the analytical results, users can determine whether a web page is credible or not. Other researchers focus on spam web pages detection for filtering low quality web pages [4][5]. But incorrect information may be presented on non-spam pages.

Study [6] propose a method to determine the truthfulness of a piece of news. In this method, the information related to the piece of news is collected from reputable news web site. Analyzing the consistence between the news and the related information, the truthfulness of the news can be determined. In [7], the focus is the truthfulness determination of an event. The relatedness between the event and its related information captured from certain web sites is measured, and the truthfulness of the event is determined. Studies in [6] and [7] are used for domain-dependent information. In [8], Honto?search1.0 is proposed to help users to determine an uncertain fact. In Honto?search1.0, sentiment distribution analysis and popularity evolution analysis are the key factors in helping users to determine the target uncertain fact. Honto?search2.0 is proposed in [9]. The objective of the system is to help users to efficiently judge the credibility by comparing other facts related to the input uncertain. Verify[10] can determine a fact statement through finding alternative fact statements and ranking these fact statements, the one on the highest position is the trustful one.

3 Proposed Solution

The goal of this paper is to determine whether a fact statement is trustful or not, even if the fact statement is a multi-answer one.

3.1 Similarity Measurement

In this section, we discuss how to measure the similarity between a piece of related information (derived from the search engine) and the target fact statement.

Necessary Sentence Generation. Not all words are necessary for the fact statement truthfulness determination. We call the words, which contributes to the truthfulness determination *necessary words*. Given a piece of information r_i related to the fact statement fs, we use N_i and ns_i to denote the collection of necessary words and the necessary sentence of r_i respectively. We use the following steps to find the necessary words: **I.** Find the consecutive sentences c_i of r_i. N_i can be extracted from c_i, since c_i is the shortest consecutive sentences which include r_i's keywords set K_{r_i}. **II.** Use Stanford Parser[1] to find the grammatical relationships between words in c_i. We divide the 52 grammatical relationships in Stanford Parser into two categories based on their importance to sentence skeleton: *essential grammatical relationships*, represented by R_e, and *unessential grammatical relationships* represented by R_o. **III.** Set N_i to K_{r_i} and use the following heuristic rules to find N_i. We use $D_i = \{d_{i1}, \ldots, d_{im}\}(1 \leq m)$ to denote the collection of the grammatical relationships of c_i. The heuristic rules used for N_i extraction are as follows: (i) If $d_{ij}(1 \leq j \leq m)$ is essential, $d_{ij}.dependent$ and $d_{ij}.governor$ are put into N_i. (ii) If $d_{ij}(1 \leq j \leq m)$ is unessential and $d_{ij}.dependent \in N_i$, $d_{ij}.governor$ are put into N_i. We sort the words in N_i by their positions in c_i and get a order. According to this order, we assemble the words in N_i, thus, the necessary sentence ns_i is generated.

Similarity Computation. The similarity between r_i and fs can be replaced by the similarity of ns_i and fs. We refine the method in [11] to calculate the similarity. First, we construct semantic vectors and order vectors by finding *best matching word* of the target word for ns_i and fs, and calculate the semantic similarity and order similarity respectively; then, combining the semantic similarity and order similarity, we get the overall similarity between ns_i and fs. We adopt a word similarity computing method in [11].

Semantic Similarity Computation. The semantic similarity between ns_i and fs is calculated by the cosine similarity of their semantic vectors. We first delete the stop words in ns_i and fs, and then get the words collections of ns_i and fs. $W_1 = \{w_{11}, \ldots, w_{1n_1}\}$ and $W_2 = \{w_{21}, \ldots, w_{2n_2}\}$ denote the two word collections of ns_i and fs respectively. We set $W = W_1 \cup W_2$ and $W = \{w_1, \ldots, w_k\}$. We use $V_1 = \{v_{11}, \ldots, v_{1k}\}$ and $V_2 = \{v_{21}, \ldots, v_{2k}\}$ to denote the semantic vectors of ns_i and fs respectively. The rules to work v_{1i} out are as follows: (i) If $w_i \in W_1$, $v_{1i} = 1$. (ii) If $w_i \notin W_1$, we find the best matching word(w_{bm}) of w_i from W_1 and set $v_{1i} = S_w(w_i, w_{bm})$. Especially, if w_{bm} does not exist, $v_{1i} = 0$. In semantic vector construction, the value of ζ is 0.2. The semantic similarity between ns_i and fs is calculated by Equation 1.

$$S_s(ns_i, fs) = \frac{V_1 \cdot V_2}{\parallel V_1 \parallel \cdot \parallel V_2 \parallel} \qquad (1)$$

Order Similarity Computation. We measure the order similarity of ns_i and fs based on their order vectors. $O_1 = \{o_{11}, \ldots, o_{1k}\}$ and $O_2 = \{o_{21}, \ldots, o_{2k}\}$ denote the order vectors of ns_i and fs. We use the following rules to work out o_{1i}: (i) If $w_i \in W_1$, o_{1i} is the position of w_i in ns_i. (ii) If $w_i \notin W_1$, we find the best matching word(w_{bm}) of w_i from W_1 by Algorithm 2. If w_{bm} exists, v_{1i} is the position of w_{bm} in ns_i; if not, $o_{1i} = 0$. Especially, in constructing order vector, the value of ζ in Algorithm 2 is 0.4. We use equation.2 to get the order similarity between ns_i and fs.

$$S_o(ns_i, fs) = 1 - \frac{\parallel O_1 - O_2 \parallel}{\parallel O_1 + O_2 \parallel} \qquad (2)$$

---

[1] http://nlp.stanford.edu/software/stanford-dependencies.shtml

Overall Similarity Computation. The overall similarity $S(ns_i, fs)$ between ns_i and fs can be calculated by combination of $S_s(ns_i, fs)$ and $S_o(ns_i, fs)$ through equation.3. The optimal value of θ is 0.85 in equation.3.

$$S(sn_i, fs) = \begin{cases} \theta S_s(sn_i, fs) + (1-\theta) S_o(sn_i, fs) & \text{if } r_i \text{ is positive on } fs \\ -(\theta S_s(sn_i, fs) + (1-\theta) S_o(sn_i, fs)) & \text{if } r_i \text{ is negative on } fs \end{cases} \quad (3)$$

3.2 Credibility Ranking

Generally, credible sources are likely to present trustful information. In addition, if an information source is important and popular, it may be credible[12]. In the following, we first introduce the importance ranking and the popularity ranking, then we merge the two rankings to get the credibility ranking.

Importance Ranking. The information related to the target fact statement is captured by a search engine, it appears in order. We combine the order and the pagerank level values to capture the importance ranking of the related information. Given two pieces of information r_i and r_j related to the target fact statement, we use pl_i and pl_j to denote the pagerank level values of the web pages from which r_i and r_j are derived. *Irank* is used to denote the importance ranking of the related information. $Irank_i$ is the position of r_i in *Irank*. If $pl_i \gtrsim pl_j$, $Irank_i \lesssim Irank_j$; If $pl_i = pl_j$ and $i \lesssim j$, $Irank_i \lesssim Irank_j$.

Popularity Ranking. Alexa ranking[2] is introduced to measure the popularity of the web sites from which the related information is derived. Given a fact statement fs and the related information collection R, $Alexa_i$ is used to denote the position of the web site, which $r_i \in R$ is derived from, in Alexa ranking. However, Alexa ranking is an absolute ranking. Given r_i and r_j, the gap between $Alexa_i$ and $Alexa_j$ may be very large. We adopt two methods to get different popularity rankings. (i) We get the popularity ranking, represented by *Prank*, by sorting $Alexa_i (1 \leq i \leq n)$ on ascending order; (ii) The popularity ranking, represented by *GRrank*, is captured by linearly mapping $Alexa_i (1 \leq i \leq n)$ into the range from 1 to n.

Credibility Ranking. We make use of the classical ranking merging algorithms (*Borda* and *Footrule*) to get the credibility ranking. Borda[13][14] is a *positional* algorithm. Given *Irank*, *Prank* and the related information collection R, $B_{Irank}(r_i)$ denotes the *Borda* scores of $r_i \in R$ on *Irank* and $B_{Irank}(r_i)$ is the number of the related information which is below r_i in *Irank*. Similarly, $B_{Prank}(r_i)$ can be derived. $B(r_i)$ is the total scores of r_i on the two rankings, which is the sum of $B_{Irank}(r_i)$ and $B_{Prank}(r_i)$.

Footrule[13] is a merging algorithm based on the distance. Given *Irank*, *Prank* and the related information collection R. We construct an complete bipartite graph $G(V, E, W)$. V is composed of R and P. P is the collection of positions and $P = \{1, \dots, |R|\}$. E is the collection of edges. W is the collection of the weights of the edges. Given a edge $< r_i, p > (r_i \in R, p \in P)$, the weight $w(r_i, p) = |Irank_i - p| + |Prank_i - p|$. By finding the complete matching at minimal cost on the graph, the merging result of

---
[2] https://www.alexa.com

Irank and *Prank* can be derived. We get an importance ranking (*Irank*) and two popularity rankings (*Prank* and *GPrank*), four credibility rankings according to the merging algorithm *Borda* and *Footrule*, they are *CBrank* (merge *Irank* and *Prank* using *Borda*), *CBGrank* (merge *Irank* and *PGrank* using *Borda*), *CFrank* (merge *Irank* and *Prank* using *Footrule*), *CFGrank* (merge *Irank* and *PGrank* using *Footrule*).

3.3 Fact Statement Determination

The information related to the target fact statement can be divided into three categories: *positive*, *negative* and *neutral*, according to the similarity between the related information and the target fact statement. Given a fact statement fs and the related information collection R, we use R_{pos}, R_{neg} and R_{neu} to denote the collections of positive, negative and neutral related information respectively. With the help of κ, we get R_{pos}, R_{neg} and R_{neu}. If $S(r_i, fs) \geq \kappa$, $r_i \in R_{pos}$; if $|S(r_i, fs)| \lesssim \kappa$, $r_i \in R_{neu}$; if $S(r_i, fs) \leq 0$ and $|S(r_i, fs)| \geq \kappa$, $r_i \in R_{neg}$. The optimal value of κ is evaluated by experiments.

Combining the similarity and the credibility ranking, we measure *contributions* of a piece of related information to the truthfulness determination. Given a piece of information r_i related to the target fact statement fs, the contributions of r_i to the truthfulness determination of fs is defined as $S(r_i, fs)/Crank_i$. Here, $Crank_i$ is the credibility ranking value of r_i. Then we propose two ways to determine the truthfulness of the target fact statement: *baseline determination method* and *SVM-based determination method*.

Baseline Determination Method. We believe that if a fact statement is trustful, the contributions of the positive related information should be larger than that of the negative related information. According to this idea, we propose baseline determination method (*BMD*). The procedure of determining the truthfulness of the target fact statement are: **I.** The positive and negative contributions of the related information are worked out respectively. The positive contributions are the sum of the contributions of the related information in S_{pos}. Similarly, the negative contributions can be worked out. **II.** The sum of the contributions of positive and negative related information is worked out. If it is larger or equal to δ, we think the target fact statement is trustful; if not, it is untruthful. Here, δ is a constant and the optimal value of δ is evaluated by experiments.

SVM-Based Determination Method. Classification method can be used to determine the truthfulness of a fact statement. In this section, we make use of SVM model to predict the classification of a fact statement and propose SVM-based determination method (*SVM-DM*). In this method, some fact statements, whose truthfulness is certain, are chosen as train set and the classification model is obtained. Using the classification model, we predict the classification of the fact statement whose truthfulness is needed to be determined. Contributions of positive, neutral, and negative to a fact statement are considered as classification features; and the classification vector of the fact statement is composed of the contributions of positive, neutral, and negative related information. In order to avoid features in larger numeric ranges dominating those in smaller numeric ranges, we linearly scale each feature to the range [-1,1]. We chose **RBF** as the kernel of SVM classification.

4 Experiments

We generate a synthetic dataset according to [10]. The dataset is composed of 50 trustful fact statements and 50 untruthful fact statements. These fact statements are fetched from Trec2007[4]. Among trustful fact statements, 30 are unique-answer ones and the rest are multi-answer ones. For each fact statement, we use Yahoo boss 2.0[5] to collect the top-150 search results as the related information. 11 experienced users, who are graduate students and experienced Internet users, help to mark the search results.

4.1 Evaluations of Key Parameters

Distribution of Related Information. We use P_f to denote the percent of the related information including the meanings of the target fact statements. fs_u and fs_m denote trustful unique-answer fact statement and trustful multi-answer fact statement respectively. Fig.1 shows P_f values when n changes. Here, n is the number of the related information considered in the truthfulness determination of a fact statement. It can be seen that, P_f for fs_u decreases with the increase of n. For a multi-answer fact statement, there are more than one correct answer to the question corresponding to the multi-answer fact statement. Thus, P_f for fs_u is always larger than P_f for fs_m. From the experiment, we can see when n is larger, more neutral or negative related information comes out.

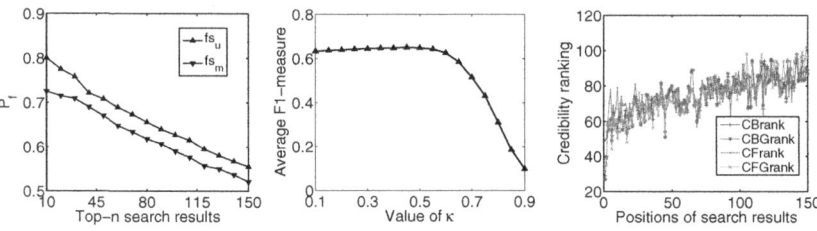

Fig. 1. Distribution of the related information

Fig. 2. F1-measure on κ

Fig. 3. Distributions of the credibility rankings

The Optimal Value of κ Detection. We measure the F1-measure values for each category at different values of κ. Then, we calculate the average of the F1-measure values for the three categories at different κ values. We believe the optimal value of κ is the value which makes the average F1-measure value greatest. Fig.2 shows the average F1-measure value when the value of κ varies. It can be seen that, the average F1-measure value reaches the peak (near 0.65) when $\kappa = 0.5$. Thus, the optimal value of κ is set to 0.5, also the default value in following experiments.

Distribution of Credibility Ranking. Fig3 show the distributions of the four credibility rankings when $n = 150$. The x-axis is the positions of the related information, and y-axis is the average of credibility ranking values of the related information at corresponding positions. From this figure, the rank values are thickly located on [20-120].

---
[3] http://trec.nist.gov
[4] http://boss.yahoo.com

It means the importance ranking and the popularity ranking can not replace each other. And the importance ranking or the popularity ranking can not replace credibility ranking. In addition, we can see some related information, which is at higher positions in the appearing order, is at lower positions in credibility ranking. It means that some related information at higher positions in the appearing order may be not credible. That is consistent with our observation. Since the gaps of Alexa ranking is considered in *CBGrank* and *CFGrank*, the ranges of *CBGrank* and *CFGrank* are larger than those of *CBrnak* and *CFrank*.

4.2 Evaluation of Determination Methods

Baseline Determination Method. The portion of the considered related information δ and the adopted credibility ranking influence the determination precision. By experiments, we find when *CFGrank* is adopted, the method has the best performance. Fig.4 shows the precision on δ and n, when the credibility ranking is *CFGrank*. When $\delta = 0.9$ and $n = 60$, the precision reaches the peak (0.74). Fig.5 shows the precision on δ on different credibility rankings, when $n = 60$. *CFGrank* and *CBGrank* can bring higher precision than *CFrank* and *CBrank*. With the increase of δ when $\delta \lesssim 0.9$, the precision increases and the precision decreases with the increase of δ, when $\delta \gtrsim 0.9$. Fig.6 shows the precision on n and four credibility rankings when $\delta = 0.9$. It can be seen that, *CFGrank* and *CBGrank* bring higher precision peak than *CFrank* and *CBrank*.

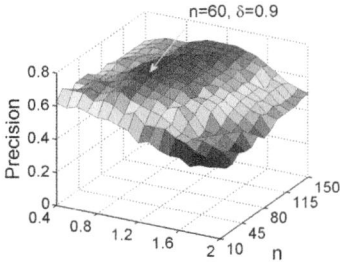

Fig. 4. Precision of *BDM* on n and δ

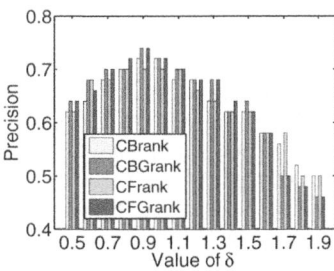

Fig. 5. Precision of *BDM* on δ

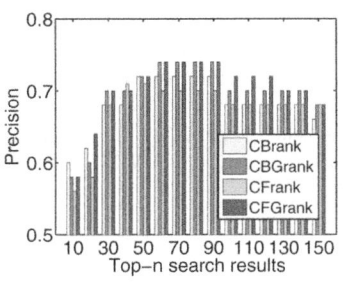

Fig. 6. Precision of *BDM* on n

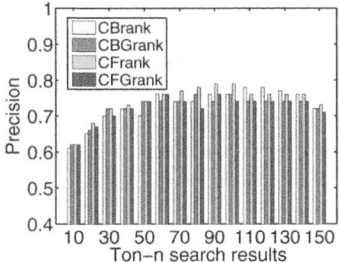

Fig. 7. Precision of *SVM − DM*

SVM-Based Determination Method. We adopt libsvm[5], RBF kernel function and three-fold cross validation to get the precision. Fig7 shows the precision on n and the four credibility rankings, it can be seen that the precision reaches the peak (0.79), when *CFrank* is adopted. Regardless of which one is adopted in the four credibility rankings, the precision first increases and then decreases, with the increase of n. Especially, when $n = 90$ and *CFrank* is adopted, we have the highest precision.

5 Conclusion and Future Work

In this paper, we propose a new method *MFSV* to determine the truthfulness of a fact statement. The results of experiments show *MFSV* is available and can be used for multi-answer fact statements. However, we just focus on domain-independent fact statements and ignore the domain knowledge. In the future, we will focus on the truthfulness determination of domain-dependent fact statements and we believe the usage of domain knowledge can help to determine a fact statement more accurately.

Acknowledgments. We would like to thank Prof. Weiyi Meng from Binghamton University for his help on this work.

References

1. McKnight, D.H., Kacmar, J.: Factors and effects of information credibility. In: ICEC 2007, pp. 423–432 (2007)
2. Schwarz, J., Morris, M.R.: Augmenting web pages and search results to scport credibility assessment. In: CHI 2011, pp. 1245–1254 (2011)
3. Lucassen, T., Schraagen, J.M.: Trust in Wikipedia: how users trust information from an unknown source. In: WICOW 2010, pp. 19–26 (2010)
4. Gyongyi, Z., Garcia-Molina, G., Pedersen, J.: Combating web spam with TrustRank. In: VLDB 2004, pp. 576–587 (2004)
5. Ntoulas, A., Najork, M., Manasse, M., Fetterly, D.: Detecting spam web pages through content analysis. In: WWW 2006, pp. 83–92 (2006)
6. Nagura, R., Seki, Y., Kando, N., Aono, M.: A method of rating the credibility of news documents on the web. In: SIGIR 2006, pp. 683–684 (2006)
7. Lee, R., Kitayama, D., Sumiya, K.: Web-based evidence excavation to explore the authenticity of local events. In: WICOW 2008, pp. 63–66 (2008)
8. Yamamoto, Y., Tezuka, T., Jatowt, A., Tanaka, K.: Supporting Judgment of Fact Trustworthiness Considering Temporal and Sentimental Aspects. In: Bailey, J., Maier, D., Schewe, K.-D., Thalheim, B., Wang, X.S. (eds.) WISE 2008. LNCS, vol. 5175, pp. 206–220. Springer, Heidelberg (2008)
9. Yamamoto, Y., Tanaka, K.: Finding Comparative Facts and Aspects for Judging the Credibility of Uncertain Facts. In: Vossen, G., Long, D.D.E., Yu, J.X. (eds.) WISE 2009. LNCS, vol. 5802, pp. 291–305. Springer, Heidelberg (2009)

---

[5] http://www.csie.ntu.edu.tw/~cjlin

10. Li, X., Meng, W., Yu, C.: T-verifier: verifying truthfulness of fact statements. In: ICDE 2011, pp. 63–74 (2011)
11. Li, Y., McLean, D., Bandar, Z., O'Shea, J., Crockett, K.: Sentence similarity based on semantic nets and corpus statistics. IEEE TKDE 18(8), 1138–1150 (2006)
12. Schwarz, J., Morris, M.R.: Augmenting Web pages and search results to support credibility assessment. In: CHI 2011, pp. 1245–1254 (2011)
13. Dwork, C., Kumary, R., Naorz, M., Sivakumarx, D.: Rank aggregation methods for the Web. In: WWW 2001, pp. 613–622 (2001)
14. Young, H.P.: An axiomatization of Borda's rule. J. Economic Theory 9, 43–52 (1974)

A Mechanism for Stream Program Performance Recovery in Resource Limited Compute Clusters

Miyuru Dayarathna[1] and Toyotaro Suzumura[1,2]

[1] Department of Computer Science, Tokyo Institute of Technology, 2-12-1 Ookayama,
Meguro-ku, Tokyo 152-8552, Japan
dayarathna.m.aa@m.titech.ac.jp, suzumura@cs.titech.ac.jp
[2] IBM Research - Tokyo

Abstract. Replication, the widely adapted technique for crash fault tolerance introduces additional infrastructural costs for resource limited clusters. In this paper we take a different approach for maintaining stream program performance during crash failures. It is based on the concepts of automatic code generation. Albatross, the middleware we introduce for this task maintains the same performance level during crash failures based on predetermined priority values assigned to each stream program. Albatross constructs different versions of the input stream programs (sample programs) with different levels of performance characteristics, and assigns the best performing programs for normal operations. During node failure or node recovery, potential use of a different version of sample program is evaluated in order to bring the performance of each job back to its original level. We evaluated effectiveness of this approach with three different real world stream computing applications on System S distributed stream processing platform. We show that our approach is capable of maintaining stream program performance even if half of the nodes of the cluster has been crashed using both Apnoea, and Regex applications.

Keywords: stream computing, data-intensive computing, reliability, highly availability, performance, auto-scaling, autonomic computing, automatic code generation.

1 Introduction

Highly availability is a key challenge faced by stream processing systems in providing continuous services. Crash faults such as operating system halts, power outages, virtual machine crashes, etc. may paralyze or take an entire application out of service. Crash faults take more time to recover since some of those need direct intervention from system administrators [18]. The widely adapted solution for recovering from crash faults in stream computing systems has been physical replication [8]. These techniques require k replicas to tolerate up to $(k - 1)$ simultaneous failures [15]. Maintaining such large number of backup nodes costs a lot in terms of electricity, rack space, cabling, and ventilation. These are the key problems faced by cloud data centers.

Balancing local stream processing load with public cloud (i.e., use of Hybrid Cloud [5]) would be a solution for maintaining stream program performance [13][11]. Yet such approaches require access to public clouds which makes it impossible to use such solutions in certain applications. For example, if the stream application deals with sensitive information (e.g., health care, national defense, etc.) it would be very difficult to follow such solution. Moreover, certain stream processing systems have license, and software issues [11] which makes it impossible to deploy them in hybrid cloud environments.

We have observed that data flow graphs tend to maintain similar shaped performance curves (i.e., similar performance characteristics) within similar stream environment conditions [4]. This indicates that relative performance of such data stream graphs is a predictable quantity.

Considering the obstacles associated with replication for maintaining performance, and characteristics of data stream programs; we introduce a different, more efficient solution for maintaining stream program performance which is applicable in resource limited stream computing clusters. Our approach is based on automatic code generation. Specifically, we generate a variety of data flow graphs (we call these "sample programs") with each giving different performance characteristics for a set of stream programs that are to be deployed in the node cluster. We select sample programs (one per input program) with consistent high throughput performance compared to the input programs, and run them in the cluster. During a node failure (e.g., system crash/shutdown) which affects a sub set of nodes in the cluster, we evaluate feasibility of performance maintenance using the existing stream applications and introduce different sample applications which produce better performance in the new environment. Our approach assigns priority to each input program and tries to maintain performance of high prioritized programs.

Albatross, the middleware on which we implemented the above approach monitors the performance status of each node, stream processing jobs and conducts the switching of sample programs appropriately to respond to the changes happen in the node cluster.

1.1 Contributions

1. *Code generation for performance* - We propose a new method for maintaining system performance during node crashes using automatic code generation. This approach reduces the requirement of keeping additional backup nodes.
2. *Switching between different versions* - We describe a method for swapping different versions of data stream programs with minimal effect on their runtime performance and without loss of integrity of data.

We implemented and evaluated our approach on System S [1] which is a large-scale, distributed stream processing middleware developed by IBM Research. In a cluster of 8 nodes we observed that our method is able to partially restore (complete restoration for some applications) performance compared to naive deployment. Our method was successful in maintaining performance of two different stream applications even when the number of nodes were reduced by half due to crash faults.

2 Background

We briefly describe operator-based data stream processing systems [12][4] and what we mean by resource limited compute clusters below.

2.1 Resource Limited Stream Compute Clusters

While our approach can be applied in public compute clusters, dynamic resource allocation has become a great problem in resource limited small (typically private) compute clusters. What we mean by use of the term "Resource Limited Stream Compute Clusters" is a compute cluster with fixed set of nodes which cannot be expanded dynamically. Our emphasize is on the number of available nodes rather than the amount of available resources in a particular node of such cluster because we believe that it does not matter how much resources a node has if it crashes suddenly. Such compute clusters are widely run by financial, academic, and health care [3] institutions that operate variety of stream processing applications. An important feature of these clusters is that it takes considerable amount of time to replace a defected node in such private clusters compared to public clusters (e.g., Amazon EC2) which can easily provision nodes on demand. We believe that our approach for performance recovery is best applicable to such resource limited clusters.

2.2 System S and SPADE

We use System S which is an operator-based [12][4], large-scale, distributed data stream processing middleware for implementing Albatross prototype [1][17]. System S uses an operator-based programming language called SPADE [6] for defining data flow graphs. SPADE depends on a code-generation framework instead of using type-generic operator implementations. SPADE has a set of built-in operators (BIOP), and also supports for creating customized operators (i.e., User Defined Operators (UDOP)) which allow for extending the SPADE language. Communication between operators is specified as streams. SPADE compiler fuses operators into one or more Processing Elements (PEs) during compilation process. A PE is a software component which spawns a unique process during the running time of its corresponding program. System S scheduler makes the connection between PEs during runtime. Out of the BIOPs used for implementing the sample programs described in this paper (abbreviations we use are shown in parenthesis), Source (S) creates a stream from data coming from an external source, Sink (SI) converts a stream into a flow of tuples that can be used by external components. Functor (F) performs tuple-level manipulations (e.g., filtering, mapping, projection, etc.). Aggregate (AG) groups, and summarizes incoming tuples. Split (SP) splits a stream into multiple output streams. Join (J) operator correlates two streams using join predicates.

2.3 Automatic Sample Program Generation with Hirundo

On a mechanism called Hirundo [4], we developed techniques for automatically generate data flow graphs with varied performance characteristics for a given stream application. Hirundo analyses and identifies the structure of a stream program, and transforms the program's data flow graph in to many different versions. While an in depth description of the Hirundo's code generator is out of the scope of this paper, we provide a brief description of its sample program generation. We explain sample program generation by Hirundo below by taking two stream applications called Regex, and Apnoea. We do not use VWAP in the following explanation because Regex and VWAP both have similar structure. Note that our aim in Hirundo was to produce optimized stream program(s) for a given stream application which is different from the aim of this work.

Stream Applications. In this paper we use three real world stream applications as examples. The first application is called "Regex" is a regular expression based data transformation application (The data flow graph is shown in Figure 1(a)). It consists of five operators and it converts date portion of `datetime` tuples represented as 2011-07-11 to 11-JUL-2011. Moreover, all the "00"s in the time portion of the tuples are changed to "22"s.

The second application is a Volume Weighted Average Price (VWAP) application. VWAP is calculated as the ratio of the value traded and the volume traded within a specified time period. The application we used (illustrated at [1]) is part of a larger financial trading application. The data flow graph of VWAP application consists of five operators (see Figure 1(b)).

The third application (shown in Figure 1(c)) is called Apnoea [3] which is part of a framework for real time analysis of time series physiological data streams extracted from a range of medical equipments to detect clinically significant conditions of premature infants.

Program Transformation. As shown in Figure 2(a), the original Regex program consists of five operators and is a linear Directed Acyclic Graph (DAG). This program is represented as S_F1_F2_F3_SI by Hirundo. Hirundo's program transformation algorithm traverses this program three operator blocks at a time and produces transformed operator blocks as shown in Figure 2(b). We call this process as Tri-Operator (Tri-OP) transformation [4]. During the transformation number of operators of some (or all) operators in the original data flow graph are changed. E.g., Number of F1, and F2 operators in the input program are increased by 4, and 2 respectively. Maximum operator count a particular operator could have is represented as depth (d). Number of sample programs generated increases with increase of depth value. E.g., When used for transforming Regex application, for a depth of 4 Hirundo generated 32 sample programs; where as a depth of 6 resulted in 136 sample programs. The generated operator blocks are stitched together in a process called "Operator Blocks Fusion". This results in sample programs like the example shown in Figure 2(c). This entire process takes only few seconds.

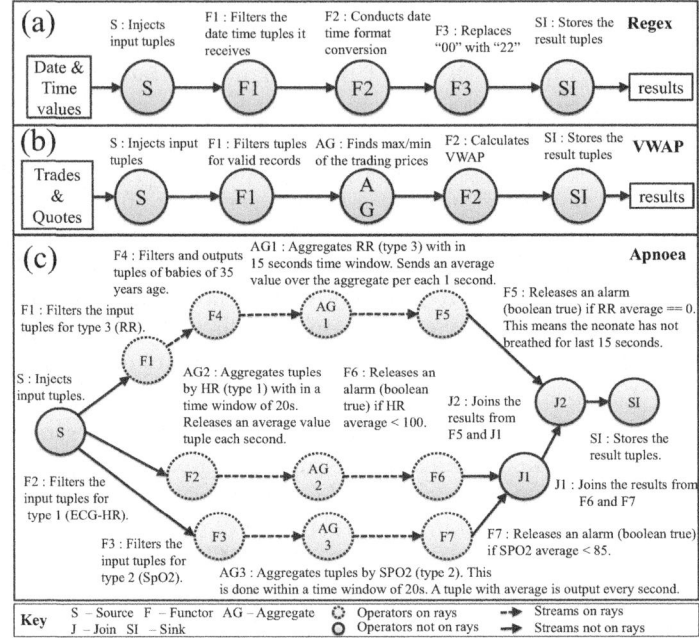

Fig. 1. Data flow graphs of example input stream applications

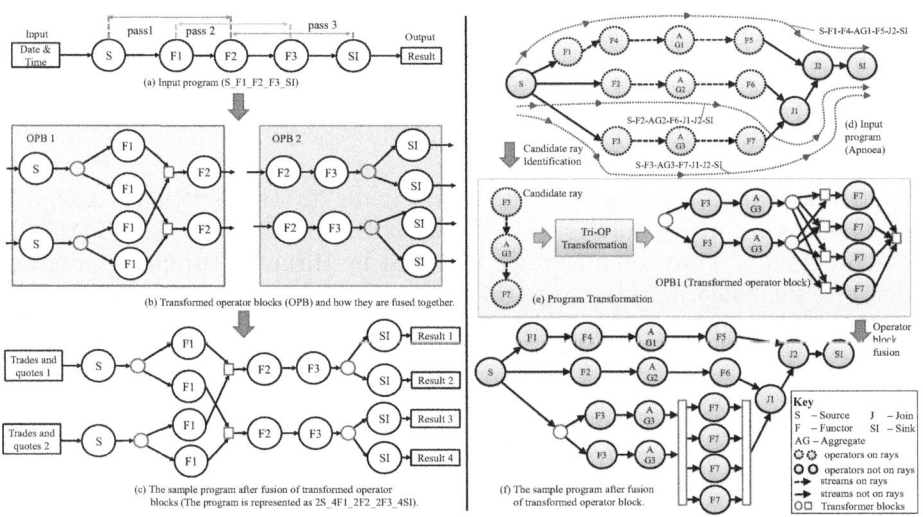

Fig. 2. An example for data flow graph transformation in Hirundo. (a) Data flow graph of Regex application. (b) Transformed operator blocks output by Hirundo for Regex application. (c) Sample program generated for Regex. (d) Data flow graph of Apnoea. (e) Transformation of the ray F3-AG3-F7. (f) A Sample program generated for Apnoea.

Since Apnoea is a multipath DAG its most expensive path is identified by profiling the Apnoea input application for a short period in the intended cluster. Then the ray (high lighted in dotted lines) of the most expensive path (S-F3-AG3-F7-J1-J2-SI) is transformed using Tri-OP transformation. The transformed operator block (shown in the right side of the Figure 2(e)) is stitched with the original DAG through operator block fusion which yields the sample application shown in Figure 2(f).

3 Related Work

Stream program performance maintenance is closely related with works conducted on highly availability, and fault tolerance of stream processing systems. Hwang et al. [9], and Gu et al. [7] have described two main approaches for highly availability called Active Standby (AS), and Passive Standby (PS). In AS two or more copies of a job are run independently on different machines. In PS a primary copy periodically checkpoints its state to another machine and uses that copy for recovery during failures. Both these approaches involve replication. Our approach is completely different from AS since, ours is based on automatic code generation and does not require backup machines. However, we share a common feature with PS since we keep incoming data tuples in main memory during the crash recovery period to avoid data losses.

Recently, MapReduce has been used for creating streaming applications. Hadoop Streaming [16] is one such implementation where Unix standard streams have been used as the interface between Hadoop, and user programs. However, it should be noted that standard Unix streams (e.g., stdin, sdout, stderr, etc.) represent only few examples of use of streams compared to the wide variety of application scenarios addressed by dedicated stream processing middleware such as System S. Logothetis et al. describe an in-situ MapReduce architecture that mines data (i.e., logs) on location where it appears [14]. They employ load shedding techniques to improve fidelity under limited CPU resources. In contrast to such load shedding techniques [2] currently our approach does not discard parts of the incoming data to recover the lost performance. Instead, Albatross transforms the stream application to a different form which could process the incoming data as it is.

Khandekar et al. [12], and Wolf et al. [17] discuss optimizing stream job performance in the context of operator fusion (COLA) and operator scheduling (SODA) respectively. However, their approaches do not focus on maintaining an agreed level of performance. Instead their focus is on performance improvement. Furthermore, COLA works on finding the optimal fusion of processing elements (using compiler outputs) of a single program, whereas Albatross constructs and uses many versions of input program(s) during its operation. This is because Albatross needs different versions with different performance levels during performance maintenance process.

4 Approach for Performance Maintenance

We formerly define our approach below. The notation used in our description is explained in Table 1.

Table 1. Notation

Notation	Description
E_N	Stream processing environment with N nodes. (N > 1)
S	Set of input stream programs. ($|S| = n$, n>0)
u	Performance margin for stream processing environment E. ($u \in \{0, ..., 100\}$, $u \in \mathbb{R}+$). This value is calculated by Albatross using current performance information of E (As described in Section 4.2).
m	A performance window set by user. ($m \in \{0, ..., 100\}$, $m \in \mathbb{R}+$)
M_i	Input stream program priority margin. ($\forall i,j$ where $i,j \in \mathbb{N}$, $M_i, M_j \in \mathbb{R}+$, $i,j \in \{0,...,(n-1)\}$, $i,j \in \mathbb{N}$, $M_i \neq M_j$, $\sum_{i=0}^{n-1} M_i = 100$). Priority margin is used for ranking input programs based on their importance. This value needs to be specified by user prior running Albatross.
r	Calibration run. If $r = 0$ it is a normal mode run. When Albatross is deployed in its usual operation its called normal mode run.
$P_S^{(r)}$	Sample program set generated for S during calibration run r. ($r \in \mathbb{N}$, $r \neq 0$). A calibration is a running of entire sample program space with the intention of obtaining the performance information.
$P_S^{(0)}$	Sample program set generated for S during normal mode run. Here, 0 in $P_{S_i}^{(0)}$ represents a normal mode run of Albatross.
X_N	Selected sample program set. ($X_{N_i} \in P_{S_i}^{(0)}$, $|X_N| = n$, $|X_{N_i}| = 1$)
perf(x)	Predicate for performance (e.g., throughput, elapsed time, etc.) of sample program x

Given an E_N which receives a steady flow of input data streams, a set of input programs S ($|S| = n$) each having a priority margin M_i, a set of sample programs generated during past calibration sessions (each represented as r) of S programs denoted by $P_S^{(r)}$, Albatross generates $P_{S_i}^{(0)}$ for each S_i ($i \in \{0,...,(n-1)\}$, $i \in \mathbb{N}$). Then, it selects sample program set X_N considering the empirical performance information of each $P_{S_i}^{(r)}$, compiles and runs them in E_N. The algorithm that selects each sample program X_{N_i} is described in next section.

During a sudden node failure which results in a different environment $E_{N'}$ with N' nodes ($N'<N$; $N'>0$), Albatross selects $X_{N'}$ ($X_{N'_i} \subset P_{S_i}^{(0)}$, $|X_{N'}| = n$, $|X_{N'_i}| = 1$) sample programs if $M_i \geq (u+m)$ and compiles them. Then, Albatross cancels the programs X_N, and starts running both X_N and $X_{N'}$ program sets in $E_{N'}$ (with N' nodes) for a time window of W_t. For each program X_{N_i} and $X_{N'_i}$, perf($X_{N'_i}$) is compared with perf(X_{N_i}) in the context of $E_{N'}$. Programs with highest performance for each i is kept running and the other jobs are canceled.

When the crashed nodes comes back online, $E_{N'}$ changes back to E_N. Therefore, the programs $X_{N'_i}$ are switched back to X_{N_i}.

4.1 Program Ranking Algorithm

It should be noted that the sample program sets generated by Albatross are not exponential. E.g., For VWAP for a transformation depth of 4 it produces only

24 sample applications. Transformation depth (d) is a non-zero integer which denotes the extent to which the input application's data flow graph is expanded [4]. Albatross uses the same program generator from Hirundo, and more details on sample program space is available from [4].

Albatross uses empirical performance information gathered from previous calibration sessions to estimate which sample program version should be the best match for the stream processing environment E. Note that in the case of System S we assume the operator placement decisions will be the same for multiple runs of a particular program on a specific node configuration. The Algorithm 1 describes the selection process. To maintain brevity we only describe the algorithm's inputs and outputs below.

```
Algorithm 1: Selection of Replacement Sample Program
Input : Input application's name (appname), structure of the input stream
        application (G), transformation depth (d), number of nodes (nodes)
Output : Replacement sample program (selectedLabel)
Description :
 1: optrunDict ← getLatestThreeOptrunIDs(appname, G, d, nodes)
 2: for all optrun in optrunDict do  /* Get sample program performance details */
 3:     perfDict[optrun] ← getPerfInfoForOptrun(optrun)
 4: end for
 5: labelDiffTable ← {}
 6: for all optrun in perfDict do /* Aggregate throughput information for each label */
 7:     for all label in optrun do
 8:         labelDiffTable[label].append(optrun[label])
 9:     end for
10: end for
11: inputStat ← ∅
12: labelPerfStat ← {}                /* Find range, average, */
13: for all label in labelDiffTable do  /* min, max throughput */
14:     labelPerfStat[label].range ← range(labelDiffTable[label])   /* values for each group */
15:     labelPerfStat[label].average ← average(labelDiffTable[label])
16:     labelPerfStat[label].min ← minimum(labelDiffTable[label])
17:     labelPerfStat[label].max ← maximum(labelDiffTable[label])
18:     if label is Input App then
19:         inputStat ← labelPerfStat[label]
20:     end if
21: end for
                                    /* Sort labels ascending order using range values */
22: sortByRangeAsc(labelPerfStat)
23: selectedLabel ← ∅               /* Select sample program with higher average */
24: for all label in labelPerfStat do:  /* throughput compared to the Input app */
25:     if label.min > inputStat.max then
26:         selectedLabel ← label
27:         break
28:     end if
29: end for
30: if selectedLabel = ∅ then        /* If could not find a suitable app, */
31:     for all label in labelPerfStat do:  /* select using average throughput */
32:         if label.average > inputStat.average then
33:             selectedLabel ← label
34:             break
35:         end if
36:     end for
37: end if
38: if selectedLabel = ∅ then /* If no suitable app found then return the input app*/
39:     selectedLabel ← appname
40: end if
41: return selectedLabel
```

Input to the algorithm is the application's name (this is the name assigned to a SPADE application under its Application meta-information tag [`Application`]), the structure of the input stream application G (G is a directed graph where each operator is represented as a vertex, and each stream that connects two operators is represented by an edge), transformation depth of the calibration run (i.e., d) and the number of nodes that are currently available on the stream processing environment. As the output, algorithm selects the first sample program which has higher average throughput compared to input stream application and has a higher minimum throughput relative to maximum throughput of input application (lines 24 - 29). If it could not find a suitable label it reduces the restrictions, and tries to select the sample program label which has higher average throughput compared to the input application (lines 30 - 37). If this attempt also fails the algorithm returns the input application's name (appname) as the sample application label since the algorithm needs to specify at least one application label (X_{N_i}) that should be run in the environment. Note that a label is an identification string. For Linear DAGs it represents arrangement of operator blocks in the sample program. E.g., Label S_4F_8F_4F_4SI means the sample program has one source, four F1 functors, eight F2 functors, four F3 functors, and four sinks. However, for multipath DAGs (E.g., Apnoea) the label serves only as a unique ID.

We select only the latest three calibration run results for sample program selection algorithm since a sample program produces similar performance behavior across multiple runs in the same stream computing environment. We make the selection of the sample program label in two steps (in lines 24 - 29 and in lines 30 - 37). The programs selected by considering the fact that having higher minimum throughput than the input application's maximum throughput, do have a higher probability of producing higher throughput compared to a decision made considering only the average throughput values (lines 24 - 29).

4.2 Program Switching Model

We employ a program switching model to remove programs (which are registered with low priority) during drastic node failures that makes the remaining nodes fully or close to fully utilized. The elimination model is formed by Cartesian product of *resource availability functions* (u) which can be defined as follows. If availability of a resource such as total amount of memory, total amount of CPU, etc. is denoted by p(y) and program switching function (binary) is given by $\phi(x)$,

$$p(y) = \frac{(current\ level\ of\ resource\ y)}{(initial\ level\ of\ resource\ y)} \tag{1}$$

$$u = 100 \times \prod p(y) \tag{2}$$

$$\phi(x) = \begin{cases} 1 & [M_i - (u + m)] > 0 \\ 0 & otherwise \end{cases} \tag{3}$$

Current implementation of Albatross uses only two resource availability variables: RAM availability (p(R)), and Node availability (p(N)). Hence, Equation 3 can be simplified to,

$$\phi(x) = \begin{cases} 1 & [M_i - ((p(R) \times p(N) \times 100) + m)] > 0 \\ 0 & otherwise \end{cases} \tag{4}$$

The choice of p(RAM) and p(Node) was made because successful operation of stream processing systems largely dependent on main memory availability. Node availability was also introduced to the program switching model because p(Node) directly reflects not only p(RAM), but also other resource availability metrics such as CPU availability, network availability, etc. on a homogeneous cluster. This results in simplified parabola shape u for which users of Albatross can easily specify the priority value m before running Albatross.

5 Implementation

Albatross prototype was developed using Python. Architecture of Albatross is shown in Figure 3. The input to Albatross is a collection of directories each

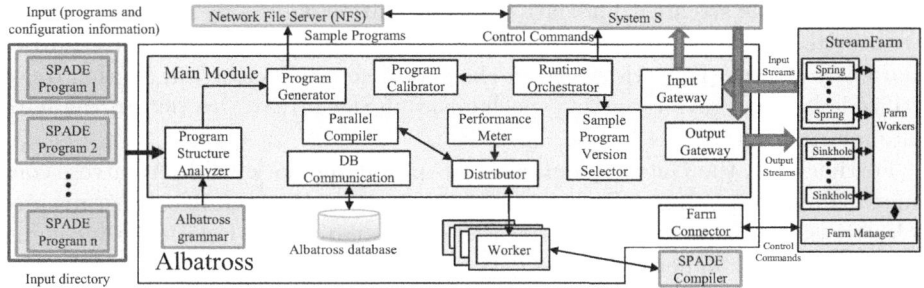

Fig. 3. System Architecture of Albatross

containing an input stream program. Each program is associated with a configuration file which lists information such as transformation depth (d) of the input program, the priority margin (M_i), etc.

Albatross has been developed targeting stream programs written in SPADE language. Therefore, Albatross depends on System S and SPADE compiler during its operations. System S is dependent on a shared file system such as Network File System (NFS), General Parallel File System (GPFS), etc. [10]. However, Albatross does not use NFS as a secondary storage to avoid potential performance bottlenecks. The experiments we conducted using Albatross were supported by a stream workload synthesis tool that we developed called "StreamFarm". Hence, Albatross does not depend on any significant file I/O during its operations. Albatross utilizes an SQLite database to store its information. Brief descriptions of important components of Albatross are described in below subsections.

We make several assumptions in creating the fault-tolerance model of Albatross. In current version of Albatross we do not employ any stream summarization/load shedding [2] techniques during the crash recovery period. All the incoming tuples are buffered in memory of the node which holds the gateway component during the crash recovery. We assume that the stream data is not bigger than what could be stored in memory of that node during that time period. Furthermore, we assume that the node which keeps Albatross and the gateway component processes does not crash during a fault recovery session.

Runtime Orchestrator module monitors health of the cluster using periodic heartbeat messages (i.e., ping messages). If a node crash was detected, the Input Gateway is informed to start the tuple buffering process. Next, the System S runtime is stopped, and the defected node is removed from the node list of System S instance configuration. After this System S is restarted on the remaining nodes. The sample programs selected by Albatross's program ranking algorithm, and the original programs are run in the environment for a short period to select the best versions to be deployed. Next, the programs not selected are stopped, and removed from the environment. Finally, the buffered data tuples are directed to the chosen applications.

All the data streams that go in/out to/from the sample applications travels through the module called Gateway. The Gateway is used for measuring the data rates of the streams. The data rate information is used by Program Version Selector to switch between the sample programs that are suitable for a particular environment (E_N). However, in the experiments described in Section 6, we utilized the data rate reporting mechanism of StreamFarm to obtain the data rate information.

Furthermore, the Gateway buffers the tuples in memory, while Albatross conducts sample programs switching. After the appropriate sample programs are selected the buffered tuples are released to System S jobs before the incoming tuples are served. When the buffer gets emptied the incoming tuples are directed to System S jobs rather than adding them to the tail of the tuple queue. In the current version of Albatross the gateway is located on a single node.

6 Experimental Evaluation

6.1 Experimental Setup

We used two clusters (lets call them A and B) of Linux Cent OS release 5.4 installed with IBM Infosphere Streams Version 1.2 and Python 1.7. Each node in cluster A had a Quad-Core AMD Phenom^{TM} 9850 processor, 512 KB L2 cache per core, 8GB memory, 160GB hard drive, 1 Gigabit Ethernet. Each node of cluster B had a dual core AMD Opteron^{TM} Processor 242, 1MB L2 cache per core, 8GB memory, 250GB hard drive, 1Gigabit Ethernet. We used SQLite version 3 as Albatross's database, and cluster B had JRE 1.6.0 installed. Both the A and B clusters were reserved for running only these experiments during the experiment period.

6.2 Evaluation of Stream Program Performance Variation

We modified Albatross not to respond to crash failures, and ran two sample applications of Regex and VWAP on Cluster A. Then we crashed two nodes which left only 6 operational nodes. The two applications showed different characteristics after the crash (See Figure 4(a)). In the case of VWAP the throughput dropped from an average of 84.16KB/s to an average of 21.38KB/s, a reduction of average data rate by 74.6%. In the case of regex application it completely stopped outputing data. Therefore, we observed that for certain applications, crash faults result in no output of data from System S. Some other applications output data at a reduced rate, but might produce different outcome than what is expected. In both the scenarios we need an explicit intervention like done by Albatross.

We ran Albatross in the Calibration Mode with VWAP application in cluster A with 8 nodes. The results of six calibration runs is shown in Figure 4(b). We observed that certain applications produce similar performance behavior across the six experiments (E.g., S_6F_AG_F_SI) and some applications produce higher performance compared to the input application (E.g., 8S_6F_AG_F_SI). This indicates that our program transformation method generates sample programs with consistent yet different levels of performance.

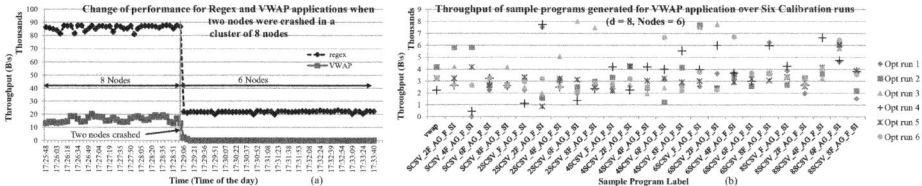

Fig. 4. (a) How the sample programs' performance change when two nodes were crashed. (b) Throughput of sample applications produced for VWAP application for Six calibration runs.

6.3 Evaluation of Performance Recovery Process

The VWAP, Regex, and Apnoea applications were calibrated by Albatross for three times prior the experiment for each node configuration (i.e., 8 nodes, 6 nodes, etc.). After the calibrations were complete, the three applications were submitted to Albatross in three separate runs. Albatross and System S were run in the cluster A, while StreamFarm was ran on cluster B to avoid potential interferences. StreamFarm allowed us to maintain a steady, high data rates throughout the experiments which gave us great support compared to file based methods that we used in our previous work [4].

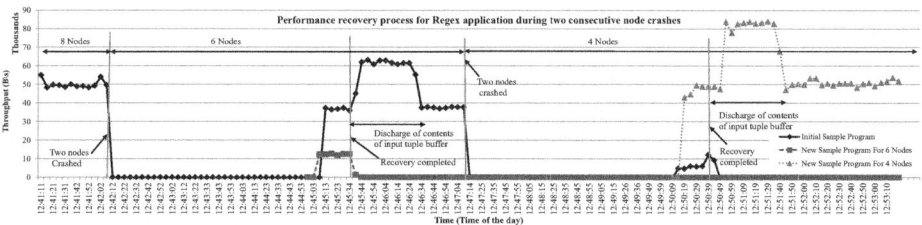

Fig. 5. Recovering performance of Regex application during two consecutive node crashes

While in the middle of the experiment run, two nodes were crashed. The resulted throughput curves are shown in Figures 5 and 6. We observed that for Regex application (See Figure 5) the new sample program introduced for 6 nodes environment (S_4F_8F_4F_4SI) was unable to produce a higher throughput compared to initial sample application (4S_4F_4F_2F_2SI), hence it's execution was canceled. The previous sample application (4S_4F_4F_2F_2SI) was run in the environment since Albatross could not find a better version of the Regex sample applications to run, yet we get only partial degradation of the Regex stream job's performance despite loss of two nodes. However, in the case of VWAP application Albatross's choice was much accurate, and performance of the new sample

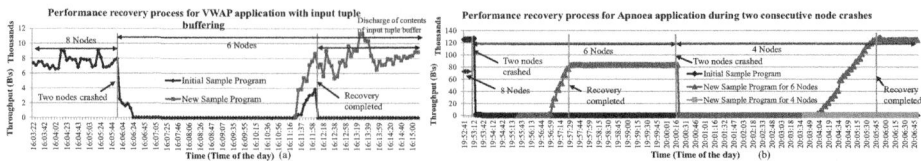

Fig. 6. Performance recovery process when (a) two nodes were crashed for VWAP. (b) two consecutive node crashes for Apnoea.

application (8S_48F_AG_F_SI) resembled almost the same performance of initial sample application (8S_64F_AG_F_SI). Note that Regex application consists of a chain of Functors where as VWAP application has an Aggregate operator that outputs data after gathering a group of tuples. Hence, the wavy curve of VWAP application compared to Regex application is formed by the aggregation operation done by VWAP. The temporary high throughput rise in both the graphs are due to the excess tuples during the release of buffered tuples just after completion of recovery process. For Apnoea, the new sample application S_F_AG_F_F_4AG_4F_F_AG_F_J_J_SI was able to restore the performance (with a partial degradation) while the initial sample application could not produce any output.

During the rest of the experiment we used only Regex and Apnoea applications. We used the full functionality of Albatross as with the previous experiment. After the first crash we allowed sufficient time for the recovery process and then crashed another two nodes. The results are shown in Figure 5 and in Figure 6(b). After the second crash we observed that for the Regex application, the initial sample program (4S_4F_4F_2F_2SI) was unable to restore the performance with 4 nodes, and the job was undertaken by 4S_8F_4F_4F_4SI which was able to produce almost the same throughput of the initial program with 8 nodes. In the case of Apnoea application, the new sample application introduced (S_F_AG_F_2F_2AG_2F_F_AG_F_J_J_SI) could not restore the performance. Hence, the previous sample application was deployed to maintain performance with 4 nodes cluster.

7 Discussion

From the evaluation results it was clear that Albatross is able to restore the operations back to the normal level (e.g., VWAP on 6 nodes, Regex on 4 nodes, and Apnoea on 4 nodes) or at least run the jobs with relatively lesser performance yet with a guarantee of the correctness of the execution (e.g., Regex on 6 nodes, and Apnoea on 6 nodes). Furthermore, Albatross was able to restore performance even when half of the nodes in the cluster were crashed (e.g., Apnoea, and Regex each on 4 nodes).

There are many further work, and limitations of the current prototype. The types of input stream applications that the current prototype can support are limited since Albatross's grammar covers a subset of the SPADE language constructs. Furthermore, the approach is infeasible for a large number of applications each demanding maintenance of higher performance. Our current program switching model is designed to avoid this. Data rate might not produce the correct picture of performance of certain stream applications. E.g., An application that aggregates tuples may emit one tuple per minute irrespective of the number of nodes that serve data for the aggregate operator. Moreover, it takes considerable amount of time (Rounded average recover times : 6 minutes for VWAP, 3 minutes for Regex application, and 5 minutes for Apnoea) to restore the normal operations of all the stream jobs. Most of this time is spent for orchestrating the System S runtime (start/stop System S runtime, reschedule PEs), and sample application compilation which accounts for significant time compared to Albatross's scheduling algorithm. However, the time period might be different for some other stream processing system. The release of tuples buffered by Albatross temporarily increases the data rate which might not be expected by some applications which receive data from System S jobs.

8 Conclusions and Future Work

In this paper we introduced a technique for maintaining performance during crash failures of stream computing systems. Our approach is widely applicable for resource limited stream processing clusters. It is based on automatic code generation. To this end we introduced Albatross, a python based middleware that monitors the status of the node clusters and strives to maintain the performance via swapping the sample programs generated for each input program. We observed that Albatross can maintain the same performance of the Regex (with 8 nodes, 4 nodes crashed), VWAP (with 8 nodes, 2 nodes crashed), and Apnoea (with 8 nodes, 4 nodes crashed) stream jobs despite loss of nodes from the stream processing environment. Therefore, we came to conclusion that our approach is capable of maintaining stream program performance even if 50% of the nodes in the cluster has been crashed in stream applications such as Regex, and Apnoea.

In future we hope to devise a sophisticated scheduling algorithm for Albatross's stream job control process to reduce the time taken for recovery process. We are also investigating on use of load shedding techniques to improve the stability of the performance recovery process.

Acknowledgments. This research was supported by the Japan Science and Technology Agency's CREST project titled "Development of System Software Technologies for post-Peta Scale High Performance Computing".

References

1. Andrade, H., Gedik, B., Wu, K.-L., Yu, P.S.: Scale-up strategies for processing high-rate data streams in systems. In: IEEE 25th International Conference on Data Engineering, ICDE 2009, March 29-April 2, pp. 1375–1378 (2009)

2. Babcock, B., Datar, M., Motwani, R.: Load shedding in data stream systems. In: Data Streams, vol. 31, pp. 127–147. Springer US (2007)
3. Catley, C., et al.: A framework to model and translate clinical rules to support complex real-time analysis of physiological and clinical data. In: Proceedings of the 1st ACM International Health Informatics Symposium, IHI 2010, pp. 307–315. ACM, New York (2010)
4. Dayarathna, M., Suzumura, T.: Hirundo: a mechanism for automated production of optimized data stream graphs. In: Proceedings of the Third Joint WOSP/SIPEW International Conference on Performance Engineering, ICPE 2012, pp. 335–346. ACM, New York (2012)
5. Furht, B., Escalante, A.: Handbook of Cloud Computing. Springer-Verlag New York, Inc. (2010)
6. Gedik, B., et al.: Spade: the system s declarative stream processing engine. In: SIGMOD 2008, pp. 1123–1134. ACM, New York (2008)
7. Gu, Y., Zhang, Z., Ye, F., Yang, H., Kim, M., Lei, H., Liu, Z.: An empirical study of high availability in stream processing systems. In: Middleware 2009, pp. 23:1–23:9. Springer-Verlag New York, Inc., New York (2009)
8. Hwang, J.-H., Cetintemel, U., Zdonik, S.: Fast and highly-available stream processing over wide area networks, pp. 804–813 (April 2008)
9. Hwang, J.-H., et al.: High-availability algorithms for distributed stream processing. In: Proceedings of the 21st International Conference on Data Engineering, ICDE 2005, pp. 779–790. IEEE Computer Society, Washington, DC (2005)
10. IBM. Ibm infosphere streams version 1.2.1: Installation and administration guide (October 2010)
11. Ishii, A., Suzumura, T.: Elastic stream computing with clouds. In: 2011 IEEE International Conference on Cloud Computing (CLOUD), pp. 195–202 (July 2011)
12. Khandekar, R., Hildrum, K., Parekh, S., Rajan, D., Wolf, J., Wu, K.-L., Andrade, H., Gedik, B.: COLA: Optimizing stream processing applications via graph partitioning. In: Bacon, J.M., Cooper, B.F. (eds.) Middleware 2009. LNCS, vol. 5896, pp. 308–327. Springer, Heidelberg (2009)
13. Kleiminger, W., Kalyvianaki, E., Pietzuch, P.: Balancing load in stream processing with the cloud. In: Data Engineering Workshops (ICDEW), pp. 16–21 (April 2011)
14. Logothetis, D., Trezzo, C., Webb, K.C., Yocum, K.: In-situ mapreduce for log processing. In: USENIXATC 2011, Berkeley, CA, USA, p. 9. USENIX Association, Berkeley (2011)
15. Tanenbaum, A.S., Steen, M.V.: Distributed Systems. Pearson Education, Inc. (2007)
16. White, T.: Hadoop: The Definitive Guide. O'Reilly Media, Inc. (2010)
17. Wolf, J., et al.: Soda: an optimizing scheduler for large-scale stream-based distributed computer systems. In: Middleware 2008, pp. 306–325. Springer-Verlag New York, Inc., New York (2008)
18. Zhang, Z., et al.: A hybrid approach to high availability in stream processing systems. In: Distributed Computing Systems (ICDCS), pp. 138–148 (June 2010)

Event Relationship Analysis for Temporal Event Search

Yi Cai[1,*], Qing Li[2], Haoran Xie[2], Tao Wang[1], and Huaqing Min[1]

[1] School of Software Engineering, South China University of Technology, Guangzhou, China
ycai@scut.edu.cn
[2] Department of Computer Science, City University of Hongkong, Hongkong, China

Abstract. There are many news articles about events reported on the Web daily, and people are getting more and more used to reading news articles online to know and understand what events happened. For an event, (which may consist of several component events, i.e., episodes), people are often interested in the whole picture of its evolution and development along a time line. This calls for modeling the depéndent relationships between component events. Further, people may also be interested in component events which play important roles in the event evolution or development. To satisfy the user needs in finding and understanding the whole picture of an event effectively and efficiently, we formalize in this paper the problem of temporal event search and propose a framework of event relationship analysis for search events based on user queries. We define three kinds of event relationships which are temporal relationship, content dependence relationship, and event reference relationship for identifying to what an extent a component event is dependent on another component event in the evolution of a target event (i.e., query event). Experiments conducted on a real data set show that our method outperforms a number of baseline methods.

1 Introduction

With the development of the Internet, news events are reported by many news articles in the form of web pages. People are getting more and more used to reading news articles online to know and understand what events happened. For a composite/complex event, it may consist of several component events, i.e., episodes. There are some interrelationships among these component events as they may be dependent on each other. For example, the event of "Toyota 2009-2010 vehicle recalls" contains several interrelated component events, e.g., the event "Toyota recall due to safety problems from 2009 to 2010" causes the happening of the event "NHTSA conduct investigations for Toyota recall" and the event "US congressional hearings hold for Toyota recall", and so on. Also, the event "US congressional hearings hold for Toyota recall" has a strong relationship with the event "Toyota's president to testify in US congressional hearings".

Quite often, what people interested in is not just a sole news article on an event, but also the related events reported by other news articles. Indeed, they are often interested in the whole picture of an event evolution or development along a timeline. This calls for modeling the dependence relationships between component events, and identifying

---

* Corresponding author.

which component events play important roles in the entire event evolution or development. Unfortunately, the current news web sites do not facilitate people in finding out relevant news articles easily, and people may need to go through all these news articles in order to find out the interrelationships between component events. Current prevailing search engines (such as Google, Yahoo and so on) allow users to input event keywords as a query and return a list of news web pages related to the query. However, instead of organizing the result by events and relationships between events, these engines just provide users with a ranking list of news web pages. It is difficult and time consuming for users to view of the huge amount of news articles and to obtain the main picture of an event. Therefore, it is necessary to provide an effective way for users to efficiently search events they are interested in, and organize the search results in an easily understandable manner syntactically, so that users can obtain the main pictures of their interested events easily and meaningfully from the semantic perspective.

Although there have been some previous works attempting to find and link incidents in news [2] [3] or discover the event evolution graphs [17] [15], they only focus on time sequence and content similarity between two component events in identifying their dependence relationships. However, using these two factors only is inadequate in identifying dependence relationships among the component events in order to form the main picture of a big event evolution or development. For example, event "Toyota recall due to safety problems from 2009 to 2010" shares little similar content with events "NHTSA conduct investigations for Toyota recall" and "US congressional hearings hold for Toyota recall". It is obvious that the first event has a strong effect on the latter two events as it caused them to happen. Unfortunately, previous works do not analyze the event relationships well and cannot find out the dependence relationships between any two events which do not share enough similar content. As a result, the main picture of a "big event" discovered by previous works often is incomplete and several significant relationships are missing.

In this paper, to satisfy the user needs mentioned above in finding and understanding the whole picture of a complex event effectively and efficiently, we conduct an in-depth event relationship analysis for event search and propose a framework to search events based on user queries. The new characteristics of the proposed framework and the contributions of our work are as follows.

- In previous works, to discover dependence relationships between events, content similarity of events is measured by matching the keywords (terms) of events. However, there may be some keywords (in two events) which are actually related/dependent but not identical. For example, "hospital" and "doctor" are dependent, but previous methods treat them as no relationship. To avoid this limitation, we adopt mutual information to measure the dependence between two terms (features), and then aggregate all mutual information between features in events to measure the content dependence degree between events. Such a process is named as *content dependence (CD) analysis* and the dependence relationship discovered based on dependence features of two events is named as *content dependence relationship*.
- As mentioned in paragraph 4, only content dependence analysis on events is inadequate to detect all event dependence relationships. According to the studies in Journalism [11] and our observation, it is not unusual for authors (reporters) to write

news articles on an event by referring to other events, when the authors consider there is a dependent relationship between them. For instance, some news articles about the event "Toyota recall due to safety problems from 2009 to 2010" refer to the event "NHTSA conduct investigations for Toyota recall". Motivated by this prevalent phenomena, We explore *event reference (ER) analysis* to detect whether there is an inter-event relationship specified by authors. The relationship between two events discovered by ER analysis is named as *event reference (ER) relationship*, which has not been explored by previous works.

– In contrast to previous works which only consider temporal relationship and content similarity, we adopt three kinds of event relationships (viz, temporal relationship, *CD* relationship obtained by *CD* analysis and *ER* relationship obtained by *ER* analysis) to identify the dependence relationship between two events. Note that *CD* and *ER* relationships are essentially event dependence relationships which are discovered by two different ways respectively. We name them by two different names with respect to the different ways for discovering them. *CD* relationships and *ER* relationships can be complementary to each other in identifying event dependence relationships.

– The search results are organized by a *temporal event map* (TEM) which constitutes the whole picture about an event's evolution or development along the timeline. Figure 1 shows an example TEM of the event "Toyota 2009-2010 vehicle recalls". A TEM provides a way to organize and represent the events search results by showing the interrelationships between/among the events. It provides an easier and more efficient means for users to know and understand their interested events in a comprehensive way.

– To evaluate the performance of our proposed approach, we conduct experiments on a real data set by comparing with a number of baseline methods. Experiment results show that our method outperforms baselines in discovering event dependence relationships.

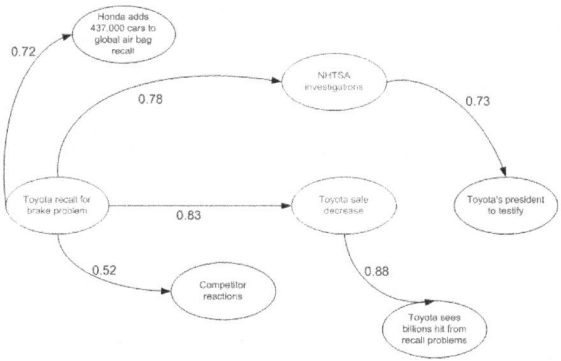

Fig. 1. An example of the *TEM* about the event "Toyota 2009-2010 vehicle recalls"

The rest of this paper is organized as follows. In section 2, we formulate the event search problem. Section 3 introduces the (temporal) event search framework. We conduct experiments on a real data set to evaluate the proposed methods for event search in section 4. In section 5, to further illustrate and evaluate our method, we study a query case about the event "SARS" happened in 2003. Related works are studied in section 6. We conclude the paper and introduce potential future works in section 7.

2 Problem Formulation

According to [9] and [17], an event is something that happens at some specific time and place. In reality, events often are reported by some documents, such as news articles in web pages. Formally, for an event a, there is a set of documents talking about a, and such a set of documents, denoted as $R_a = \{d_1^a, d_2^a, \cdots, d_n^a\}$, is named as **related document set** of a. Each document is about one event and an event can be reported by multiple documents. A document introducing an event includes the start time, place(s) and content of the event. Thus, for each document d_x^a, there should be a timestamp $\tau_{d_x^a}$, a set of place names $q_{d_x^a} = \{t_{x,1}, \cdots, t_{x,n}\}$ and a set of terms $h_{d_x^a} = \{f_{x,1}, \cdots, f_{x,n}\}$ about the event's content. We define an event as follows.

Definition 1. *An* ***event*** *a is a tuple (L_a, P_a, F_a) where L_a is the life cycle of a, P_a is the set of places where a happens, and F_a is the set of features describing a.*

The **life cycle** L_a of event a is the period (time interval) from the beginning time St_a to the end time Et_a of a, i.e., $L_a = [St_a, Et_a]$, where St_a is the earliest timestamp among all the timestamps of related documents of a, and Et_a is the latest timestamp among all that of related documents of a. The **place set** P_a of an event a is a set of terms denoted by P_a where $P_a = \{t_{a,1}, t_{a,2}, \cdots, t_{a,m}\}$ and each $t_{a,x}$ is a term which represents a place. For an event, it may consist of several component events, i.e., episodes.

For example, for the event of SARS epidemic which happened in 2002 among some 37 countries around the world, the life cycle of this event is from November 2002 to May 2006. The place of the event includes China, Canada, Singapore and so on. There are many reported news on the event on the Web. To describe the event, we can extract from the set of documents the set F_a of features (i.e., keywords), such as "SARS", "flu-like", "fever" and so on. The event of SARS epidemic consists of several component events such as "Experts find disease infect and SARS outbreaks", "China informs and cooperates with WHO", "SARS has great impact on economy" and so on.

Definition 2. *For each event a, it contains a set of* ***component events*** *denoted as $CE_a = \{a_1, a_2, \cdots, a_n\}$ where $1 \leq n$, a_x is a component event of a and $R_a = R_{a_1} \cup R_{a_2} \cup \cdots \cup R_{a_n}$.*

Definition 3. *Among all the component events of an event a, the* ***seminal component event*** *of a is the one whose start time is the same as that of a, i.e., the start time of the seminal component event is no later than those of the other component events of a.*

Definition 4. *Among all the component events of an event a, the* ***ending component event*** *of a is the one whose end time is as the same as that of a, i.e., the end time of the ending component event is no earlier than those of the other component events of a.*

For an event a and its component events, it is obvious that $St_a = Min_{i=1}^{n}(St_{a_i})$, $Et_a = Max_{i=1}^{n}(Et_{a_i})$, $P_a = P_{a_1} \cup P_{a_2} \cup \cdots \cup P_{a_n}$ and $F_a = R_{a_1} \cup F_{a_2} \cup \cdots \cup F_{a_n}$. We observe that there is a temporal requirement for two events to have a dependence relationship between them, as follows.

Observation 1. *If there is a dependence relationship from event a to event b, i.e., a is dependent on b, then there is a temporal relationship between a and b such that $St_b <= St_a$, i.e., b happens earlier than or at the same time as a.*

Definition 5. *A **Temporal Event Map** is a weighted directed graph, denoted by $TEM = (N, E, W_d)$, which consists of events as nodes, relations as edges, and weights on the edges as strength degrees of dependence relationships. In particular, each vertex $v \in N$ is an event, each edge $e_x \in E$ is a dependence relationship between two events, and $w_y \in W_d$ is a weight which indicates the strength degree of a dependence relationship.*

An example of temporal event map of the event "Toyota 2009-2010 vehicle recalls" is shown in Fig. 1.[1]

We formulate the problem of temporal event search as follows. The input of the search problem is a tuple (I_t, I_p, I_f) where I_t is a time interval, I_p is a set of terms of places, and I_f is a set of keywords about an event content. The event which is relevant to (corresponding to) the input is named as the **target event**, i.e., the event happens in the places in I_p during I_t, and the feature set of the target event contains I_f. The output of the search problem is a TEM constituting all the component events of the target event.

The problem of temporal event search can thus be regarded as a function ϕ:

$$\phi : I \times D \to T'$$

where I is the set of input, D is the set of documents and T' is the set of TEMs.

For the example of Fig. 1, we may have the following input:

$$I_t = [1/11/2009, 23/2/2010]; I_p = (USA); I_f = (Toyota, recall)$$

then the temporal event map for such a search task is the one shown in Figure 1.

3 Event Relationship Analysis

In this section, we propose a framework of event relationship analysis to support temporal event search. In our method, we first identify a set of related documents for the target event and extract component events from the related documents. We conduct content dependence (CD) and event reference (ER) relationship analysis to identify dependence between events.[2]

---

[1] We use the width of a line to indicate the strength of a dependence relationship.
[2] In the rest of the paper, we use the term "event" to denote "component event" for convenience wherever there is no ambiguity.

3.1 Preliminaries

A user query can be considered as search requirements corresponding to a target event which satisfies all the needs from the user. The related document set of the target event can be obtained by a function θ:

$$\theta : I \times D \to R$$

where I is the set of input, D is the set of documents and R is the set of related document sets.

In general, we consider (I_t, I_p, I_f) as three kinds of (not all are compulsory) user search requirements. In some cases, users may only input one or two of the (I_t, I_p, I_f). For such special cases, we only take the user input requirements into consideration, i.e., subset of (I_t, I_p, I_f).

For each target event a corresponding to an input I and its related document set R_a, we can detect several component events from R_a. All component events of a should happen during I_t, and their places are contained in I_p and features contained in I_f. The component event detection of a target event is a function φ:

$$\varphi : R \to E$$

where R is the set of related documents and E is the set of the component events.

For the problem of event detection, there have been many existing works published such as [1] [12] [14]. In this paper, we adopt the topic-model based method [14] as the preferred method to detect events.

3.2 Content Dependence Analysis

In analyzing content dependence (CD) relationships for temporal event search, we notice that features of an event a may have various degrees of importance in representing a. Some features are more representative than others for the event. An event can be represented by a *feature vector*, denoted by $\vec{F}_a$, which is a set of feature:value pairs.

$$\vec{F}_a = (f_{a,1} : v_{a,1}, f_{a,2} : v_{a,2}, \cdots, f_{a,n} : v_{a,n}), \forall i, 0 < v_{a,i} \leq 1$$

where $f_{a,i}$ is a feature and $v_{a,i}$ is the importance degree of $f_{a,i}$ for the event a. Hence, $v_{a,i}$ is the *NTF-IEF* (normalized term frequency-inverse event frequency) value of $f_{a,i}$, i.e.,

$$v_{a,i} = \frac{tf_{a,i}}{MAX_u(tf_{a,u})} \log \frac{N}{ef_i} \quad (1)$$

where $tf_{a,i}$ is the frequency of term i in R_a, N is the total number of component events, $MAX_u(tf_{a,u})$ is the maximal value among all $tf_{a,u}$ and ef_i is the number of component events containing term $f_{a,i}$.

As mentioned before, previous works use content similarity (most works adopt cosine similarity) to identify dependence relationships between events. However, two events may have some keywords which are dependent but not identical, which causes the previous works to be inadequate in measuring how relevant these two events are.

According to [10], variables (i.e., keywords) which are not statistically independent suggest the existence of some functional relation between them, and mutual information provides a general measure of dependencies between variables. Thus, we adopt mutual information to measure the dependence between features, and further use an aggregation of all mutual information between the feature sets in two events to measure the content dependence degree between them.

Formally, for two events a and b, the content dependence degree, denoted by $Cd(a,b)$, is an aggregation of all mutual information between all features in $\overrightarrow{F}_a$ and that in $\overrightarrow{F}_b$, as follows:

$$Cd(a,b) = \frac{\sum_{f_x \in \overrightarrow{F}_a} \sum_{f_y \in \overrightarrow{F}_b} I(f_x, f_y)}{|\overrightarrow{F}_a||\overrightarrow{F}_b|} \quad (2)$$

where $|\overrightarrow{F}_a|$ ($|\overrightarrow{F}_b|$) is the cardinality of the set $\overrightarrow{F}_a$ ($\overrightarrow{F}_b$), and $I(f_x, f_y)$ is the dependence degree between features f_x and f_y, measured as follows:

$$I(f_x, f_y) = P(f_x, f_y) \log \frac{P(f_x, f_y)}{P(f_x)P(f_y)} \quad (3)$$

where $P(f_x, f_y)$ is the probability of f_x and f_y co-occurring in the same document among all the related documents, and $P(f_x)$ is the probability of f_x occurring in a document among all documents, and $P(f_y)$ is the probability of f_y occurring in a document among all the documents.

By measuring all mutual information between two component events, we can obtain a **component content dependence matrix** of an event a, denoted as M_a^c, as follows:

$$M_a^c = \left\{ \begin{array}{c} Cd(1,1), Cd(1,2), \cdots, Cd(1,m) \\ \cdots \\ Cd(n,1), Cd(n,2), \cdots, Cd(n,m) \end{array} \right\}$$

where each entry is a content dependence degree between two component events.

3.3 Event Reference Analysis

Although content dependence measurement can address the limitation of content similarity measurement, it may still miss some dependence relationships between events. In particular, the existence of a dependence relationship between two events does not necessarily mean that there exists a content dependence relationship between them. In many cases, although the contents of two events are very different and even of different topics, people may still regard that there is a dependence relationship between them. For instance, "Experts find disease infect and SARS outbreaks" has an impact on "SARS has great impact on economy" and "SARS has a great impact on Tourism". The latter two events are dependent on the first one even though their content dependence degree is indeed very small.

According to the studies of Journalism [11], when authors of news articles about an event a find and regard that there exists a dependence relationship between a and b (e.g., b triggers the happening of a, or a is evolved from b and so on), their articles may actually refer event b. This is in line with our observation on our collected data set. For instance, some news articles about the event "Toyota recall due to safety problems from 2009 to 2010" refer to the event "NHTSA conduct investigations for Toyota recall". Such an explicit reference relationship made by authors in their news articles reflect their viewpoints and consideration on the inter-event relationships [11]. Therefore, we may regard such event reference relationships as more meaningful and reliable than content dependence relationships, and ER relationship analysis provides a way to discover those event dependence relationships missed by CD analysis and obtain a more complete temporal event map (TEM).

We can also observe that when a news article of an event c refers to another event a, there are usually some phrases that identify event a in the documents of event c, and we name such phrases as *core features* of a. The definitions of core feature set of an event is defined below.

Definition 6. *The **core feature set** of an event a, denoted by F_a^c is a set of features which are salient in the event, distinguishable from those of other events, and jointly can identify the event.*

For two events a and b, if there exists a related document of b, denoted as d_x^b, such that $\exists f_i \in F_a^c, f_i \in d_x^b$, and $\tau(d_x^b) > St_a$ (i.e., a happens earlier than b), then we say there is a **reference relationship** from b to a, i.e., a **is a reference of** b or b **refers to** a. Such a reference relationship is a fuzzy relationship, and the more core features of a are mentioned in b, the more strength degree of the relationship. For example, for the event "US congressional hearings hold for Toyota recall" denoted by a and the event "Toyota's president to testify in US congressional hearings" denoted by b, we find that the core feature set of event a is $F_a^c = \{congress, hearing, safety\}$ while the core feature set of b is $F_b^c = \{Akio, Toyoda, testify, apologize\}$. For event a, some of its core features also exist in some documents (news articles) of b, (e.g., "congress", "hearing" and "safety" all appear in the news titled as "Toyota's president to testify before Congress" on Feb 19, 2010), so we say b refers to a.

The strength degree of a reference relationship from b to a is determined by a function $Cr(a, b)$ which is to be defined below. For event b referring to event a, it should follow the temporal restriction of Observation 1. For two events a and b, in order to find out whether b refers to a, we need to discover the core feature set of a first and then check whether the core features of a exist in the related documents of b.

According to our observation, the core feature set of an event has the following properties.

Property 1. The core features of an event a are the most salient and representative features of a, i.e., the features appear in the related documents of a with a high frequency.

Property 2. The core features of an event a are distinguishable from those of other events, i.e., the core features should facilitate us in identifying event a from all other events easily.

Based on the above properties, we propose the following function to select core features of an event a:

$$u(f_i, a) = p(f_i|a) \cdot p(a|f_i) \tag{4}$$

where $p(f_i|a)$ is the probability of feature f_i to exist in the related documents of event a, and $p(a|f_i)$ is the probability of a document (in which f_i is a feature) being on event a. Note that $p(f_i|a)$ and $p(a|f_i)$ reflect the properties 1 and 2 respectively.

We select top-k core features based on equation 4. For two events a and b, the more related documents of b refer to more core features of a, the stronger is the reference relation from b to a. We propose a function to measure the strength degree of a reference relation from a to b as follows:

$$Cr(a,b) = \frac{\sum_{i=0}^{N_b} M_{b,i}^a}{|F_a^c|} \times \frac{1}{N_b}, \forall M_{b,i}^a > 1 \qquad (5)$$

where N_b is the number of related documents of b, $M_{b,i}^a$ is the number of core features of a existing in the document d_i^b, $|F_a^c|$ is the cardinality of F_a^c. Note that there is a restriction for $M_{b,i}^a$ in $Cr(a,b)$ where $M_{b,i}^a > 1$, highlighting that a reference relationship from b to a should refer more than one core feature of a.

For the reason that the values of $Cr(a,b)$ and $Cr(b,a)$ may be greater than zero, a could refer to b and also b could refer to a, and the strength degrees of reference relationship from a to b could be different with that from b to a. An event can be refereed by many other events. Besides, one event can also refer to many other events.

By measuring all component event reference degree between any two events, we can obtain an **component event reference matrix** of a target event a, denoted by M_a^r, as follows:

$$M_a^r = \left\{ \begin{array}{c} Cr(1,1), Cr(1,2), \cdots, Cr(1,m) \\ \cdots \\ Cr(n,1), Cr(n,2), \cdots, Cr(n,m) \end{array} \right\}$$

Each entry is a reference degree between two events.

3.4 Temporal Event Map Construction

We adopt content dependence (CD) analysis and event reference (ER) analysis to identify event dependence relationships. In cases when users are only interested in the ER relationships between events, we can do a projection on the TEM and obtain an *event reference TEM*, which is a sub-graph of the entire TEM. Similarly, if users are only interested in the CD relationships between events, we also can do a projection on the TEM and obtain a *content dependence TEM*. Besides, it is easy to show all the CD, ER and event dependence relationships in a TEM. While there are many interesting issues related to the visualization of TEM, we omit further discussion here since are our focus in this paper is on event relationship analysis.

4 Evaluation

In this section, we conduct experiments on a real data set to evaluate our approach by comparing it with a number of baseline methods.

4.1 Experiment Setting

To evaluate our method for temporal event search, we collect 5063 English news articles (i.e., web page documents of news) from some mainstream news websites such as CNN News and BBC News. We select ten queries about major events to test our method, such as "Toyota 2009-2010 vehicle recalls", "2010 Copiap mining accident", "SARS in 2003" and so on. Among these, the event "SARS in 2003" contains the most number of related news articles (i.e., 231 articles), and the event "Christchurch Earthquake in 2010 in New Zealand" contains the fewest number of related news articles (i.e., 39 articles). According to our observation on the data set, when an event a refers to another event b, the number of referred core features of b is often around five.[3] Thus, we select top-5 core features to measure event reference relationships in our experiment.

To compare with our method, we adopt three baseline methods. The first one is the state-of-the-art method of discovering event evolution relationship proposed by Yang [17], which is similar to the method in [2] and we denote it as EEG. The second baseline, denoted as CDM, only considers content dependence analysis and does not use event reference analysis to judge event dependence relationships. Different from CDM, the third one, denoted as ERM, only considers event reference analysis instead of using content dependence analysis to judge event dependence relationships.

We have invited five human subjects to annotate the dependence relationships between events. All the annotated relationships are combined synthetically to obtain a set of relationships, i.e., the union of all the relationships annotated. Such a set of relationships given by the annotators is considered as a standard answer set (ground truth) of event dependence relationships. For the reason that different people may have different viewpoints on the event relationships due to, e.g., their knowledge and background, not every annotator came up with the same set of dependence relationships. Therefore, the standard answer set is an aggregation of the annotations given by all the annotators.

For the evaluation we use $Precision$, $Recall$ and $F-measure$ as the metrics. We denote the set of event dependent relationships (i.e., edges in a TEM) annotated by annotators as R_A, and the set of event dependence relationships discovered by machine as R_M. The metrics are defined as follows:

$$Precision = \frac{R_A \cap R_M}{R_M}; Recall = \frac{R_A \cap R_M}{R_A}$$

$$F - measure = \frac{2 \times Precision \times Recall}{Precision + Recall}$$

4.2 Experiment Results

In constructing TEM, there is a parameter α which is used to prune the "weak" event dependency relationships. So first, we test different values of α to evaluate the effect of α on *Precision*, *Recall* and *F-measure* for setting the best value of parameter α for the following experiment. In our testing, we use two query events, one is "SARS in 2003"

---

[3] Such an observation is only based on our collected data set. It could be different for other data sets.

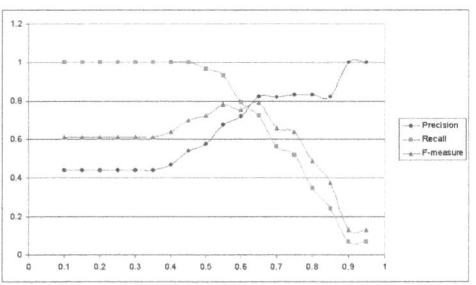

Fig. 2. The effect of α on Precision, Recall and F-measure

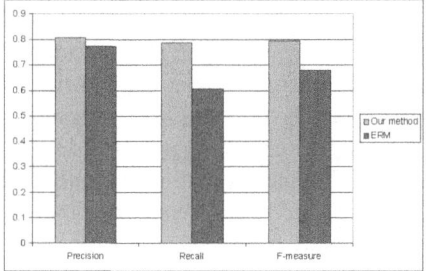

Fig. 3. Our method vs. EEG

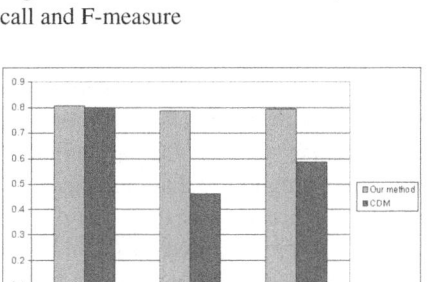

Fig. 4. Our method vs. CDM

Fig. 5. Our method vs. ERM

which contains the most number of related news articles and the other is "Christchurch Earthquake in 2010 in New Zealand" which contains the fewest number of related news articles. Figure 2 shows the effect of α on *Precision*, *Recall* and *F-measure*. According to Fig. 2, we find that as α increases, the *Precision* and *F-measure* increase while *Recall* decreases. The reason is that when the value α is small, there are many event dependence relationships whose dependence degree is great than α (but the dependence relationship is still actually "weak"), so the $Recall$ is high and the $Precision$ is low. As α increases, more and more event dependence relationships of which dependence degree is lower than α are pruned, so the $Recall$ becomes lower and the $Precision$ becomes higher. When $\alpha = 0.65$, we obtain the highest value of *F-measure*. Thus, we set $\alpha = 0.65$ for all the test queries subsequently.

After setting the value of α, we conduct all test queries and average the results of them on different metrics. Figures 3-5 show the comparison of our method with all the three baseline methods on $Precision$, $Recall$ and $F - measure$. According to Figures 3-5, it is obvious that our method outperforms all the baseline methods on $Precision$, $Recall$ and $F-measure$. The $Precision$ and $Recall$ values of our method are around 0.8, meaning that not only most event dependence relationships discovered by our method are correct, but also our method can discover more event dependence relationships than the baselines. Our method's $F - measure$ score is also around 0.8 since it is a combination of $Precision$ and $Recall$. Note that CDM outperforms EEG a little on all the metrics, indicating that using mutual information to measure feature

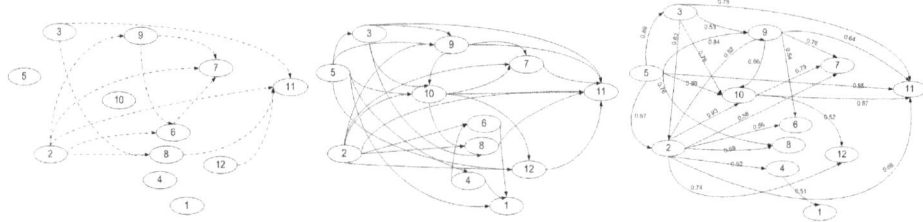

Fig. 6. The Result of EEG **Fig. 7.** The Result of Annotators **Fig. 8.** The Result of Our Method

dependence is better than just matching keyword similarity (as done by previous works). ERM outperforms EEG and CDM on all three metrics. It indicates that using event reference analysis (i.e., ERM) to identify event dependence relationships is more effective than using content dependence relationship analysis (i.e., CDM) and content similarity analysis (i.e., EEG). Besides, it is quite interesting to see that the *Recall* of CDM is greater than that of ERM, while both of them are smaller than that of our method. This means that the event dependence relationships identified by CDM and ERM are indeed different and complementary, and our method being a combination of CDM and ERM has the strength of both methods'. In other words, taking both content dependence and event reference analysis into consideration in identifying event dependence relationships can perform better than taking just one of these.

4.3 Case Study

To illustrate the performance of our proposed method more clearly, we further show a specific search case on the query event "SARS happened from 1/3/2003 to 30/6/2003 around the world" denoted by Q_{SARS}. The test query is $I_p = (China)$, $I_t = [1/3/2003, 30/6/2003]$, $I_f = (SARS)$.

Table 1. Component Events for the query about SARS event from 1/3/2003 to 30/6/2003

Component Event	Summary
1	SARS has great impact on Tourism
2	SARS cases are reported and updated regularly to reflect the disease seriousness
3	Experts treat patients with medicine in hospital
4	SARS has great impact on transportation especially airline
5	Experts find disease infect and SARS outbreaks
6	SARS has great impact on economy
7	Other countries donate and offer help for China for SARS
8	Scientists' find coronavirus and conduct animal test for vaccine
9	China informs and cooperates with WHO on fighting SARS
10	China makes effort on prevent disease spread
11	Beijing has made SARS under control
12	Quarantine probable cases and close schools for disinfecting

Table 2. Comparison on discovered event relationships of our method and EEG for Q_{SARS}

	Correct	Missed	Incorrect	New	Total
Our method	20	7	2	2	24
EEG	10	17	2	0	12

Table 1 shows all the component events which are related to Q_{SARS}. Table 2 shows the statistics of the discovered event relationships by our method and EEG based on the results of human annotators for this case. Figures 6-8 show the relationship graph (or TEM) obtained by EEG method, given by the human annotators and our method for Q_{SARS}, respectively. Our method can find more and miss less correct inter-event relationships. In addition, our method can discover not only the inter-event relationships but also the strength degrees of such relationships. More interestingly, our method can find some new relationships which were not found by EEG and even human annotators. Such new relationships are confirmed and approved by the annotators as meaningful ones (e.g., the relationship from event 5 to event 2 and the relationship from event 3 to event 2).

5 Background and Related Works

There are many works about processing events which may include news event or system events, although most of these work focus on news event.

To the best of our knowledge, there is no work on temporal event search before. A related work is done by Jin et al. [6] who present a temporal search engine supporting temporal content retrieval for Web pages called TISE. Their work supports Web pages search with temporal information embedded in Web pages, and the search relies on a unified temporal ontology of Web pages. TISE handles Web pages search only, and it cannot handle event search nor discover the event relationships.

Topic detecting and tracking (TDT) is a hot research topic related to our work. Given a stream of constantly generated new documents, TDT groups documents of the same topic together and tracks the topic to find all subsequent documents. There are several techniques on detecting news topics and tracking news articles for a new topic. For instance, Allan et al. [1] define temporal summaries of news stories and propose methods for constructing temporal summaries. Smith [12] explores detecting and browsing events from unstructured text. Some techniques are proposed to detect particular kinds of events. For example, Fisichella et al. [7] propose a game-changing approach to detect public health events in an unsupervised manner. Modeling and discovering relationships between events as generally out of the scope of current TDT research.

Mei and Zhai [8] study a particular task of discovering and summarizing the evolutionary patterns of themes in a text stream. A theme in an interval may be part of an event or a combination of several events that occur in the interval. Their work does not however capture the interrelationships of major events. Fung et al. [4] propose an algorithm named Time Driven Documents-partition to construct an event hierarchy in a text corpus based on a user query.

Some other works focus on discovering stories from documents and representing the content of stories by graphs. For example, Subasic et al. [13] investigate the problem of discovering stories. Ishii et al. [5] classify extracted sentences to define some simple language patterns in Japanese so as to extract causal relations, but their work cannot handle cases which are not defined in their patterns.

An event evolution pattern discovery technique is proposed by Yang et al. in [16]. It identifies event episodes together with their temporal relationships. They consider temporal relationships instead of evolution relationships. Although the temporal relationships can help organize event episodes in sequences according to their temporal order, they do not necessarily reflect evolution paths between events. An extended work of them occurs in [15]. Yang et al. [17] define the event evolution relationships between events and propose a way to measure the event evolution relationships. In their work, identifying an event evolution relationship between two events depends on the similarity of the features of the two events. Based on a small number of documents and events in a news topic, Nallapati et al. [9] define the concept of event threading. Their definition of event threading is a content similarity relationship from previous event to a later event. The event threading is organized as a tree structure rather than a graph. In order to identify event threading, they employ a simple similarity measure between documents to cluster documents into events and the average document similarity to estimate the content dependencies between events. Feng and Allan [2] extend Nallapati's work to passage threading by breaking each news story into finer granules, and propose a model called incident threading in [3].

6 Conclusions and Future Works

In this paper, we have defined three kinds of event relationships which are temporal relationship, content dependence relationship and event reference relationship, and have applied them to measure the degree of inter-dependencies between component events to support temporal event search. We have also formalized the problem of event search and proposed a framework to search events according to user queries. Experiments on a real data set show that our proposed method outperforms the baseline methods, and it can discover some new relationships missed by previous methods and sometimes even human annotators.

Admittedly, several possible future extensions can be made to our work. In our current method, only top-5 core features are selected for event reference analysis over the collected data set. How to choose the "right" number of core features for event reference analysis automatically for different data sets is an open issue for further study. Another potential extension is to implement a visualization tool with a sophisticated user interface based on our current method.

Acknowledgements. This work is supported by Foundation for Distinguished Young Talents in Higher Education of Guangdong, China (NO. LYM11019); the Guangdong Natural Science Foundation, China (NO. S2011040002222); the Fundamental Research Funds for the Central Universities,SCUT (NO. 2012ZM0077); and Guangdong Province University Innovation Research and Training Program (1056112107).

References

1. Allan, J., Gupta, R., Khandelwal, V.: Temporal summaries of new topics. In: SIGIR 2001: Proceedings of the 24th Annual International ACM SIGIR Conference on Research and Development in Information Retrieval, pp. 10–18. ACM, New York (2001)
2. Feng, A., Allan, J.: Finding and linking incidents in news. In: CIKM 2007: Proceedings of the Sixteenth ACM Conference on Conference on Information and Knowledge Management, pp. 821–830. ACM, New York (2007)
3. Feng, A., Allan, J.: Incident threading for news passages. In: CIKM 2009: Proceeding of the 18th ACM Conference on Information and Knowledge Management, pp. 1307–1316. ACM, New York (2009)
4. Fung, G.P.C., Yu, J.X., Liu, H., Yu, P.S.: Time-dependent event hierarchy construction. In: Proceedings of KDD 2007, pp. 300–309. ACM, New York (2007)
5. Ishii, H., Ma, Q., Yoshikawa, M.: Causal network construction to support understanding of news. In: Proceedings of HICSS 2010, pp. 1–10. IEEE Computer Society, Washington, DC (2010)
6. Jin, P., Lian, J., Zhao, X., Wan, S.: Tise: A temporal search engine for web contents. In: IITA 2008: Proceedings of the 2008 Second International Symposium on Intelligent Information Technology Application, pp. 220–224. IEEE Computer Society, Washington, DC (2008)
7. Fisichella, K.D.M., Stewart, A., Nejdl, W.: Unsupervised public health event detection for epidemic intelligence. In: CIKM 2010: Proceeding of the 19th ACM Conference on Information and Knowledge Management (2010)
8. Mei, Q., Liu, C., Su, H., Zhai, C.: A probabilistic approach to spatiotemporal theme pattern mining on weblogs. In: WWW 2006: Proceedings of the 15th International Conference on World Wide Web, pp. 533–542. ACM, New York (2006)
9. Nallapati, R., Feng, A., Peng, F., Allan, J.: Event threading within news topics. In: CIKM 2004: Proceedings of the Thirteenth ACM International Conference on Information and Knowledge Management, pp. 446–453. ACM, New York (2004)
10. Steuer, J.S.R., Daub, C.O., Kurths, J.: Measuring distances between variables by mutual information. In: Proceedings of the 27th Annual Conference of the Gesellschaft für Klassifikation: Cottbus, pp. 81–90. Springer (2003)
11. Rich, C.: Writing and Reporting News: A Coaching Method, 6th edn. Wadsworth Publishing Company (2009)
12. Smith, D.A.: Detecting and browsing events in unstructured text. In: SIGIR 2002: Proceedings of the 25th Annual International ACM SIGIR Conference on Research and Development in Information Retrieval, pp. 73–80. ACM, New York (2002)
13. Subasic, I., Berendt, B.: Web mining for understanding stories through graph visualisation. In: ICDM 2008: Proceedings of the 2008 Eighth IEEE International Conference on Data Mining, pp. 570–579. IEEE Computer Society, Washington, DC (2008)
14. Tang, J., Zhang, J., Yao, L., Li, J., Zhang, L., Su, Z.: Arnetminer: extraction and mining of academic social networks. In: KDD, pp. 990–998 (2008)
15. Yang, C.C., Shi, X.: Discovering event evolution graphs from newswires. In: WWW 2006: Proceedings of the 15th International Conference on World Wide Web, pp. 945–946. ACM, New York (2006)
16. Yang, C.C., Shi, X.-D., Wei, C.-P.: Tracing the Event Evolution of Terror Attacks from On-Line News. In: Mehrotra, S., Zeng, D.D., Chen, H., Thuraisingham, B., Wang, F.-Y. (eds.) ISI 2006. LNCS, vol. 3975, pp. 343–354. Springer, Heidelberg (2006)
17. Yang, C.C., Shi, X., Wei, C.-P.: Discovering event evolution graphs from news corpora. Trans. Sys. Man Cyber. Part A 39(4), 850–863 (2009)

Dynamic Label Propagation in Social Networks

Juan Du, Feida Zhu, and Ee-Peng Lim

School of Information System,
Singapore Management University
{juandu,fdzhu,eplim}@smu.edu.sg

Abstract. Label propagation has been studied for many years, starting from a set of nodes with labels and then propagating to those without labels. In social networks, building complete user profiles like interests and affiliations contributes to the systems like link prediction, personalized feeding, etc. Since the labels for each user are mostly not filled, we often employ some people to label these users. And therefore, the cost of human labeling is high if the data set is large. To reduce the expense, we need to select the optimal data set for labeling, which produces the best propagation result.

In this paper, we proposed two algorithms for the selection of the optimal data set for labeling, which is the *greedy* and *greedyMax* algorithms according to different user input. We select the data set according to two scenarios, which are 1) finding top-K nodes for labeling and then propagating as much nodes as possible, and 2) finding a minimal set of nodes for labeling and then propagating the whole network with at least one label. Furthermore, we analyze the network structure that affects the selection and propagation results. Our algorithms are suitable for most propagation algorithms. In the experiment part, we evaluate our algorithms based on 500 networks extracted from the film-actor table in freebase according to the two different scenarios. The performance including input percentage, time cost, precision and f1-score were present in the results. And from the results, the greedyMax could achieve higher performance with a balance of precision and time cost than the greedy algorithm. In addition, our algorithm could be adaptive to the user input in a quick response.

1 Introduction

The problem of label propagation has in recent years attracted a great deal of research attention [12, 17, 4], especially in the setting of social networks where an important application of it is to better understand the elements of the network, such as user profiles [8]. As user profiles are often represented by node labels denoting their interests, affiliations, occupations, etc, it is therefore desirable to know the correct labels for as many nodes as possible. However, in real-life social network applications, complete label information of the entire network is rare due to users' privacy concern and unwillingness to supply the information. Consequently, label propagation has been widely used to derive from the known labels of a subset of nodes the unknown ones of the other nodes for the rest of the network [1]. The underlying assumption is the well-observed phenomenon of "homophily" in social networks, i.e., users with strong social connections tend to share similar social attributes.

To trigger the label propagation process over a social network, we need to first acquire the correct labels for an initial set of nodes, which we called a *seed set*. As the acquisition cost of the labels is usually high, e.g., by human labeling and verification, the goal for label propagation in these settings is usually to find as small a seed set as possible such that the knowledge of these node labels would maximize the label propagation. A seemingly similar problem is the classic influence maximization problem, the goal of which is to find as small a set of nodes as possible to initiate certain adoption (e.g., products, innovation, etc.) such that it will trigger the maximum cascade, which has been the focus of many influential research works including Kleinberg's [6].

However, it is important to note the critical difference between our problem and the influence maximization problem. Our label propagation problem has an extra dimension of complexity as a result of the uncertainty of the labels assigned to the seed set. In the influence maximization problem, the labels to be assigned to the seed set are mainly for status indication which are known a priori — if a node is chosen as a seed to initiate the adoption, its label is set as "active", otherwise, its label remains as "inactive". The challenge is to find the right set to assign the initial "active" labels to maximize the cascade. On the other hand, in our label propagation problem, labels represent categorical attributes the values of which remain undecided until specified by users, i.e., for each node in the seed set, technically, users can specify any chosen label from the label universe. The challenge in identifying the right set is not only to study the network structure but to consider all possibilities of label assignment as well.

This important difference between the two problems also suggests that, in our problem setting, a dynamic model of seed set computation based on step-wise user input could be more suitable. Instead of computing the seed set all at once, we in fact should compute the seed set one node at a time based on user input for the next label. As shown in Example 1, different label revealed at each step could lead to drastically different propagation result.

Example 1. In Figure 1, suppose in Step 1 the network is initialized with only node 4 labeled as "A" and the propagation method is the majority voting algorithm such that each node gets the label of the majority of its neighbors. Depending on which node and its label is known in Step 2, we would get entirely different final propagation result. If in Step 2 we know node 6 with label "B", the propagation can not proceed, and if node 5 is further known with the same label "B", the result will be as shown in the right-upper network. On the other hand, if in Step 2 node 1 is revealed with label "A", the network will be fully propagated with label "A". Yet, if in Step 2 node 1 is labeled as "B" instead, then more nodes' labels need to be known in order to continue the propagation.

Therefore, in this paper, we propose the dynamic label propagation problem, which is to find, incrementally based on user input, the optimal seed set to

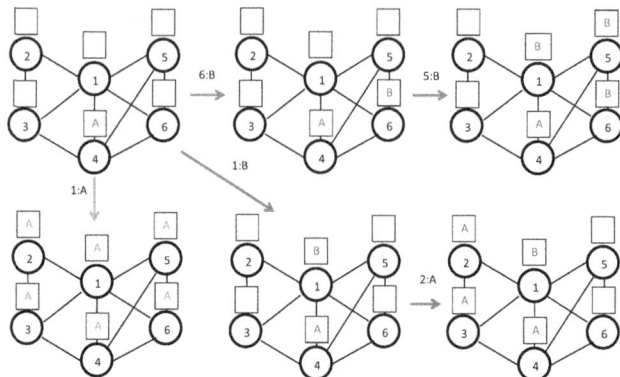

Fig. 1. Examples of different propagation results by dynamic label input

propagate the entire network. A closely related problem is to find the optimal k-size seed set where k is a user-specified parameter, e.g., budget constraint. We show that both problems are NP-hard, and present a greedy algorithm which gives good empirical results. We propose four evaluation criteria and compared different propagation models. To explore the connection between the actual label distribution and the network structure properties, we show the propagation results for some widely-used network measures including density, modularity and single node number. Our empirical results on a real-world data set demonstrate the effectiveness and efficiency of our proposed greedy algorithm.

The rest of the paper is organized as follows. We first introduce some popular propagation algorithms and the relation to our algorithm of finding the optimal given label set in Section 2. And in Section 3 we provide the details of our algorithms. Some network structure analysis that will affect the selection and propagation are shown in Section 4. And we evaluate the algorithms in Section 5. The related work is introduced in Section 6 and finally our work is concluded in Section 7.

2 Problem Formulation

2.1 Problem Definition

We denote a labeled network as $G = (V, E, L)$, where V, E and L represent the non-empty sets of nodes, edges, labels respectively. Given a labeled network $G = (V, E, L)$, there exists an Oracle function $\mathcal{O}_G : V \to L$ such that given a query of any node $v \in V$, $\mathcal{O}_G(v) \in L$, which simulates user input on the node labels. We assume initially no labels are know for any node of G, and each node could obtain a label of L during the label propagation, which could get updated during the process. However, we also assume that labels obtained from the Oracle will never change.

We begin by defining the notion of a *propagation scheme* as follows. The idea is that, given a set of nodes whose labels are known initially, a propagation scheme defines the set of the nodes each of which would obtain a label by the end of the label propagation process. The propagation scheme is defined as a function to achieve the greatest generality since the exact choice of the propagation algorithm would depend on the nature of the application. We leave the detailed discussion of the propagation scheme to subsequent parts of the paper.

Definition 1. *[Propagation Scheme] Given a labeled network $G = (V, E, L)$, an Oracle function $\mathcal{O}_G$ and a $S \subseteq V$ such that for each $v \in S$, $\mathcal{O}_G(v)$ is known, a* propagation scheme *is a function $P : 2^V \to 2^V$ such that $P(S) \subseteq V$ and for each $v \in P(S)$, v would obtain a label by the end of the label propagation process.*

The question of the greatest interest to users is the *Minimum Label Cover (MLC)* problem which is to find the smallest node set to obtain labels initially such that the subsequent propagation could cover the whole network, i.e., assign labels for every node. A closely related problem is the *K-set Label Cover (KLC)* problem in which we are interested in how much of the network we can at most cover if we know the labels of K nodes, which is useful for applications in which a budget is given to acquire the initial labels. These two problems are related in that a solution to the KLC problem would also give a solution to the MLC problem. Notice that in both problem settings, the Oracle to reveal the node labels is not available to the algorithm to find the seed set. In contrast, in our dynamic problem definitions later, the Oracle is available at each step to answer label queries.

Definition 2. *[Minimum Label Cover (MLC)] Given a labeled network $G = (V, E, L)$ and a propagation scheme $P(.)$, the* Minimum Label Cover *problem is to find a node set $\mathcal{S}$ of minimum cardinality, such that the label propagation as defined by $P(.)$ would cover the entire network, i.e., $\mathcal{S} = \operatorname*{argmin}_{|S|}\{S|P(S) = V\}$.*

Definition 3. *[K-set Label Cover (KLC)] Given a labeled network $G = (V, E, L)$, a propagation scheme $P(.)$ and a positive integer K, the* K-set Label Cover *problem is to find a node set $\mathcal{S}^K$ of cardinality K such that the label propagation as defined by $P(.)$ would achieve the maximum coverage of the network, i.e., $\mathcal{S}^K = \operatorname*{argmax}_{|P(\mathcal{S})|}\{\mathcal{S}||\mathcal{S}| = K\}$.*

We are now ready to define our dynamic label propagation problem, which essentially is to solve MLC and KLC incrementally given user input at each step.

Definition 4. *[Dynamic Minimum Label Cover (DMLC)] Given a labeled network $G = (V, E, L)$, an Oracle function $\mathcal{O}_G$ and a propagation scheme $P(.)$, the* Dynamic Minimum Label Cover *problem is to find a node sequence of minimum length, $\mathcal{S} = (v_1, v_2, \ldots, v_{|\mathcal{S}|})$, such that the label propagation as defined by $P(.)$ would cover the entire network, i.e., $\mathcal{S} = \operatorname*{argmin}_{|S|}\{S|P(S) = V\}$.*

Definition 5. *[**Dynamic K-set Label Cover (DKLC)**]* *Given a labeled network $G = (V, E, L)$, an Oracle function $\mathcal{O}_G$, a propagation scheme $P(.)$ and a positive integer K, the* Dynamic K-set Label Cover *problem is to find a node sequence of length K, $\mathcal{S} = (v_1, v_2, \ldots, v_K)$, such that the label propagation as defined by $P(.)$ would achieve the maximum coverage of the network, i.e.,*
$$\mathcal{S}^K = \underset{|P(\mathcal{S})|}{\operatorname{argmax}}\{\mathcal{S}||\mathcal{S}| = K\}.$$

2.2 Complexity Analysis

In this section we give some complexity analysis of the varied problem settings, mostly based on known hardness results with some quite straightforward problem reductions. The detailed proofs are omitted due to space limit. First it is not hard to see the NP-hardness of the MLC problem as a result of the following theorem from [6].

Theorem 1. *[6] The influence maximaization problem is NP-hard for the Linear Threshold model.*

In our definition of the MLC problem, if we set the propagation scheme to be the function which corresponds to the Linear Threshold model as described in [6], and our label set L to be the set containing only a single label, then the status of a node whether or not it has acquired this label would map exactly to the status of being "active" or "inactive" as in the Linear Threshold model in [6]. Therefore, the influence maximization problem is indeed a sub-problem of the MLC problem. Due to Theorem 1, we have the following theorem for the MLC problem.

Theorem 2. *The MLC problem is NP-hard.*

As we can solve the MLC problem in polynomial time by systematically try a sequence of increasing values of K for the corresponding KLC problem, Theorem 2 implies that the KLC problem is also NP-hard.

Corollary 1. *The KLC problem is NP-hard.*

By similar argument, if we set our label set L to be the set containing only a single label to match exactly the status of a node being "active" or "inactive" as in the Linear Threshold model in [6], the having the Oracle available will not lend additional information as in this case the label, which is actually status, is known a priori. As such, the static versions of the problem are actually subproblems of the dynamic versions. We therefore also have the following results by similar argument.

Theorem 3. *The DMLC problem is NP-hard.*

Corollary 2. *The DKLC problem is NP-hard.*

Since both versions of the dynamic label propagation problems are NP-hard, we resort to heuristic algorithms. In particular, we develop a greedy algorithm which will be detailed in Section 3. In [6], it has been shown that such a greedy hill-climbing algorithm would give an approximation to within a factor of $(1-1/e-\epsilon)$ for Linear Threshold model. It is worth noting that in this paper we are not limited to the Linear Threshold model, as we will discuss in the following. Unfortunately, the approximation bounds are not known for the greedy algorithm in models with other propagation methods, e.g., K-nearest neighbor algorithm, which we would like to explore in our future study.

2.3 Propagation Models

We present a discussion of some widely-used propagation models focusing on their applicability in our problem setting.

K-nearest Neighbor Algorithm. K-nearest neighbor algorithm (KNN) is a method for classification, while in label propagation, it is also widely used. The idea of KNN is that the node will be labeled as the same label as his nearest top-K nodes' labels. The distance of two nodes could be measured by different factors like SimRank [5], which measures the structural-context similarity. In this case, the selection prefers the nodes that are more similar to others.

Linear Threshold Model. Linear threshold model is widely used in information diffusion. Given a set of active node, and a threshold θ for each node, at each step, an inactive node will become active if the sum of the weights of the edges with active neighbors exceeds the threshold θ. Similar to this process, during the label propagation, a node will accept the label if the sum of the weights of the edges with neighbors by this label exceeds the threshold θ. In the linear threshold model, the selection prefers the nodes with higher degree and higher edge weights. We call it majority voting in the following sections to differentiate the propagation with information diffusion.

Independence Cascading Model. Independence cascading model is another widely used model in information diffusion and was also deeply discussed in [6] along with linear threshold model. When a node v becomes active in step t, it is given a single chance to activate each currently inactive neighbor w with a predefined probability. In addition, if v succeeds, then w will become active in step $t+1$; but whether or not v succeeds, it cannot make any further attempts to activate w in subsequent rounds. Obviously, in the label propagation scenario, node v should be able to propagate its labels out at any steps rather than only once. And therefore, this model is not suitable for label propagation.

Supervised Learning Algorithm. Supervised learning algorithms use the nodes with existing labels to train the classifier and then propagate to the unlabeled nodes, like Support Vector Machine(SVM) and Hidden Markov Model.

These algorithms need a certain number of labeled nodes as training dataset to train the model first. However, in our case, the labeling of the nodes is unknown and need to be adaptive to the user input in a quick response, and thus the supervised learning algorithms are not quite suitable.

3 Seed Node Detection Algorithm

3.1 Design Ideas

The complexity of the formation of set $\mathcal{S}$ is $O(2^n - 1)$, where n is vertex number. In addition, the selection also needs to consider the situation of nodes with different labels, which consequently will decrease the performance. So before propagation on the incomplete network, we need to employ some techniques to simplify it first. And according to the different characteristics of various networks, the techniques might be varies. Here, we introduce two approaches: pruning and clustering.

Pruning. In social network like twitter, there are some users who have many followers such as celebrities, film stars and politicians. We call these users as "Hub users". When the label stands for affiliations rather than interests, the propagation will fail due to the existence of these "Hub users". In addition, the normal users who do not know each other off-line will decrease the performance of propagating affiliations as well. And thus, we need to prune some users before propagation under different circumstances.

Besides, in social network, some nodes are isolated due to a lot of reasons such as they are puppets or new-comers. In this situation, the degree of these nodes in a certain target network is usually small. If these spam nodes are not essential in the specific scenario, then it could be pruned. Since different pruning techniques will be employed according to different label propagation scenarios, so the modification of the network will not affect the propagation result significantly.

By pruning techniques, we could not only remove noise nodes to increase precision, but also decrease the number of nodes in the network. And thus the computation time according to $O(2^n - 1)$ will be reduced.

Clustering. To reduce the complexity, another step is to divide the network into several subgraphs. However, a question is that if the network could be clustered well and then the minimum set $\mathcal{S}$ will be inferred by randomly choose a node in each cluster directly. Actually, the clustering approach of previous research works cannot achieve best results, which consequently leads the wrong labeling by choosing only one node. And therefore, the idea is that, before propagation, we just do a roughly clustering on the network. For each cluster, we select a minimum set $\mathcal{S}'$, and then union all the $\mathcal{S}'$. During the combination, the nodes those could be propagated by the others in $\bigcup \mathcal{S}'$ will be deleted. Actually, the selection after clustering might not be the optimal one compared to that on the original network. However, to deal with large networks, it works when considering time cost.

Finding Set $\mathcal{S}$. To select set $\mathcal{S}$, there are two approaches. The first one is to give the final result $\mathcal{S}$ directly in off-line mode, and the other one is to add the node to $\mathcal{S}$ online. The main difference between these two selection processes is that the second one is more dynamic. In the off-line modes, as long as the prediction of one node's label is different from the actual one, the selection according to the propagation result might be changed, as shown in Example 1. And thus, it needs to pre-compute all the situations for the nodes with different labels, which is impractical for large and complex networks. On contrary to the off-line mode, the online one picks up the nodes dynamically according to the user input. Once a node's label was given by users, the selection considers the current network states. In another words, in each iteration i, the selection depends on the network structure and the labels constructed in $(i-1)$'s iteration rather than the initial network state. The details of the algorithm will be shown as follows.

3.2 Algorithm

We propose a greedy algorithm to select the set $\mathcal{S}$ dynamically, which is called *G-DS*. In each iteration, we pick up a node which maximizes the propagation coverage. The measurement of the maximization considers different labels. Suppose in the i's iteration, the existing labels in the network are the set $\mathcal{L} = l_1, l_2, ..., l_k$. And then, for each unknown vertex v, we calculate the increase coverage "Cov" by v labeled as l_x as

$$Cov(v, l_x) = \frac{\#known\ label\ nodes}{\#total\ nodes}, \qquad (1)$$

and the probability "P" that v to be labeled as l_x according to the current network status in the $(i-1)$'s iteration is

$$P(v, l_x) = \frac{\#nodes\ labeled\ as\ l_x\ in\ v's\ neighbors}{\#v's\ neighbors}, \qquad (2)$$

and finally sum up $Cov * P$ for each label to get the average coverage $AvgCov$ as Equation 3. In each iteration, we pick up the vertex with the largest $AvgCov$ score. The *G-DS* algorithm is shown in Algorithm 1.

$$Score = AvgCov(v) = \sum_{l_x \in \mathcal{L} \cup l_{new}} Cov(v, l_x) * P(v, l_x). \qquad (3)$$

However, the performance of the *G-DS* algorithm is low. The time complexity is $O(n^3)$ in the worst case, where n is the node number in the network. Each time to choose a node, it needs to calculate the score for each unlabeled nodes with different existing labels. In order to improve the efficiency, we propose a semi-greedy algorithm, called *GMax-DS*. *GMax-DS* algorithm is similar to the *G-DS* algorithm. However, instead of calculating the $AvgCov$ score for all the situations, it only considers the most possible label for node v according to the current network states as

Algorithm 1. *G-DS Algorithm*

Require: $G = (V, E, L), k$
Ensure: $|\mathcal{K}| == k$ or $|\mathcal{K}| == |V|$
1: $\mathcal{S} = \emptyset, \mathcal{K} = \emptyset, L = \emptyset$
2: **while** $|\mathcal{K}| < k$ or $|\mathcal{K}| < |V|$ **do**
3:     start the i's iteration
4:     **for** each $v \in (V - \mathcal{K})$ **do**
5:        **for** each $l_x \in L$ **do**
6:           compute $AvgCov(v)$ according to the $(i-1)$'s iteration
7:        **end for**
8:     **end for**
9:     $\mathcal{S}.add(max(AvgCov(v)))$
10:    input the label $\mathcal{O}_G(v)$ for the node with the $max(AvgCov(v))$ score
11:    propagate the network by $\mathcal{S}$, update $\mathcal{K}$
12:    add l' to L if L does not contain it
13: **end while**
14: **return** $\mathcal{S}$

in Equation 4. For example, for vertex v, its neighbors have labels like l_1, l_2 and l_3, among which, l_1 occurs most, and thus the score is calculated as $Cov(v, l_1)$.

$$Score = Cov(v, l_{max}) = Cov(v, l_{max}) * P_{max}. \quad (4)$$

In the *GMax-DS* algorithm, we replace the score in Algorithm 1 line 6 with the one in Equation 4. Since we just consider the label with the highest probability during the calculation, the computation cost will be significantly decreased. And the time complexity of *GMax-DS* in the worst case is $O(n^2)$.

4 Social Network Structure

In real cases, the label distribution is related to the network structures. And therefore, the network structure will also affect the performance of our algorithm. So in this section, we present some structure features in social networks which might influence the performance of the selection. Actually, the performance is related to the propagation method as well. And thus, all the comparisons are based on the same propagation method. We randomly extracted 15 networks from film-actor table in FreeBase[1], and compared the performance based on two simple propagation methods, KNN and majority voting algorithms. The results based on different network attributes are shown in Figure 2.

4.1 Graph Density

If the graph is less dense, then it indicates that the nodes are not well connected to others. Usually, the nodes with the same labels are more coherent. The propagation methods propagate the labels to a node from its neighbors or similar

---
[1] http://www.freebase.com/view/film/actor

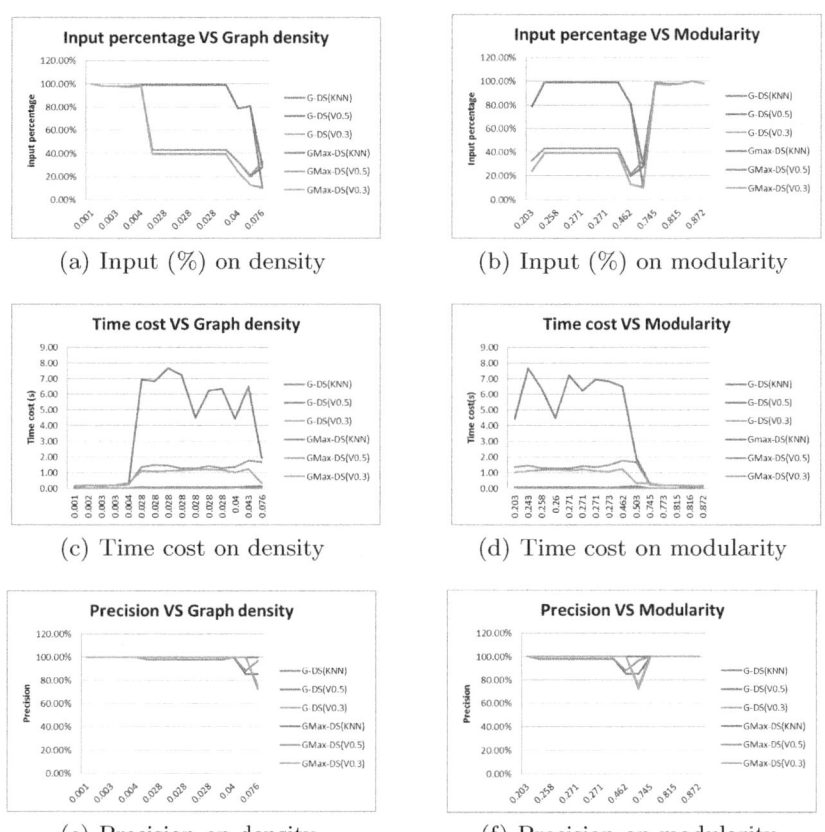

Fig. 2. Performance according to the *G-DS* and *GMax-DS* algorithms by KNN and Majority voting propagation methods under different network structures (The different methods are shown in different colors. Note that some networks are with similar properties and thus their points meet on the graph.)

nodes. However, in a sparse network, it is hard to propagate the labels. As in Figure 2.(a), (c) and (e) shows, the performance will arise linearly when the density increases.

4.2 Modularity

Modularity[10] measures the strength of a division of a network into modules. Networks with high modularity have dense connections between the nodes within the modules but sparse connections between nodes in different modules. So according to the definition of modularity, a network with higher modularity requires less input for labels. In addition, it is much easier for labels to be propagated within the modules rather than across the modules, and therefore increases the precision. The results are shown in Figure 2.(b), (d) and (f), which indicates that the performance will also arise when modularity increases. (The modularity score is calculated based on [2].)

4.3 Single Node Number

Actually, according to the analysis of graph density and modularity, the tendency of input percentage, time cost and precision shown in Figure 2 should be linearly. However, there is some exceptions. We further looked into these networks and found that these graphs include many one-degree nodes, which we also mentioned in Section 3. And here, if we just propagate these nodes from the only neighbor they connect to is unsafe. So here we will just pick up these one-degree nodes and add them to the input set, which increases the input percentage and the final precision in our result.

The reasons why we choose density and modularity as the attributes we further looked into is that: 1) the network structures they present affect the propagation performance directly, and 2) they are related to some other attributes like average degree, cluster coefficient, etc. However, there might be some other factors. And due to the limitation of the pages, we do not enumerate all the attributes here.

5 Experiment

5.1 Dataset

We utilized the film-actor table from FreeBase. In a network, the nodes indicate the actors and the edges stand for the relations that these two actors appeared in the same film. The labels for the node are the films that the actor performed within the network. We randomly extracted 500 networks from FreeBase for different actors' networks. The descriptions of the 500 networks are shown in Table 1. Furthermore, in Figure 3, we present a propagation result for a 131-vertex network from the set we select by *GMax-DS* and KNN, where the color indicates different labels. The size of the seed set is 13 and the precision is around 84.2%, which strongly illustrates that our algorithm works in real case.

Table 1. Description for the 500 networks extracted from freebase film-actor table

	Node	Edge	Density
#Minimum	2	81	0.94
#Medium	13	148	2.15
#Max	458	6937	6.21
#Average	24.50	377.51	1.84

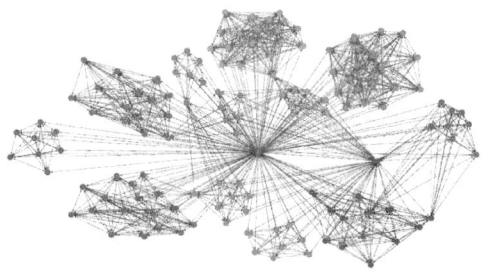

Fig. 3. Case study of the propagation result: the size of set S is 13, and the vertex number is 131. The precision is 84.2%.

5.2 Experiment Setup

We compared the *G-DS* and the *GMax-DS* algorithm according to different performance measurements. And based on the two scenarios we discussed in section 1, the comparison also included the two scenarios, which are the selection of the minimal set and the size-k set. The propagation algorithms we chose here are KNN and majority voting. In addition, we also compare our algorithm with the naive off-line one, which is to check all the possible seed sets and pick up the best one. "TK" stands for the selection of top-k nodes while "A" is to cover the whole network. And "GA" indicates the *G-DS* algorithm while "GM" means the *GMax-DS* algorithm. In addition, "K" and "V" indicate the propagation algorithms respectively, "K" is KNN and "V" is the majority voting with different thresholds as 0.3 and 0.5.

5.3 Experiment Result

Time Cost. Since our algorithm needs to be adaptive to the user input, so the time cost for the selection should be limited. Once the labels for a node are decided, our algorithm needs to pick up another node into S for input quickly. So, in our experiments, we evaluate the time cost in both scenarios. The result is shown in Figure 4, and notice that the time is normalized to log. Mostly our algorithm spent only less than 0.0001 seconds to select the data set for input. The only one extreme case is larger than 1. Since *G-DS* algorithm need to consider all the possible labels in the selection, it takes more time than *GMax-DS* algorithm when the network is with more labels.

F1-score under KMLC. Considering the scenario of selecting the size-k set for input, we evaluate the f1-score where k is equal to 3, 5 and 10. The result is given in Figure 5. When using the KNN propagation algorithm, the f1-score could be mostly beyond 90%. On contrary, by majority voting algorithms with threshold as 0.3, the f1-score is around 70% in average. In addition, comparing *G-DS* with *GMax-DS* algorithms, *GMax-DS* outperforms *G-DS* algorithm in both propagation algorithms.

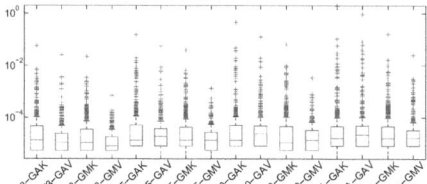
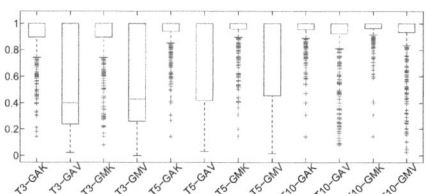

Fig. 4. Time cost (normalized to log) of the selection

Fig. 5. F1-score of selecting top-K nodes

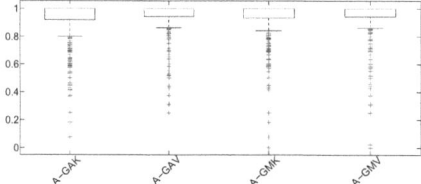
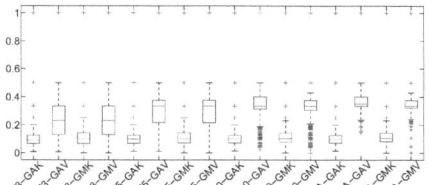

Fig. 6. Precision of covering all the nodes with at least one label

Fig. 7. Input percentage of covering all the nodes with at least one label

Precision under MLC. To pick up the minimal set, we evaluate the precision score and the results are shown in Figure 6. The median number mostly reaches 100% and the lower bound of the precision is around 80%, which indicates that our selection performs well to ensure the precision independence of the propagation algorithms.

Input Percentage under MLC. Actually, the precision is also related to the input percentage. When the input percentage is higher, then the precision will consequently be higher. So we further looked into the input percentage by different propagation methods with *G-DS* and *GMax-DS* algorithms. The evaluation result is illustrated in Figure 7. In general, the input percentage is less than 40%, and the average value for KNN and majority voting is 10% and 30% respectively. Some values are even smaller as 1 or 3. In this experiment, we could find that according to different propagation methods, the input percentage could varies, which has already been discussed in Section 2. In addition, by KNN, the input percentage is around 10% in average and the maximum value is around 25%. In most real cases, this number of input is acceptable.

Compared to the Naive Selection. Furthermore, we also compared our *GMax-DS* algorithm to the force brute selection. The results are shown in Table 2. The propagation method here is KNN. We might see from the table that the precision and the input percentage of our algorithms are mostly the same as the naive one. However, considering the time cost, different with our algorithm, the naive one will increase exponentially when the vertex number increases. On contrary, ours grow linearly and is under control.

Table 2. Performance comparison between naive algorithm and GMax-DS algorithm based on the networks with different vertex number

		5	10	15	19	20
Input percentage	Naive	20%	10%	7%	5%	10%
	GMax-DS	20%	10%	7%	5%	15%▲
Precision	Naive	100%	100%	100%	100%	100%
	GMax-DS	100%	100%	100%	100%	95%▼
Time cost(s)	Naive	0.1	8	**789**	**7882**	**18379**
	GMax-DS	0.015	0.124	0.063	0.156	0.327

From the above experiments, we might infer that the time cost of the *GMax-DS* algorithm is less than that of the *G-DS* algorithm. And in general, the time cost is limited to an acceptable value, which could be adaptive to the user input. In addition, to find the size-k set $\mathcal{S}$, even the size is quite small as 3, some network could also be propagated well and achieve higher f1-score. However, it would be better if k increases. To select the set $\mathcal{S}$ to cover the whole network, the precision could achieve higher even the input percentage is small.

6 Related Work

To our knowledge, there is no work on *dynamic* label propagation in social network. However, there is some researches in information diffusion to find the most influential user sets, which is similar to our problem to some extents. Both are propagated from neighbors. But the difference is that, in information diffusion, the status of a node is usually active or in active[9]; while in label propagation, the node might have multiple labels. In information diffusion, the status is not intrinsic like retweeting the posts [11], while for label propagation, a node's label like affiliation is intrinsic and will not changed according to different network structures. And thus, the problem in label propagation is more complicated than that in information diffusion.

In information diffusion, one of the most widely used algorithms to find the most influential nodes is the greedy algorithm. David Kempe [6] proposed a greedy algorithm to maximize the spreading of influence through social network first. He proved that the optimization problem of selecting the most influential nodes is NP-hard and provided the first provable approximation guarantees for efficient algorithm. The algorithm utilized the submodular functions to ensure finding the most influential nodes in each iteration. Later, based on Kempe's work, Yu Wang [13] proposed a new algorithm called Community-based Greedy algorithm for mining top-K influential nodes to improve the efficiency in large graphs.

In addition, there are some other attributes to measure the role of the nodes in the network, like the degree centrality, closeness centrality, betweenness centrality, eigenvector centrality, etc. [14] measured the node's importance in the network respectively in different aspects. And some papers also compared different measures. For example, Kwak et al. [7] looked into three measurements - number of followers, page-rank and number of retweets, and drew a conclusion that the finding of influential nodes differs by different measurements. As well as Kwak's work, [3] and [15] also compared different measures of influence like number of mentions, a modified page-rank accounting for topics, also found that the ranking of the influential nodes depends on the influence measures. In our problem of label propagation, the selection of the data set for input also differs by utilizing different propagation methods. And our selection algorithm should be adaptive to various propagation methods.

7 Conclusion

In this paper, we proposed the *G-DS* and *GMax-DS* algorithms to select the optimal seed set to maximize the propagation performance. Due to the label complexity, our algorithm could adjust itself dynamically according to the various user inputs. In addition, we further analyzed various network structure attributes since they are related to the label distribution and will affect the selection directly. Our empirical evaluations on real-world FreeBase data set demonstrated the effectiveness and efficiency of our algorithm in terms of input percentage, time cost, precision and f1-score.

Acknowledgments. This research is supported by the Singapore National Research Foundation under its International Research Centre @ Singapore Funding Initiative and administered by the IDM Programme Office.

References

[1] Bakshy, E., Hofman, J., Mason, W., Watts, D.: Identifying influencers on twitter. In: Fourth ACM International Conference on Web Seach and Data Mining, WSDM (2011)
[2] Blondel, V., Guillaume, J., Lambiotte, R., Lefebvre, E.: Fast unfolding of communities in large networks. Journal of Statistical Mechanics: Theory and Experiment 2008(10), P10008 (2008)
[3] Cha, M., Haddadi, H., Benevenuto, F., Gummadi, K.: Measuring user influence in twitter: The million follower fallacy. In: 4th International AAAI Conference on Weblogs and Social Media (ICWSM), vol. 14, p. 8 (2010)
[4] Gregory, S.: Finding overlapping communities in networks by label propagation. New Journal of Physics 12(10), 103018 (2010)
[5] Jeh, G., Widom, J.: Simrank: a measure of structural-context similarity. In: Proceedings of the Eighth ACM SIGKDD International Conference on Knowledge Discovery and Data Mining, pp. 538–543. ACM (2002)
[6] Kempe, D., Kleinberg, J., Tardos, É.: Maximizing the spread of influence through a social network. In: Proceedings of the Ninth ACM SIGKDD International Conference on Knowledge Discovery and Data Mining, pp. 137–146. ACM (2003)

[7] Kwak, H., Lee, C., Park, H., Moon, S.: What is twitter, a social network or a news media? In: Proceedings of the 19th International Conference on World Wide Web, pp. 591–600. ACM (2010)
[8] Lampe, C., Ellison, N., Steinfield, C.: A familiar face (book): profile elements as signals in an online social network. In: Proceedings of the SIGCHI Conference on Human Factors in Computing Systems, pp. 435–444. ACM (2007)
[9] Myers, S., Zhu, C., Leskovec, J.: Information diffusion and external influence in networks. In: Proceedings of the 18th ACM SIGKDD International Conference on Knowledge Discovery and Data Mining, pp. 33–41. ACM (2012)
[10] Newman, M.: Modularity and community structure in networks. Proceedings of the National Academy of Sciences 103(23), 8577–8582 (2006)
[11] Sun, E., Rosenn, I., Marlow, C., Lento, T.: Gesundheit! modeling contagion through facebook news feed. In: Proc. of International AAAI Conference on Weblogs and Social Media, p. 22 (2009)
[12] Wang, F., Zhang, C.: Label propagation through linear neighborhoods. IEEE Transactions on Knowledge and Data Engineering 20(1), 55–67 (2008)
[13] Wang, Y., Cong, G., Song, G., Xie, K.: Community-based greedy algorithm for mining top-k influential nodes in mobile social networks. In: Proceedings of the 16th ACM SIGKDD International Conference on Knowledge Discovery and Data Mining, pp. 1039–1048. ACM (2010)
[14] Wasserman, S., Faust, K.: Social network analysis: Methods and applications, vol. 8. Cambridge University Press (1994)
[15] Weng, J., Lim, E., Jiang, J., He, Q.: Twitterrank: finding topic-sensitive influential twitterers. In: Proceedings of the Third ACM International Conference on Web Search and Data Mining, pp. 261–270. ACM (2010)
[16] Xie, W., Normal, E., Li, C., Zhu, F., Lim, E., Gong, X.: When a friend in twitter is a friend in life. In: Proceedings of the 4th International Conference on Web Science, pp. 493–496 (2012)
[17] Zhu, X., Ghahramani, Z.: Learning from labeled and unlabeled data with label propagation. Tech. rep., Technical Report CMU-CALD-02-107, Carnegie Mellon University (2002)

MAKM: A MAFIA-Based *k-Means* Algorithm for Short Text in Social Networks

Pengfei Ma and Yong Zhang

Research Institute of Information Technology, Tsinghua University, Bejing 100084, China
mpf07@mails.tinghua.edu.cn, zhangyong05@tsinghua.edu.cn

Abstract. Short text clustering is an essential pre-process in social network analysis, where k-means is one of the most famous clustering algorithms for its simplicity and efficiency. However, k-means is instable and sensitive to the initial cluster centers, and it can be trapped in some local optimums. Moreover, its parameter of cluster number k is hard to be determined accurately. In this paper, we propose an improved k-means algorithm MAKM (MAFIA-based k-means) equipped with a new feature extraction method TT (Term Transition) to overcome the shortages. In MAKM, the initial centers and the cluster number k are determined by an improved algorithm of Mining Maximal Frequent Item Sets. In TT, we claim that co-occurrence between two words in short text represents greater correlation and each word has certain probabilities of spreading to others. The Experiment on real datasets shows our approach achieves better results.

Keywords: Short Text Clustering, K-Means, Feature Extraction, MAKM, TT.

1 Introduction

Social networks have been very popular in recent years, people can post the messages of what they hear, see or think, and most of them are short texts. Short text clustering is an essential pre-process in social network analysis, due to incredible growth of data, the efficiency and scalability of the clustering algorithm become more and more important. K-means is a widely used method in various areas, such as data mining, machine learning and information retrieval [2-4]. Because of its simplicity and efficiency, the k-means algorithm is accepted and widely used in many different applications.

However, there are two main problems in k-means algorithm. First, the algorithm is extremely likely to be trapped in some local optimums. Second, the algorithm needs to fix the parameter k, because different k values can cause great affection to the clustering result. Although there are some studies on the choices of initial centers to avoid local optimums [5-8], these methods have not made notable gains comparing with the normal k-means algorithm, and less of them consider how to fix the parameter k.

The main contribution of this paper is an improved k-means algorithm MAKM (MAFIA-based k-means) for short text clustering, in which the optimized initial centers and number k of clusters can be determined. In order to support MAKM, we also propose a new feature extract method TT (Term Transition) for short text.

MAKM is based on the fact that there are some key words frequently occurring in the same clusters, while the key words of different clusters are not exactly the same. Hence, the procedure for mining maximal frequent item sets is actually a course of clustering, with which the initial centers and cluster number k can be fixed. However, before this, there are two preconditions.

Considering a cluster with key words T_1 and T_2, a text may contain only T_1 or T_2, or both, and we may get two maximal frequent item sets $\{T_1\}$, $\{T_2\}$, that will result in a split of the cluster. In TT, each word has certain probabilities of spreading to others. The more frequently two words co-occur, the larger probability of spreading to each other they have. So, the text simply containing T_1 could also be very similar to the text where only T_2 appears, thus the maximal frequent item set will be $\{T_1, T_2\}$.

Secondly, there hasn't been a suitable minimum support for the algorithm of Mining Maximal Frequent Item Sets when the clusters' sizes are greatly different. Specifically, large cluster will be split into several small clusters with small minimum support, and small cluster will be lost with large minimum support. In MAKM, we find all maximal frequent item sets with different minimum support in different iterations.

The rest of the paper is organized as follows. In Section 2, we describe the new feature extraction method TT. In Section 3, we present the improved k-means algorithm MAKM. In Section 4, we show the experiment results on real datasets. Finally, we conclude this paper in Section 5.

2 TT Algorithm

TT (Term Transition) is a new feature extraction algorithm for short text, which contains two basic core ideas: feature transition and dimension reduction.

2.1 Feature Transition

In TT, each word has certain probabilities of spreading to others. The more frequently the words co-occur, the larger probability they will have. In order to derive the law of this diffusion, some auxiliary data structures are defined as follows.
- Transition Probability Matrix (TPM): TPM is a m × m matrix which describes the transition probability from one word to another. Especially, we treat TPM as P2 when only considering two words, and P2 can be formally defined as

$$P2[i,j] = P(i|j) \times P(j|i) \times \frac{P(j|i)}{P(i|j)+P(j|i)} \tag{1}$$

In order to generate TPM with multiple words, two rules should be satisfied. Rule 1: The sum of any word's transition probabilities to all words (including the word itself) should be one; Rule 2: The ratio of transition probabilities between two words each other should be equal to the ratio with multiple words. So we have

$$\begin{cases} \sum_{k=0}^{m-1} \text{TPM}[i,k] = 1 \\ \frac{\text{TPM}[i,j]}{1-\sum_{\substack{k=0 \\ k \neq i}}^{m-1} \text{TPM}[i,k]} = \frac{P2[i,j]}{1-P2[i,j]} \end{cases} \quad (2)$$

Then we can obtain TPM from equation (3)

$$\text{TPM}[i,j] = \begin{cases} \frac{\frac{P2[i,j]}{1-P2[i,j]}}{1+\sum_{\substack{k=0 \\ k \neq i}}^{m-1} \frac{P2[i,k]}{1-P2[i,k]}} & , i \neq j \\ 1 - \sum_{\substack{k=0 \\ k \neq i}}^{m-1} \text{TPM}[i,k] & , i = j \end{cases} \quad (3)$$

- Variation Probability Matrix (VPM): VPM is a m × m matrix which denotes the final transition probability matrix. Let α be a feature vector of a short text, then $\hat{\alpha} = \alpha \cdot \text{VPM}$ is the final feature vector after word transition. The final transition probability from word i to j depends on the weights and transition probabilities of all paths that arrive at it. For example, the word i can transfer to j through different paths, such as $\{i \to j\}, \{i \to k_1 \to j\}, \{i \to k_1 \to k_2 \to j\}, \dots, \{i \to k_1 \to \dots \to k_m \to j\}$, It's something like but different from Markov Chain, because we consider that the paths of different lengths have different weights. So we can define VPM as

$$\text{VPM} = \frac{1}{\Lambda}\sum_{k=1}^{\infty}(\lambda \cdot \text{TPM})^k = (1-\lambda) \cdot \text{TPM} \cdot (I - \lambda \cdot \text{TPM})^{-1},$$

$$\text{where } \Lambda = \sum_{k=1}^{\infty}\lambda^k = \frac{\lambda}{1-\lambda}, \text{and } 0 < \lambda < 1 \quad (4)$$

Note that, the matrix $(I - \lambda \cdot \text{TPM})$ is certainly invertible in equation (4). Because each value in TPM is between 0 to 1 and the summation of each row is 1, according to the Gershgorin Circle Theorem, the moduli of TMP's eigenvalues are less than 1, so $|I - \lambda \cdot \text{TPM}| = \lambda \left|\frac{1}{\lambda}I - \text{TPM}\right| \neq 0$ ($0 < \lambda < 1$), and it's invertible. We call λ the decay factor which corresponds to the weight of different transitions. Through experiments on real datasets we found that it can achieve better results when $\lambda = 0.75$.

2.2 Dimension Reduction

The standard deviation of different dimensions reflects the ability to distinguish clusters. The dimensions with lower standard deviation are helpless or even harmful to clustering. In TT algorithm, it calculates the feature vectors in higher dimensional space using TF-IDF and updates each feature vectors based on equation (4). Then it calculates and sorts the standard deviation of each dimension, and prunes the dimensions with lower standard deviation using MDL (Minimal Description Length) method. Finally, it normalizes all feature vectors. Through experiments we found that most standard deviations are very small, and these words will bring negative effect to the clustering, such as extra time consumption and the fuzzy edges of different clusters. So the pruning is beneficial to the clustering algorithm.

The TT algorithm is summarized as follows:

```
Algorithm-TT
Input: The dataset Φ of short text.
Output: The feature vectors of each short texts.
1: Sampling of Φ to get a sub dataset φ;
2: Calculate VPM on the sub dataset φ by equation (4);
3: Initialize feature vectors of each short texts on the sub
    dataset Φ using TF-IDF;
4: Update feature vector by α=α·VPM;
5: Dimension reduction by MDL pruning;
6: Normalize all feature vectors in the new dimensional space.
```

3   MAKM Algorithm

MAKM (MAFIA-based k-means) is an improved k-means algorithm for short text clustering. The algorithm can find the suitable number k of clusters and choose the optimized initial centers automatically with only one input parameter Grain Factor (γ) that ranges from zero to one and describes the clustering rough degree. If γ is approaching to one, the similar small clusters will be combined into one cluster, while the big one will be split to several small clusters when γ is approaching zero.

In MAKM, we choose the optimized initial centers and find the number k of clusters through an improved MAFIA [1] algorithm, a method of Mining Maximal Frequent Item Sets. MAFIA has one input parameter, the minimum support (minsup), which directly determines the result. We will find all maximal frequent item sets (clustering centers) with different minimum support in different iterations.

3.1   The Conversion between Short Text Dataset and Transaction Set

In order to use the method of MAFIA, we need convert a short text dataset to a transaction set. Each short text will be converted to a transaction as follows:

Let $\alpha = (x_1, x_2, \ldots, x_m)$ be a feature vector of a short text, and suppose M is the root mean square of α, i.e. $M = \sqrt{\frac{\sum_{j=1}^{m} x_i}{m}}$, then the transaction T can be obtained as

$$T = \{t | t = t_i, t_i > M\} \tag{5}$$

Vice versa, if we get a transaction T, the feature vector α can be obtained as

$$\alpha = \frac{\beta}{|\beta|} \cdot VPM, \text{ where } \beta_i = \begin{cases} 1, t_i \in T \\ 0, t_i \notin T \end{cases} \tag{6}$$

3.2   The Initialization of Minimum Support

Let $I = \{i_1, i_2, \ldots, i_m\}$ be a set of words, $T = \{t_1, t_2, \ldots, t_n\}$ be a set of transactions, where each transaction t_k is a set of words such that $t_k \subset I$, and i_k.count is the number of transactions that contain i_k. Then we obtain the support of i_k and average support as

$$\sup_k = \frac{i_k.count}{\sum_{u=0}^{m-1} i_u.count} \quad (7)$$

$$\sup_{ave} = \frac{1}{\sum_{u=0}^{m-1} i_u.count} \times \sum_{k=0}^{m-1} (i_k.count \times \sup_k) \quad (8)$$

The average support is the weighted average of each word's support, where the weight is the support count. In order to make the minimum support change slowly when the grain factor approaching 0.5, while change quickly when the grain factor approaching 0 or 1, we can define the initial minimum support as

$$\sup_{min} = \begin{cases} \min\{\sup_k\} + \exp\left(\frac{\gamma}{0.5} \times \ln(\sup_{ave} + 1 - \min\{\sup_k\})\right) - 1, \gamma < 0.5 \\ \sup_{ave} + \exp\left(\frac{\gamma - 0.5}{0.5} \times \ln(\max\{\sup_k\} + 1 - \sup_{ave})\right) - 1, \gamma \geq 0.5 \end{cases} \quad (9)$$

3.3 The Update of Minimum Support and the Iterative Termination Condition

The maximal frequent item sets are found by invoking MAFIA iteratively until convergence. Let T be all transaction sets, T_{cur} be the transaction set during current iteration, and T_{del} be the deleted transaction set, it's easy to see that they satisfy $T = T_{cur} \cup T_{del}$ and $T_{cur} \cap T_{del} = \emptyset$.

At the beginning of each iteration, the minimum support will be recalculated by equation (9) on transaction set T_{cur}, and then we can get a new local maximal frequent item set (LMFI) by invoking MAFIA.

For each item set in LMFI, if the count in T_{cur} is greater than the count in T_{del}, the item set will be added to global maximal frequent item set (GMFI), and the transaction contained the item set in T_{cur} will be moved to T_{del}. The iterations continue until there is no item set moved.

The MAKM algorithm is summarized as follows:

```
Algorithm-MAKM
Input: The clustering grain factor γ, and the data set.
Output: The clusters.
1: Sampling the data points as dataset D;
2: Convert the dataset D to a transaction set T by equation (5);
3: Initialize the Global Maximal Frequent Item Set, GMFI = ∅;
4: Initialize the transaction sets, T_cur = T and T_del = ∅;
5: REPEAT
6:   Calculate minimum support by equation (9);
7:   Invoke MAFIA to get the Local Maximal Frequent Item Set(LMFI);
8:   Update GMFI, T_cur, and T_del described in 3.3;
9: UNTIL (T_del is no longer changed)
10: Convert GMFI to feature vectors described in 3.1, and update
    feature vectors by α = α·VPM to obtain the initial cluster
    centers, and the size of GMFI is the number of clusters;
11: Invoke K-Means to get the cluster result.
```

4 Experiments

In this section we present an evaluation of the performance of MAKM and TT on real datasets.

4.1 Experimental Framework

Datasets. We gathered data from Sina Microblog, and we used the provided API to gather information about a user's social links and tweets. We launched our crawler in September 2011, and the crawler started at a specified user to the users he followed. We gathered information about a user's follow links and all tweets ever posted. In total, there were 48,401,561 tweets of 84753 users.

We selected seven specific topics, and partitioned them into two completely different distribution datasets. Table1 shows the distribution of the datasets.

Table 1. The Distribution of Datasets

	WENZHOU's train crash	Nuclear leakage of Japan	The centennial anniversary	Li Na won the France Open	World Expo Shangha	Willian Kate: Royal	and The	The credit downgrade of the U.S.
Dataset 1	80,000	40,000	20,000	15,000	8,000	8,000		7,000
Dataset 2	8,000	8,000	8,000	8,000	8,000	8,000		7,000

Preprocess. Before the clustering, we should perform some preprocessing works on the datasets, such as stop words removing and Chinese words segmentation. In our work, we use ICTCLAS, a Chinese words segmentation tool developed by intelligent science laboratory of institute of computing technology, Chinese academy of science.

4.2 Results

We conducted extensive experiments on the two datasets, some results are reported below.

Fig. 1 shows the difference of category centers between the feature extraction method TT and TF-IDF for each category on dataset 1 and dataset 2. We found that, most of D-values (*Difference value*) are approaching zero in a category, and only a few D-values are a little bigger or smaller than zero. It shows that there exist transitions among the key tokens.

Fig. 2 shows the change of clustering average F1 score and the number of clusters found against different value of grain factor γ of MAKM on dataset 1 and dataset 2. We can see that the clustering average F1 score is not sensitive to γ when γ changed from 0.5 to 1.0. However, the average F1 score decreases when γ becomes smaller. This is because larger γ makes similar small clusters merged to one bigger cluster, and smaller γ results in big cluster being split into multiple smaller clusters.

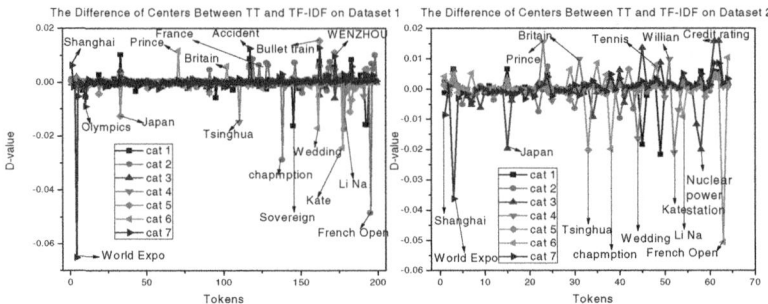

Fig. 1. The centroid difference of each category between TF-IDF and TT in feature extraction

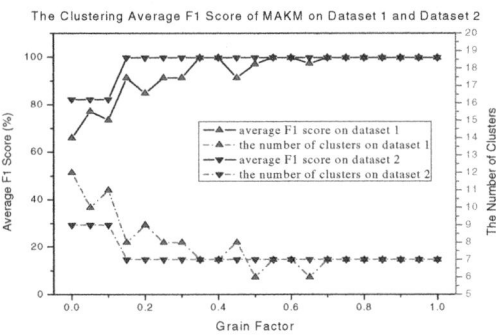

Fig. 2. The effect of grain factor on MAKM

Table 2. The Clustering Result of MAKM when $\gamma = 0.0$

Cluster#	Accuracy(%)	Recall(%)	F1(%)	Category
1	99.99	78.81	88.15	WENZHOU's train crash
2	100.00	42.96	60.10	Nuclear leakage of Japan
3	99.98	84.74	91.73	The centennial anniversary of Tsinghua
4	99.99	99.67	99.83	Li Na won the France Open
5	99.80	17.01	29.06	WENZHOU's train crash
6	99.15	30.01	46.07	Nuclear leakage of Japan
7	100.00	26.91	42.41	Nuclear leakage of Japan
8	99.72	99.67	99.69	World Expo Shanghai
9	99.97	99.78	99.88	Willian and Kate: The Royal Wedding
10	99.91	99.98	99.95	The credit downgrade of the U.S.
11	98.78	4.05	7.78	WENZHOU's train crash
12	99.77	15.26	26.47	The centennial anniversary of Tsinghua

Table2 shows the detailed clustering result of MAKM when $\gamma=0.0$. We can see that the original seven categories are clustered to twelve clusters, and clustering accuracies are mostly larger than 99%. Some clustering recalls are quite low, because of the split of the big clusters.

We run the standard K-Means (KM), Hierarchical K-Means (HKM) [8] with different computation times (p) and MAKM 25 times on dataset 1 both equipped with equipped with TT feature extraction method, and we found the average number of iterations of MAKM is six, while the k-means is sixteen. The result is reported below.

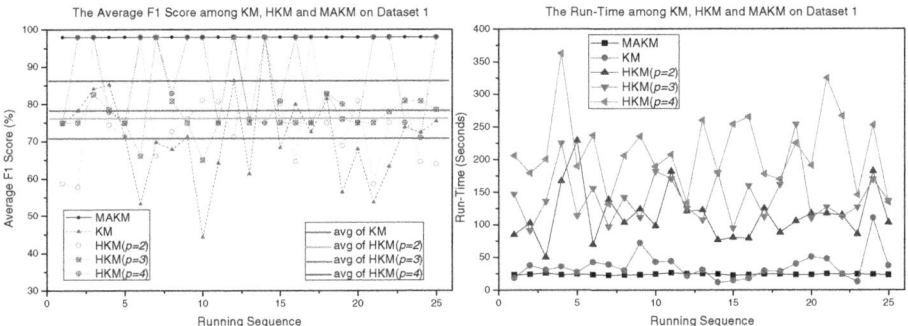

Fig. 3. The average F1 score and running time comparisons among KM, HKM, and MAKM

Fig. 3 shows the comparison results of average F1 score and running time among KM, HKM and MAKM on dataset 1. In the experiment, all algorithms use TT as the feature extraction algorithm. We set MAKM sample half of the dataset to choose the initial clustering centers. The grain factor γ is set to 0.8 in MAKM, and k is set to the actual number of categories in KM and HKM. For HKM, the computation times is set to 2, 3, and 4, respectively.

After 25 times repeats, we can see that our algorithm, MAKM has great advantages both in accuracy and efficiency. The average F1 score of MAKM is 97.94% with average running time 24.21s, KM is 70.96% with 35.86s, HKM (p=2) is 76.43% with 114.64s, HKM (p=3) is 78.72% with 141.99s, and HKM (p=4) is 86.13% with 212.96s. Besides, MAKM is a stable clustering algorithm when the sampling rate is 0.5.

5 Conclusion

In this paper, we proposed an improved k-means algorithm (MAKM) for short text clustering. This algorithm can determine the most suitable number k and optimized initial centers before the k-means procedure, which overcomes the shortage that the standard k-means algorithm may be trapped in some local optimums for its sensitivity to cluster number k and initial clustering centers. We also came up with a new feature extraction method TT for short text which is efficient and beneficial to support MAKM. We carried out experiments on the real datasets and the results show out MAKM can solve the problem of local optimums very well with high accuracy which confirms MAKM is a more stable algorithm for short text clustering.

Acknowledgment. The work described in this paper was supported by grants from National Basic Research Program of China (973 Program) No.2011CB302302, National Natural Science Foundation of China under Grant No. 61170061 and Tsinghua University Initiative Scientific Research Program.

References

1. Goil, S.: MAFIA: efficient and scalable subspace clustering for very large (1999)
2. Elkan, C.: Using the triangle inequality to accelerate k-means. In: ICML, pp. 147–153 (2003)
3. Huang, Z.: Extensions to the k-Means Algorithm for Clustering Large Data Sets with Categorical Values. Data Mining and Knowledge Discovery 2, 283–304 (1998)
4. Hassan-Montero, Y., Herrero-Solana, V.: Improving Tag-Clouds as Visual Information Retrieval Interfaces, Spain, October 25-28 (2006)
5. Bradley, P.S., Fayyad, U.M.: Refining Initial Points for K-Means Clustering. In: ICML (1998)
6. Lu, J.F., Tang, J.B., Tang, Z.M., Yang, J.Y.: Hierarchical initialization approach for K-Means clustering. Pattern Recognition Letters (2008)
7. Aggarwal, A., Deshpande, A., Kannan, R.: Adaptive Sampling for k-Means Clustering. In: Dinur, I., Jansen, K., Naor, J., Rolim, J. (eds.) APPROX 2009. LNCS, vol. 5687, pp. 15–28. Springer, Heidelberg (2009)
8. Arai, K., Barakbah, A.R.: Hierarchical K-means: an algorithm for centroids initialization for K-means. In: Reports of the Faculty of Science and Engineering, vol. 36(1), Saga University, Japan (2007)

Detecting User Preference on Microblog*

Chen Xu, Minqi Zhou**, Feng Chen, and Aoying Zhou

Institute of Massive Computing, East China Normal University
{chenxu,fengchen}@ecnu.cn, {mqzhou,ayzhou}@sei.ecnu.edu.cn

Abstract. Microblog attracts a tremendous large number of users, and consequently affects their daily life deeply. Detecting user preference for profile construction on microblog is significant and imperative, since it facilitates not only the enhancement of users' utilities but also the promotion of business values (e.g., online advertising, commercial recommendation). Users might be instinctively reluctant to exposure their preferences in their own published messages for the privacy protection issues. However, their preferences can never be concealed in those information they read (or subscribed), since users do need to get something useful in their readings, especially in the microblog application. Based on this observation, in this work, we successfully detect user preference, by proposing to filter out followees' noisy postings under a dedicated commercial taxonomy, followed by clustering associated topics among followees, and finally by selecting appropriate topics as their preferences. Our extensive empirical evaluation confirms the effectiveness of our proposed method.

1 Introduction

Personalized applications are rather important and remain to be solved imperatively, especially in the web 2.0 era, with their ultimate intentions for online advertising, i.e., recommending appropriate services (e.g., online games, movies, music and commodities) for the right users [1]. Here, right means users might be interested in the recommended services with high probability.

As one of the foundations of online advertising, user profile leverages the system to recommend appropriate services to right users in terms of their specific needs in a great deal. It is widely used in many personalized application scenarios, e.g., adaptive web search [2] based on dedicated user profile to improve the search results. Herein, user profile includes both his demographic information - e.g., name, age, country and education level - and individual preferences (or interests) [3], i.e, the preference levels on items such as digital products and sport products. User preference is harder to detect but poses more value for online advertising, compared to the demographic information. In this research, we focus on preference detection to construct user profile.

---

* This work is partially supported by National Science Foundation of China under grant numbers 61003069 and 61232002, and National High-tech R&D Program (863 Program) under grant number 2012AA011003.
** Corresponding author.

Our goal is to detect user preference in the new environment, i.e., microblog, such as Twitter and Sina Microblog. Detecting user preference is not easy in such online social network, and one challenge is useful information for preference detection is limited within a small portion, surrounded by a large volume of noises (useless information). For example, over 85% topics in Twitter are headline news or persistent news [4] which are useless in preference detection.

In microblog application, a user is able to post messages, denoted as *tweets*, which are propagated to others via two following relationships (i.e., *followee* and *follower*). Microblog is different from other social networks (e.g., Facebook), in both friendship authorization and linkages. Here, a user is allowed to *follow* any user without the permission barrier [5]. Hence, the relationship linkages on microblog are unilateral, unlike the bilateral ones in the other social networks such as Facebook. This speciality leads to our unique observations: 1) reading the tweets from his followees reflects user's real preferences indeed, even if he is reluctant to exposure preference on his own tweets; 2) the tweets from user's followees do belong to many topics, but only part (not all) of those topics, i.e., the most common ones, reveals his preference; 3) a user prefers the tweets from his followees at different levels and inversely followees' influences over user are not the same with each other.

Existing works such as [6] employ homophily to leverage user's behaviours from social neighbours, but it only covers part of the followees in microblog. The method in [7] simply treats users' followees fairly to extract user preference, which is oblivious to the difference on followees' influence. Hence, these works do not exploit those observations above sufficiently.

In this study, we detect the preference for any user, by filtering out and clustering tweets from his followees, etc., based on the observations listed above and our contributions include:

– Detailed analysis on our collected Sina Microblog dataset is provided, especially the statistics of noisiness, which shows the speciality of microblog.
– We filter out relatively useless followees, and the remainings are the high-quality ones for user preference detection.
– We incorporate social linkages for influence evaluation, which integrates the structure with content on social media context.

The extensive empirical evaluation confirms our *two-step* approach is well suited for preference detection based on our observations and highlights the effectiveness of this method compared to the existing work.

The rest of this paper is organized as: Section 2 describes our dataset used in this paper; Section 3 illustrates the problem formally; Section 4 shows our approach to detect user preference and Section 5 gives empirical evaluation; Section 6 summarizes the related work and Section 7 concludes our paper.

2 Dataset Description

We have crawled a dataset consisted of users, relationship linkages and tweets from Sina MicroBlog for the purpose of preference detection. This dataset is obtained in the following manner. Firstly, we choose 45 seed users, denoted as a set U, from the Sina MicroBlog platform. Then, we crawl the followees and followers of each user in U to form another user set U'. All the tweets of users in $\mathbb{U} = U \cup U'$, involving 11,380 users, are crawled and inputed into a set T.

The period of the tweets in T ranges from October 2010 to March 2012, with the number of tweets per month shown in Figure 1. We perform a semantic grounding of the tweets against a tree-structured reference *taxonomy* [3], with the leaf node as keyword and non-leaf node (except root) as topic (Figure 3). This taxonomy is derived from the homepage of Taobao which is more suitable for Chinese context compared with the translation of Google Taxonomy. Here, the tweet involving any keyword in the taxonomy is considered as an *interesting tweet* since it could be used for preference detection potentially. Figure 1 depicts the number of interesting tweets as well as the one of total tweets per month. For any user, the *utilization* of his tweets is defined as the percentage of the number of his interesting tweets vs. the total tweets of this user. Figure 2 shows the frequency distribution, as well as cumulative distribution, of the percentage of users in $\mathbb{U}$ given utilization. The result indicates the utilization of most users are limited between 20% and 40%, which brings a challenge to preference detection.

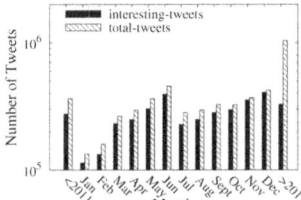

Fig. 1. Tweets Distribution

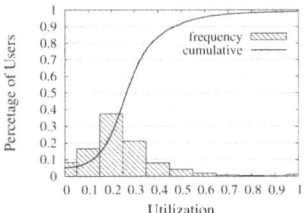

Fig. 2. Tweets Utilization

3 Problem Statement

Formally, $\mathbb{U} = \{u_0, u_1, \cdots, u_{|\mathbb{U}|-1}\}$ represents all the users in a social network and u_i is the $(i+1)^{th}$ user. The social network is depicted by a directed graph, G_f, with users as nodes and following linkages as edges. The edge from u_i to u_j indicates the following relationship between them, i.e., u_j is u_i's followee and u_i is u_j's follower meanwhile. In this study, we use E_{u_i}, R_{u_i} and D_{u_i} to denote u_i's followees, followers and friends (i.e., both followees and followers) respectively.

The problem is to find out what u_i is interested in, i.e., his preference, based on the given information. It includes the following relationship and u_i's tweets as well as the tweets of E_{u_i} and R_{u_i}. The outcome is user preference organized as a vector, $\boldsymbol{P}_{u_i}$, where each element is the preference level on its corresponding item [8]. The items in preference vector come from the top level topics (except root) of the taxonomy in Figure 3 and Table 1 illustrates an example of this vector.

4 User Preference Detection

This section shows our proposed method to detect user preference on microblog. We argue that user's following actions reflect his preferences, since most of the users are reluctant to exposure interests on his own tweets but his reading materials, i.e., followee's tweets, reveal the preferences. Herein, we propose our *two-step* approach to address this issue by extracting the information from followee's tweets (Section 4.1) firstly and then detecting preference with user's following linkages (Section 4.2).

4.1 Tweet Signature Extraction

We start off by extracting the information from the tweets of user's followees, since those are what this user wants to read on the microblog application. Here, u_{ij} is one of u_i's followees and u_{ij}'s tweets are summarized as *tweet signature*. In this work, u_{ij}'s tweet signature is represented as a vector:

$$\boldsymbol{\Theta}_{u_{ij}} = (\theta_{1,u_{ij}}, \theta_{2,u_{ij}}, \cdots, \theta_{n,u_{ij}}, \cdots)$$

where $\theta_{n,u_{ij}}$ is u_{ij}'s preference level on the item θ_n. Items in this vector are a set of the top level topics in the aforementioned taxonomy, such as sport, digital and food in Figure 3.

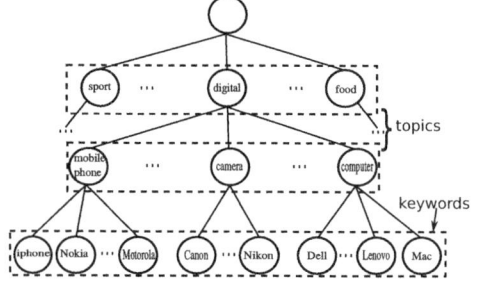

Table 1. Preference Vector

topic	sport,	⋯	digital,	⋯	food	total
vector	(0.2,	⋯	0.36,	⋯	0.01)	1

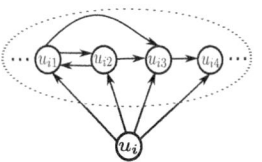

Fig. 3. Tree-structured Taxonomy   **Fig. 4.** An Example of $\mathbb{G}_{u_i}$

The preference level $\theta_{n,u_{ij}}$ is the sum of weighted term frequency of all the descendants of $\boldsymbol{\theta}_n$, i.e., $desc(\boldsymbol{\theta}_n)$, in that taxonomy tree. Formally,

$$\theta_{n,u_{ij}} = \sum_{\boldsymbol{\omega}_x \in desc(\boldsymbol{\theta}_n)} \sum_{0 \leq d < W} e^{-\alpha d} tf(\boldsymbol{\omega}_x, T_{u_{ij}}^{(d)}) \qquad (1)$$

where $T_{u_{ij}}^{(d)}$ is the tweets generated d days ago (less than W days) by u_{ij} and α is a decay factor (we set it to 0.025 as in [9]) to simulate the interest decline.

4.2 Capturing Preference with Following Linkage

The previous section shows how to extract the information from followees' tweets. In this section, we illustrate the method to detect user preference with following linkages. This issue is addressed from the two aspects: 1) in general, a user prefers several main topics on what his followees talk about rather than all the topics of his followees; 2) a user is interested in his followees at different levels which means different followees have different influences on him. The former shows user's preferences on different topics whereas the latter indicates the preference on different people. This idea drives us to filter out useless information and evaluate the followees' influence over him, in order to capture user preference. The following discussions are derived from the two views above.

Followee Filtering. In microblog, a user does need to read something useful from his followees who would post rich and colorful tweets. However, the user is usually interested in several main topics of his followees rather than all the topics. In this work, we exploit clustering to find those topics. That is, the followees are clustered into several clusters by their tweet signatures and then each cluster involves the followees with preference on some distinctive topics.

Clearly, cosine distance is appropriate to measuring the distance between tweet signatures for clustering. In addition, we incorporate *screen name* within similarity comparison, given the particularity of Chinese name. The similarity between two screen names can be evaluated by Jaccard index of q-gram. Hence, the distance function (Eq. 2) between two of u_i's followees, i.e., u_{ix} and u_{iy}, is the combination of cosine distance, in term of the tweet signature, and Jaccard similarity, in term of the screen name.

$$Dist(u_{ix}, u_{iy}) = (1 - \frac{\Theta_{u_{ix}} \cdot \Theta_{u_{iy}}}{|\Theta_{u_{ix}}||\Theta_{u_{iy}}|})[1 - \frac{q(SN_{u_{ix}}) \cap q(SN_{u_{iy}})}{q(SN_{u_{ix}}) \cup q(SN_{u_{iy}})}] \qquad (2)$$

where $SN_{u_{ix}}$ denotes u_{ix}'s screen name and $q(SN_{u_{ix}})$ indicates the q-gram set of $SN_{u_{ix}}$. In particular, we set $q = 2$. Then, we use DBSCAN to cluster those followees. The bigger cluster indicates u_i has a higher preference on the common topics of the followees in this cluster. In other words, the bigger clusters are more useful to detect user preference than smaller ones. Motivated by that, clusters are ranked by size and the followees in top-k clusters, denote as E'_{u_i}, are selected.

Followee Influence. The key point now is to evaluate followee's influence over u_i, since we have acquired the tweet signatures of his followees and filtered out some useless followees. The simplest way is to treat each followee in E'_{u_i} fairly, i.e., *equivalent* influence, as in [7]. That is, u_{ij}'s *influence over u_i, i.e., $Inf(u_i, u_{ij})$, is $\frac{1}{|E'_{u_i}|}$, where $|E'_{u_i}|$ is the number of followees in E'_{u_i}.

It means each followee in E'_{u_i} has an equivalent influence over u_i. However, a user is influenced by followees at different levels generally. As an example shown in Figure 4, u_i follows u_{i1}, u_{i2}, u_{i3} and u_{i4}. Besides, there are other following linkages among followees, e.g., u_{i1} follows u_{i2} and u_{i3}. From the aspect of u_i, u_{i3}'s tweets can not only be pulled to u_i (u_i follows u_{i3}) directly, but also reposted by u_{i1} and u_{i2} (if they do) which might be pulled to u_i further. Intuitively, the influence of u_{i3} over u_i is higher than the one of both u_{i1} and u_{i2} over u_i, in $\mathbb{G}_{u_i}$, i.e., the social network between u_i and his followees in E'_{u_i}.

This example calls for an alternative scoring function to evaluate the importance of each user in $\mathbb{G}_{u_i}$. Let $s^{(t)}(u_{ij})$ denote the score of u_{ij} after t-th iteration and $0 < \mu < 1$ a dumping factor. This scoring function is:

$$s^{(t)}(u_{ij}) = (1-\mu) \sum_{v \in E'_{u_i} \cap E_{u_{ij}}} \frac{s^{(t-1)}(v)}{|E'_{u_i} \cap E_{u_{ij}}|} + \mu \frac{1}{|\mathbb{G}_{u_i}|} \quad (3)$$

where $E'_{u_i} \cap E_{u_{ij}}$ is u_{ij}'s followees in $\mathbb{G}_{u_i}$ and $|\mathbb{G}_{u_i}|$ is the total number of the users in $\mathbb{G}_{u_i}$. Here, we propose the normalized *PageRank-like* influence. That is, u_{ij}'s influence over u_i is $\frac{s(u_{ij})}{\sum_{v \in E'_{u_i}} s(v)}$, where $s(u_{ij})$ is u_{ij}'s score by Eq.3 above.

Hybrid Model. After the followee filtering and influence evaluation, we propose a *hybrid model* which combines u_i and his followees' tweet signatures without loss of generality. In this model, user preference is the weighted sum of his own and followees' tweet signatures. Formally, u_i's preference vector $\boldsymbol{P}_{u_i}$ is:

$$\boldsymbol{P}_{u_i} = (1-\beta)\boldsymbol{\Theta}_{u_i} + \beta \sum_{u_{ij} \in E'_{u_i}} Inf(u_i, u_{ij})\boldsymbol{\Theta}_{u_{ij}} \quad (4)$$

where $\boldsymbol{\Theta}_{u_i}$ is u_i's tweet signature, $\boldsymbol{\Theta}_{u_{ij}}$ is tweet signature of u_{ij}, one of u_i's followees in E'_{u_i}, and β is a parameter tuning the weight between u_i and his followees in E'_{u_i}. Clearly, this model ranks the topics intrinsically to detect the most common ones as user preference.

In particular, different β leads our hybrid model to be with different implications. Firstly, $\beta = 1$ makes u_i's tweet signature no sense. In other words, only the tweets from followees are used to detect u_i's preferences. Secondly, $\beta = 0$ means only user's own tweets are used. Hence, $\boldsymbol{P}_{u_i}$ is equal to $\boldsymbol{\Theta}_{u_{ij}}$. Finally, $0 < \beta < 1$ is the trade-off between two cases above. Here, user preference is generated by both his own and followees' tweet signatures.

5 Empirical Evaluation

In this study, we invite the seed users in our dataset (Section 2) to evaluate the effect of our proposed method. They are asked to judge the top-5 topics in preference vector generated from different sources. DCG is used to evaluate the quality of user preference detected by our approach. We use gain value of 5, 3, 1, and 0 for judgement grade 1, 2, 3 and 4, respectively. In the following, Section 5.1 shows the impact of factor β and relationship in our proposed method and Section 5.2 highlights the effectiveness of our approach.

5.1 Impact of β and Relationship

Under the framework of hybrid model, we evaluate the ability of followee's tweets for preference detection in comparison to 1) user's own tweets, 2) friends' tweets, and 3) followers' tweets. In particular, followees' and friends' influences over user are calculated by PageRank-like influence, whereas followers' influences over user are calculated by equivalent influence.

Table 2. Average DCG(λ)

λ	1	2	3	4	5
$\beta = 0$	2.50	5.22	6.90	8.15	9.59
$\beta = 0.5$, followee	2.77	6.49	7.64	8.94	10.19
$\beta = 1.0$, followee	**2.92**	**6.70**	7.85	**9.35**	**10.64**
$\beta = 0.5$, follower	2.31	5.08	6.59	8.16	9.68
$\beta = 1.0$, follower	2.38	5.05	6.77	8.15	9.76
$\beta = 0.5$, friend	2.65	5.71	7.36	8.34	9.38
$\beta = 1.0$, friend	2.81	6.47	**8.01**	8.95	10.08

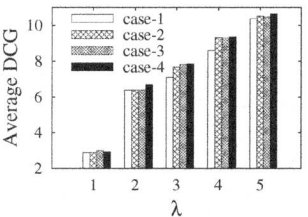

Fig. 5. Diff. Combinations

Impact of Factor β. Here, β is set to be 0, 0.5 and 1 respectively to tune the weight between himself and the others. Firstly, we consider the different β with followee in Table 1. It shows that the case with $\beta = 1$ gets the highest gain, whereas $\beta = 0$ acquires the lowest gain. The gain of $\beta = 0.5$ is between the value of two cases. This result shows that the tweets from followees are more useful than his own tweets. As our observation before, the tweets from followees are pulled to user actively so that it potentially has what user is interested in, whereas the tweets generated by himself might reluctant to show preference. Secondly, the result on different β with friend shows friends' tweets is better than user's own tweets to detect preference. Finally, the result on different β with follower shows the ability of followers' tweets to find user preferences is approximate to the one of his own tweets.

Impact of Relationship. We compare the results with different relationships, including followees, followers and friends. At first, we focus on DCG gain of the preference generated via different relationships with $\beta = 1$ in Table 2 which means user's own tweets are excluded here. Hence, the ability of information from the followees to find user preference is much more than the one from the followers. It verifies our observation that the following action implies user's preferences. It also shows the followees' tweets do a little better than friends' tweets. In general, user's friends on the social network might be his classmates, friends or relatives in the real life. He follows those real friends whatever they talk about on the virtual social network, which does not show a strong preference. Hence, only the friends' information as in [6] is not enough to detect user preferences. Additionally, the result on different relationships with $\beta = 0.5$ illustrates the followees' tweets are most useful for detection.

5.2 Effectiveness Comparison

In general, the combination of followees' tweets and $\beta = 1$ (as shown in Talbe 2 by bold font) gets the highest DCG . Under this setting, we compare four cases further, in order to highlight the effectiveness of our filtering and PageRank-like influence. 1) *no filtering + equivalent influence*: the followees are not filtered and followee's influences are equivalent; 2) *filtering + equivalent influence*: the followees are filtered and followee's influences are equivalent; 3) *no filtering +*

PageRank-like influence: the followees are not filtered and followee's influences are evaluated by PageRank-like influence; 4) *filtering + PageRank-like influence*: the followees are filtered and followee's influences are evaluated by PageRank-like influence. In particular, case 1 is the method used in [7].

The result in Figure 5 shows that case 2 has a higher score than case 1, which highlights our followee filtering improves the quality of detecting user preferences. The comparison between case 3 and 4 also concludes it. In addition, case 3 and case 4 outperforms case 1 and case 2 respectively, which indicates the effect of our PageRank-like influence evaluation.

6 Related Work

This section shows the work related to ours in two categories, including social media and recommender system.

[10] shows initial but significant statistical results on exploring user interests in the microblogs. Interest propagation along with the friendship in social network is discussed in [11]. However, we try to find user preference by the social network topology rather than study the pattern of interest propagation on social media. [7] studies the profile generated from different sources, including followees, followers, etc., but it simply treats them fairly. Besides, [6] combines users' behavioural and social data based on homophily. This work addresses the social networks with bilateral relationships, whereas our work focuses on the ones with unilateral linkages such as microblogs.

Our work is considered as one of the fundamental and critical steps in recommender system. [12] summarizes state-of-the-art recommender system. User preference, the main topic focused in our paper, is an indispensable basic for recommendation. Search engine usually employs user preference to improve personalized web search [2] which is a classical application of recommender system. User preference can be extracted from a large number of logs on search engine. However, in microblog environment, the useful information is limited as illustrated before.

7 Conclusion

In this paper, we explore the issue to find user preference for profile construction in microblog application. To achieve that goal, our proposed approach here makes full use of followees' tweets rather than user's own tweets, since a user might be reluctant to exposure preference actively but his following actions reveal the preferences. Also, we explore followee filtering and PageRank-like influence evaluation to improve the quality of preference detection. Empirical evaluation shows our approach together with followees' tweets is well suited for preference detection on social network with unilateral linkages such as microblog.

References

1. Liu, Y., et al.: Finding the right consumer: optimizing for conversion in display advertising campaigns. In: WSDM, pp. 473–482 (2012)
2. Sugiyama, K., et al.: Adaptive web search based on user profile constructed without any effort from users. In: WWW, pp. 675–684 (2004)
3. Gauch, S., Speretta, M., Chandramouli, A., Micarelli, A.: User profiles for personalized information access. In: Brusilovsky, P., Kobsa, A., Nejdl, W. (eds.) Adaptive Web 2007. LNCS, vol. 4321, pp. 54–89. Springer, Heidelberg (2007)
4. Kwak, H., et al.: What is twitter, a social network or a news media? In: WWW, pp. 591–600 (2010)
5. Weng, J., et al.: Twitterrank: finding topic-sensitive influential twitterers. In: WSDM, pp. 261–270 (2010)
6. Liu, K., et al.: Large-scale behavioral targeting with a social twist. In: CIKM, pp. 1815–1824 (2011)
7. Hannon, J., et al.: Recommending twitter users to follow using content and collaborative filtering approaches. In: RecSys, pp. 199–206 (2010)
8. Agarwal, D., et al.: Targeting converters for new campaigns through factor models. In: WWW, pp. 101–110 (2012)
9. Guy, I., et al.: Social media recommendation based on people and tags. In: SIGIR, pp. 194–201 (2010)
10. Banerjee, N., et al.: User interests in social media sites: an exploration with microblogs. In: CIKM, pp. 1823–1826 (2009)
11. Yang, S.H., et al.: Like like alike: joint friendship and interest propagation in social networks. In: WWW, pp. 537–546 (2011)
12. Adomavicius, G., et al.: Toward the next generation of recommender systems: A survey of the state-of-the-art and possible extensions. IEEE Trans. Knowl. Data Eng. 17(6), 734–749 (2005)

MUSTBLEND: Blending Visual Multi-Source Twig Query Formulation and Query Processing in RDBMS

Ba Quan Truong and Sourav S Bhowmick

Singapore-MIT Alliance, Nanyang Technological University, Singapore
School of Computer Engineering, Nanyang Technological University, Singapore
{bqtruong,assourav}@ntu.edu.sg

Abstract. Recently, in [3, 9] a novel XML query processing paradigm was proposed, where instead of processing a visual XML query *after* its construction, it *interleaves* query formulation and processing by exploiting the latency offered by the GUI to filter irrelevant matches and prefetch partial query results. A key benefit of this paradigm is significant improvement of the *user waiting time* (UWT), which refers to the duration between the time a user presses the "Run" icon to the time when the user gets the query results. However, the current state-of-the-art approach that realizes this paradigm suffers from key limitations such as inability to correctly evaluate certain visual query conditions *together* when necessary, large intermediate results space, and inability to handle visual query modifications, limiting its usage in practical environment. In this paper, we present a RDBMS-based *single* as well as *multi-source* XML *twig* query evaluation algorithm, called MUSTBLEND (**MU**lti-**S**ource **T**wig **BLEND**er), that addresses these limitations. A key practical feature of MUSTBLEND is its portability as it *does not* employ any special-purpose storage, indexing, and query cost estimation schemes. Experiments on real-world datasets demonstrate its effectiveness and superiority over existing methods based on the traditional paradigm.

1 Introduction

Formulating XML queries using XPath or XQuery languages often demand considerable cognitive effort from the end users and require "programming" skills that is at least comparable to SQL [1, 7]. The traditional approach to address this challenge of query formulation is to build an intuitive and user-friendly visual framework [4] on top of a state-of-the-art XML database. Figure 1 depicts an example of such a visual interface. Although query formulation now becomes significantly easier, evaluation of XQuery queries (especially over multiple data sources) on existing XML supports provided by commercial RDBMSs is often slow. To get a better understanding of this problem, we experimented with the datasets and queries in Figure 2[1]. Figure 2(c) shows the query evaluation times on XML-extended relational engines of two popular commercial RDBMS. Due to legal restrictions, these systems are anonymously identified as *XSys-A* and *XSys-B* in the sequel. Observe that most queries either take more than 30 minutes to evaluate (denoted by DNF in the paper) or are not supported by (denoted by NS in the paper) the underlying RDBMS. Note that the query evaluation time in a visual querying framework

---

[1] For the time being, the reader may ignore the bold underlined text and the identifiers in braces.

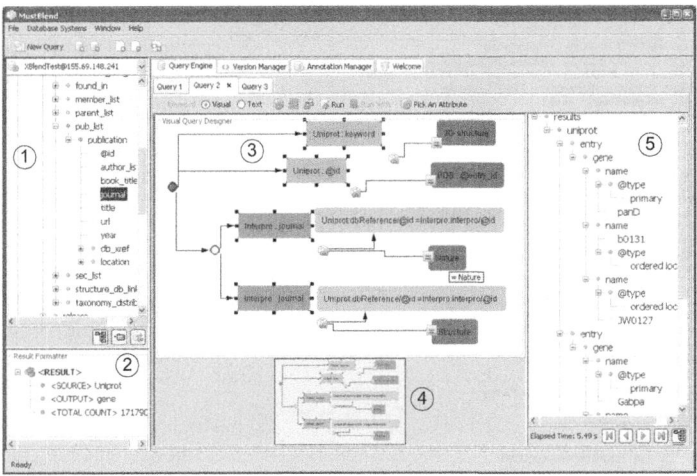

Fig. 1. Visual interface of MUSTBLEND

is identical to the *user waiting time* (UWT), which refers to the duration between the time a user clicks on the "Run" icon to the time when she gets the query results.

A Novel Visual Querying Paradigm. To resolve the issue of unusually long UWT of many XML queries, in [3,9] we took the first step towards exploring a novel XML query processing paradigm on top of a relational framework by *blending* the two traditionally orthogonal steps, namely visual query formulation and query processing. Let us illustrate this paradigm with an example. Consider the XML document in Figure 3(a). Suppose a user wishes to retrieve the name elements of entries (entry/name) that are related to the *"human"* organism (organism/name) and are created (@created) in *"2001"*. Using the visual interface in Figure 1, one can formulate the query as follows. (a) *Step 1:* Select the entry/name from Panel 1 to Panel 2 as *output expression*. Note that Panel 1 depicts the *structural summary* of the XML data sources. (b) *Step 2:* Select the created attribute from Panel 1, drag it to Panel 3, and add the value predicate *"2001"*. (c) *Step 3:* Select the name of organism from Panel 1, drag it to Panel 3, and add the predicate *"human"*. (d) *Step 4:* Click on the "Run" icon.

If we rely on traditional query processing paradigm, then the query evaluation is only initiated *after* Step 4. Although the final query that a user intends to pose is revealed gradually in a step-by-step manner during query construction (Steps 1 to 3), it is not exploited by the query processor *prior* to clicking of the "Run" icon. In contrast, in the new paradigm query construction and query processing are interleaved to prune false results and prefetch partial query results by exploiting the latency offered by the GUI-based query formulation (processing starts immediately after Step 1).

The key benefits of the new paradigm are as follows. First, since a complex XQuery query is evaluated by a set of smaller queries (to retrieve partial results), this new paradigm is less likely to stress the query optimizer compared to a single complex XQuery in traditional paradigm. Second, it significantly improves the UWT for many

Id	Queries	Result Size	Avg QFT (in sec)
Q1	for $entry in doc('UNIPROT.BIOXML')/uniprot/entry, $interpro in doc('INTERPRO.BIOXML')/interprodb/interpro, $cellCategory in doc('PDB.BIOXML')/PDBx:datablock/PDBx:cellCategory where $intepro/pub_list/publication/year > "1950" {3} and $entry/keyword = "3D-structure"{1} and $interpro/@id = $entry/dbReference/@id {2} and $PDBx:cellCategory/PDBx:cell/@entry_id = $entry/dbReference/@id {4} return $entry/name;	8	46.0
Q2	for $entry in doc('UNIPROT.BIOXML')/uniprot/entry, $interpro in doc('INTERPRO.BIOXML')/interprodb/interpro, $publication in $entry/pub_list/publication where **$publication/journal** = "Structure" {4} and **$publication/year** = "2002" {5} and $entry/@created[contains(., "2001")] {1} and $entry/organism/name = "Human" {2} and $interpro/@id = $entry/dbReference/@id {3} return $entry/name;	23	67.7
Q3	for $entry in doc('UNIPROT.BIOXML')/uniprot/entry, $interpro in doc('INTERPRO.BIOXML')/interprodb/interpro where $interpro/pub_list/publication/journal[contains(., "Cell")] {4} and ($entry/organism/name[contains(., "Mouse")] {1} or $entry/organism/name[contains(., "Human")]) {2} and $interpro/@id = $entry/dbReference/@id {3} return $entry/gene;	8871	44.8
Q4	for $entry in doc('UNIPROT.BIOXML')/uniprot/entry where $entry/organism/name = "Mouse" {1} or $entry/organism/name = "Human" {2} return $entry/protein;	17161	27.4

(a) Representative queries

Source	Size	No. of files	No. of Attributes	No. of Elements
UniProt	1.5GB	1	38,380,645	20,836,316
Interpro	69MB	1	1,427,234	988,079
PDB	287MB	30	692,583	5,578,498

(b) Datasets

Id	XSys-A	XSys-B
Q1	NS	NS
Q2	DNF	NS
Q3	DNF	NS
Q4	68.6	269.9

(c) Query performance (in sec.)

Fig. 2. Query evaluation times of representative queries

queries. Since we initiate query processing during query construction, UWT is the time taken to process a part of the query that is yet to be evaluated (if any).

Related Work and Motivation. Despite these appealing benefits of the new paradigm, the approach presented in [3,9] suffers from the following limitations. Firstly, it was designed only for queries in which every condition a user draws on the query canvas need to be processed *independently*. For example, the conditions drawn in Steps 2 and 3 in the above query can be independently matched against the database and the final query results can be computed by identifying common nodes in the partial results of these two conditions. However, this framework fail to correctly handle queries where conditions may need to be evaluated *together*. Consider the XML document in Figure 3(b). Suppose we wish to retrieve the names of proteins (interpro/name) that appear in the *"Nature"* journal (journal element) in *"2000"* (year). Independent evaluation of these two conditions as above will return the rightmost interpro/name element (*"Carboxyl transferase"*). However, it is associated with two *different* publication elements instead of a single one containing *"Nature"* and *"2000"*. Hence in order to retrieve correct results, these two conditions must be evaluated *together*. Secondly, [3,9] retrieves and materializes entire subtrees satisfying matching conditions drawn by users. However, this may adversely affect the overall prefetching performance in many cases due to the size of intermediate results. Thirdly, the new paradigm should be efficient and robust even when modifications (e.g., deletion or update of conditions) are committed by users during query formulation. Systematic investigation of how it handle such query modifications was beyond the goal of the aforementioned study. In this work, we seek to overcome these central limitations by proposing a novel algorithm called MUSTBLEND on top of a relational framework.

2 Visual Twig Query Model

We begin by introducing the twig query model which we support in this paper and the visual interface to formulate such queries.

2.1 Multi-Source Twig (MUST) Pattern

Most XML processors, both native and relational, have overwhelmingly focused on *single-source* AND-twig queries modeled as a twig pattern tree [6]. A *single-source* twig query is evaluated on a set of documents represented by a single XML schema or DTD. Jiang et al. [8] extended the notion of such AND-twig queries to process twigs with both AND and OR operators. Hence, at the very least, our query model should support such queries. Additionally, as discussed in Section 1, our query model should support queries over multiple data sources using joins. We refer to such twig queries as *multi-source twig (MUST) patterns*.

A MUST pattern Q is a graph with four types of nodes: location step query node (QNode), logical-AND node (ANode), logical-OR node (ONode), and return node (RNode). Each Q has a single node of type RNode which represents the output node. While labels of ANode and ONode are always "AND" and "OR" respectively, QNodes' and RNodes' labels are tags. An edge in Q can be of two types, namely, *axes edge*

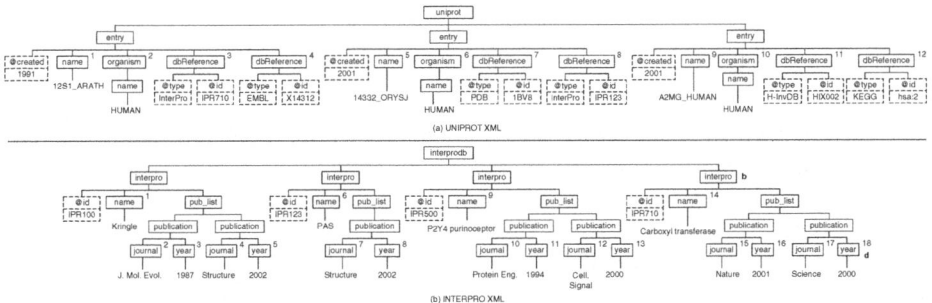

Fig. 3. XML representations of UNIPROT and INTERPRO data sources

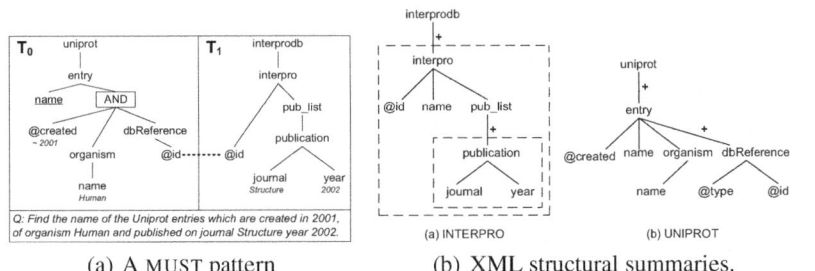

(a) A MUST pattern   (b) XML structural summaries.

Fig. 4. Twig query model and structural summary

and *join edge*. The former represents parent-child or attribute relationship [2] between a pair of nodes belonging to the same source whereas the latter connects two nodes from two different sources. Specifically, a *join edge* (q_1, q_2) asserts that q_1 and q_2 have equal value[3]. For example, Figure 4(a) shows the MUST pattern representation of the query $Q2$ in Figure 2(a). We denote the RNode by underlined tag (*e.g.*, name); and axes and join edges as direct and dashed lines, respectively.

Representing MUST Pattern Using XQuery. Observe that the aforementioned MUST pattern can be represented as an XQuery query. A MUST query Q is a 3-tuple $(\mathcal{F}, \mathcal{W}, \mathcal{R})$ where $\mathcal{F}$ is a set of for clause items, $\mathcal{W}$ is a set of predicates in DNF in the where clause, and $\mathcal{R}$ contains the output expression specified in the return clause. Specifically, the syntax of Q is as follows.

$$\begin{array}{ll} \text{FOR} & \$x_1 \ in \ p_1, \ldots, \$x_n \ in \ p_n \\ \text{WHERE} & (a_1 \wedge a_2 \wedge \ldots \wedge a_k) \vee \ldots \vee (c_1 \wedge c_2 \wedge \ldots \wedge c_m) \\ \text{RETURN} & r \end{array}$$

We categorize the *where-expressions* in $\mathcal{W}$ into two types, namely *join expressions* and *non-join expressions*. A *join expression* captures the join edge in a MUST pattern and involves predicates expressing join conditions over two document sources. On the other hand, a *non-join expression* expresses a filtering condition on a single document source. In the sequel, we refer to each expression in $\mathcal{W}$ as *condition*. Finally, the return clause has a single *output expression* r (RNode).

Extension of Query Model. The MUSTBLEND framework can easily support a variety of XPath axis and qualifiers as long as the underlying XML engine can support their evaluation. For instance, if a user visually specifies a path expression containing AD and preceding axis at a particular formulation step, then this visual action will be translated to a corresponding SQL statement by MUSTBLEND and forwarded to the underlying query engine for execution. Having said this, we would like to stress that a wide variety of XML queries are not easy to formulate even visually as it requires a deep understanding of the language which many end-users do not possess. It is of paramount importance to balance expressiveness and usability in MUSTBLEND as compromising the latter will render it impractical to end-users in a wide variety of domains [7].

2.2 Visual Query Interface

Figure 1 depicts the screen dump of the current version of the *path-based* visual interface of MUSTBLEND. The left panel (Panel 1) displays the XML *structural summary* (discussed later) of different XML data sources. When the users drag a node from Panel 1, the path expression corresponding to this node is automatically built. To formulate a query, the users first specify the output expression r (return node) by dragging that path expression from Panel 1 and dropping it to Panel 2. The *Visual Query Designer*

---

[2] We consider XPath navigation only along the child (/) and attribute (/@) axes. Extension to other navigation axis is orthogonal to the proposed technique.

[3] MUSTBLEND only supports equality join condition but inequality join condition can be supported easily.

panel (Panel 3) depicts the area for formulating query conditions. To build a non-join condition, the users drop a path expression in this panel. A *Condition Dialog* will appear for users to fill in all remaining information (op, val). If the dropped expression's data source is different from the output expression's data source, another dialog will appear for users to build the join edge between the two data sources. The user may drop a new condition on an existing condition in Panel 3 in order to indicate her intention to consider these two conditions together. Otherwise, she may drop the new condition on a blank space to indicate that it is independent of existing conditions. Two or more conditions can be combined using AND/OR (default is AND) connectives. The circular nodes in Figure 1 are color coded to represent AND (red) or OR (yellow) connectives. A satellite view (Panel 4) is provided with zooming functionality for more user-friendliness. The user can execute the query by clicking on the "Run" icon. The *Results View* (Panel 5) displays the query results.

3 Blending Visual Query Formulation and Processing

We now discuss how we can facilitate blending of query formulation and processing. We assume that a user *does not modify* previously constructed query fragments during formulation (no deletion or updates). In the next section, we shall relax this assumption.

Recall that MUSTBLEND GUI provides the flexibility to users to impose constraints on a set of conditions *together* (e.g., conditions on journal and year elements). However, this feature introduces two challenges. First, it is not always necessary that the underlying query processor need to *evaluate* these conditions together (twig). Hence we need a mechanism to detect automatically when a set of conditions should be evaluated together. Second, in order to facilitate evaluation of these conditions together it is often necessary to identify a common ancestor node (e.g., publication element for conditions on journal and year). It is unrealistic to assume that the end-users should explicitly specify them as it requires understanding of the XML structure. We introduce the notion of *inner structure tree* (IST) and *user actions tree* (UAT) to automatically resolve these two issues. We begin by introducing some auxiliary concepts.

An XML document is modeled as ordered directed trees, denoted by $\mathbb{D} = (\mathcal{N}, \mathcal{S})$, where $\mathcal{N}$ is a set of nodes (elements and attributes) and $\mathcal{S}$ is a set of edges (hierarchical relationships). Given an XML tree $\mathbb{D} = (\mathcal{N}, \mathcal{S})$, a *path* of a node $n \in \mathcal{N}$ in $\mathbb{D}$, denoted as $path(n)$, is a concatenation of dot-separated labels $\ell_1.\ell_2\ldots\ell_k$, such that $\ell_i (1 \leq i \leq k-1)$ is the label of n's ancestor at level i. ℓ_1 is the label of the root node and ℓ_k is the label of n itself.

We adopt the DataGuide [5] as our XML *structural summary*. Intuitively, a DataGuide *structural summary*, denoted by $\mathbb{S}$, is a tree representing all unique paths in $\mathbb{D}$. That is, each unique path p in $\mathbb{D}$ is represented in $\mathbb{S}$ by a node whose path from the root node to this node is p. An edge may have a label "+" *iff* the target node of the edge has cardinality "+" with respect to $\mathbb{D}$. Further, every unique label path of $\mathbb{D}$ is described exactly once, regardless of the number of times it appears in $\mathbb{D}$. Figure 4(b) depicts the structural summary of the XML document in Figure 3(b). Observe that the edges incident on interpro and publication nodes have label "+" as interprodb and pub_list nodes in Figure 3(b) have multiple occurrences of these child nodes, respectively. A subtree of $\mathbb{S}$ is a *Plus-tree* (*P-tree* in short) if its root is the target node of an "+"

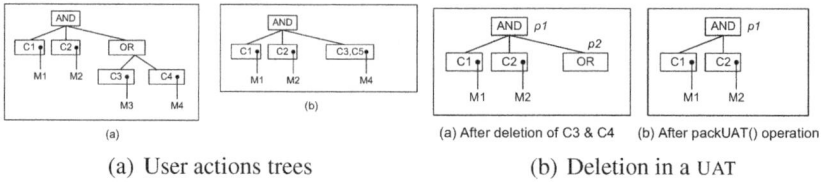

(a) User actions trees (b) Deletion in a UAT

Fig. 5. User action trees and deletion operation on it

edge. For example, in Figure 4(b) the subtree $g_{publication}$ rooted at publication node is a P-tree of $\mathbb{S}$ while the subtree g_{pub_list} rooted at pub_list node is a subtree but not a P-tree of $\mathbb{S}$. We denote a set of all P-trees of $\mathbb{S}$ by $ptree(\mathbb{S})$.

Inner Structure Tree (IST). Let $g_1, g_2 \in ptree(\mathbb{S})$ and $root(g_1) \neq root(g_2)$. Then g_1 is an *inner structure tree* (IST) of g_2, denoted by $g_1 \sqsubset g_2$, if and only if g_1 is a subtree of g_2. Note that $path(root(g_2))$ is a prefix of $path(root(g_1))$. For example, in Figure 4(b), $g_{publication} \sqsubset g_{interpro}$ (highlighted by dashed rectangles).

User Actions Tree (UAT). A *user actions tree* (UAT), denoted as U, describes how a set of conditions that are connected by AND or OR connectives are to be processed by MUSTBLEND to generate the final query results. Each internal node of U represents an AND or OR connective. Each leaf node of U is a 2-tuple $v = (\mathcal{C}_a, M)$, where $\mathcal{C}_a$ is a set of non-join conditions that are processed together and M is the temporary relation that stores the prefetched data satisfying $\mathcal{C}_a$. For example, Figure 5(a) depicts two UATs. Observe that M_4 in Figure 5(a)(b) is generated by evaluating C_3 and C_5 together whereas M_2 is generated by processing C_2 independently from rest of the conditions. When do we process a set of conditions together? We elaborate on this now.

Given a condition C, let $target(C.S.exp)$ refers to the *target node* (rightmost node) in the path expression S of C. When the exp is obvious from the context, we denote it as $target(C)$. If $target(C)$ is contained in $g \in ptree(\mathbb{S})$ then we say that g includes C, denoted by $C \vdash g$. For example, consider C_3 in $Q2$. Here $target(C_3)$ is the journal node that is contained in $g_{publication}$. Hence, $C_3 \vdash g_{publication}$. Given a set of conditions $\mathcal{C}_a$ and $g \in ptree(\mathbb{S})$, g *minimally includes* $\mathcal{C}_a$, denoted by $\mathcal{C}_a \vdash_m g$, iff $\forall C_i \in \mathcal{C}_a, C_i \vdash g$ and there $\nexists g' \in ptree(\mathbb{S})$ such that $g' \sqsubset g, \forall C_i \in \mathcal{C}_a, C_i \vdash g'$.

Let $\mathcal{C}$ be a set of conditions and r be the output expression on $\mathbb{S}$. Then the conditions in $\mathcal{C}$ are processed together *iff* (a) the label of the parent node of $\mathcal{C}$ in U is AND and (b) $g_1 \sqsubset g_2$ where $(g_1, g_2) \in ptree(\mathbb{S})$, $\mathcal{C}_a \vdash_m g_1$ and $(\mathcal{C}_a \cup \{r\}) \vdash_m g_2$. Note that the $root(g_1)$ is the common ancestor satisfying all conditions of $\mathcal{C}$. For example, consider Figure 5(a). Suppose $target(r.exp)$ is the name node. If we formulate two conjunctive conditions on nodes journal and year, then they can be evaluated together to find publication nodes satisfying these conditions. This is because $g_{publication}$ is an IST of $g_{interpro}$, $g_{publication}$ minimally includes the conditions on journal and year, and $g_{interpro}$ minimally includes the conditions on journal, year, and name.

Algorithm 1. The MUSTBLEND algorithm

Input: Actions on the user interface
Output: Query results M

1 Initialize M and user actions tree U;
2 Initialize queue $\mathbb{Q}$;
3 $M_o \leftarrow$ **fetchOutputExp**(r);
4 $\mathcal{A} \leftarrow$ **getGUIAction**();
5 **while** ($\mathcal{A} \neq$ "Run") **do**
6   **if** ($\mathcal{A} ==$ "Add") **then**
7   C_{add} is the new condition;
8   C_{target} is the drop target;
9        $\_SQL \leftarrow$ **fetchCondMatch**($r, C, C_{add}, C_{target}$);
10   U.**insert**(C_{add});
11       $\mathbb{Q}$.**insert**($\_SQL$);
12   **else**
13   **if** ($\mathcal{A} ==$ "Delete") **then**
14   C_{del} is the deleted condition;
15          $U \leftarrow$ **deleteHandler**($C_{del}, U, \mathbb{Q}$);
16   **else**
17   **if** ($\mathcal{A} ==$ "Update") **then**
18   C_{old} and C_{new} are old and new conditions;
19   Initialize $upFlag = \emptyset$;
20             $U \leftarrow$ **updateHandler**($C_{old}, C_{new}, upFlag, U, \mathbb{Q}$);
21   C.**insert**(C_i);
22    $\mathcal{A} \leftarrow$ **getGUIAction**();
23 **if** ($\mathbb{Q} \neq \emptyset$) **then**
24   Wait for materializing all partial results;
25 **else**
26   Modify U by removing unnecessary internal nodes;
27    $M \leftarrow$ **retrieveFinalResults**(U, r);
28 **return** M

3.1 Algorithm MUSTBLEND

We now present the Algorithm MUSTBLEND (Algorithm 1). Importantly, for the sake of generality, we present a generic approach that is independent of any specific relational approaches. The reader may refer to [11] for an example of how various subroutines in the algorithm can be realized on a specific tree-unaware XML storage system. First, when the output expression r is dragged into Panel 2, it materializes the *identifiers* of the elements/attributes in the XML tree that satisfy r by invoking the *fetchOutputExp* procedure (Line 03). It generates an SQL query for this task. An *identifier* of an element n in an XML tree $\mathbb{D}$ is one or more attributes of n that can uniquely identify n in $\mathbb{D}$. Note that we materialize the identifiers instead of entire subtrees because it is more space-efficient. It is worth mentioning that the identifier scheme is not tightly coupled to any specific system as any numbering scheme (*e.g., region encoding, dewey number-based* [6]) that can uniquely identify nodes in an XML tree can be used as an identifier.

Next, Lines 05–22 are executed repeatedly until the "Run" icon is clicked. When a user drags a new query condition C_{add} and drops it on an existing condition C_{target}, Lines 07–11 are executed. The algorithm invokes the *fetchCondMatch* procedure to materialize the identifiers in M_o that satisfy C_{add} (Line 9). Then, it adds C_{add} into the UAT U (Line 10). Figure 5(b) depicts the UAT generated after the conditions in the running query are visually formulated. MUSTBLEND detects that C_3 and C_5 need to be processed together and identifies the common ancestor. Lines 13–20 are executed if the user modifies a portion of the query that have already been constructed. We shall elaborate on these steps in Section 4. Note that the translated SQL queries generated by these steps are inserted into a queue $\mathbb{Q}$. These queries are then processed sequentially in another process thread. The node identifiers retrieved by the above steps are materialized in a set of relations where the schema of each relation contains only document identifier and node identifier attributes.

Once all the conditions are visually formulated, the user may click on the *"Run"* icon to retrieve the query results. Once the materialization of all partial results are completed, the algorithm invokes the *retrieveFinalResults* (Line 27) which traverses U to retrieve the query results from the temporary tables storing the partial results.

fetchCondMatch **Procedure.** Let $\mathbb{S}_1$ and $\mathbb{S}_2$ be the structural summaries of the data sources of the output expression r and the new condition C_{cur}, respectively. First, this procedure retrieves the P-trees g_1 and g_2 that minimally includes the two input condition sets ($\{r, C_{cur}, C_{target}\} \vdash_m g_1$, $\{C_{cur}, C_{target}\} \vdash_m g_2$). Next, it determines if join across data sources is needed by comparing the data sources of r and C_{cur}. If join is not required ($\mathbb{S}_1 = \mathbb{S}_2$) then it first retrieves the set of conditions C_a that have already been formulated by the user and $C \vdash g_2 \: \forall C \in C_a$ and $C_{target} \in C_a$. If $g_2 \sqsubset g_1$ then it generates an SQL statement that processes the conditions in C_a and C_{cur} together due to reasons discussed earlier. Otherwise, it first generates SQL statements to retrieve the node identifiers satisfying C_{cur} and then it appends statements for determining subtrees that contain these identifiers as well as satisfy r. Note that when the user drops C_{cur} on a blank space, then $C_{target} = \emptyset$. In this case, g_2 is set to $\mathbb{S}_2$. Consequently, the condition $g_2 \sqsubset g_1$ is not satisfied and the above step is followed to process the new condition independently. When join across data sources is required, this procedure first updates g_1 where $\{C_j, C_{cur}, C_{target}\} \vdash_m g_1$. Then an SQL query is generated for prefetching portion of data satisfying C_{cur} using the join condition C_j. Due to space constraints, the formal description of the algorithm is given in [11].

retrieveFinalResults **Procedure.** This procedure can be divided into two main steps: (a) processing of the UAT and (b) retrieval of *complete* subtrees from the database satisfying the query [11]. The objective of the first step is to retrieve all identifiers of instances of r that satisfy the set of conditions in the UAT. After that, the second step is used to build an SQL query to extract all subtrees satisfying the identifiers extracted from the first step. While the second step is straightforward, we propose a *disk-based* and *main memory-based* strategies called DISKRETRIEVE and MEMRETRIEVE, respectively, to realize the first step. The DISKRETRIEVE strategy processes U recursively and *returns an SQL query* to retrieve the identifiers from the materialized relations. Given the node *root* in U, the algorithm first identifies whether it is an "AND" node or an "OR" node. If it is an "AND" node, then it adds the "INTERSECT" operator into the SQL statement. Otherwise, the "UNION" operator is used. Then, it retrieves the child nodes of *root* and processes them one by one. If the child node is a leaf node, then the algorithm adds corresponding SQL statement. Otherwise, it recursively process the internal nodes and finally returns an SQL query for execution. For example, reconsider the UAT in Figure 5(b). The SQL query generated by this procedure is as follows: select * from M1 INTERSECT select * from M2 INTERSECT select * from M4.

While the DISKRETRIEVE strategy requires the partial results to be materialized in the database and retrieved the final results using SQL queries, the MEMRETRIEVE approach reduces I/O cost by storing the identifiers in memory. In particular, it stores the partial results in the main memory[4] and use a similar procedure to the aforementioned algorithm except that it retrieves the intermediate relations directly from the memory instead of building SQL queries.

Remark. Observe that Algorithm MUSTBLEND does not exploit predicate selectivities to optimize prefetching performance. Unfortunately, this strategy is ineffective here as users can formulate low and high selective conditions in any arbitrary sequence of actions. Consequently, it is not advantageous to speculate an end-user's subsequent actions in order to take full advantage of selectivity estimates.

4 Visual Query Modifications

In this section, we address the issue of modification to a visual query. We consider two types of modification, namely *delete* and *update*. Deletion enables a user to delete a query condition $C_{del} \in \mathcal{C}$ that has been constructed by him. The update operation allows a user to update a previously formulated condition or change the default AND connective to OR. Specifically, we allow the following updates types (a) Update of the value of a condition. (b) Update of the operator of a condition. (c) Update of AND/OR connectives. Note that the path expression of a condition is not allowed to be updated visually as it often demands syntactic knowledge of XPath expressions from the users. To modify the path expression, one must delete the condition and add a new one.

Handling Deletions. The *deleteHandler* procedure handles deletion of a condition C_{del} in the following way. First, it checks if the translated SQL query for C_{del} is still in the query queue $\mathbb{Q}$. If it is, then it indicates that the query has not been executed yet. Hence, the algorithm will remove it from $\mathbb{Q}$. Otherwise, the results of C_{del} have already been materialized in a temporary table M_C. Consequently, the algorithm will drop M_C. Next, it updates the UAT U by deleting C_{del} from it. Finally, it checks if an internal node of U has become a leaf node due to the deletion of C_{del} and modify U accordingly (*packUAT* procedure). For example, consider the UAT in Figure 5(a). Figure 5(b)(b) depicts the structure of the UAT after deleting C_3 and C_4. Note that if $C_{del} \in \mathcal{C}_a$ (where $1 \leq i \leq |\mathcal{C}_a|$) is deleted, then all conditions in $(\mathcal{C}_a - C_{del})$ shall be reevaluated. The algorithms in Section 3.1 can be exploited for this purpose.

Handling Updates. We first discuss updates on conditions (leaf nodes in the UAT) and then present the effect of updates on the internal nodes. Suppose that a user updates the condition $C_{old} \in \mathcal{C}$ to C_{new}. Let M_{old} and M_{new} be the materialized tables satisfying C_{old} and C_{new}, respectively. There are four possible cases as follow for such update operation. **(a) *Case 1:*** $M_{old} \subset M_{new}$. In this case the results in M_{old} also satisfy C_{new}. However, not all nodes satisfying C_{new} have been retrieved. Hence, it is necessary to retrieve these additional nodes and merge them with M_{old}. **(b) *Case 2:*** $M_{old} \supset M_{new}$. Nodes satisfying C_{new} are already in M_{old}; however M_{old} also contains nodes that do

---

[4] The intermediate relations are implemented using *HashMap*.

not match C_{new}. Consequently, these nodes need to be deleted. **(c) *Case 3:*** $(M_{old} \cap M_{new}) \neq \emptyset$ and $(M_{old} \neq M_{new})$. This case represents the scenario where some of the nodes in M_{old} are part of the result matches for C_{new}. Note that M_{old} also contains nodes that are not relevant to C_{new}. Hence, we need to delete non-matching nodes from M_{old} and retrieving matching nodes that are not in M_{old}. **(d) *Case 4:*** $(M_{old} \cap M_{new}) = \emptyset$. We delete M_{old} and retrieve matching nodes for M_{new}.

The *updateHandler* procedure first determines whether the update operation is on a query condition or on an AND/OR node. If the former is true then it determines the *update code* (1, 2, and 0 for Cases 1, 2, and 3 and 4, respectively) based on C_{old} and C_{new} only. In case, it is not possible to determine the code (e.g., the value of a condition is a string) then by default it is considered as Case 4. If the *update code* is 0, then it considers this modification as deletion of C_{old} and insertion of C_{new}. Note that if $C_{old} \in \mathcal{C}_a$ then we execute these two steps as well. If the *update code* is greater than 0, then the algorithm first checks if the SQL query for C_{old} is still in $\mathbb{Q}$. If it is still in $\mathbb{Q}$, then C_{new} will be translated into an SQL query by using the algorithms discussed in the preceding section and it will replace the old query in $\mathbb{Q}$ with the new one. On the other hand, if the SQL query for C_{old} has already been executed, then the algorithm will generate an INSERT SQL statement (for Case 1) or a DELETE statement (for Case 2). Note that the former statement retrieves additional nodes from the database that satisfy C_{new} and inserts them in M_{old}. Similarly, the latter statement deletes nodes in M_{old} that do not satisfy C_{new}.

Now consider the update of an AND node (recall that it is created by default) to an OR node. If each child leaf node n of an updated AND node represents a single condition C then there is no modification to the prefetching process during query formulation. However, if at least one of the child node n contains two or more conditions ($\mathcal{C}_a$) that need to be processed together then n needs to be modified along with its prefetched relation (if any). Consequently, the algorithm first removes unnecessary internal nodes (if any) from U that may have resulted due to the update operation. If n is updated to OR node, then it is decomposed into a set of leaf nodes where each node represents a single query condition $C_i \in \mathcal{C}_a$. The prefetched partial results (if any) associated with n is deleted and SQL queries for each C_i where $1 < i \leq |\mathcal{C}_a|$ are generated to prefetch partial results matching each of these conditions. Otherwise, if an OR node is restored back to an AND node then the original leaf nodes are restored. Due to space constraints, the formal description of *updateHandler* is reported in [11].

5 Performance Study

MUSTBLEND is implemented in Java on top of a recently proposed *path materialization-based (PM)* [6] XPath processor[5] on relational backend called ANDES [10]. We create two variants of MUSTBLEND (see *retrieveFinalResults* procedure), namely one having DISKRETRIEVE strategy (denoted by MB-H) and another MEMRETRIEVE strategy (denoted by MB-M). All experiments were conducted on an Intel Core 2 Quad 2.66GHz processor and 3GB RAM. The operating system was Windows XP. The RDBMS used was MS SQL Server 2008 Developer Edition.

---

[5] PM approach has advantages over node-based approach when XML data are schemaless [6].

Query	MB-M	MB-H	XSys-A	XSys-B	Zorba
Q1	0.12	0.24	NS	NS	NS
Q2	0.13	0.25	DNF	NS	1495.8
Q3	26.1	0.73	DNF	NS	171.9
Q4	134.4	0.83	68.6	269.9	0.45
Q5	0.16	0.20	3.2	16.0	0.45
Q6	0.25	0.24	2.0	18.0	0.64
Q7	1.34	0.61	72.6	449.2	0.47
Q8	0.13	0.45	165.5	NS	145.8
Q9	2.16	0.51	DNF	NS	1400.6
Q10	0.16	0.86	NS	NS	NS

(a) User Waiting Times (in sec.)

Query	MB-M	ANDES
Q1	2.14	DNF
Q2	9.39	18.3
Q3	2.18	DNF
Q4	1.26	12.7
Q5	1.83	2.1
Q6	1.29	68.1
Q7	1.66	22.0
Q8	1.38	18.5
Q9	2.91	19.6
Q10	3.37	7.1

(b) TPT vs complete query execution times (in sec.)

Fig. 6. Performance results (DNF – Did Not Finish in 30min; NS – Not Supported)

Query	Approach	Step Out	Step 1	Step 2	Step 3	Step 4	Step 5	TPT
Q1	MB-M	0.10 (179430)	0.28 (6123)	0.32 (156172)	0.24 (9)	0.09	-	2.03
	MB-H	-	-	-	-	0.20	-	2.14
Q2	MB-M	0.09 (179430)	0.32 (5850)	0.34 (9595)	7.28 (30294)	1.14 (4317)	0.10	9.27
	MB-H	-	-	-	-	-	0.22	9.39
Q4	MB-M	0.12 (179430)	0.23 (7566)	0.13 (9595)	134.41	-	-	134.89
	MB-H	-	-	-	0.78	-	-	1.26
Q6	MB-M	0.03 (18093)	0.18 (35)	0.17 (12479)	0.28 (18064)	0.43 (682)	0.21	1.30
	MB-H	-	-	-	-	-	0.20	1.29
Q9	MB-M	0.08 (171790)	0.15 (5601)	1.14 (130318)	0.96 (70327)	2.23	-	4.56
	MB-H	-	-	-	-	0.58	-	2.91

Fig. 7. Running times of materialization of partial results (in sec.) for representative queries

We compare our ANDES-based MUSTBLEND implementation with two popular commercial XML-extended relational engines, *XSys-A* and *XSys-B* (see Section 1), realizing traditional query processing paradigm. Appropriate indexes were created for all approaches and prior to our experiments, we ensure that statistics had been collected. The bufferpool of the RDBMS was cleared before each run. We also compare *Zorba* (try.zorba-xquery.org), an open-source XQuery processor written in C++ which adopted latest optimization techniques [2]. We do not compare it with [9] as the latter does not correctly support queries that require a set of conditions to be evaluated together (e.g., Q_2).

Experimental Setup. We use the XML representations of UNIPROT, PDB, and INTERPRO downloaded from their official websites. The features of these datasets are given in Figure 2(b). Since *Zorba* fails to handle large datasets (UNIPROT), we reduce the UNIPROT dataset by a factor of 50 (28MB) so that we can study its performance.

We chose ten single and multi-source twig queries that join up to three data sources. Q_1 to Q_4 are shown in Figure 2(a) and the remaining queries are given in [11] (due to space constraints). These queries are selected based of several features such as result size, number of conditions in the where clause, number of data sources, existence of ISTs with minimally inclusive conditions (highlighted in underlined bold), and existence of AND/OR connectives. The subscripts of the labels in curly braces in the where clause represent the default sequence of steps for formulation of conditions in MUSTBLEND. Note that if a join and a non-join condition have same subscript then it means that the join condition is formulated immediately after its non-join counterpart and are evaluated together in MUSTBLEND. For example, the sequence of steps of Q_1 is

Query	Sequence	Time	Time	Time	Time	MB-M	MB-H
Q5	[C2, C1]	0.09	0.10	-	-	0.14	0.26
Q3	[C3, J3, C2, C1]	0.89	0.24	0.34	-	25.6	0.77
	[C2, C1, C3, J3]	0.18	0.34	0.82	-	26.5	0.70
	[C3, J3, C1, C2]	0.81	0.15	0.34	-	26.5	0.78
Q10	[C2, C1, J3, J4]	0.17	0.67	1.21	0.18	0.15	0.92
	[C1, C2, J3, J4]	0.79	0.32	0.17	1.21	0.15	0.90
	[C1, J3, J4, C2]	0.80	1.21	0.17	0.32	0.16	0.93

Fig. 8. Effect of query formulation sequence (in sec.)

depicted in Figure 2(a). Note that the join condition J_2 and the non-join condition C_2 share same subscript. That is, J_2 is specified immediately after the formulation of C_2 and are processed together in one step. Unless mentioned otherwise, we shall be using the default sequence for formulating a query.

In order to formulate visual queries, fifteen unpaid volunteers with no prior knowledge of XQuery query language participated in the experiments. Details related to participants' profile is given [11]. Each query was formulated six times by each participant (using the default sequence unless specified otherwise) and reading of the first formulation of each query was ignored. The average query formulation time (QFT) for a query by all participants is shown in the right-most column in Figure 2(a).

Experimental Results. We now present performance results of MB-H and MB-M.

User Waiting Times (UWT). Figure 6(a) shows the average *user waiting time* (UWT) of all approaches. It is computed by taking the average of the UWTs of all participants. In *XSys-A*, *XSys-B*, and *Zorba*, UWT refers to the query execution times. Clearly, disk-based and memory-based variants of MUSTBLEND are significantly faster than approaches based on traditional paradigm in most queries. In particular, MB-M and MB-H are at least two orders of magnitude faster than *XSys-A* or *XSys-B* for queries that join multiple data sources ($Q1 - Q3, Q8 - Q10$). Also, MB-H typically has superior performance compared to MB-M especially for queries with larger result size (e.g., $Q3$, $Q4$). Note that UWT of MB-H is less than a second for all queries. Lastly, although we use a much smaller UNIPROT dataset for *Zorba*, surprisingly, MB-M and MB-H are still significantly faster than *Zorba*.

Materialization of partial results. We now report the execution times for materialization of partial results of a set of conditions in a visual query. Figure 7 reports the performance of five representative queries (results for all benchmark queries are available in [11]). Each column labeled $Step\ i$ represents the running time associated with the materialization of corresponding query condition(s) of i-th step (subscript in the label inside curly braces) in the sequence. The last step in MB shown in bold refers to retrieval of entire subtrees satisfying the complete query. The values in parenthesis represent the size of the materialized relations. $Step\ out$ refers to the output expression selection step. In response to this action, MUSTBLEND retrieve all nodes (identifiers) in the database satisfying the output expression. Notably, the only difference between MB-M and MB-H is the last step where final results are retrieved (*retrieveFinalResults* procedure). The preceding steps are identical in both approaches.

We can make the following observations. First, the large size of intermediate results does not adversely affect the UWT. Additionally, retrieving all the node identifiers (can

MUSTBLEND: Blending Visual Multi-Source Twig Query Formulation   241

Id	Original query	Modified query
MQ1	for $entry in doc('UNIPROT.XML')/uniprot/entry where ($entry/keyword = "3D-structure" {C₁} or $entry/keyword = "Calcium" {C₂}) and ($entry/organism/name[contains (., "Cell")] {C₃} or $entry/organism/name = "Mouse" {C₄}) return $entry/gene	for $entry in doc('UNIPROT.XML')/uniprot/entry where ($entry/keyword = "3D-structure" {C₁} or $entry/keyword = "Calcium") {C₂} and ($entry/organism/name[contains (.,"virus")] {C₅} or $entry/organism/name = "Human") {C₆} and $entry/feature/@description[contains(.,"protein")]{C₇} return $entry/gene
MQ2	for $entry in doc('UNIPROT.XML')/uniprot/entry, $interpro in doc('INTERPRO.XML')/interprodb/interpro where $entry/organism/name[contains(., "Human")] {C₁} and $interpro/pub_list/publication/journal = "Structure"{C₂} and $interpro/pub_list/publication/year = "2002" {C₃} and $interpro/@id = $entry/dbReference/@id {J₂} return $entry/name	for $entry in doc('UNIPROT.XML')/uniprot/entry, $interpro in doc('INTERPRO.XML')/interprodb/interpro, $publication in $interpro/pub_list/publication where ($entry/organism/name[contains(., "Human")] {C₁} and $entry/protein/name[contains(., "protein")]) {C₄} and ($interpro/pub_list/publication/journal = "Structure" {C₂} or $interpro/pub_list/publication/journal = "Cell") {C₅} and $publication/year > "1980" {C₆} and $publication/year <= "2000" {C₇} and $interpro/@id = $entry/dbReference/@id {J₂} return $entry/name
MQ3	for $entry in doc('UNIPROT.XML')/uniprot/entry, $interpro in doc('INTERPRO.XML')/interprodb/interpro, $cell in doc('PDB.XML')/datablock/cellCategory/cell where $entry/keyword = "3D-structure" {C₁} and $entry/organism/name[contains(., "Human")] {C₂} and $interpro/@id = $entry/dbReference/@id {J₃} and $cell/@entry_id = $entry/dbReference/@id {J₄} return $entry/name	for $entry in doc('UNIPROT.XML')/uniprot/entry, $interpro in doc('INTERPRO.XML')/interprodb/interpro, $publication in $interpro/pub_list/publication, $cell in doc('PDB.XML')/datablock/cellCategory/cell where $entry/keyword = "3D-structure" {C₁} and $entry/organism/name[contains(., "Mouse")] {C₂} and $publication/year > "1956" {C₆} and $publication/year <= "2000" {C₇} and $interpro/@id = $entry/dbReference/@id {J₃} and $cell/@entry_id = $entry/dbReference/@id {J₄} return $entry/name

Fig. 9. Effect of query modifications

be large) satisfying an output expression is feasible as it does not affect the prefetching operations and UWT adversely. Second, MB-M is faster than MB-H when the final result set is small (e.g., $Q1, Q2, Q10$) whereas MB-H is faster when the final result set is large (e.g., $Q3, Q4$). Finally, for the majority of the queries in MB, interestingly, the *total prefetching times* (the total time taken for all prefetching operations, denoted by TPT) are significantly less than the query execution times in *XSys-A*, *XSys-B*, and *Zorba* (Figure 6(a)). This is due to benefits of the new paradigm mentioned in Section 1.

TPT vs complete query execution times. The aforementioned experiments do not demonstrate whether the performance benefit of MUSTBLEND is due to the visual querying paradigm instead of the efficiency of underlying storage scheme of ANDES. In this experiment, we shall shed light on this issue. Specifically, we measure the TPT of $Q1 - Q10$ using MB-H and the execution time of each query in its entirety on ANDES. Note that we did not undertake similar experiments on *XSys-A*, *XSys-B*, and *Zorba* as these systems do not allow us to retrieve and materialize node identifiers as partial result matches. Recall that in MUSTBLEND we only materialize node identifiers in order to minimize intermediate results size. Figure 6(b) reports the performance results. Clearly, in most cases the TPT is significantly lower than the cost of executing an entire query on ANDES. Observe that the UWT (Figure 6(a)) is also significantly smaller than the evaluation time of an entire query on ANDES.

Effect of query formulation sequence. A visual query can be formulated by following different sequence of steps. We now assess the effect of these different sequences on the UWT in MB. Figure 8 lists different formulation sequences for three representative queries (results for all benchmark queries are available in [11]), average times (all participants) to retrieve partial results, and the average UWT. Note that $Q5, Q3$ and $Q10$ are on one, two and three data sources. Notably, there are hardly any significant changes in both the prefetch times and the UWT. This is primarily due to the following reasons. Firstly, GUI latency can always be exploited by MUSTBLEND at each step irrespective of the ordering

Query	Type	Sequence of Modifications	Prefetch Time	Avg. UWT	Result Size
MQ1	IM	1. Update C4 to C6 (Case 4)	0.23	1.2	1928
		2. Update C3 to C5 (Case 4)	0.13	1.46	2135
		3. Insert $entry/protein/name = "Protein"	0.17	0.05	0
		4. Delete $entry/protein/name = "Protein"	-	0.31	2135
		5. Insert C7	0.38	0.41	866
	BM	Same sequence as IM	-	0.42	866
MQ2	IM	1. Update the value of C1 to "Mouse" (Case 4)	0.46	0.42	972
		2. Update C3 to C6 (Case 1)	7.36	6.57	1653
		3. Insert C7	1.31	2.76	1638
		4. Update the value of C1 to "Human"	0.41	0.69	1401
		5. Insert C5	5.96	3.66	0
		6. Update to OR node for C2 and C5	7.92	8.35	2748
		7. Insert C4	0.60	0.97	1740
	BM	Same sequence as IM	-	0.98	1740
MQ3	IM	1. Insert C6	1.33	0.20	2
		2. Insert C7	1.38	0.55	2
		3. Update C2 to C5 (Case 4)	0.40	0.10	1
	BM	Same sequence as IM	-	0.12	1

Fig. 10. Effect of query modification in MUSTBLEND (in sec.)

of the visual steps. Secondly, in any query formulation sequence, each visual step results in evaluation of a simple XPath fragment, which is much faster to evaluate compared to a large chunk of complex XQuery as the former stresses the underlying query processor less.

Effect of query modifications. Figure 9 depicts three representative queries on one, two, and three data sources before and after modifications (denoted by $MQ1$, $MQ2$, and $MQ3$, respectively). In this figure, we highlight the changes in underlined bold. For ease of reference, all unique conditions in the original and modified versions of a query are given unique identifiers (e.g., C_1). In order to simulate real-world scenario, we consider two types of modification scenario, namely *incremental* and *bulk* modifications. In *incremental modification* (denoted by IM), after each modification action we execute the query by clicking on the "Run" icon. Hence, if there are n modifications performed by a user then the query is evaluated n times. On the other hand, in *bulk modification* (denoted by BM), all modifications to a query is first formulated before it is executed. Hence a modified query is executed *only once*.

Figure 10 reports the performances of IM and BM in MB-M. Since deletion of a condition does not require retrieval of new matches, we mainly focus on updates. To simulate real-world scenario, we mix update operations with insertion of new conditions. The sequence of operations performed by a user for a query is recorded in the second column. As a user may insert/update a condition and restore it back later (this modification will not appear in the final modified query) after realizing his mistake, we represent this scenario by inserting (resp. updating) and deleting (resp. update back) query conditions that do not appear in Figure 9 (e.g., the third and fourth modification actions for $MQ1$, first and fourth updates for $MQ2$). We can make the following observations from the results in Figure 10. First, all prefetching activities in IM due to the modifications are completed within few seconds. Second, the UWTs for both IM and BM are significantly faster than traditional approaches. The modified $MQ2$ and $MQ3$ do not return any results in 30 minutes or they are not supported by *XSys-A* and *XSys-B*. The UWTs of $MQ1$ for *XSys-A* and *XSys-B* are $173.7s$ and $866.6s$, respectively. *Zorba* takes $22.1s$ and $2957.7s$ for modified $MQ1$ and $MQ2$, respectively. However, it does not support $MQ3$. These results clearly demonstrate that MUSTBLEND's performance is not adversely affected by query modifications, highlighting again its strength.

6 Conclusions

Our research sought to understand and provide insights to a new XML query processing paradigm where the latency offered by visual query formulation is utilized to prefetch partial results. We have presented MUSTBLEND - an algorithm to realize this paradigm over relational framework by addressing some of the central limitations of [3,9]. Specifically, it can handle richer variety of queries and only stores *synopsis* of intermediate results to make the overall process space-efficient. As MUSTBLEND does not employ special-purpose storage, indexing, and cost estimation schemes to improve UWT, it can easily be built on top of any off-the-shelf RDBMS. Further, the proposed algorithm ensures that the prefetching activities are completely transparent to the users and their interaction behaviors are not affected by this paradigm. MUSTBLEND has excellent performance for a wide variety of queries. It can also gracefully accommodate modifications to a query during construction. All these features are important for deployment of MUSTBLEND in real-world environment.

References

1. Abiteboul, S., Agrawal, R., Bernstein, P., et al.: The Lowell Database Research Self-Assessment. In: CACM, vol. 48(5) (2005)
2. Bamford, R., Borkar, V.R., et al.: XQuery Reloaded. In: VLDB (2009)
3. Bhowmick, S.S., Prakash, S.: Every Click You Make, I Will be Fetching It: Efficient XML Query Processing in RDBMS Using GUI-driven Prefetching. In: ICDE (2006)
4. Braga, D., et al.: XQBE (XQuery By Example): A Visual Interface to the Standard XML Query Language. In: ACM TODS, vol. 30(2), pp. 398–443 (2005)
5. Goldman, R., Widom, J.: DataGuides: Enabling Query Formulation and Optimization in Semistructured Databases. In: VLDB (1997)
6. Gou, G., Chirkova, R.: Efficiently Querying Large XML Data Repositories: A Survey. In: IEEE TKDE, vol. 19(10) (2007)
7. Jagadish, H.V., Chapman, A., Elkiss, A., et al.: Making Database Systems Usable. In: ACM SIGMOD (2007)
8. Jiang, H., Lu, H., Wang, W.: Efficient Processing of XML Twig Queries with OR−Predicates. In: SIGMOD (2004)
9. Prakash, S., Bhowmick, S.S., Widjanarko, K.G., Dewey Jr., C.F.: Efficient XML Query Processing in RDBMS Using GUI-Driven Prefetching in a Single-User Environment. In: Kotagiri, R., Radha Krishna, P., Mohania, M., Nantajeewarawat, E. (eds.) DASFAA 2007. LNCS, vol. 4443, pp. 819–833. Springer, Heidelberg (2007)
10. Soh, K.H., Bhowmick, S.S.: Efficient Evaluation of NOT-Twig Queries in Tree-Unaware Relational Databases. In: Yu, J.X., Kim, M.H., Unland, R. (eds.) DASFAA 2011, Part I. LNCS, vol. 6587, pp. 511–527. Springer, Heidelberg (2011)
11. Truong, Q.B., Bhowmick, S.S.: MustBlend: Blending Visual xml Query Formulation with Query Processing in RDBMS. Technical Report, http://www.cais.ntu.edu.sg/~assourav/TechReports/MustBlend-TR.pdf

Efficient SPARQL Query Evaluation via Automatic Data Partitioning

Tao Yang[1], Jinchuan Chen[2], Xiaoyan Wang[1], Yueguo Chen[2], and Xiaoyong Du[1]

[1] School of Information, Renmin University of China
[2] Key Laboratory of Data Engineering and Knowledge Engineering
(Renmin University of China), MOE, China
{yangtao2007,jcchen,wxy,chenyueguo,duyong}@ruc.edu.cn

Abstract. The volume of RDF data increases very fast within the last five years, e.g. the Linked Open Data cloud grows from 2 billions to 50 billions of RDF triples. With its wonderful scalability, cloud computing platform like Hadoop is a good choice for processing queries over large data sets. Previous works on evaluating SPARQL queries with Hadoop mainly focus on reducing the number of joins through careful split of HDFS files and algorithms for generating Map/Reduce jobs. However, the way of partitioning RDF data could also affect the performance. Specifically, a good partitioning will greatly reduce or even totally avoid cross-node joins and significantly reduce the cost of query evaluation. Based on HadoopDB, this work processes SPARQL queries in a hybrid architecture where Map/Reduce takes charge of the computing tasks and an RDF query engine, RDF-3X, stores the data and evaluates join operations over local data. Based on analysis of query work-loads, we propose a novel algorithm for automatically partitioning RDF data. We also present an approximate solution to physically place the partitions in order to reduce data redundancy. All the proposed approaches are evaluated by extensive experiments over large RDF data sets.

1 Introduction

RDF, an abbreviation for *Resource Description Framework*, is a model recommended by W3C for data interchange on the Web. Basically, RDF represents each fact as a triple $< s, p, o >$. RDF dataset is essentially a graph with each vertex per entity and each edge per relationship between two entities. The SPARQL query is a widely accepted query language for accessing RDF triples. A SPARQL query contains a set of *triple patterns*, i.e. at least one element of s, p, and o is a variable. It can also be represented as a graph, with some vertexes or edge labels (predicates) as variables. The results of a SPARQL query are sub-graphs of the RDF graph. Hence a SPARQL query is basically a sub-graph pattern matching task. As a running example, Fig. 1 illustrates the statement and corresponding query graph of a SPARQL query, which tries to find all the persons who obtained his/her degree from the same university which he/she currently belongs to.

In recent years, with the quick proliferation of RDF data, it is often infeasible to store all RDF triples in a single node, which motivates the interests of processing SPARQL queries in a distributed environment, especially within the Hadoop platform [10,12]. Benefiting from the Map/Reduce framework, these works obtain high scalability of evaluating SPARQL queries over billions of RDF triples. However, SPARQL queries usually contain multiple joins and these join operations may be conducted in multiple worker nodes, which is not favored by Map/Reduce because cross-node communications are not permitted in the map phase. Thus a SPARQL query may need multiple M/R jobs which is quite expensive since each such job requires several seconds to fire up, not to speak of the time cost of communication between multiple nodes.

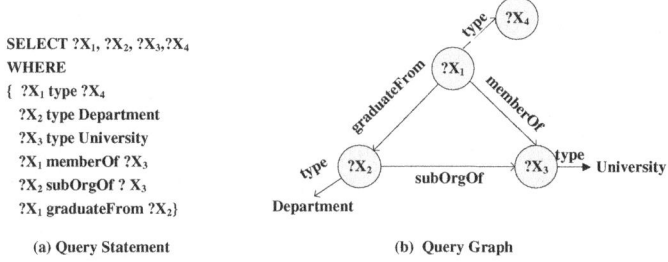

Fig. 1. An example of a SPARQL Query

In the distributed database community, a classical technique to reduce cross-node communication is *data partitioning*. The basic idea is to put the tuples which may be involved in a join in the same worker node [4,14,6]. For this purpose, usually we need to analyze previous query workloads and identify which tuples or rows are probably appear in the same queries [14,6]. Taking this idea, this work aims at facilitating scalable and efficient processing of SPARQL queries via automatic data partitioning. Our work is based on the HadoopDB project [3], which proposes a hybrid architecture by combining Map/Reduce and databases. The principle idea is to execute M/R jobs over a database cluster. In this way, the hybrid system could inherit the scalability and fault-tolerance from Map/Reduce framework while obtaining high efficiency from the powerful capability of processing complex operators like joins and aggregations in traditional databases. In our work, each worker node is equipped with a RDF-3X query engine [13], a state-of-art single-node system for processing SPARQL queries.

Our partitioning approach is inspired by the observation that in many applications there usually exist some frequent query patterns. A *query pattern* is a special SPARQL query, which essentially defines a code such that a group of similar SPARQL queries can be compiled according to it. For example, the query shown in Fig. 1 can be regarded as a query pattern. The variable X_4 can be replaced by different constants like *Student, Professor, and Staff* and correspondingly generate different queries. The general idea of our partitioning approach is to divide the RDF graph into twigs, or tiny sub-graphs, according to the frequent query patterns[1]. We can then ensure that no cross-node joins are needed when processing SPARQL queries complied with any query pattern. What should also be mentioned is that, thanks to the powerful capability of RDF-3X for processing triple joins, even for queries not in any identified patterns, the query performance of our system would also defeat those works based on Hadoop systems [10].

The work most similar to ours is [9], where the authors also propose to evaluate SPARQL queries over HadoopDB. In [9], the whole RDF graph is divided into several huge sub-graphs based on a graph partitioner METIS [2]. These subgraphs would be stored at different worker-nodes, with triples near the division boundaries replicated to multiple nodes. Based on this partitioning, most queries could be answered based on the triples inside a single node.

Compared with [9], our solution has several significant advantages. First of all, [9] does not consider the dynamic properties in query workloads, and cannot guarantee that there are no cross-node joins for frequent query patterns. Suppose a query pattern happens to involve the triples on the partition boundaries, querying the queries compiled with this pattern have to coordinate triples in different nodes. Secondly, [9] may result in many duplicated triples and does not mention how to alleviate this redundancy. In our solution, the partitioning contains two steps. The first step is exactly a logical partitioning, i.e. it will divide the original dataset into many small parts but does not really move them. At the second step, we will place these partitions into different worker nodes. During this placement phase, we will try to reduce the data redundancy by putting partitions with large overlapping into the same worker node. Finally, the partitioning in [9] is based on graph partitioning, a known NP-complete problem [5], which will cost lots of computational efforts.

The contributions of our work are summarized as follows.

- We propose a query-driven data partitioning approach and based on it develop an efficient solution for processing SPARQL queries over large scale RDF data.
- We prove that the placement problem of reducing data redundancy is NP-hard.
- We present an approximate algorithm for reducing data redundancy, which is based on the LNS (Large Neighborhood Search) solution [14].
- We conduct extensive experiments over two large datasets, i.e. LUBM [8] and BTC [1], to evaluate the efficiency and effectiveness of our proposed approaches.

Next we will illustrate the architecture of our system in Sec. 2. We then discuss the partition and placement approaches in Sec.3. Sec. 4 will report our experimental results. We will discuss related works in Sec. 5 and conclude this paper in Sec. 6.

---

[1] The query patterns are assumed to be available, and the efforts of analyzing query workloads and identifying frequent patterns exceed the scope of this paper.

2 System Architecture

Our system architecture is illustrated in Figure 2. This system contains three modules including *Data Pre-Processor*, *Query Engine* and a hybrid platform combining Map/Reduce and RDF-3X.

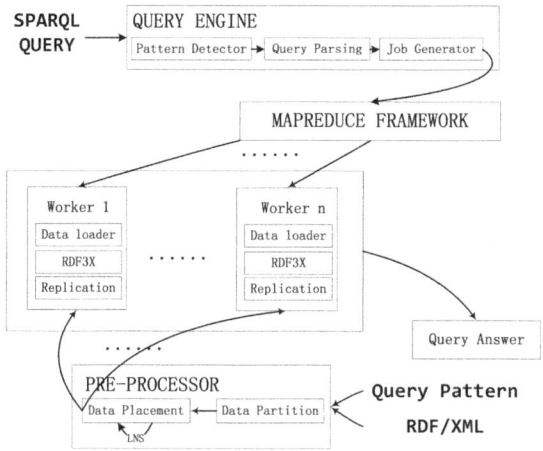

Fig. 2. System Architecture

In the data pre-processor, RDF data are partitioned according to query patterns. The *data partition* procedure guarantees that there are no cross-node joins when evaluating any queries compiling to any registered pattern. Each query pattern would have an independent partitioning. All the partitions generated by all the frequent query patterns need to be put into the nodes, through the *data placement* procedure. Note that a triple may appear in multiple partitions since we perform an independent partitioning for each query pattern. Hence the major concern of the placement procedure is to reduce the data redundancy. Once obtaining the partitions, the *data loader* procedure on each worker node will load all triples to the RDF-3X database installed in that machine.

In the query engine, after receiving a SPARQL query, the *pattern detector* figures out whether this query matches any query pattern. If YES, because the triples of each sub-graph matching this query have been placed to a single worker, we can pushdown the whole query to RDF-3X and simply generate one M/R job to retrieve the results. For those queries matching no patterns, we just generate M/R jobs according to the algorithm in [10]. In practice, we can design query patterns to accommodate as many queries as possible, e.g. by replacing more constants with variables. Finally, the query engine submits the jobs to MapReduce framework.

Queries are executed using MapReduce with RDF-3X as a local engine. MapReduce checks configuration files and locates data replications with job parameters.

Then each task sends SPARQL queries to RDF-3X installed in each worker node. Query results are returned back as the *InputFormat* of `Mapper`.

3 Data Partitioning

We now illustrate the data partitioning process, which contains two steps, i.e. query-driven partitioning and placement.

3.1 Query-Driven Partitioning

Since each SPARQL query pattern can be regarded as a directed graph, we will use "edge" and "triple pattern" interchangeably when the meaning is clear from the context. For each such sub-graph matching a given query pattern, if all triples in this sub-graph are placed into the same worker node, there would be no cross-node joins when evaluating all queries within this pattern. This property will be guaranteed by our partitioning algorithm. Before going into the algorithm details, we first illustrate a framework to present the principle idea of our partitioning algorithm.

Algorithm 1. PartitioningFramework(RDF dataset D, Query Q)

1 $Q' \leftarrow$ an empty graph;
2 Randomly choose an edge e_0 from Q and insert it into Q' ;
3 $S \leftarrow$ all triples matching e_0;
4 Randomly choose a partition for each triple in S;
5 Record the partition information;
6 **while** $E(Q') < E(Q)$ **do**
7 $\quad$ Select an edge e such that $e \in E(Q) \setminus E(Q')$ and e is connected to Q';
8 $\quad$ Insert e into Q';
9 $\quad$ $S \leftarrow$ all triples matching e;
10 $\quad$ **foreach** *triple t in S* **do**
11 $\quad\quad$ Find the set of triples S' that can be joined with t according to Q and have already been partitioned;
12 $\quad\quad$ Put t in each of the partition containing at least one triple in S';
13 $\quad$ Record the partition information;

Algorithm Framework. Alg. 1 illustrates a framework of our partition algorithm. We first initialize an empty graph Q', which will be used to record the progress of partitioning. Next, Step 2 tries to randomly choose an edge from Q and add it to Q'. All triples matching this edge will be loaded by scanning the whole dataset (Step 3), and each matched triple will be assigned to a random partition (Step 4). We will store the partitioning informations, e.g. the partition which a triple is assigned to, in a table (Step 5). After processing this initial edge, we then add other edges in Q to Q' one by one. Step 7 claims that the edge which we choose at each step must be connected to Q', which is a critical requirement to ensure the algorithm target. Again, we retrieve all triples matching this new edge (Step 9). Now we need to decide the partitions for these matched triples. For this purpose, we basically conduct a join according to the structure of Q between these new coming triples and those have already been partitioned (Step 11).

Lemma 1. *For any query pattern Q, all sub-graphs matching Q will have their triples assigned to the same partition if we divide the original dataset according to Algorithm 1.*

Proof. Without lose of generalization, suppose Q contains m edges. We change the indices of these m edges and get $e_1, \cdots, e_m$ such that e_i is the ith edge inserted into Q' according to Algorithm 1. Any sub-graph matching Q also contains m triples. Again, we change the indices of the m triples in a specific sub-graph to obtain $t_1, \cdots, t_m$ so that t_i has e_i as its corresponding triple pattern in Q. It is not hard to see that t_i must be placed into the same partition as t_{i-1} for $i = 2, \cdots, m$. Thus all these m triples must appear in the same partition which is initially decided by t_1.

The Partition Algorithm. The framework in Algorithm 1 cannot be directly applied in practice. This algorithm requires a table or index to record the partitioning decision for each processed triple, which is not feasible due to the large volume of RDF dataset. Moreover, the RDF dataset is usually quite skew. Some triple patterns may have huge number of matched triples. This skewness must be handled carefully otherwise the performance could be very bad. We now illustrate several techniques utilized for overcoming these shortcomings and, based on these techniques, present the partitioning algorithm.

A Compact Data Structure for Storing the Partitioning Results. In Alg. 1, for each e chosen to be inserted into Q', all the triples matching e should be checked to see whether they could join with previous partitioned triples. This triple-to-triple join is quite costly in terms of both computation and memory with the existence of billions of triples. In order to improve the performance, we now present a compact data structure to store the partitioning results.

This structure is based on the following observation. Any triple t which could be joined with existing ones must satisfy the following two requirements: (1) its corresponding triple pattern e must share at least one common variable with one or more edges in Q', and (2) there exists at least one partitioned triple t' such that both t and t' are assigned with the same value for this common variable. Therefore, instead of recording all the partitioned triples, we just need to keep all the distinct values of each variable that have shown in any partitioned triples. Specifically, we construct a Hash table for each variable v in Q' containing a set of (var_key, pos) entries with one var_key per distinct value of v, and the corresponding pos be the identifier no. of the partition storing all the triples having the same value, i.e. var_key, to this variable v.

Thus every time we join a candidate triple with previous allocated ones, our algorithm will first look up the Hash table with its value on the joined variable and then assign this triple to the corresponding partition indicated by the returned *pos* value. In the meanwhile, if this triple contains another variable, its value should be added to the index for the follow-up operations.

Algorithm 2. Partitioning(TripleSet D, Query Q, Int n)

1 $TP \leftarrow$ estimate(Q, D);
2 $Q' \leftarrow$ an empty graph;
3 **while** $|E(Q')| < |E(Q)|$ **do**
4     $e \leftarrow$ chooseEdge(Q, Q', TP);
5     $S_{Temp} \leftarrow$ loadTriples(D, e);
6   **foreach** $t \in S_{Temp}$ **do**
7   **if** $E(Q')=0$ **then**
8          $i \leftarrow$ a random value in $[1, \cdots, n]$;
9   **foreach** v *in* $Var(e)$ **do**
10   putIndex(Γ,hash(t,e,v),i);
11   **else**
12   **foreach** v *in* $Var(e) \cap Var(Q')$ **do**
13           $i \leftarrow$ readIndex(Γ,hash(t,e,v));
14   **if** $i < 0$ **then**
15   continue;
16   **else**
17   add t into S_i;
18   **foreach** v *in* $Var(e) \cap (Var(Q) \setminus Var(Q'))$ **do**
19           $i \leftarrow$ readIndex(Γ,hash($t,e,Var(e) \setminus v$));
20   putIndex(Γ,hash(t,e,v),i);
21   insert e to Q';

Choosing Edges Based on Selectivity Estimation. In practice, the selectivities of the triple patterns could be quite skew. For LUBM dataset containing 1 billion triples, if we first choose the edge $?X_1$ *type* $?X_4$ in the example query in Fig.1, about 10^9 of records need to be retrieved and recorded in the index. Clearly, not all the records can satisfy the query. Many retrieved triples have no contributions to the query results. In this work, we adopt the *selectivity estimation* technique to improve the performance. As a classical method in the database community, the principle of *selectivity estimation* is to evaluate predicates with low selectivities first in order to reduce the number of tuples involved in joins. We utilize a simple heuristic to estimate the selectivity of each triple pattern. Suppose the number of triples contained in a predicate is num_p, and the number of distinct values of the variable in this triple pattern is num_v, the selectivity of this triple pattern is estimated by num_p/num_v. For details, please refer to our technical report [18].

The Algorithm. Now, we are ready to present the partitioning algorithm, which is listed in Alg. 2. Step 1 is to construct a list TP storing the selectivities of each triple pattern by analyzing D. The *chooseEdge* function in Step 4 is to select a new edge which should be connected to Q'. This selection is based on the priorities stored in TP. Steps 7-10 is to process the first edge. Each triple matching the first edge will be assigned to a random partition (Step 8). A *hash* function computes a Hash code for the variable value of this triple with this triple pattern, and this code would be used as the key to store the partition result in an index Γ(Step 10). For subsequent edges, a variable on the edges may be *join variable*, i.e. appearing in Q', or non-joinable variable, i.e. not in Q'. The partition of the current triple is decided by its *join variables* through checking the index(Step 13). Note that a triple cannot be joined if we cannot find its key in the index(Steps 14-15). We also store the partition information for the non-joinable variables (Steps 18-20), which may be used to link with follow-up edges. Note that there is at most one non-joinable variable for each edge.

3.2 Placement

According to the partitioning process shown in Sec. 3.1, each query pattern will generate n partitions. We now discuss how to physically place these partitions in different worker nodes. Note that when partitioning the original RDF dataset according to a query pattern Q, those triples which could not satisfy this pattern will not be assigned to any partition. Thus, after processing all the m query patterns, there are still a large portion of triples satisfying no patterns and therefore are not partitioned. In practice, such triples will be seldom accessed and we call them *cold* triples. Each of these cold triples would be assigned to a randomly chosen worker node. In this section, we focus on the placement of the *hot* triples, i.e. those probably satisfying at least one query pattern. Since our partitioning algorithm guarantees no cross-node joins for queries compiled to frequent patterns, the major concern of placement is to reduce the data redundancy, i.e. the number of replicated triples among different worker nodes. We will discuss how to estimate data redundancy and give the definition of the placement problem. Then we will prove that the placement problem is NP-hard. Finally we will illustrate an efficient approximate solution for this problem.

Problem Definition. During the partitioning process, we totally generate $m \cdot n$ partitions. Hence each reasonable placement solution, denoted by P, needs to arrange these $m \cdot n$ partitions in a $m \cdot n$ matrix such that: i) the ith column of P contains all the partitions corresponding to the ith query pattern ($i = 1, \cdots, m$) and, ii) all partitions in the jth row would be put into the jth worker node ($j = 1, \cdots, n$). Thus a placement solution is exactly a $m \cdot n$ matrix and there could be n^m different possible solutions. The data redundancy of a placement solution P could be evaluated by the following equation.

$$\gamma_P = \sum_{\forall i,j (i \neq j)} \gamma_{i,j} - \sum_{i=1}^{n}\sum_{j=1}^{m} \gamma_{i,j} \qquad (1)$$

Here $\gamma_{i,j}$ is the number of replicated triples in the ith and j^{th} partitions, and γ_P means the overall redundancy of this solution. The value of each $\gamma_{i,j}$ can be estimated based on the index described in Sec. 13. The details are skipped due to page limit. Interested readers are recommended to read our technical report [18]. The first part on the right side is basically the total number of replicated triples among all the $m \cdot n$ partitions. The value of $\sum_{j=1}^{m} \gamma_{i,j}$ is the overall redundancy among all partitions in the jth row. The second part is just to compute the sum of all redundancy in each row. When evaluating γ_P, the row-level redundancy should be subtracted since all partitions in a row would be put in the same worker node. Note that the first part is a constant for all different placement solutions. Hence in order to minimize γ_P, we just need to maximize the value of the second part.

Definition 1. *Given the partitioning results of the m query patterns, i.e. a set of $m \cdot n$ partitions,* **the placement problem** *is to minimize the data redundancy as defined in Equation 1 by arranging these $m \cdot n$ partitions into the n worker nodes.*

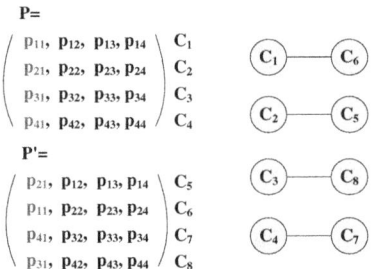

Fig. 3. An example of placement solution

Complexity Analysis. The *placement problem* could be transformed to the *maximum weight independent set* problem (MWIS in short). For any graph where each vertex is attached with a positive weight, a maximum independent set (MIS) is a set of vertexes in this graph which are pairwise disconnected and, all other vertexes should have at least one neighbor in this set. A MWIS is just the heaviest MIS. Let us define a *reasonable combination*, denoted as C, as a set of m partitions each of which corresponds with a distinct query, i.e. a row in a placement solution. Hence there could be totally n^m reasonable combinations. We then build a graph $\mathcal{G}$ by adding a node for each reasonable combination, and adding an edge between two nodes if and only if their underlying combinations contain overlapping partitions. Each node is attached with a weight which is equal to the overall redundancy of this combination. Fig. 3.2 illustrates a simple example for partitioning a dataset into four nodes according to four queries. On the left part, there are two possible placement solutions, P and P'. Each element of the two matrixs, e.g. p_{11}, is a partition generated in the partitioning process. The difference between P and P' are highlighted in red color. Each row represents a reasonable combination with their labels, i.e. $C_1, \cdots, C_8$, listed on the right. The graph on the right part of this figure contains eight nodes for these combinations, with edges connecting combinations with overlapped elements. For example, the nodes C_1 and C_6 are connected since both nodes contain the same element p_{11}.

Algorithm 3. Placement(Solution P)
1 $SearchArray \leftarrow$ Relax(P);
2 $best \leftarrow$ Evaluate(P);
3 $scoreOfOneSearch \leftarrow$ LocalSearch($SearchArray, best$);
4 **while** $scoreOfOneSearch > best$ **do**
5     $best \leftarrow scoreOfOneSearch$;
6     $ideal \leftarrow$ array stored in LocalSearch;
7     $SearchArray \leftarrow$ Relax($ideal$);
8   LocalSearch($SearchArray, best$);
9 **return** $ideal$;

Note that in any placement solution, there will be no repeated partitions and all the n^m partitions should show up. Clearly, each maximum independent set will constitute a solution P, and the optimal solution is exactly the one with the largest sum-of-weight. The MWIS problem is known to be NP-hard [16], and we have to resort to some approximate approaches.

A LNS-Based Approximate Algorithm. In this paper, we adopt the LNS (Large Neighborhood Search) algorithm [14]. The principle of LNS is an iterative process. Starting from an initial solution, each iteration will search the nearby solutions of the previous optimal solution, and repeat this procedure if finding a better solution. Otherwise, if no better solutions are found in the neighborhood, the current optimal solution will be output as the result. The algorithm is illustrated in Algorithm 3. Firstly, we takes the output of the partitioning algorithm as the initial solution. Then in relaxation part, some placed parts in the initial solution will be selected to be relaxed according to proportion ratios given as parameters (Step 2). In the core phrase of LNS, the LocalSearch function will explore nearby solutions and try to find better solutions (Step 4,9). The process will be repeated if a better solution is met during the local search process (Steps 5-9).

4 Experimental Analysis

4.1 Experiment Setup

Hardware. We perform our experiments on a 8 node cluster. Each node has the following configuration: two 2.4GHz Intel(R) Xeon(R) E5654 processors, 48GB main memory and 250G disk space. We run the partitioning and placement algorithms on one of the machine in cluster with an extra 4.5T hard disk.

Software. We modify the source codes of Hadoop-0.20.2 in order to adapt with RDF storage. The version of the RDF-3X engine used in our experiments is 0.3.7. In the *pre-processor* module, we parse each SPARQL query with jena-2.6.2 together with arq-2.8.3 [17]. In comparison with the state-of-art, we also implement the system in [10], called *TKDE11* in this paper. We do not compare the performance with the system in [9], since it usually evaluates SPARQL queries over database clusters instead of the Map/Reduce platform.

DataSets and Queries. Throughout our experiments, we use one synthetic dataset, the Lehigh University Benchmark(LUBM) [8] and one realistic dataset, the Billion Triple Challenge 2010(BTC) [1]. LUBM generates synthetic data about universities on a university domain ontology. Besides data generator, LUBM also has 14 standard queries focusing on both scalability and inference testing. In our experiments, we generate a dataset of 10,000 universities with default parameters. The LUBM dataset contains around 1.1 billion triples. BTC contains the triples from twelve sources such as Yago, DBPedia, and Freebase [7]. We choose BTC to test effectiveness and efficiency of our solutions on large scale and real-life RDF data. After cleaning the noisy or duplicated data, we obtain about 1.28 billion triples. Before the query execution, we also encode each field of these triples into integers to facilitate the query processing.

Query patterns are obtained by rewriting the SPARQL queries. We simply unbound some concrete values as variables in some triple patterns. For example, in LUBM query1, we turn triple pattern ?X ub:takesCourse <http://www.Department0.University0.edu/GraduateCourse0> into ?X ub:takesCourse ?Y to generate a query pattern. The complete list of all queries used in our experiments can be found in our technical report [18].

Table 1. Query running time in seconds of LUBM 10000 dataset

	Q1	Q2	Q3	Q4	Q5	Q6	Q7
Our Solution	16.3	235.2	16.3	16.3	16.3	255.4	22.4
TKDE11	116.3	687.6	174.4	757.8	371.1	342.8	289.2

	Q8	Q9	Q10	Q11	Q12	Q13	Q14
Our Solution	28.3	165.2	16.4	15.3	17.5	74.5	213.3
TKDE11	1320.3	1371.8	184.5	103.3	56.0	91.0	325.9

4.2 Evaluation

Partitioning Time. In our system, the time for partitioning the LUBM dataset is about 20 minutes. The RDF-3X needs about 40 minutes for loading these triples and building the index.

Query Performance. Next, we will compare the performance between our solution and TKDE11, and evaluate the effectiveness of our placement algorithm for reducing data redundancy.

Table 1 illustrates the query performance of our solution and TKDE11. We can observe that our solution always performs better than TKDE11. Moreover, for most queries such as Q1, Q3-Q5, and Q7-Q11, the performance of our solution is about one order of magnitude faster than its competitor. These results validate the principle idea that good data partitioning can significantly improve query performance. Next we will analyze the results in detail.

1. For the queries with simple semantic and small result set like Q1, Q3-Q5, Q7-Q8, Q10, and Q11-Q12, our solution performs about 10 times faster than TKDE11. The reasons are two folds: 1) there are only one or two join variables in those queries and their triple patterns usually contain constants, which ensure RDF-3X to utilize its index for selecting candidate triples very fast and, 2) their result sets are quite small, less than 10^3 triples, and little time cost is needed in data transformation between the map and reduce processes.
2. Q2 and Q9 are queries with complex structure, and small result set. Specifically, there is a triangle relationship among three variables, say $?X$ $?Y$ and $?Z$. When dealing with these queries, TKDE11 will generate two M/R jobs with job1 evaluating two joins on $?X$ and $?Y$ separately, and job2 joining the output of job1 with $?Z$. The time cost is huge with large middle files written to and read from HDFS files. On the contrast, our solution can benefit from the powerful capabilities of RDF-3X for evaluating joins. Moreover, our system needs only M/R job based on our partitioning approach.
3. The performance of our method is slightly better than TKDE11 in Q6, Q13 and Q14. These queries all have huge results, larger than 10^8 triples. Lots of triples should be scanned and large amount of triples need to be transferred between RDF-3X engine and Map/Reduce framework. Hence both the query execution and result retrieval phases will cost a lot of time compared with those queries in the above cases.

For the queries matching no frequent patterns, our method has similar performance as TKDE11. The only difference between our solution and TKDE11 method for processing these queries is where the data comes from. We evaluate this case using some queries in LUBM query set by processing each triple pattern at a time. There are no evident difference in the time costs of both methods. Thus we do not report the numbers here.

It should be noted that the queries above are tested using cold runs, which means that the main memory and file system cache were cleared before execution.

Also, because RDF-3X does not support inferencing, we rewrite the reasoning-needed queries in LUBM query set to equivalent ones using union operations before query execution.

The time of query processing tested on BTC is presented in Table 2. Due to the feature of large number of distinct predicates for BTC dataset, the result set for queries over BTC is much smaller than that of the queries over LUBM. Relatively simple query and smaller results explains the resemblance of the query time for our solution in Table 2. It should be noted that there is predicate variable in Q1 and Q4, in which case our solution performs far better than TKDE11 strategy because we do not need to scan the whole triples to get the result.

Table 2. Query running time in seconds of BTC dataset

	Q1	Q2	Q3	Q4	Q5	Q6	Q7
Our Solution	16.1	16.2	16.2	15.2	15.9	16.1	16.3
TKDE11	296.1	45.1	95.6	330.3	22.3	74.5	22.8

Placement. We test the effectiveness of our placement algorithm with various column adjust number K and row adjust number M, i.e. at each iteration of Algorithm 3, K columns and M rows of the previous matrix are relaxed which define a neighborhood containing $(M!)^K$ different solutions. The y-axis in Figure 4 and Figure 5 is the ratio of redundancy decrease, which is computed by $1 - \gamma'/\gamma$ with γ' and γ are the redundancies of the final solution and initial solution respectively.

As illustrated in Figure 4, it is obvious that the higher K and M are, the better performance is achieved. Figure 5 illustrates the redundancy decrease ratio obtained after each iteration of searching neighborhood with different K and M. Clearly, the redundancy becomes smaller after each iteration, and after several runs the redundancy becomes stable.

5 Related Work

SPARQL Query Processing. Most previous works on evaluating SPARQL queries over RDF data are based on a single node [13,7,17]. The RDF-3X [13] is widely accepted as the state of art for SPARQL query engine, which stores all triples in B+ tree, and builds exhausted indexes of all SPO permutations. Due to the centralized mode, these works cannot scale to handle huge volume of RDF triples which are still increasing in high velocity.

In order to process such huge RDF datasets, [12,10,11] suggest to store RDF triples in HDFS files and evaluate SPARQL queries by rewriting them as a series of Map/Reduce jobs. [10] presents a method for generating Map/Reduce jobs and heuristics of dividing RDF triples into separate HDFS files. Myung et. al. [12] propose an algorithm for basic graph pattern matching, and they will process a

SPARQL query by a sequence of Map/Reduce jobs. None of these works consider to improve query performance through better data partitioning. The work in [9] proposes to partition a RDF dataset according to its graphical features and try to avoid cross-node communications by replicating some vertexes near the boundaries of each partition. However, as illustrated in Sec.1, our solution has several evident advantages compared with [9].

Data Partitioning. In distributed databases, data partitioning is one of the most important technologies for achieving platform scalability. The partitioning solutions are realized mainly in horizontal partitioning or vertical partitioning [4]. In brief, horizontal partitioning, such as Hash, Round-robin, Range, etc., is to divide a relational table into multiple groups of rows, whereas vertical partitioning is tries to divide a table into several clusters of columns. Recently, along with the quick development of practical applications, researchers tend to use nested horizontal or vertical partitioning methods [15], or even hybrid approaches [4], to achieve better performance. The idea of improving performance through clever partitioning gives us a good inspiration, but all of these works focus on relational databases and cannot be applied directly to partition RDF datasets. The works in [6] design partitioning based on elaborative analysis on query workloads such that frequent queries could be answered more quickly, which is also adopted in this paper in our data partitioning solution.

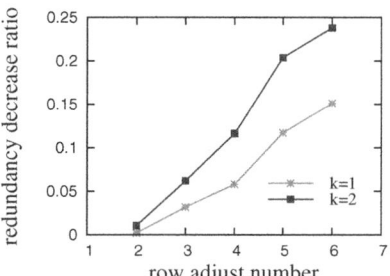

Fig. 4. Redundancy decrease

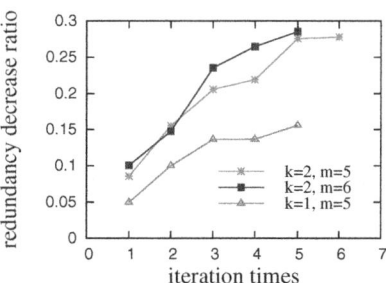

Fig. 5. Iteration

6 Conclusion and Future Works

In this paper, we propose a automatic data partitioning approach in order to improve the performance of processing SPARQL queries in a Map/Reduce framework. Compared with previous works, our system can completely avoid cross-node joins for frequent queries and reduce data redundancy. According to the simulation results, our method could accelerate the query processing time by up to two orders of magnitude. Our on-going projects include how to perform update or even migration when there are significant changes on the dataset and/or workload, e.g. large number of new-coming tiples, new identified query patterns, and skewed query accessing.

Acknowledgements. This work is funded by the National Science Foundation of China under Grant No. 61170010, National Basic Research Program of China (973 Program) No. 2012CB316205, and HGJ Project 2010ZX01042-002-002-03. We also want to express our thankfulness to Sa-Shixuan International Research Centre for Big Data Management and Analytics (hosted in Renmin University of China) which is partially funded by a Chinese National "111" Project "Attracting International Talents in Data Engineering and Knowledge Engineering Research".

References

1. Btc 2010 (2010), http://www.hpi.uni-potsdam.de/naumann/sites/btc2010
2. Metis, http://glaros.dtc.umn.edu/gkhome/views/metis/index.html/
3. Abouzeid, A., Bajda-Pawlikowski, K., Abadi, D., Silberschatz, A., Rasin, A.: Hadoopdb: An architectural hybrid of mapreduce and dbms technologies for analytical workloads. PVLDB 2(1), 992–933 (2009)
4. Agrawal, S., Narasayya, V., Yang, B.: Integrating vertical and horizontal partitioning into automated physical database design. In: SIGMOD 2004, pp. 359–370 (2004)
5. Andreev, K., Räcke, H.: Balanced graph partitioning. In: SPAA, pp. 120–124 (2004)
6. Chang, C., Kurç, T.M., Sussman, A., Çatalyürek, U.V., Saltz, J.H.: A hypergraph-based workload partitioning strategy for parallel data aggregation. In: PPSC (2001)
7. Du, F., Chen, Y., Du, X.: Partitioned indexes for entity search over rdf knowledge bases. In: Lee, S.-g., Peng, Z., Zhou, X., Moon, Y.-S., Unland, R., Yoo, J. (eds.) DASFAA 2012, Part I. LNCS, vol. 7238, pp. 141–155. Springer, Heidelberg (2012)
8. Guo, Y., Pan, Z., Heflin, J.: Lubm: A benchmark for owl knowledge base systems. Web Semantics: Science, Services and Agents on the World Wide Web 3(2-3), 158–182 (2005)
9. Huang, J., Ren, D.J.K.: Scalable sparql querying of large rdf graphs. PVLDB 4(11), 1123–1134 (2011)
10. Husain, M., McGlothlin, J., Masud, M.M., Khan, L., Thuraisingham, B.: Heuristics based query processing for large rdf graphs using cloud computing. IEEE TKDE 23(9), 1312–1327 (2011)
11. Kim, H., Ravindra, P., Anyanwu, K.: Scan-sharing for optimizing rdf graph pattern matching on mapreduce. In: IEEE CLOUD, pp. 139–146 (2012)
12. Myung, J., Yeon, J., Lee, S.-G.: Sparql basic graph pattern processing with iterative mapreduce. In: Proc. of the 2010 Workshop on Massive Data Analytics on the Cloud, MDAC 2010, pp. 6:1–6:6 (2010)
13. Neumann, T., Weikum, G.: Rdf-3x: a risc-style engine for rdf. PVLDB 1(1), 647–659 (2008)
14. Pavlo, A., Curino, V., Zdonik, S.: Skew-aware automatic database partitioning in shared-nothing, parallel oltp systems. In: SIGMOD 2012, pp. 61–72 (2012)
15. Rao, J., Zhang, C., Megiddo, N., Lohman, G.: Automating physical database design in a parallel database. In: SIGMOD 2002, pp. 558–569 (2002)
16. Sanghavi, S., Shah, D., Willsky, A.S.: Message passing for maximum weight independent set. IEEE Trans. on Information Theory 55(11), 4822–4834 (2009)
17. Wilkinson, K., Sayers, C., Kuno, H.A., Reynolds, D.: Efficient RDF Storage and Retrieval in Jena2. In: ISWC 2003, pp. 131–150 (2003)
18. Yang, T., Chen, J., Wang, X., Chen, Y., Du, X.: Efficient sparql query evaluation via automatic data partitioning, technical report (2012), http://iir.ruc.edu.cn/~jchchen/rdfpartition.pdf

Content Based Retrieval for Lunar Exploration Image Databases

Hui-zhong Chen[1,2], Ning Jing[1], Jun Wang[3], Yong-guang Chen[4], and Luo Chen[1]

[1] School of Electronic Science and Engineering,
National University of Defense Technology, Changsha, China
[2] Shanghai Branch, Southwest Electronic and Telecommunication
Research Institution, Shanghai, China
[3] The Third Research Institute of Ministry of Public Security, Shanghai, China
[4] Ordnance Engineering College, Shijiazhuang, China

Abstract. Being a novel research aspect following the recent new round of lunar explorations, content-based lunar image retrieval provides a convenient and efficient way for accessing relevant lunar remote sensing images by their visual contents. In this paper, we introduce a novel method for mining relevant images in lunar exploration databases. A novel feature descriptor derived from relationships of salient craters in lunar images and a compound feature model organizing different features are proposed. Based on the features, similarity measurement rules and a retrieval algorithm are proposed and described in detail. Both theoretical analysis and experimental results of our method are provided, verifying that our features and model are effective and the method can get a good relevant retrieval results in lunar image databases.

Keywords: Content-based lunar image retrieval, Feature extraction, Similarity measurement, Relevant retrieval, Image mining.

1 Introduction

With the new round of lunar exploration recently, several programs by different countries have been developed or in plan, such as the CLEMENTINE, LRO of USA, SMART-1 of ESA, SELENE of Japan, Luna-Glob of Russia, MoonLITE of UK, Chandrayaan-1 of India and Chang'e of China et al. Large quantities of lunar remote sensing images are obtained and available to the public soon after the mission. The databases containing massive and diverse images are conventionally organized by keys like mission Sol. or spacecraft clock time, which makes it difficult for users to retrieve and take advantage of the full potential of the image data [9].

Following the current lunar exploration missions, content-based lunar image retrieval has been studied as a quite novel research aspect. C. Meyer et al.[9] of NASA Ames research center proposed a method for content-based retrieval of images for planetary exploration.

Being a relevant research domain, for earth remote sensing images, there are various contributions that focus on content-based retrieval and, oftentimes, considered as image information mining[8]. For example, in GeoIRIS system[14], a content retrieval system architecture is designed, covering the function of automatic feature extraction, visual content mining from large-scale image databases. Recently, [13] presented a novel indexing structure that was developed to efficiently and accurately perform content-based shape retrieval of objects from a large-scale satellite imagery database. [2] described a framework for modeling directional spatial relationships among objects and using this information for contextual classification and retrieval. [3]proposed a knowledge-discovery algorithm that links low-level image features with high-level visual semantics to automate the process of retrieving semantically similar images. More earlier relevant contributions can be found in [1][12][11][4] et al.

Lunar images are lack of color information and the shape of objects contained is about the same. Therefore it brings new challenges to us and further researches are required for the content-based lunar image retrieval.

2 Problem Description

To retrieve relevant images, content features are extracted and indexes are built in the preprocessing stage and stored after that. The images whose features belong to the top k best matches are returned as the resulting. Figure 1 gives out the overall structure of content-based lunar image retrieval.

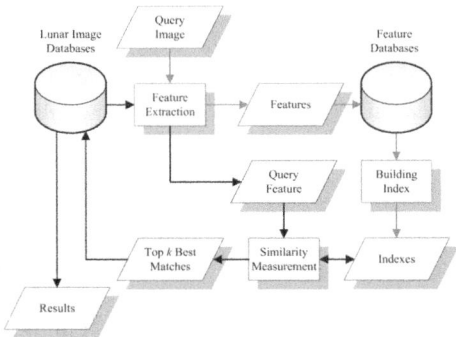

Fig. 1. Overall structure of content-based lunar image retrieval, the grey arrows demonstrate the preprocessing steps and the black arrows demonstrate the retrieval steps

Firstly, several notations are defined: I: a lunar remote sensing image; N: number of images in the database; **I**: image set in the database, $\mathbf{I} = \{I_i|\ i = 1, 2, \ldots N\}$; **F**: content feature of I; I_q: sample query image; $\mathbf{I}_{result}$: image set of query results; k: number of resulting image sets; S: similarity value.

Then the problem can be defined as follows:

Definition 1. (problem of feature extraction): Given a lunar remote sensing image I, find a method $Extract(\cdot)$, s.t. $\mathbf{F} = Extract(I)$.

Definition 2. (problem of similarity measurement): Given two lunar remote sensing images I_1 and I_2, find a method $S(\cdot, \cdot)$, s.t. $Sim_{12} = S(\mathbf{F}_1, \mathbf{F}_2)$, the higher Sim_{12} is, the more similar I_1 and I_2 are in content.

Definition 3. (problem of content-based lunar image retrieval): Given a query sample I_q, find a set of k lunar images $\mathbf{I}_{result}$ from the database, s.t. $\forall I_i \in \mathbf{I} \setminus \mathbf{I}_{result}$, $\forall I_j \in \mathbf{I}_{result}$, $Sim_{jq} \geq Sim_{iq}$.

3 Feature Extraction

We have developed a compound feature model that supports different content features. And we also propose a feature descriptor particularly for lunar images based on the salient regions.

3.1 Compound Feature Model

Most lunar images looks similar, because their visual contents are about the same. In this condition, to organize different features including global descriptors and local descriptors, a compound feature model is defined as follows:

$$\mathbf{F} = \left(\vec{gf}, \mathbf{MF}, \mathbf{LF}\right) \quad (1)$$

Here, $\vec{gf} = \{gf_1, gf_2, \ldots, gf_D\}$ is a D-dimensional global feature vector, and each image has only one $\vec{gf}$. $\mathbf{MF}$ is the set of D_m-dimensional mid-level local feature descriptors, $\mathbf{MF} = \{\vec{mf_i} | i = 1, 2, \ldots, N_m\}$, $\vec{mf_i} = \{mf_{i1}, mf_{i2}, \ldots, mf_{iD_m}\}$. And $\mathbf{LF}$ denotes the set of low-level local features: $\mathbf{LF} = \{\vec{lf_i} | i = 1, 2, \ldots, N_l\}$, $\vec{lf_i} = \{lf_{i1}, lf_{i2}, \ldots, lf_{iD_l}\}$. The "mid-level" means the descriptors are calculated on the low-level features, The detail will be described in the next section.

We use HU moment[5] as shape feature, Tamura[15] descriptor as texture feature, and they are combined into one feature vector as the global compound feature. SURF local descriptors are used as low-level features.

3.2 LIFBS Feature

LIFBS is based on the salient regions of a lunar image. Here, the saliency means the most visually attentive parts in a lunar image, usually containing big and conspicuous impact craters.They are set of circle regions, which can be calculated by automated Crater Detection Algorithms (CDAs)[10][6] or labeled by users manually. Below are some notations to be used in the description of LIFBS generating algorithm: c: $c = (x, y)$, center position of a salient region; r: radius of a salient region; H: visual strength of a salient region; SR: a salient region, $SR = (c, r, H)$; N_s: number of salient regions in a lunar image; **SRS**: set of all salient regions in a lunar image, **SRS** $= \{ SR_i | i = 1, 2, \ldots, N_s \}$.

Then the detailed LIFBS generating algorithm is as follows: ***Step 1.*** Input **SRS**, $\forall SR_i, SR_j \in$ **SRS** and $i \neq j$, calculate the distance between centers:

$$D_{ij} = \text{Dist}(SR_i, SR_j) = \sqrt{(x_i - x_j)^2 + (y_i - y_j)^2} \qquad (2)$$

for each $SR_i \in$ **SRS**, repeat ***Step 2- Step 6***.

Step 2. Sort all the $SR_j, j \neq i$ in ascend order according to D_{ij}, mark the rank order number as o, and $j = \text{rank}^{-1}(o)$.

Step 3. Let $o = 1$, set SR_i's nearest neighboring salient region $SR_j, j = \text{rank}^{-1}(o)$ as the base neighboring salient region $SR0$, and its center $c_j = (x_j, y_j)$ as the base point $c0$. Let the line from c_i to $c0$ is the base line, denoted as $L0$.

Step 4. Let c_i be the origin, $L0$ be the 1st quadrant angular bisector, then-coordinates divides the image to 4 quadrants: $QR_n, n = 1, 2, 3, 4$.

Step 5. If c_j is in QR_n, check if there already exists feature components of QR_n, if not, calculate following 3 components: $s_n = r_j/r_i$, $h_n = H_j/H_i$, $d_n = \text{Dist}(SR_j, SR_i)/r_i = D_{ij}/r_i$ and make s_n, h_n, d_n as the feature components of QR_n. If components of all the four QR exist, go to ***Step 6***; else let $o = o + 1$, $j = \text{rank}^{-1}(o)$, repeat ***Step 5***.

Step 6. Combine components of all the QR to a 12-dimensional LIFBS feature vector for SR_i: $\vec{fv}_i = (s_1, h_1, d_1, s_2, h_2, d_2, s_3, h_3, d_3, s_4, h_4, d_4)$

Step 7. Let **FV** $= \left\{ \vec{fv}_1, \vec{fv}_2, \ldots \vec{fv}_N \right\}$ and return.

4 Similarity Measurement

Similarity measurement compares content features between the sample query image and images in lunar image gallery. The k most similar images are outputted as results. The visual features of a lunar image are defined as 3 types: a global feature vector, mid-level local features and low-level local features. we propose a compound feature model based similarity measurement, which consists of a filtering phase and a refining phase. That is:

$$S(\mathbf{F}_q, \mathbf{F}_i) = \begin{cases} S_{filter}(\mathbf{F}_q, \mathbf{F}_i), \text{Filtering} \\ S_{refine}(\mathbf{F}_q, \mathbf{F}_i), \text{Refining}, I_i \in \mathbf{I}_c \end{cases} \qquad (3)$$

$S_{filter}(\mathbf{F}_q, \mathbf{F}_i)$ denotes the filtering phase similarity measurement by the sum of weighted global feature and mid-level local features.

$S_{refine}(\mathbf{F}_q, \mathbf{F}_i)$ denotes the refining phase similarity measurement. The low-level local features between query image and each one in $\mathbf{I}_c$ will be used for calculation and sorted to get the final query result. The similarity measurement is calculated on weighted global, mid-level and low-level features.

5 Retrieval Algorithm

5.1 Algorithm Description

Based on compound feature model and similarity measurement, a content based lunar image retrieval algorithm is proposed. The input of the algorithm is a query lunar image I_q, output is the resulting set of images $\mathbf{I}_{result} = \{I_i | i = 1, 2, \ldots k, I_i \in \mathbf{I}\}$. The algorithm consists of three phases: preprocessing, filtering, and refining. In preprocessing phase, HU moments, Tamura texture feasure, SURF descriptors and LIFBS descriptors are extracted to construct the compound feature model; In filtering phase, according to filtering similarity measurement, global and middle-level local similarity are calculated. In refining phase, the refining similarity measurement is calculated based on all these types of feasues including the low-level. The detailed algorithm is as follows.

Algorithm 1. Relevant Retrieval
Input:
  I_q: Query Sample Image, k: Number of Results, k_c: Number of Candidates;
Output:
  $\mathbf{I}_{result}$: Set of Result Images.
1 let set of candidates $\mathbf{I}_c \leftarrow \emptyset$;
2 let set of results $\mathbf{I}_{result} \leftarrow \emptyset$;
3 Cal_HU (I_q), Cal_Tamura (I_q)
4 let set of detected SURF points: $\{SURF\} \leftarrow$ Detect_SURF_Points (I_q)
5 Cal_LIFBS (I_q);
6 Cal_SURF_Descriptors $(\{SURF\})$;
7 for each $I_i \in \mathbf{I}$ do
8     $Sim_{iq} \leftarrow S_{filter}(\mathbf{F}_q, \mathbf{F}_i)$;
9 end for
10 SortDescend $(Sim_{1q}, Sim_{2q}, \ldots Sim_{Nq})$;
11 for the top k_c Sim_{iq} do]
12    $\mathbf{I}_c \leftarrow \mathbf{I}_c \cup I_i$;
13 end for
14 for each $I_i \in \mathbf{I}_c$ do
15    $Sim_{iq} \leftarrow S_{refine}(\mathbf{F}_q, \mathbf{F}_i)$;
16 end for
17 SortDescend $(Sim_{1q}, Sim_{2q}, \ldots Sim_{k_cq})$;
18 for the top k Sim_{iq} do]
19    $\mathbf{I}_{result} \leftarrow \mathbf{I}_{result} \cup I_i$;
20 end for
21 return $\mathbf{I}_{result}$;

6 Experimental Results

In this Section, the proposed method is evaluated by experiments upon 2 datasets. One is 11591 remote sensing lunar images and their transformations from

Chang'e-2 and CLEMENTINE databases; another is the primitive 1960 images that are divided from Chang'e-2 entire lunar remote sensing map. The environment for evaluation is: Intel Core i3, 2.93 GHz, 4-core processor, 2GB RAM, Microsoft Visual C++ 2008 Compiler, OpenCV 2.0 Image Processing Library.

First of all, we tested the LIBFS feature with three sample lunar images (shown in Figure 2). (a) is a Chang'e lunar image containing 6 salient regions, (b) is an image of the same area from CLEMENTINE dataset, (c) is another Chang'e image also containing 6 salient regions used for comparison.

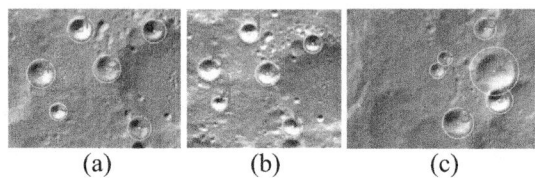

Fig. 2. Lunar image samples: the circles indicate the salient regions and the number in a circle is region ID

Figure 3 shows the extracted LIFBS of the sample images in Figure 2. The x-axis in Figure 3 denotes the 12-dimensional feature descriptor and the y-axis is the value. The matching results tell that all the six feature vectors are matched between (a) and (b), but only one match between (a)-(c) and (b)-(c). From experimental results, the LIFBS we proposed is able to distinguish the images that contain the similar objects and with similar distribution.

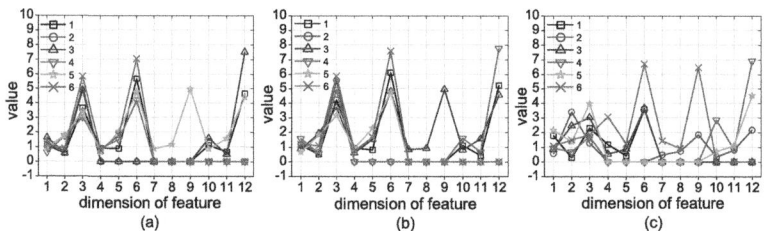

Fig. 3. The illustration of LIFBS

Then we test the retrieval results of our content-based lunar remote sensing images retrieval algorithm. Firstly, query images, which are processed by transforms of rotation, moving, scaling, brightness, contrast, noise, fuzzy transformations (see Figure 4 for an example), are tested on the first dataset. Here, we use $Precision - Recall$ curves for evaluation of query results, defined as follows,

$$Precision = \frac{|\mathbf{I}_{similar}| \cap |\mathbf{I}_{result}|}{|\mathbf{I}_{result}|}, Recall = \frac{|\mathbf{I}_{similar}| \cap |\mathbf{I}_{result}|}{|\mathbf{I}_{similar}|} \qquad (4)$$

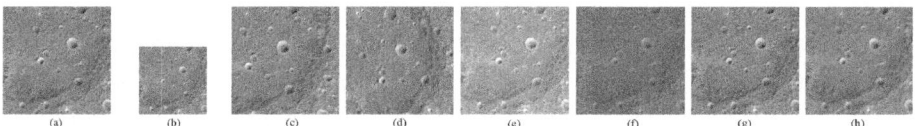

Fig. 4. Transform of Lunar image: (a) original image (b) scaling (c) moved (d) rotation (e) brightness (f) contrast (g) noise (h) fuzzy.

Figure 5 shows the comparisons of filtering phase results(average value of 100 different queries). final query results of our method using global features (HU moment + Tamura texture), and LIBFS feature. From the figure, the final query result of our method is the best in all conditions, its *Precision − Recall* curve is closer to the ideal curve that recall is always 1.0.

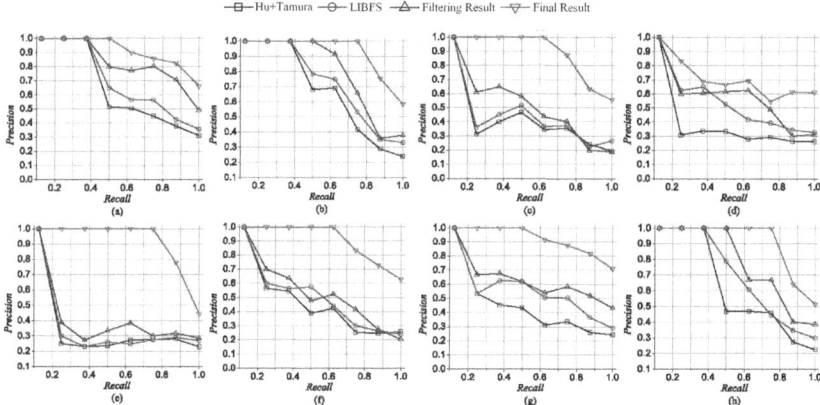

Fig. 5. Precision-Recall curves of query results: (a) original image (b) scaling (c) moved (d) rotation (e) brightness (f) contrast (g) noise (h) fuzzy.

To conclude, our retrieval algorithm based on the compound feature model and different content features can find out lunar remote sensing images from the gallery which are visually similar with the query sample. Facing various transformations, our approach maintains good *precision* and *recall*.

7 Conclusion

In this paper, a method has been proposed for retrieving the relevant images in lunar remote sensing image databases. Unlike the existing methods, our method takes the domain knowledge into consideration and our research covers most key points of content-based lunar image retrieval including the feature extraction, similarity measurement and the retrieval algorithm. The method is based on LIFBS feature descriptor proposed according to the distribution of salient craters and a compound feature model. The similarity measurement and retrieval algorithm are described in detail. Experiments show that our method is able to find out relevant images from different data sources and by the compound feature model it can get better retrieval results in lunar image databases than using simple combination of shape and texture features. For better performance, our future work will be emphasized on implementing our method to distributed parallel computing platforms.

References

1. Agouris, P., Carswell, J., Stefanidis, A.: An environment for content-based image retrieval from large spatial databases. ISPRS Journal of Photogrammetry and Remote Sensing 54(1), 263–272 (1999)
2. Aksoy, S., Cinbis, R.G.: Image mining using directional spatial constraints. IEEE Geoscience and Remote Sensing Letters 7(1), 33–37 (2010)
3. Barb, A.S., Shyu, C.-R.: Visual-semantic modeling in content-based geospatial information retrieval using associative mining techniques. IEEE Geoscience and Remote Sensing Letters 7(1), 38–42 (2010)
4. Datcu, M., Seidel, K., Walessa, M.: Spatial information retrieval from remote-sensing images part i: Information theoretical perspective. IEEE Transactions on Geoscience and Remote Sensing 36(5), 1431–1445 (1998)
5. Hu, M.-K.: Visual pattern recognition by moment invariants. IRE Transactions on Information Theory, 179–187 (1962)
6. Hui-Zhong, C., Yong-Guang, C., Jing, N., et al.: Roi detection method for lunar imagery based on surf. J. lnfrared Millim. Waves 30(6), 561–566 (2011)
7. Hui-Zhong, C., Yong-Guang, C., Jing, N., et al.: Rpcpf: A parallel index for matching the high-dimensional vectors in multimedia databases. Chinese Journal of Computers 34(10), 2009–2017 (2011)
8. Li, J., Narayanan, R.M.: Integrated spectral and spatial information mining in remote sensing imagery. IEEE Transactions on Geoscience and Remote Sensing 42(3), 673–685 (2004)
9. Meyer, C., Deans, M.: Content based retrieval of images for planetary exploration. In: IEEE/RSJ International Conference on Intelligent Robots and Systems, San Diego, CA, USA, pp. 1377–1382 (2007)
10. Salamuniccar, G., Loncaric, S.: Open framework for objective evaluation of crater detection algorithms with first test-field subsystem based on mola data. Advances in Space Research 42(1), 6–19 (2007)
11. Schröder, M., Rehrauer, H., Seidel, K.: Spatial information retrieval from remote-sensing images part ii: Gibbscmarkov random fields. IEEE Transactions on Geoscience and Remote Sensing 36(5), 1446–1455 (1998)
12. Schroder, M., Rehrauer, H., Seidel, K.: Interactive learning and probabilistic retrieval in remote sensing image archives. IEEE Transactions on Geoscience and Remote Sensing 38(5), 100–119 (2000)
13. Scott, G.J., Klaric, M.N., Davis, C.H.: Entropy-balanced bitmap tree for shape-based object retrieval from large-scale satellite imagery databases. IEEE Transactions on Geoscience and Remote Sensing 49(5), 1603–1616 (2011)
14. Shyu, C.-R., Klaric, M., Scott, G.J.: Geoiris: Geospatial information retrieval and indexing system-content mining, semantics modeling, and complex queries. IEEE Transactions on Geoscience and Remote Sensing 45(4), 2839–2852 (2007)
15. Tamura, H., Mori, S., Yamawaki, T.: Textural features corresponding to visual perception. IEEE Transactions on Systems. Man, And Cybernetics 8(6), 460–473 (1978)

Searching Desktop Files Based on Access Logs

Yukun Li, Xiyan Zhao, Yingyuan Xiao, and Xiaoye Wang

Key Laboratory of Intelligence Computing and Novel Software Technology, Tianjin
Key Laboratory of Computer Vision and System, Ministry of Education
Tianjin University of Technology, 300384, Tianjin, China
{liyukun,yyxiao,wangxy}@tjut.edu.cn, zhaoxiyan322@sina.com

Abstract. People often meet trouble in searching a desktop file when they can not remember exact words of its filename. In this paper, we firstly propose an algorithm to generate access logs by monitoring desktop operations and implement a prototype. By running it in several computers of selected participants we collected a data set of access logs. Then we propose a graph model to represent personal desktop files and their relationships, and highlight two file relationships(content relationship and time relationship) to help users search desktop files. Based on the graph model, we propose a desktop search method, and the experimental results show the feasibility and effectiveness of our methods.

1 Introduction

When people want to re-find a desktop file and can not remember its location, they often choose desktop search tools to do it. Because most desktop search tools are based on keyword search technology, people often meet trouble in searching a desktop file when forgetting exact words of the filename. Because of the limitation of human memory, it is unreasonable to ask each person to exactly remember words of every filename of desktop. For example, if a user wants to search the file "An draft on dataspace framework.pdf" with existing desktop search tools, he/she has to remember one or some words of set { "draft", "dataspace", "framework"}. Because most existing desktop search tools do not distinct accessed files from a great number of system files, they often work at low performance. This paper focuses on helping people efficiently search desktop file when they lose memory about exact file information.

1.1 Related Work

Chirita and Nejdl [1] proposed to connect semantically related desktop items by exploiting analysis information about sequences of accesses. Peery et al. [2] presented a multi-dimension query method in personal dataspace, which individually grades each dimension(content, structure and metadata), then combines the three dimension scores into a meaningful unified score. All the works above do not refer to how to get the access logs and how to search desktop files based on them. In [3], an idea about identifying personal tasks based on user's operations was proposed and demonstrated. In [4], a method on querying personal file based on user's working context was proposed. Some researchers of database area studied about managing personal data set, and the work involves personal dataspace model [5,6], pay-as-you-go integration [7,8], index [9] and query. Some interesting prototypes were developed like iMemes [10], Semex [7], MyLifeBit [11], HyStack [12] and so on. The works listed above didn't efficiently solve the problem on how to search desktop file based on access logs.

1.2 Contribution Summary

The contributions of this paper can be summarized as below:(1) Propose an algorithm to generate access logs by monitoring user's operations on desktop and implement a prototype system, and by running it on several computers of selected participants we collect a data set of access logs from eight persons.(2)Propose a graph model to represent personal desktop files and their relationships, and highlight two file relationships(content relationship and time relationship) to help users re-find personal desktop files, furthermore propose a desktop search method based on the graph model.

The rest is organized as follows: In section 2, we describe our desktop search method. Section 3 is about experiments. Section 4 concludes this paper.

2 Searching Methods Based on Access Logs

As most accesses to desktop files are re-finding [13], we propose that (1)it should be enough for desktop search to scan only the accessed files, (2)the accessed files can be identified by monitoring desktop operations and (3)the access logs can provide additional methods about desktop search.

2.1 Generating Access Logs

We propose to generate user access logs by monitoring the recently-accessed folder of operating system like Windows XP, and take a 3-ary tuple { *OperationTime, OperatedFileName, OperatedDirectory*}to represent the schema. The steps include: (1) If a change of the latest accessed desktop file is detected, a new access record will be generated, and the attributes OperationTime, OperatedFileName and OperatedDirectory can be identified through APIs provided by operating system; (2) If the latest accessed file is not involved in the log table, it means the user is accessing a new file. By this method we developed a prototype and collected logs of eight persons about one year.

Table 1 shows a part of an author's access logs, which includes 8 records and refers to 5 different desktop files. Except "A proposal for applying an award.doc", all the files are related to the activity "submitting to DASFAA 2013", although some filenames are not similar, like "figure1.vsd" and "submission to DASFAA

2012". When the user wants to re-find "figure1.vsd" and only remembers it relates to the activity "submitting to DASFAA 2013", instead of remembering the filename "figure1.vsd", it will be difficult for user to do by existing desktop search tools. But if the time relation between the two files is highlighted, which will provide the user additional ways for searching "figure1.vsd". Therefore besides content similarity, we propose to highlight time relationship to help users search desktop files more efficiently.

Table 1. Overview of a part of a user's access logs

No	User	File name	Access time
1	U1	Submission to DASFAA 2012.tex	2012-10-01 14:00
2	U1	figure1.vsd	2012-10-01 14:02
3	U1	Experimental data for DASFAA submission.xls	2012-10-01 14:05
4	U1	Comments from a Coauthor.doc	2012-10-01 14:20
5	U1	A proposal for applying an award.doc	2012-10-01 14:30
6	U1	Submission to DASFAA 2012.tex	2012-10-01 14:35
7	U1	figure1.vsd	2012-10-01 15:30
8	U1	Experimental data for DASFAA submission.xls	2012-10-01 15:40

2.2 Desktop File Graph Model and Construction

We propose a graph model to describe desktop files and their relationships, and name it *DFG(Desktop File Graph)*. A *DFG* is described as *G(F,R,n)*, where *F* is a set of desktop files accessed by user, *n* is the number of files in *F*, and *R* is a set of file relationships. Based on the observations mentioned in section 2.1, we take the following two relationships into consideration: *content similarity(Co)* and *time relationship(Ti)*, where *Co* means the similarity of two files in content, and *Ti* means the possibility that two files are accessed together. The *DFG* model provides an additional method for users to search desktop files. How to identify the relationships is the key problem. In this section, we propose methods to identify the two relationships.

As to content relationship, we propose to take filename similarity to approximately represent the content relationship of two files. For each file, we take a set of tokens included in the filename to denote its content. By computing the similarity of token sets of two files, we can work out the content similarity of them. In our work we take the formula 1 to compute Jaccard similarity [14] of the two token sets of the files F_i and F_j, and regard it as the content relationship of the two files.

$$Co(F_i, F_j) = \frac{|F_i.S_{token} \cap F_j.S_{token}|}{|F_i.S_{token} \cup F_j.S_{token}|} \quad (1)$$

In formula 1, $F_i.S_{token}$ means the token set of file F_i, and $F_j.S_{token}$ means the token set of file F_j. If $Sim(F_i, F_j)$ is bigger than 0, we add an edge between F_i

and F_j in the graph to denote the content relationship, and the weight of the edge (F_i, F_j) equals to the value of $Co(F_i, F_j)$.

As to the time relationship, Our algorithm is based on the position below: If two files are often accessed at the same time, we think they have time relation. The hard problem is how to decide "at the same time". In this work we propose to take "accessed sequentially" to approximately evaluate "accessed at the same time". For example, let A and B be two files, the more times they are accessed sequentially, the closer time relationship they have.

How to compute the time relationship is a challenging problem. Firstly, to two given files, the times they are accessed sequentially is dynamic; Secondly, it needs a method to increasingly update the time relation value based on its existing value. We propose a simple method to compute it as formula 2.

$$Ti_n(F_i, F_j) = \frac{Ti_{n-1}(F_i, F_j) + 1}{2} \qquad (2)$$

In formula 2, $Ti_{n-1}(F_i, F_j)$ means the existing time relation value between the two files denoted by F_i and F_j, and the initial value $Ti_0(F_i, F_j)$ is 0. When a new sequential access to F_i and F_j is found during monitoring user accesses, their time relation will be updated based on formula 2. The new value will be bigger than the old one, and its maximum value will not exceed 1. For example, when their first sequential access is found, $Ti_1(F_i, F_j) = (0+1)/2 = 0.5$, and when the second sequential access is found, $Ti_2(F_i, F_j) = (0.5+1)/2 = 0.75$.

Algorithm 1 shows the process of constructing desktop file graph. It supposes there exists a desktop file graph G_s, and shows how the file set and the two file relationship sets will be updated when a new access to a desktop file is found.

2.3 Searching Method

we propose a simple interface to perform the graph-based search, whose format is "$keyword_1, keyword_2, ..., keyword_n \setminus [C|T]$", where $keyword_i$ is a keyword user input, C and T are options which are set by users when they plan to search desktop files, where C means searching based on content relationship and T means searching based on time relationship. For example, "database, index \ C" means searching the files including keywords "database" and "index" based on content relationship, "database, index \ T" means searching the files whose filename includes keywords "database" and "index" based on time relationship. Based on the input keywords, we take Jaccard [14] method to compute the similarity between the input keywords($In.S_{keywords}$) and each file's token set($F_i.S_{token}, 1 \leq i \leq n$) by formula 3, and get a n-ary vector V_s as the primary results, which is taken to generate final results based on the desktop file graph.

$$V_s(i) = \frac{|F_i.S_{token} \cap In.S_{keywords}|}{|F_i.S_{token} \cup In.S_{keywords}|}. \qquad (3)$$

Assume the desktop file space is a graph $G(F, Co, Ti, n)$, where F is the set of desktop files accessed by user, n is the number of files in F, Co is the edge set

Algorithm 1. Constructing desktop file graph
Input: A new accessed file f and a graph $Gs(F, Co, Ti, m, n)$, where F is a set of files, Co is the content relation set, Ti is the time relation set, n is the total number of files, and m is the ID number of the file accessed last time.
Output: An updated graph $Gs(F, Co, Ti, m, n)$.

1: **procedure** *Constructing Desktop File Graph*$(f, Gs(F, Co, Ti, m, n))$
2:   **if** $f \in F$ **then**
3:     find the ID number of f in $Gs.F$ and store it into k
4:   **else**
5:     add a new file $Gs.F_{n+1}$
6:     n = n + 1, k = n
7:     **for** (int i = 1, i $\leq$ n, i++) **do**
8:       $S_{co} = |Gs.F_i.Tokens \cap Gs.F_k.Tokens| \,/\, |Gs.F_i.Tokens \bigcup Gs.F_k.Tokens|$
9:       **if** $S_{co} > 0$ **then**
10:        $Co(Gs.F_i, Gs.F_k) = S_{co}$
11:       **end if**
12:     **end for**
13:   **end if**
14:   $Ti(Gs.F_m, Gs.F_k) = (Ti(Gs.F_m, Gs.F_k) + 1)/2$
15: **end procedure**

of content relationship, Ti is the edge set of time relationship. In our method, we imagine $G(F, Co, Ti, n)$ as two virtual graphs $Gc(F, Co, n)$ and $Gt(F, Ti, n)$, and take two $n \times n$ adjacency matrixes to present them, where the nondiagonal element a_{ij} is the weight of the edge from vertex i to vertex j, and the diagonal element a_{ii} is set 1 here. Let M be the adjacency matrixes of selected graph view ($Gc(F, Co, n)$ or $Gt(F, Ti, n)$), based on V_s we can compute the result file set by the formula $V_r = V_s \times M$, and the result V_r is a n-ary vector. Based on V_r, we can compute the final result Rs by the formula $Rs = \{F_i | V_r(i) \neq 0, 0 \leq i \leq n\}$. Naturally, based on the values of V_r, the searching results can be ranked easily.

3 Experiments

Table 2 shows the participants' attributes(age, sex and position) and data sets. The parameters of data set include time length of data collection(Time), access times, accessed files, re-access times, and the ratio of re-access times to access times(Re-accessRatio). From the table we can discover most operations of desktop are re-accesses.

3.1 Experimental Design

We create a benchmark with the help of the participants. To the best of our knowledge there is no existing benchmark on evaluating desktop re-finding methods. Based on the number of the files a user wants to search, we classify the searching cases into two categories: single file search and multiple file search.

Table 2. Overview the statistics on access log collection

User	Age	Sex	Position	Time (day)	Access Times	Accessed Files	Re-access Times	Re-accessRatio (%)
U1	26	Female	Master	351	9514	1836	7678	80.70
U2	25	Male	Master	351	5994	2291	3703	61.78
U3	27	Female	Master	223	1005	393	612	60.90
U4	36	Male	PhD	355	7320	1894	5426	74.13
U5	25	Male	Master	354	15040	3829	11211	74.54
U6	29	Male	PhD	183	3021	813	2208	73.09
U7	22	Female	Undergraduate	213	6064	1522	4542	74.90
U8	23	Female	Undergraduate	233	6587	1755	4832	73.36

Single file search means relocating a specific file, and multiple file search means searching multiple files. We ask each participant to design some searching cases according to their searching experience, and give the correct answer for each search based on what they want to find. We let each user U_i design 10 search samples respectively for single file search and multiple file search, and ask them to give a file or a file set to every search sample as right answer.

We take the popular measures recall, precision and F-score [15] to evaluate our methods. Because we have not found existing work about helping users search desktop files based on monitoring user access logs, and desktop search tools are popular ways for users to re-find desktop files, we select two popular desktop tools MS desktop search and Google desktop search engine as baseline to evaluate our method. To each search, we perform it with different methods and take top-k files returned as the final results, and set $k = 30$ in our experiments. By comparing the final results with the benchmark for each search sample, we can compute the recall, precision and F-score of each search, then we can work out the average value of recall, precision and F-score of each method.

3.2 Experimental Results

Figure 1 illustrates the advantages of our method: (1)Either to single file search or to multiple file search, our access log-based method's F-score is the best; (2) The recall of our method equals to 1 approximately, which is much better than other tools; (3) Precisions of all methods are not high, which is in accord with our expects because there exist some unrelated files whose names share some same words. Totally our method has better precision than other desktop search tools.

Like desktop search engine, our method also has two types of cost: off-line cost and online cost. (1)As to online cost, MS desktop search tool shows the lowest performance, it always takes several minutes to handle a search and returns a great number of files which often include many system files. Our log-based method and Google Desktop search engine show a better online performance, especially the log-based method's average response time is less than one second,

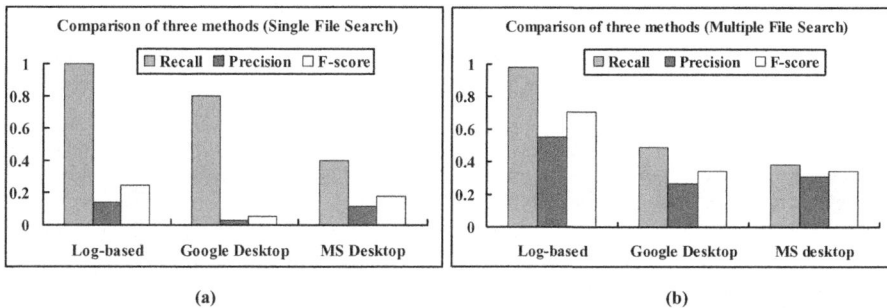

Fig. 1. Comparison of log-based method, Google desktop and MS desktop

which can satisfy most users' needs. (2)As to off-line cost, the cost of our method is much lower than the selected desktop search tools. Take google desktop search for example, it always takes several hours to build the initial index in some cases, and the update of index is also delayed much more, which sometimes results in search failure. To MS desktop search, it has little additional cost for updating. Totally, our access log-based method's performance is comprehensively better than other desktop search tools, and can satisfy users' requirements.

We also have the following observations in experiments. (1)Sometimes users do not name a desktop file according to its content for some reasons like "download it from a web site and keep its original filename", "get it from other persons", and so on; (2)People archive personal desktop files with folders according to different rules. For example, some folders are created based on user activities, like "Submission to DASFAA 2013", which includes the files related to the submission to DASFAA 2013, and sometimes based on the file categories, like "dataspace paper", which includes the papers related to dataspace topic.(3)Access logs provide users additional facets to search desktop files like access frequency, access time, operation types and so on. The observations discover some interesting research topics and we will study them in the future.

4 Conclusions

In this paper, we firstly propose a method to generate access logs by monitoring users' operations on desktop and build a data set of access logs of eight persons. Then we propose a desktop search method based on access logs. The experimental results show the effectiveness of our method.

Acknowledgments. This research was supported by the Natural Science Foundation of China under grant number 61170027, 61170174; Natural Science Foundation of Tianjin under grant number 11JCYBJC26700.

References

1. Chirita, P.-A., Nejdl, W.: Analyzing User Behavior to Rank Desktop Items. In: Crestani, F., Ferragina, P., Sanderson, M. (eds.) SPIRE 2006. LNCS, vol. 4209, pp. 86–97. Springer, Heidelberg (2006)
2. Peery, C., Wang, W., Marian, A., Nguyen, T.D.: Multi-Dimensional Search for Personal Information Management Systems. In: 11th International Conference on Extending Database Technology, pp. 464–475. ACM Press, Nantes (2008)
3. Li, Y., Zhang, X., Meng, X.: Exploring Desktop Resources Based on User Activity Analysis. In: 33rd International ACM SIGIR Conference on Research and Development in Information Retrieval, p. 700. ACM Press, Geneva (2010)
4. Li, Y., Meng, X.: Supporting Context-based Query in Personal DataSpace. In: 18th ACM Conference on Information and Knowledge Management, pp. 2–6. ACM Press, Hong Kong (2009)
5. Franklin, M.J., Halevy, A.Y., Maier, D.: From databases to dataspaces: A new abstraction for information management. SIGMOD Record (SIGMOD) 34(4), 27–33 (2005)
6. Dittrich, J.P., Antonio, M., Salles, V.: iDM:A unified and versatile data model for personal dataspace management. In: 32nd International Conference on Very Large Data Bases, pp. 367–378. ACM Press, Seoul (2006)
7. Dong, X., Halevy, A.: A platform for personal information management and integration. In: 2nd Biennial Conference on Innovative Data Systems Research, pp. 119–130. Online Proceedings, Asilomar (2005)
8. Dong, X., Halevy, A., Yu, C.: Data integration with uncertainty. In: The 33rd International Conference on Very Large Data Bases, pp. 687–698. ACM Press, Vienna (2007)
9. Dong, X., Halevy, A.: Indexing dataspaces. In: The ACM SIGMOD International Conference on Management of Data, pp. 43–54. ACM Press, Beijing (2007)
10. Blunschi, L., Dittrich, J.P., Girard, O.R., Karakashian, S.K., Salles, M.A.V.: A Dataspace Odyssey: The iMeMex Personal Dataspace Management System. In: 3rd Biennial Conference on Innovative Data Systems Research, pp. 114–119. Online Proceedings, Asilomar (2007)
11. Gemmell, J., Bell, G., Lueder, R., Drucker, S.M., Wong, C.: MyLifeBits: fulfilling the Memex vision. In: 10th ACM International Conference on Multimedia, pp. 235–238. ACM Press, Juan les Pins (2002)
12. Karger, D.R., Bakshi, K., Huynh, D., Quan, D., Sinha, V.: Haystack: A customizable general-purpose information management tool for end users of semistructured data. In: 2nd Biennial Conference on Innovative Data Systems Research, pp. 13–26. Online Proceedings, Asilomar (2005)
13. Elsweiler, D., Baillte, M., Ruthven, I.: Exploring Memory in Email Refinding. ACM Transactions on Information Systems(TOIS) 26(4), Article No.21 (2008)
14. Bayardo, R.J., Ma, Y., Srikant, R.: Scaling up all pairs similarity search. In: 16th International Conference on World Wide Web, pp. 131–140. ACM Press, Banff (2007)
15. Raghavan, V.V., Bollmann, P., Jung, G.S.: Retrieval System Evaluation Using Recall and Precision: Problems and Answers. In: 12th International Conference on Research and Development in Information Retrieval, pp. 59–68. ACM Press, Cambridge (1989)

An In-Memory/GPGPU Approach to Query Processing for Aspect-Oriented Data Management

Bernhard Pietsch

Database and Information Systems Group
Friedrich-Schiller-University Jena
B.Pietsch@uni-jena.de

Abstract. Under the paradigm of aspect-oriented data management (AODM), cross-cutting concerns in the data model – like multi-language support or functional versioning – are to be encapsulated and separated from the core aspect data. At runtime, a re-weaving of data influenced by different aspects has to be done. Previous research demonstrated that running queries directly against the referential model of AODM for relational databases via SQL is slow and inefficient. This paper presents an approach to accelerate queries by using a native storage model for aspect specific data and a specialized in-memory as well as a GPGPU query method.

1 Introduction

The aspect-oriented programming paradigm [10] was introduced to address the problem of cross-cutting concerns in object-oriented or procedural programming. But not only code is vulnerable to tangling of different aspects, similar problems occur in data modeling. In [11] a paradigm for aspect-oriented data management (AODM) in relational databases has been proposed, as a generalized solution for integrating the business object perspective with additional aspects such as multi-language or versioning requirements. It focuses on the modularization of aspects, separating data of different aspects in the data model. However, at data retrieval, aspect data reintegration (weaving) is required. The multi-dimensional nature of aspect integration is a major challenge to the relational model. Specific structures suffer from high redundancy, generic structures induce complex and often slow operations, especially for data retrieval. This paper presents an approach to circumvent overly complex querying against the relational data by using specialized in-memory structures (kept up-to-date concurrently to a referential database) that support aspect aware mass data retrieval. Speed-up potential is demonstrated by a basic performance test of CPU and GPGPU proof-of-concept implementations.

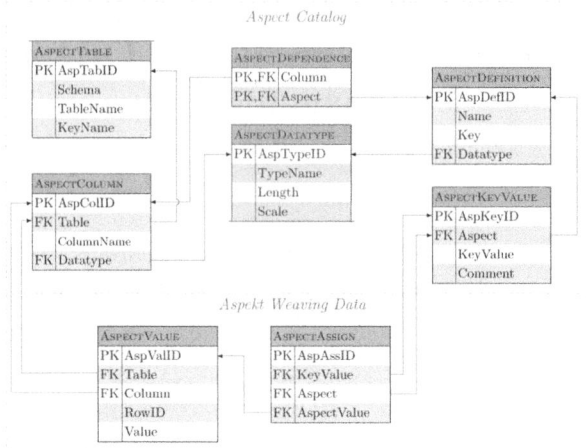

Fig. 1. AODM Referential Model (adapted from [12])

2 Related Work

2.1 Aspect-Oriented Data Management

In [12], a referential model for aspect meta data and aspect specific data under the AODM paradigm was proposed. Its relational schemata are displayed in Figure 1, a fixed set of schemata containing *all* structural meta data and aspect specific attributes, despite supporting any number of business tables, achieved by using an *entity-attribute-value* pattern (EAV/CR) [14]. This concept avoids redundancy and explosion of additional aspect tables for every functional table. However, querying EAV data in relational databases requires a transformation back to a one-column-per-attribute format via a "pivot" operation. Pivoting via multiple self-joins is not performing well in general [3]. To retrieve aspect specific data relations from the referential model, two cascading pivot operations have to be applied. As shown in [13], static SQL does not perform very well.

As a usable and efficient alternative, an API for AODM was proposed and implemented as an database-external query accelerator [16]. For query purposes, a filter language based on predicate logic was introduced. Aspect filter (AF) expressions are used as constraints for the contexts of aspect specific data. E.g.

$$Language = \text{'en' } OR \text{ } CurrencyArea \text{ } IN \text{ } (\text{'USD', 'GBP'})$$

limits the visibility of attribute values – for attributes depending on the aspects LANGUAGE and/or CURRENCY AREA – to those in the specified subspaces.

2.2 General-Purpose Computing on GPUs

Early general-purpose computing on graphics processing units (GPGPU) approaches were limited to "single instruction, multiple data" (SIMD) [4] vector-based GPU parallel processing, due to hardware limitations of graphics devices at that time. GPGPU first found its way into number crunching problem solving, exploiting the pure FLOPS power of graphics processing units. Much more elaborate instruction sets, control flow flexibility and memory management made graphics devices attractive for other fields of research and application. The availability of simultaneous random memory look-ups and multiple flow paths as opposed to classic SIMD made the NVIDIA company coin the term "single instruction, multiple threads" (SIMT) for contemporary GPGPU device architecture. GPGPU languages such as OpenCL or CUDA are still developing dynamically.

In database research, GPGPU became popular for specialized sub-tasks in query processing like sorting [5], join acceleration [7,17], or OLAP processing [9,18]. In recent years, more effort went into a broader integration of GPGPU approaches into relational DBMS (RDBMS), e.g. as full fledged co-processors [6], or as a partial implementation of a DBMS' command processor [1].

3 Implementation

In support of the referential RDBMS, a co-processor is designed and implemented to deal with data retrieval by using a redundant in-memory set of aspect-specific data.

An attribute value under the AODM paradigm may be aspect-dependent and thus not necessarily a single value. Instead, it may be considered as a (possibly sparse) multi-dimensional grid of values, the grid dimension determined by the number of aspects influencing that attribute. Different attributes in a relational schema may depend on different aspects – e.g. a *name* might depend on a LANGUAGE aspect, while *price* depends on a CURRENCY aspect, hence attributes with common aspect dependencies are grouped into sub-relations and provide the base for querying aspect-specific data in memory.

3.1 Grid-File Index

To access aspect-specific tuples in a sub-relation, an index structure based on grid files (GF) [15,8] is used. This GF index needs one dimension per aspect, plus one additional dimension to account for the assignment to the business entity. The tuple data is stored in fixed-sized buckets referenced by the index structure. The index is used for "point" and special partial match queries, especially for inserting, updating and deleting individual aspect-specific tuples. However, when using AF expressions in a mass query, after an initial interval-based evaluation, all remaining buckets must be evaluated per tuple. This can be done in parallel, of course, either by multiple CPU nodes or on a GPGPU device (after transfering the ncessary data).

3.2 GPGPU Approach

For GPGPU, bucket design was adapted to better fit CUDA requirements for data retrieval. Within a warp (32 simultaneous threads), coalesced memory access is encouraged to reduce device memory IO. For better support of this access pattern, buckets for CUDA kernel processing are column-based instead of row-based, where each column consists of one key dimension of the aspect-specific tuples. Data manipulation is still done by the CPU and is slowed down due to worse cache coherency, but OLTP through-put is limited by the ACID compliant RDBMS performance on the referential data anyway.

There are two natural approaches for AF evaluating: abstract syntax tree (AST) traversal or stack machine (SM) execution. For AST traversal, the AST of an AF expression has to be transferred to the device and recursively evaluated for each tuple. Recursion is a relatively new feature for CUDA devices, available since Fermi. The construction of pointer-rich structures in device memory is still relatively complex. Evaluation by use of a SM seems to be more suited for CUDA. The translation of an AF expression into a SM instruction sequence can be done by the CPU in a single traversal. The instruction sequence can be encoded as a pointer-free array that is straight-forward transferable to device global memory. On the CUDA device, the instruction sequence can be interpreted in a simple intra-thread loop. The interpretation of each instruction cannot be done without code branches for different opcodes, but all threads evaluating a line under a common AF expression always choose the same branch at run-time, not creating any flow divergence between those threads.

The pseudo-code procedure STACKMACHINEPARALLEL illustrates a SIMT-parallel interpretation of an AF expression for a tuple list. Each thread uses its own local bit-stack for the evaluation, the stack machine instruction sequence (*instr_list*) is stored in device constant memory as an array. Each tuple in *tuple_list* (transferred to device global memory) is evaluated in its own thread and the result written to *result_bitvector*.

procedure STACKMACHINEFILTERPARALLEL (**in** *tuple_list*, **in** *instr_list*, **out** *result_bitvector*)
    for each tuple t in *tuple_list* **in parallel**
        if t is not empty **then**
            for each instruction i in *instr_list*
                if $i.type$ == AND **then**
                      pop $i.op\_num$ bits from stack
                      if all bits are set **then** push 1 **else** push 0
                else if $i.type$ == OR **then**
                      pop $i.op\_num$ bits from stack
                      if no bits are set **then** push 0 **else** push 1
                else if $i.type$ == NOT **then**
                      $temp \leftarrow$ pop
                      push (invert $temp$)
                else if $i.type$ == PRED **then**
                      if $i.val == t.keyVal[i.key]$ **then** push 1 **else** push 0
            result_bitvector[t'th bit] $\leftarrow$ pop

The result bit-vector resides in device global memory. For queries with few matching tuples, the bit vector creates significant memory overhead inevitable due the SIMT principle. The bit vector is reduced to the actual set of matching tuples by the use of stream compaction [2]. Finally the result list has to be transferred back to the system memory.

4 Performance

Performance results were gathered on an AMD Athlon64 X2 at 2.6GHz, 8GB RAM, an NVidia GeForce 450GTS (4 multiprocessors à 48 CUDA cores), connected via 16 PCI-E lanes (2.5 GT/s each), operated by a 3.3 Linux kernel. A DB2 9.7 and a PostgreSQL 9.0.6 instance were used as RDBMS.

4.1 Data Set

A single "business" data table Module was used and four additional aspects were created – LANGUAGE, REGION, VERSION and PRICEGRADE – and "activated" for the respective attributes listed below, e.g., norm depends on LANGUAGE, REGION and VERSION, price depends on REGION, VERSION and PRICEGRADE.

```
CREATE TABLE Module (id       VARCHAR(100) NOT NULL PRIMARY KEY,
                     name     VARCHAR(70) NOT NULL,
                     variant  VARCHAR(70) NOT NULL,
                     price    NUMERIC(8,2) NOT NULL,
                     norm     VARCHAR(70),
                     material VARCHAR(70),
                     rowid    INTEGER NOT NULL UNIQUE);
```

Two data sets were created: *Small* (2000 entities with ≈100,000 aspect specific tuples) and *Large* (38000 entities with ≈1,100,000 tuples). The data was loaded into relational tables conforming to the AODM referential model, and concurrently loaded into the in-memory grid-file structure described in the previous section.

In addition to bulk load, in an OLTP test a set of transactions deleting and inserting aspect specific tuples was run through an intermediate software layer that (a) forwarded tuple manipulations to the relational database and (b) recorded tuple manipulations and applied them to the in-memory structure when a transaction commit was acknowledged by the relational database. That way, the in-memory structure can be used for accelerating queries issued by read-only transactions in a consistent way. The purpose of this OLTP test was to measure the additional computational load by the in-memory module during aspect specific data manipulation.

4.2 Test Conditions

Retrieval of selected aspect specific information has to take two different kinds of filters into account: Firstly, for selection of entity-sets, a high selective (HS) and a low selective (LS) filter for Module entities in both data sets were constructed. Secondly, as constraints for the aspect context, a highly selective and compact AF expression was used

  F1 = *Language IN ('en', 'en_US') AND Version = '7'*

as well as a more complex and less selective filter

  F2 = *(Language = 'fr' AND Version IN ('8', '9')) OR*
  (Language IN ('en', 'en_US', 'en_UK', 'zh', 'zh_CN', 'zh_HK')
  AND Version IN ('1', '2', '3', '4', '5')) OR
  (Language IN ('po', 'it', de') AND Version <> '11')

Table 1. Query times (in ms)

Data Set	Entity	Asp. Filter	DB2 view	PostgreSQL view	PostgreSQL mat.	In-Memory CPU	In-Memory GPU
Large	HS	F1	>21,600,000	231,000	308	13.1	3.47
		F2	>21,600,000	242,000	640	14.8	5.51
	LS	F1	>21,600,000	235,000	334	19.3	4.17
		F2	>21,600,000	248,000	1,180	28.7	6.66
Small	HS	F1	>21,600,000	38,500	138	1.19	0.535
		F2	>21,600,000	40,100	144	1.35	0.735
	LS	F1	>21,600,000	39,600	159	1.74	0.541
		F2	>21,600,000	40,200	211	2.71	0.834

The combination of two data sets, two entity filters and two AF expressions resulted in eight test queries. These were run against the relational database view and materialized views as well as the in-memory system with and without GPGPU usage. Query results were written into a *null* stream for precision considerations. RDBMS measurements where run with buffer pool capacities high enough to keep all relevant data in RAM and warm caches.

4.3 Results

On DB2, initial generation of materialized query tables (MQT) had to be cancelled after a run-time of more than 24 hours for *Small*. Consequently no MQTs were used on DB2. On PostgreSQL, rebuilding the materialized views for *Small* took 46 seconds and 213 seconds for *Large*. The in-memory data structures are currently not made persistent in a fully ACID complaint way, they have to be restored from the relational schemas during recovery. Loading data from the database into an empty in-memory structure took 4 seconds (*Small*), 34 seconds (*Large*) respectively. This is almost exclusively the time needed by the RDBMS to do table scans and copy the data to the in-memory process. During the OLTP test, two different conditions were used: OLTP transactions were solely forwarded to the RDBMS as well as additionally run against the in-memory structure as described in section 4.1. No significant difference in transaction throughput was found.

The central performance results of this paper are shown in Table 1, query times (in ms, rounded to 3 decimal figures) for all eight queries are listed for both RDBMS (baseline) and both the CPU and GPGPU in-memory approaches. As mentioned above, MQTs were not available on DB2, and unfortunately, queries against non-materialized views had to be cancelled after a run-time of 6 hours. On PostgreSQL, queries against views were slow, around 40 seconds for *Small* and 4 minutes for *Large*. However, when materialized views ("mat." condition) were used, query times expectably improved by orders of magnitude.

Figure 2 graphically shows the query times for the in-memory and GPGPU conditions. The left and middle bar in each chart displays the run-time for

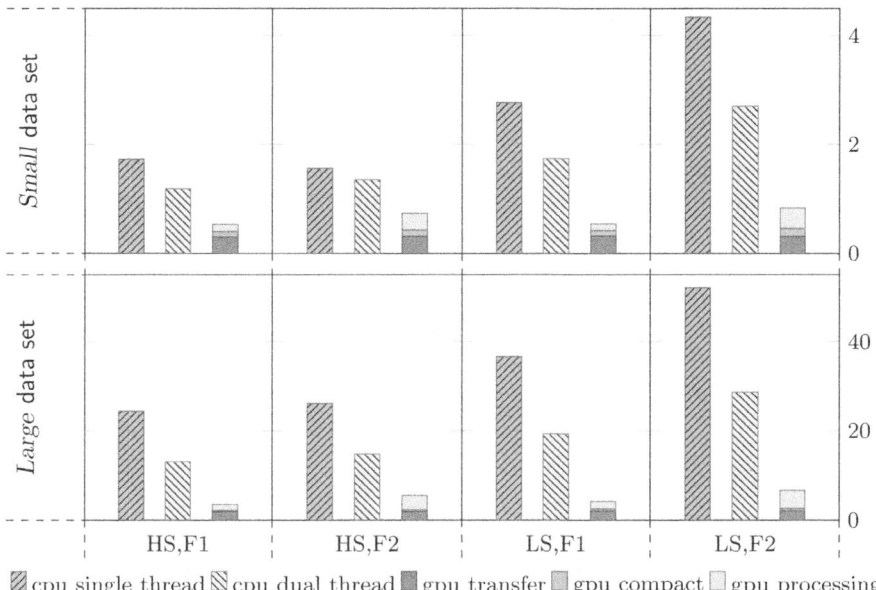

Fig. 2. Time to calculate query results (in ms)

in-memory query processing in a single thread or two threads respectively. The right bar for GPGPU query processing is segmented into portions representing transfer between host and device memory, stream compaction of the result vector, and the core GPGPU filtering and memory management activity (pooled as "*gpu processing*" in the figure). Results for the in-memory approach are in line with the expectations. Multi-threading yields a better speed-up for *Large* and more complex filtering, as the overhead for thread management and synchronisation for merging results is less important. Note that the CUDA approach results are not directly comparable to the CPU condition, as the absolute numbers are highly dependent on the selection of specific hardware.

5 Conclusion

For applications using the AODM referential model with mostly read-only aspect data, materialized views are a comfortable and fast enough solution. However, neither materialized nor ordinary views can deal with more dynamic data. This paper presented an in-memory approach promising to speed up aspect data queries significantly (by five orders of magnitude compared to static SQL in the very basic tests used in this paper), and at the same time does not impair data manipulating workloads. This proof-of-concept query co-processing system however is not yet fully integrated into a RDBMS. As an auxillary in-memory technique, it does not provide ACID compliant persistence and has only limited support for concurrency and transactional consistency on its own, it relies on the referential data managed by the RDBMS instead. Additionally, performance of real life complexity queries (including joins, aggregations and write-after-read statements) has to be tested.

Although performing better on the specific test platform than its CPU/in-memory counterpart, the GPGPU approach buys its speed-up with considerable drawbacks like limited concurrency, as CUDA devices are not able to serve more than one CPU thread at a time. Data transfer between host and device itself is another major obstacle. As visible in the results, for some test queries, transfer dominates the query times. To tackle these issues, intelligent scheduling, light weight compression and transfering concurrently with CUDA kernel execution has to be considered. Apart from technical obstacles, GPGPU devices are not yet established as features of typical database or application server hardware.

References

1. Bakkum, P., Skadron, K.: Accelerating sql database operations on a gpu with cuda. In: Proc. 3rd Workshop. GPGPU 2010, pp. 94–103. ACM, NY (2010)
2. Billeter, M., Olsson, O., Assarsson, U.: Efficient stream compaction on wide simd many-core architectures. In: Proc. Conf., HPG 2009, pp. 159–166. ACM, NY (2009)
3. Dinu, V., Nadkarni, P., Brandt, C.: Pivoting approaches for bulk extraction of Entity-Attribute-Value data. Comput. Meth. Prog. Biomed. 82, 38–43 (2006)
4. Flynn, M.J.: Some computer organizations and their effectiveness. IEEE Transactions on Computers C-21(9), 948–960 (1972)
5. Govindaraju, N., Gray, J., Kumar, R., Manocha, D.: Gputerasort: high performance graphics co-processor sorting for large database management. In: Proc. Inf. Conf. on Management of Data, SIGMOD 2006, pp. 325–336. ACM, NY (2006)
6. He, B., Lu, M., Yang, K., Fang, R., Govindaraju, N.K., Luo, Q., Sander, P.V.: Relational query coprocessing on graphics processors. ACM Trans. Database Syst. 34(4), 21:1–21:39 (2009)
7. He, B., Yang, K., Fang, R., Lu, M., Govindaraju, N., Luo, Q., Sander, P.: Relational joins on graphics processors. In: Proc. Int. Conf. on Management of Data, SIGMOD 2008, pp. 511–524. ACM, NY (2008)
8. Hinrichs, K.: Implementation of the grid file: Design concepts and experience. BIT Numerical Mathematics 25, 569–592 (1985)
9. Kaczmarski, K.: Comparing GPU and CPU in OLAP Cubes Creation. In: Černá, I., Gyimóthy, T., Hromkovič, J., Jefferey, K., Královič, R., Vukolić, M., Wolf, S. (eds.) SOFSEM 2011. LNCS, vol. 6543, pp. 308–319. Springer, Heidelberg (2011)
10. Kiczales, G., Lamping, J., Mendhekar, A., Maeda, C., Lopes, C., Loingtier, J.-M., Irwin, J.: Aspect-Oriented Programming. In: Akşit, M., Matsuoka, S. (eds.) ECOOP 1997. LNCS, vol. 1241, pp. 220–242. Springer, Heidelberg (1997)
11. Liebisch, M.: Aspektorientierte Datenhaltung – ein Modellierungsparadigma. In: GI Workshop GvD, Bad Helmstedt, Deutschland, pp. 13–17 (May 2010)
12. Liebisch, M.: Supporting functional aspects in relational databases. In: Proc. 2nd Int. Conf. ICSTE 2010, San Juan, Puerto Rico, USA (October 2010)
13. Liebisch, M.: Accessing Functional Aspects with Pure SQL - Lessons Learned. In: Proc. 15th East European Conf., ADBIS 2011, Vienna, Austria, vol. 2, pp. 85–94 (September 2011)

14. Nadkarni, P.M., Marenco, L., Chen, R., Skoufos, E., Shepherd, G., Miller, P.: Organization of heterogeneous scientific data using the EAV/CR representation. J. Am. Med. Inform. Assoc. 6(6), 478–493 (1999)
15. Nievergelt, J., Hinterberger, H., Sevcik, K.C.: The grid file: An adaptable, symmetric multikey file structure. ACM TODS 9, 38–71 (1984)
16. Pietsch, B.: Aspektorientierte Datenhaltung in relationalen DBMS – Implementierung und Bewertung einer Zugriffsschicht. Diploma thesis, Institute for Computer Science, FSU Jena (September 2011)
17. Pirk, H., Kersten, M.L., Manegold, S.: Accelerating foreign-key joins using asymmetric memory channels. In: Int. Conf. on VLDB, pp. 585–597 (2011)
18. Wittmer, S., Lauer, T., Datta, A.: Real-time computation of advanced rules in OLAP databases. In: Eder, J., Bielikova, M., Tjoa, A.M. (eds.) ADBIS 2011. LNCS, vol. 6909, pp. 139–152. Springer, Heidelberg (2011)

On Efficient Graph Substructure Selection

Xiang Zhao[1], Haichuan Shang[2], Wenjie Zhang[1], Xuemin Lin[1], and Weidong Xiao[3]

[1] The University of New South Wales, Australia
{xzhao,zhangw,lxue}@cse.unsw.edu.au
[2] The University of Tokyo, Japan
shang@tkl.iis.u-tokyo.ac.jp
[3] National University of Defense Technology, China
wdxiao@nudt.edu.cn

Abstract. Graphs have a wide range of applications in many domains. The graph substructure selection problem is to find all subgraph isomorphic mappings of a query from multi-attributed graphs, such that each pair of matching vertices satisfy a set of selection conditions, each against an equality, range, or set containment operator on a vertex attribute. Existing techniques for single-labeled graphs are developed under the assumption of identical label matching, and thus, cannot handle the general case in substructure selections. To this end, this paper proposes a two-tier index to support general selections via judiciously materializing certain mappings. Moreover, we propose efficient dynamic query processing and index construction algorithms. Comprehensive experiments demonstrate the effectiveness and efficiency of our approach.

1 Introduction

Recent decades have witnessed an explosion of structured data, which strongly demands effective management solutions. Graphs are widely used to model complex structured data in many applications, including bioinformatics, pattern recognition, etc. Hence, graph management attracts great interest from academia and industry.

Given a graph database and a query graph, *subgraph containment search* returns graphs containing the query, and has been well addressed [3, 8, 11]. In some applications, it is desirable to discover all occurrences of the query in one large graph. This problem is named *subgraph all-matching*, and has been studied in [13–15]. Containment search tells if there is a mapping of the query, while all-matching retrieves all of them.

In subgraph containment search and all-matching, current techniques work on *single-labeled* graphs for efficiently answering queries with equality conditions on vertex labels. That is, every vertex has a single label, and two matching vertices are required to have an identical label. However, in many applications, structured data is modeled with *multi-attributed* graphs, exhibiting multiple attributes on vertices, each with a corresponding value. In this sense, an aforementioned label in previous work refer a value. Consequently, more powerful operators on vertices become desirable to handle such graphs. For example, in Fig. 1(1) are two data graphs, each

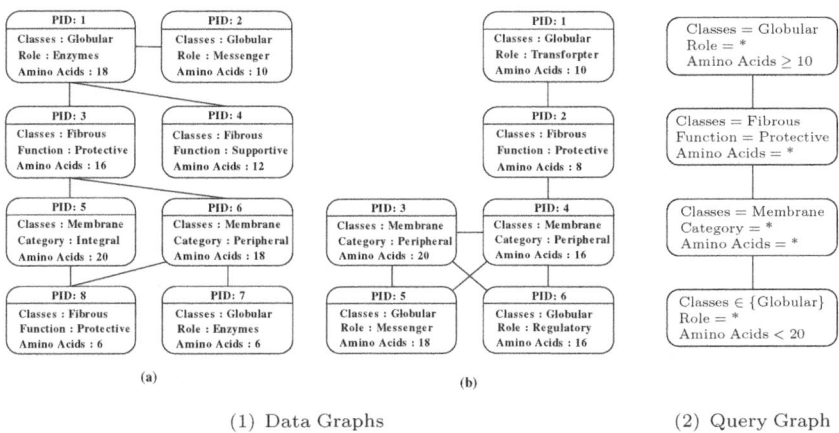

Fig. 1. Substructure Selection

depicting a protein-protein interaction network. Every vertex has several heterogeneous attributes with its own value domain. A biologist may want to find all four-protein interactions, with requirements modeled in Fig. 1(2), where '*' denotes a wildcard – arbitrary value in the domain. The query graph advises (1) the four proteins have a chain interaction; and (2) each protein possesses specific attributes; e.g., the first protein is of class 'Globular', takes an arbitrary role, and has no less than 10 amino acids. The query is issued to the database to retrieve the mappings satisfying all the requirements. We call this type of queries *substructure selection*.

In substructure selections, we have comparatively large graphs [9] in the database, and every *data* graph possesses multiple attributes on vertices. A *query* is a graph where each vertex comprises a set of selection conditions in *conjunctive normal form* such that each selection condition is against an equality, range, or set containment operator on one attribute. The problem aims to find all subgraph isomorphic mappings from the query to every data graph such that the attributes at each pair of matching vertices meet the selection conditions. Note the vertex attributes are not necessarily to be identical; in fact, they are different in most real applications. For instance, consider the database in Fig. 1(1) and the query in Fig. 1(2). In graph (a), the vertex with PID 1 possesses attributes Classes, Role and Amino Acids, whereas the vertex with PID 3 does not have Role but Function. Substructure selection returns three answers: $((a), \{1, 3, 6, 7\})$, $((b), \{1, 2, 4, 5\})$, $((b), \{1, 2, 4, 6\})$, where, for each mapping result, the first element is the provenance graph ID, and the second is the set of matching vertices.

While substructure selection is fundamental to structure oriented analysis, little has been done due to intrinsic challenges. Techniques for subgraph containment query [3, 8, 11] adopt the exclusion logic to prune false positives. Nonetheless, they do not provide sufficient support to find all subgraph isomorphic mappings. Subgraph all-matching algorithms [13–15] are designed for efficiently locating subgraph isomorphic mappings, and are yet not able to handle the general selection conditions over multi-attributed graphs. In particular, the more general requirements, e.g., multi-attributes, range selection, impose unseen challenges to the mapping computation. Individually, NOVA [15] and SPath [14] are solely based on vertex labels, and hence, the multi-attributes render them infeasible under our scenario. The distance-based pruning of GADDI [13] becomes weak due to more candidates in general selection. DELTA [12] imposes the rigid constraint of *fixed* attributes on vertices with identical values; i.e., all vertices have the same set of attributes. Thus, this model does not lend itself to the majority of real applications. By relaxing subgraph isomorphism, graph *simulation* based pattern query [4] incorporates range conditions on the single labels at vertices. Either, it does not solve our problem. We address the above challenges in this paper.

To the best of our knowledge, this is among the first attempts to study the substructure selection problem. In summary, we make the following contributions:

- We propose a new type of fundamental queries – substructure selection that handles general selections on multi-attributed graphs;
- We design a novel structure SS-index to speed up the online computation via judiciously materializing partial embeddings;
- We devise an efficient method to dynamically compose effective query execution plans to reduce the overall search cost; and
- We propose SS-search algorithm employing effective plans, as well as a scalable index construction algorithm for SS-index utilizing query logs.

In addition, extensive experiments demonstrate the effectiveness and efficiency of the proposed techniques, which significantly outperform other alternatives.

Organization. The paper is organized as follows: Section 2 states preliminaries and the problem. Section 3 presents a depth-first search (DFS) paradigm, and then introduces the design of SS-index. Leveraging the index, we propose SS-search for query processing in Section 4, and postpone the discussion on index construction in Section 5. Section 6 reports the experimental study, followed by conclusion in Section 7.

Related Work. Research on graph databases is a well-established activity, especially in pharmaceutical and chemical industries. This paper focuses on exact structural mappings. Techniques for containment queries are categorized as (1) feature-based, such as gIndex [11], cIndex [2], FG-index [3], etc; and (2) non-featured-based, represented by gString [5], GCoding [16], etc. C1-index [7] takes an initial step to support wildcard on vertices by leaving those vertices to the verification phrase. CP-index [9] employs embeddings to speed up containment search over large graphs. GBLENDER [6] presents the first visual paradigm blending subgraph query formulation and processing. These methods without exception are designed for single-labeled graphs under the assumption of single vertex label with string value, which are hence intrinsically incapable of general selections. Finding all matches of a query is also studied [13–15]. To handle multi-labeled graphs with the constraint of fixed attributes on vertices, DELTA transforms it into a spatial indexing problem. Due to the same reason, it is difficult to extend the approach to general structural analysis. In addition, it utilizes equality conditions on those specified attributes. Either, it cannot handle substructure selections with various vertex attributes taking abundant selection conditions. Among others, graph simulation based pattern query incorporating range conditions on the single vertex labels are also investigated [4]. As a result, it substantially differs from the subgraph isomorphism based mappings studied in this paper.

2 Preliminaries

This paper focuses on undirected simple graphs; i.e., self-loops or multiple edges are not allowed. A *data graph* is a multi-attributed graph, denoted by $r = (V_r, E_r, l_r)$, where V_r is the vertex set, $E_r \subseteq V_r \times V_r$ is the edge set, and l_r is an attribute function. $v \in V_r$ has an attribute set A_v, and each *attribute* $a_v \in A_v$ is assigned a *value* $l_r(a_v)$. Equivalently, a vertex v has a value $l_r(a_v)$ for attribute a_v. Nevertheless, v may not have a particular attribute a_v, in which case $l_r(a_v) = nil$. Besides, $|V_r|$ and $|E_r|$ denote the numbers of graph vertices and edges, respectively.

A *query graph* is denoted by $s = (V_s, E_s, \varphi_s)$, where V_s is the vertex set, $E_s \subseteq V_s \times V_s$ is the edge set, and φ_s is an attribute selection function. E_s enforces the *connection constraints*, while φ_s exerts the *attribute constraints*. $v \in V_s$ also has an attribute set A_v such that $a_v \in A_v$ is assigned a *selection condition* $\varphi_s(a_v)$. That is, φ_s imposes a condition $\varphi_s(a_v)$ on attribute a_v, against an equality, range, or set containment operator.

Given vertices $u \in V_r$, $v \in V_s$, u *satisfies* v on attribute a, provided

- $l_r(a) = \varphi_s(a)$, if $\varphi_s(a)$ defines an equality condition; or
- $l_r(a) \in \varphi_s(a)$, if $\varphi_s(a)$ defines a range condition; or
- $l_r(a) \subseteq \varphi_s(a)$, if $\varphi_s(a)$ defines a set containment condition; or
- arbitrary value in the domain, if $\varphi_s(a)$ is a wildcard.

u *matches* v, if u's attribute values satisfies v's corresponding constraints conjunctively.

Example 1. Consider in Fig. 1(1) graph (a) as r, the vertex with PID 1 as u; the graph in Fig. 1(2) as s, the upmost vertex as v. u has a value 'Enzymes' on attribute Role. $A_v = $ {Classes, Function, Amino Acids}. Range condition φ_s(Amino Acids) = '$\geq$ 10' requires a value no less than 10 on attribute Amino Acids. u matches v.

A data graph r' is a *subgraph* of r (denoted $r' \sqsubseteq r$), if there is an *injection* $f : V_{r'} \to V_r$ such that (1) $\forall v \in V_{r'}$, $f(v) \in V_r$, all attribute values at $f(v)$ are retained at v by $l_{r'}$; and (2) $\forall e = (u,v) \in E_{r'}$, $(f(u), f(v)) \in E_r$. Similarly, the subgraph relation between query graphs s' and s (denoted $s' \sqsubseteq s$) is an injection from $V_{s'}$ to V_s that retains the attribute constraints φ_s on corresponding vertices.

Definition 1 (Substructure Mapping). *Given a data graph r and a query graph s, a substructure mapping is an injection $f : V_s \to V_r$ such that (1) $\forall v \in V_s$, $f(v) \in V_r$, $f(v)$ matches v; and (2) $\forall (u,v) \in E_s$, $(f(u), f(v)) \in E_r$.*

Problem Statement. Given a set of data graph as database R and a query graph s, the problem of substructure selection finds all substructure mappings from the query graph s to each data graph r in R.

Example 2. Consider the graphs in Fig. 1(1) as R, the graph in Fig. 1(2) as s. The vertices with PID 1, 3, 6, 7 form a substructure mapping from s to graph (a). Substructure selection is to retrieve all the three of such mappings in graphs (a) and (b).

3 Substructure Selection

In this section, we first present a DFS paradigm for substructure selection, and then propose a novel index structure to support and boost the computation.

3.1 Algorithm Framework

We illustrate the DFS paradigm for processing query s against graph r in Algorithm 1. Assuming vertices are sorted in a given/arbitrary order, we use $s[m]$ to denote the m-th vertex in V_s, and $s[1..m]$ to denote the subgraph induced by the first m vertices. In each iteration, we match the available vertices in r with $s[m]$. Current partial mapping f is extended, if u satisfies (1) the attribute constraints on $s[m]$; and (2) all connection constraints between $s[m]$ and $s[1..m-1]$ (Line 5). Hence, f is fed to next recursion iteratively (Line 7). A substructure mapping is found when V_s are fully matched (Lines 1 – 3). The algorithm terminates after all possibilities are explored, and all valid mappings are found.

To process query s against a database R, we iterate Algorithm 1 over each graph r of R. Henceforth, the analysis will focus on processing s against a single data graph r, since the cost of processing R simply adds a constant factor.

It can be immediately verified that all substructure mappings from s to r can be found by Algorithm 1, with each vertex in the subgraph satisfying all the constraints on its matching vertex. We observe that the most costly step in Algorithm 1 is the iterative extension of *partial mappings*. While there exist an exponential number of possible extensions, many of them fail halfway. If we can render the extension

Algorithm 1. SelectSubstructure(s, r, m, f)

    Input : s is query; r is data graph; m is depth; f is mapping.
    Output : F is a set of substructure mappings, initialized as $\emptyset$
1 if $m > |V_s|$ then
2     $F \leftarrow F \cup f$;   /* found a mapping */
3   return
4 for each unmapped vertex u in r do
5   if u satisfies
       attribute constraints of $s[m] \wedge$ connection constraints with $s[1..m-1]$ then
6         $f' \leftarrow f, f'[m] \leftarrow u$;   /* extend the mapping */
7   SelectSubstructure (s, r, $m+1$, f');

in a faster and more informative manner, we are expected to discover valid mappings and suspend invalid ones more efficiently. Following presents an observation that enable us to develop effective indexing and query algorithms based on those qualified partial mappings. Let $F_s(r) = \{f_r(s)\}$ denote all substructure mappings from s to r.

We prove that every full substructure mapping is always grown from certain partial mappings. As a consequence, if we can leverage the pre-computation of partial mappings, we may start growing full mappings based on the partial mappings. Furthermore, chances are we can take a leap during the growth by jointing other partial mappings. In another word, we start with a subgraph, expand it by vertex extension, and leap with the aid of other subgraphs when possible. Thus, full mappings are to be found faster, reducing the overall response time. To this end, we will present shortly an index structure to effectively organize such partial mapping information.

Theorem 1. *If $s' \sqsubseteq s$, $F_r(s'|s) \subseteq F_r(s')$, where $F_r(s'|s) = \{f_r(s')|f(s') \sqsubseteq f(s)\}$.*

3.2 Index Structure

We propose a two-tier index structure called SS-index (Substructure Selection-index). The upper tier is a set of template graphs organized in a prefix tree; the lower tier stores mappings in the database subsumed by the corresponding templates.

First, we introduce *template graph*, denoted by $t = (V_t, E_t, \phi_t)$, where V_t is the vertex set, $E_t \subseteq V_t \times V_t$ is the edge set, and ϕ_t is a function assigning attributes to V_t. Intuitively, template graph removes attribute values (resp. selection conditions) from data (resp. query) graph such that each vertex comprises attributes only. A template t' is a *subgraph* of t (denoted $t' \sqsubseteq t$), if there is an *injection* $f : V_{t'} \to V_t$ such that (1) $\forall v \in V_{t'}$, $f(v) \in V_t \wedge \phi_{t'}(v) \subseteq \phi_t(f(v))$; and (2) $\forall e = (u,v) \in E_{t'}$, $(f(u), f(v)) \in E_t$. Furthermore, a data graph r is *subsumed* by a template t (denoted $r \trianglelefteq t$), if there is a *bijection* $f : V_r \to V_t$ such that (1) $\forall v \in V_r$, $f(v) \in V_t \wedge A_v \subseteq \phi_t(f(v))$; and (2) $\forall e = (u,v) \in E_r$, $(f(u), f(v)) \in E_t$. We also say r is an *embedding* of t. Similarly, subsumption relation between a query s and a template t is a bijection $s \trianglelefteq t$.

Example 3. Consider Fig. 1, and template graph t in Fig. 2. The subgraph of graph (b) induced by vertices with PID 1, 2, 4 and 6 is subsumed by t; and t also subsumes s.

SS-index consists of selected template graphs in a *prefix tree*, as well as corresponding embeddings. In particular, the selected templates utilize a prefix-sharing strategy for both compact storage and efficient access. Every leaf node of the prefix tree corresponds to a distinct template (the upper-tier), and it links to a *subindex* in the lower-tier comprising all the embeddings in the database subsumed by the template. These embeddings are recorded and sorted by combined search keys. Particularly, for each mapping, we keep, as an entry, its values of the indexed attributes as in the template, provenance graph ID and vertex ID's. In a subindex, nonetheless, multiple entries may have identical values for certain indexed attributes. Under this scenario, the embeddings having the same search key are folded into a single

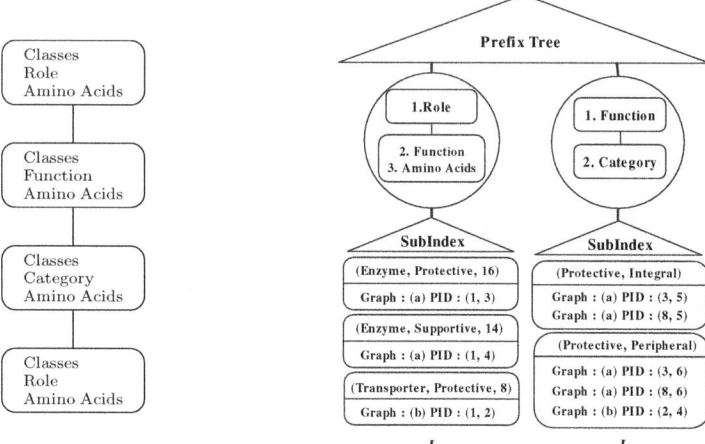

Fig. 2. Template Graph **Fig. 3.** Example of SS-index

entry, in order to reduce the memory footprint of the index, while provenance graph ID's and vertex ID's are recorded separately.

Example 4. Consider the SS-index in Fig. 3. The circles under the prefix tree indicate leaf nodes, representing two template graphs. I_a and I_b are subindices for the two templates, respectively. Entries in I_b are sorted in ascending order of their search keys – 'Function-Category'. There are five embedding entries with two distinct values 'Protective, Integral' and 'Protective, Peripheral' in I_b.

4 Query Processing

This section introduces the design of SS-search (Substructure Selection-search) algorithm. We first give an overview of SS-search and emphasize the importance of a good query execution plan, then investigate the primitive operations involved in execution plans, and finally conceive an algorithm to generate effective plans.

4.1 Algorithm Overview

We summarize SS-search algorithm into three phases: (1) index probing, (2) query plan generation, and (3) mapping discovery.

In index probing, all indexed templates subsuming a subgraph of the query (a.k.a., partial query) are obtained. Under an ideal scenario, a template subsuming the query is indexed, and hence, all substructure mappings are collected from the subindex of that template the embeddings satisfying the attribute constraints, where connection constraints are naturally preserved. Following discusses the solution to general cases, which first generates a query execution plan and then searches for full mappings.

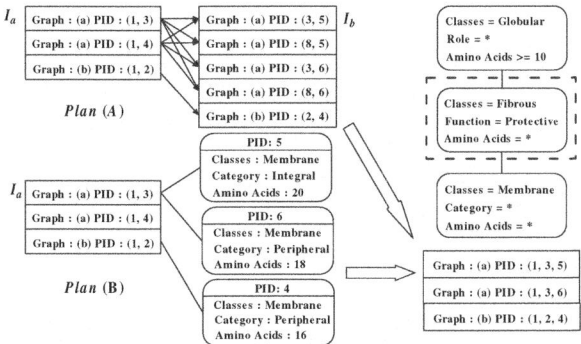

Fig. 4. Example of Query Executions

An embedding r' of template t' is *qualified* with respect to partial query s', if r' satisfies all the constraints of s', where $s' \trianglelefteq t'$. Given an indexed template t' subsuming a partial query s', qualified partial mappings of s' can be obtained from the subindex of t'. Thus, we save the online cost of computing these partial mappings, and grow them to full mappings thereafter. Note these partial mappings are inevitably explored by Algorithm 1. Moreover, we quest if the cost can be further saved during the extension process. We argue this is achievable provided there is a "wise" query execution plan that instructs how to search for full mappings. Let us take the following example.

Example 5. Consider in Fig. 4 two executions for the query in the upper right. Two partial queries are subsumed by indexed templates (see Fig. 3). To expand the mappings in I_a, Plan (A) joins them with those from I_b; Plan (B) scans the adjacency list of the second vertex (bounded by dashed line), in order to match the third vertex.

In fact, Plan (B) outperforms Plan (A). The performance gap arises from that joining results from subindices is more expensive than scanning adjacency list for extension in this case. Nevertheless, it remains unclear if join always performs worse for all cases. In addition, we observe the number of partial mappings during the extension also greatly influence the performance. More partial mapping implies larger input to be processed subsequently. With regards to memory consumption and runtime performance, we seek an appropriate plan to reduce partial mappings during the execution.

To this end, we propose an algorithm that first chooses an indexed template as seed, and then extends to a complete plan with minimum cost. Prior to the algorithm details, we look into the primitive operations constituting execution plans. The runtime cost and number of intermediate results of each operation are analyzed, so as to evaluate different plans. In the sequel, "partial mapping" is also referred as "intermediate result", since a partial mapping is an intermediate result potentially for a full mapping.

4.2 Primitive Operations

It is sufficient to consider the following six operations, as summarized in Table 1.

Presumably, all single attributes (templates of single vertex with one attribute) are included in SS-index. Hence, the vertices satisfying attribute constraint c are retrieved by accessing the subindex under the template corresponding to the attribute of c. Let n_c denote the number of entries satisfying c therein. For ease of exposition, we introduce $\rho(C) \triangleq \prod_c \frac{n_c}{\sum_r |V_r|}$, where $c \in C$, C is a set of attribute constraints, and $r \in R$.

Index Retrieval $\mathtt{IR}(I_t)$. Given a subindex I_t under template t, it return the partial mappings indexed by I_t. Denote the total number of entries under I_t as n_t. Straightforwardly, it outputs n_t intermediate results, and hence, consumes $O(n_t)$ time.

Graph Scan $\mathtt{GS}(C)$. Given a set of attribute constraints C, it retrieves the vertices satisfying C by scanning the database. Trivially, the estimated number of intermediate results is $\sum_r |V_r| \cdot \rho(C)$, and it takes $O(\sum_r |V_r|)$ time, $r \in R$.

Attribute Validation $\mathtt{AV}(F_{s'}, v, C)$. Consider partial mappings $F_{s'}$, a vertex $v \in V_{s'}$, $s' \sqsubseteq s$, and a set of attribute constraints C. For each $f_{s'} \in F_{s'}$, it verifies whether $f_{s'}(v)$ satisfies C, and retains all the qualified ones. The validation cost is $O(|C| \cdot |F_{s'}|)$, while the number of intermediate results is $|F_{s'}| \cdot \rho(C)$.

Connection Validation $\mathtt{CV}(F_{s'}, u, v)$. Consider partial mappings $F_{s'}$, and edge $(u, v) \in E_{s'}$, $s' \sqsubseteq s$. For each $f_{s'} \in F_{s'}$, it verifies if $(f_{s'}(u), f_{s'}(v)) \in f_{s'}$, and retains those having passed the validation. The possibility of an edge in a graph is $p(e) = \frac{\sum_r p_r(e)}{|R|}$, where $p_r(e) = \frac{2|E_r|}{|V_r|(|V_r|-1)}$, $r \in R$. Thus, it produces $|F_{s'}| \cdot p(e)$ intermediate results, and runs in $O(\theta \cdot |F_{s'}|)$ time, where θ is the average vertex degree of R.

Mapping Extension $\mathtt{ME}(F_{s'}, v, C)$. Consider partial mappings $F_{s'}$, a vertex $v \in V_{s'}$ and a set of attribute constraints C. For each $f_{s'} \in F_{s'}$, it explores vertices u such that u satisfies C and connects to at least one vertex of $f_{s'}$. For each valid extension, we have a new partial mapping $f_{s'} \cup \{u\}$ with v mapped to u. Thus, it costs $O(\theta \cdot |C| \cdot |F_{s'}|)$ time, where θ is the average vertex degree in the database. Hence, the number of intermediate results is approximated as $\theta \cdot |F_{s'}| \cdot \rho(C)$.

Mapping Join $\mathtt{MJ}(F_{s'}, F_{s''})$. A mapping join operation connects two sets of partial mappings $F_{s'}$ and $F_{s''}$, where $s', s'' \sqsubseteq s$. Denote as $\hat{V} = V_{s'} \cap V_{s''}$ the set of join

Table 1. Summary of Primitive Operations

Operation	Intermediate Result Number	Runtime Cost
$\mathtt{IR}(I_t)$	n_t	$O(n_t)$
$\mathtt{GS}(C)$	$\sum_r |V_r| \cdot \rho(C)$	$O(\sum_r |V_r|)$
$\mathtt{AV}(F_{s'}, v, C)$	$|F_{s'}| \cdot \rho(C)$	$O(|C| \cdot |F_{s'}|)$
$\mathtt{CV}(F_{s'}, u, v)$	$|F_{s'}| \cdot p(e)$	$O(\theta \cdot |F_{s'}|)$
$\mathtt{ME}(F_{s'}, v, C)$	$\theta \cdot |F_{s'}| \cdot \rho(C)$	$O(\theta \cdot |C| \cdot |F_{s'}|)$
$\mathtt{MJ}(F_{s'}, F_{s''})$	$\sum_{v \in \hat{V}} n_{s'}^v \cdot n_{s''}^v$	$O(|F_{s'}| \cdot |F_{s''}| \cdot |\hat{E}|)$

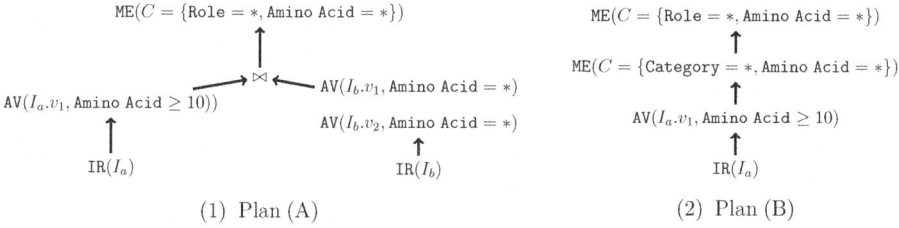

Fig. 5. Query Execution Plans

keys, and $\hat{E} \subseteq E_s$ the set of connection constraints on the join results. The join results are the combinations of $f_{s'}$ and $f_{s''}$, such that (1) $\forall v \in \hat{V}, f_{s'}(v) = f_{s''}(v)$, and (2) $\forall (u, v) \in \hat{E}, (f_{s'}(u), f_{s''}(v)) \in E_s, u \in V_{s'}, v \in V_{s''}$. The cost is in $O(|F_{s'}| \cdot |F_{s''}| \cdot |\hat{E}|)$. The number of intermediate results is estimated as $\sum_{v \in \hat{V}} n^v_{s'} \cdot n^v_{s''}$, where $n^v_{s'}$ (resp. $n^v_{s''}$) is the numbers of mappings such that $\exists u \in V_{s'}$ (resp. $u \in V_{s''}$), $f_{s'}(v) = u$ (resp. $f_{s''}(v) = u$).

GS and IR are the two primitive operations not requiring partial mappings as input. A complete query execution plan can be considered as a tree with every node representing a primitive operation. For example, two sample plans are depicted in Fig. 5 regarding Example 5. The overall cost of a query execution plan is given by summing up the costs of all nodes, where the cost of internal nodes are also dependent on the intermediate results of the antecedent operation. We compare different query execution plans in terms of their overall cost. Immediate is there exist an exponential number of plans.

Next we prove that computing the query execution plan with minimum cost is difficult, and hence, most online applications cannot afford the cost. Following subsection proposes as a remedy a practical algorithm to find effective plans efficiently.

Theorem 2. *Given a substructure selection problem, finding the query execution plan with minimum cost is NP-hard.*

4.3 Query Plan Generation

We first lay down four heuristics for composing effective query execution plans.

- **A proper execution order reduces the overall processing cost.** We generate tree-structured plans to reduce the search space. It differs from the left-deep tree in relational database in that the internal nodes are the operations requiring partial mappings as input – AV, CV, ME and MJ. The rationales behind are it (1) avoids materializing partial mappings of each operations; and (2) significantly reduces the search space of plan composition.
- **Joining overlapped mapping reduces intermediate results.** Suppose we have two sets of partial mappings $F_{s'}$ and $F_{s''}$ for partial queries $s', s'' \sqsubseteq s$, respectively. We exert MJ only when $F_{s'}$ and $F_{s''}$ overlap, i.e., have at least one common vertex and/or edge; otherwise, the intermediate results of join are exactly the Cartesian product of $F_{s'}$ and $F_{s''}$. As a consequence, we prioritize ME, and postpone MJ till the partial queries overlap.

- **Eager constraints validation reduces intermediate results.** We allocate constraints validation as early as possible in the plan so as to reduce the input size of subsequent operations. Specifically, we insert CV immediately after IR and ME to check the inner connection constraints. Similarly, we allocate AV once there are attribute constraints on extending vertices.
- **Early validation of selective constraints reduces intermediate results.** In particular, given a set of constraints, we order them in descending order of selectivity so that the partial mappings in subsequent steps are reduced. When there is a tie, we further differentiate them by selection ranges of the constraints; i.e., constraints with smaller selection ranges are ordered ahead.

Algorithm 2. GeneratePlan(s, I)

    Input   : s is a query; I is SS-index.
    Output : P is the execution plan, initialized as $\emptyset$.
1 open $\leftarrow V_s$, closed $\leftarrow \emptyset$;
2 cand $\leftarrow \{t | s' \trianglelefteq t \wedge s' \sqsubseteq s$ such that t is indexed by $I\}$;
3 remove non-maximal templates from cand;
4 $t \leftarrow$ PickSeed(cand), cand $\leftarrow$ cand $\setminus \{t\}$;
5 open $\leftarrow$ open $\setminus V_t$, closed $\leftarrow$ closed $\cup V_t$;
6 $P \leftarrow P \cup$ IR ;   /* append validations when possible */
7 **while** open $\neq \emptyset$ **do**
8   ME $\leftarrow$ PickExtension(open), add extension vertex into closed;
9      $P' \leftarrow P, P \leftarrow P \cup$ ME ;   /* append validations when possible */
10   **for each** $t \in$ cand **do**
11   **if** $V_t \cap$ closed $\neq \emptyset$ **then**
12           $P' \leftarrow P' \cup$ MJ;
13   **if** EstimateOutput(P') < EstimateOutput(P) **then**
14              $P \leftarrow P'$;   /* replace the existing plan */
15   open $\leftarrow$ open $\setminus V_t$, closed $\leftarrow$ closed $\cup V_t$;
16   cand $\leftarrow$ cand $\setminus \{t\}$;
17 **return** P

Hence, we compile the guidelines into Algorithm 2 for efficiently composing a query execution plan, which consists of two stages: (1) seed selection, and (2) plan growth.

The algorithm starts by retrieving as set cand all indexed templates subsuming a partial query, and then removing non-maximal template graphs (subgraphs of others in cand, Lines 2 – 3). The template t, having the least partial mappings satisfying all the constraints under consideration, is adopted as the plan seed (Line 4). Afterwards, the algorithm utilizes two exclusive sets – open and closed – to indicate the status of every query vertex. Vertices in open has been considered in the plan, and closed implies a vertex is not considered yet. After including V_t into closed, we put index retrieval as the first operation (Line 6), and begin to grow the seed to a complete plan.

In the second stage, we iteratively insert mapping extension operations into the current plan. In each iteration, all required attribute and connection validations are exploited, and then all possible mapping extensions of the current plan are identified. Among all the extensions, the one with minimum estimated cost is chosen via calling PickExtension (Line 8). Note only the vertices in open are considered. Each mapping extension introduces a new vertex to the plan, which is put into closed thereafter. Additionally, if there exists a candidate template of V_t subsuming a subgraph of $V_{s'} \sqsubseteq$ closed, we create an alternative plan by replacing the mapping extensions of $V_{s'}$ in that specific template with a mapping join of V_t. If this alternative is estimated by EstimateOutput to have less intermediate results, we adopt the alternative plan (Lines 11 – 14). The process repeats till all query vertices are in closed.

SS-search employs the resulting query execution plan to discover the entire mapping set eventually. We remark the cost of the first stage consists of (1) subgraph isomorphic tests for identifying indexed templates; and (2) removals of non-maximal templates. One may note the worst case complexities of both parts are NP-complete, whereas they are not significantly large in practice, as the indexed templates are normally small.

5 Index Construction

This section discusses the construction of SS-index. We judiciously index effective template graphs to strike a balance between index quality and cost. An *effective* template is both sufficiently frequent and sufficiently discriminative. Additionally, our algorithm considers to utilize query logs to pick selective templates.

Frequency of a template graph t is defined as $freq(t) = \frac{|\{r|r' \trianglelefteq t\}|}{|R|}$, and *discrimination ratio* as $disc(t) = \frac{|\{r|r' \trianglelefteq t\}|}{|\{r|r' \trianglelefteq t'\}|}$, where $r' \sqsubseteq r$, $r \in R$, and $t' \sqsubset t$. Following [8, 11], we choose the template graphs conforming to $freq(t) \geq \alpha$ and $disc(t) \leq 1 - \beta$, where α and β are frequency and discrimination thresholds, respectively.

When query logs are available, we use average number of partial mappings that can be retrieved by the historical queries to estimate the selectivity of the templates. Therefore, *selectivity* is defined as $sele(t) = \frac{\sum_{s' \trianglelefteq t} |F_{V_t}|}{|\{s|s' \trianglelefteq t\}|}$, where F_{V_t} is the set of embeddings of t, $s' \sqsubseteq s$, s is an individual from historical queries S. If there exist more than one embedding of t in s, we choose the one with minimum $|F_{V_t}|$. Given a selectivity threshold γ, a template graph is selective, if $sele(t) \leq \gamma$.

An issue brought to attention is the exponential number of possible template graphs; i.e., we need to test all possible combinations of attributes at each vertex for all graph structures. This is computationally prohibitive when there are larger number of attributes and/or structures. To resolve the issue, we make the observations: (1) a common combination of attribute is comparable to the "frequent itemset" in a transactional database; and (2) an embedding of template t' is subsumed by another t if $t' \sqsubseteq t$; we put the attribute values of an embedding of t, which are missing from t', as nil. This enables us to employ only the maximal frequent attribute sets as the attributes for indexing discriminative structures, while alleviating the exponential growth issue.

Particularly, we treat the set of attributes at a vertex as one transaction, $|R| \cdot |V_r|$ transactions in total. Maximal frequent attribute sets [1] are derived from the "transactional database"; i.e., frequent attribute sets all whose immediate supersets are infrequent. These attribute sets, denoted by $\mathcal{A}$, constitute the domain of attributes carried by template vertices; i.e., a template is extended with vertex v, only if $A_v \subseteq A \in \mathcal{A}$.

Algorithm 3. IndexTemplate(D, L, α, β, $[S, \gamma]$)

Input : R is database; L is maximum template size; α is frequency threshold; β is discrimination threshold; S is query set; γ is selectivity threshold.
Output : I is SS-index, initialized as $\emptyset$.
1 $\mathcal{A} \leftarrow$ compute maximal frequent attribute sets;
2 **for each** distinct single attribute att **do**
3 $\quad$ t is a template of att;
4 $\quad$ $I \leftarrow I \cup \{t\}$;   /* include in the upper-tier */
5 **for each** template t discovered by gSpan-like procedure using $\mathcal{A}$ **do**
6 $\quad$ **if** CheckThreshold (t, α, δ, $[S, \varepsilon]$) **then**
7 $\quad\quad$ $I \leftarrow I \cup \{t\}$;   /* include in the upper-tier */
8 **for each** template $t \in I$ **do**
9 $\quad$ $I \leftarrow I \cup$ ConstructSubindex(t) ;   /* include in the lower-tier */
10 **return** I

We propose an apriori-based template indexing algorithm (Algorithm 3). Line 1 computes the maximal frequent attribute sets $\mathcal{A}$. Then in Lines 2 – 4, we index templates for single attributes. These individual attributes avoid expensive graph scans in query processing. In Lines 5 – 7, we follow the procedure of gSpan [10] to generate template structures, and it only extend a template to a new vertex whose attribute set is a subset of an element in $\mathcal{A}$. CheckThreshold tests whether the template satisfies both frequency and discrimination thresholds; additionally, the selectivity threshold is verified when query logs are available. Meeting the thresholds qualifies a template to be organized in a prefix tree in SS-index (upper-tier). Finally, Lines 8 – 9 construct the subindices for all templates in the upper-tier by calling ConstructSubindex.

6 Experiments

The following algorithms are involved in the experimental evaluation:

- **SS-search** is the proposed algorithm for substructure selection using SS-index.
- **QuickSI** is a state-of-the-art subgraph containment search algorithm [8]. We received the source code from the authors, and modified it to support finding all substructure mappings. Essentially, the adapted QuickSI realized our Algorithm 1 following the DFS paradigm. Treating data graph templates with the first attribute on vertices (disregarding the remaining attributes) as single-labeled graphs, we mined discriminative tree features and built index for pruning.

- **GADDI** is a state-of-the-art subgraph all-matching algorithm [13]. It was reengineered to handle general selection conditions, and built index with discriminative subgraphs for every data graph using the templates as QuickSI.

All algorithms were implemented in C++, and compiled using GCC 4.4.3 with -O3 flag. Experiments were run on a machine of Intel Xeon 2.40GHz dual CPU with 4G memory running Debian Linux. We used the following default settings for all algorithms – frequency threshold $\alpha = 0.1$ and discrimination threshold $\beta = 0.1$.

We conducted experiments on both real and synthetic datasets. In the interest of space, we only show the results on real dataset AIDS. AIDS is an antivirus screen compound dataset at NCI/NIH, containing 43, 905 molecules. On average, each graph has 25.4 vertices and 27.3 edges. Each vertex represents an atom with a 3-dimensional coordinate, and edges depict the chemical bonds between atoms. For every vertex, we used the name as an attribute with the first dimensional coordinate as value. In total, there are 62 distinct first attributes, valued in [-47.3, 63.4]. Additionally, we used 'coordinate' as the other attribute with the second dimensional coordinate as value. Since coordinate is continuous in the value domain, we tested range conditions on AIDS.

Five sets of 1000 query graphs were used (denoted Q8, Q12, Q16, Q20 and Q24), average number of edges being 8, 12, 16, 20 and 24, respectively. For each query, 30% vertices have non-trivial attribute constraints with average selection range of 0.1. Q16 was used by default, if not otherwise specified. We report the average results per query.

Index Construction. We first evaluate the indexing performance of SS-index using different structural features. We enforced Algorithm 3 to grow path, tree and subgraph templates, respectively, and show the results in Table 2(1). Three resulting algorithms are shortened as Path-SS, Tree-SS and Subgraph-SS, respectively. It is clear Path-SS consumes the most space, since the number of path templates is much more than the others. Tree-SS takes the second place regarding both space and indexing time, while Subgraph-SS has the smallest index size but largest runtime due to expensive subgraph isomorphic tests. Further, Fig. 6(1) reflects the effect of various features on runtime. Subgraph-SS has a greater starting point, but decreases faster than Path-SS, as subgraphs are more selective in larger graphs than paths. Tree-SS provides the greatest performance upgrade for all query sets. Subsequently, we chose trees as index features in the remaining experiments to strike a balance between space costup and runtime speedup. Also, we suggest, if memory is not critical, one may also include subgraph features to gain further speedup when data graphs a large.

The comparison of indexing performance involving three algorithms are presented in Table 2(2). GADDI spends the greatest space and runtime on indexing due to the lack of optimization pertinent to the problem. SS-index and QuickSI have comparative indexing performance. Specifically, SS-index needs to manage the selected embeddings, and thus, has slightly larger size and greater runtime, though SS-index and QuickSI both utilizes tree features. We will see shortly this offline effort is rewarding.

Table 2. Experiment Results-I

(1) Comparison of Features			(2) Indexing Performance		
SS-index	Size (kB)	Time (s)	AIDS	Size (kB)	Time (s)
Path-SS	4867.9	175.6	QuickSI	876.5	225.6
Tree-SS	1249.8	367.4	GADDI	4834.3	3198.2
Subgraph-SS	657.1	581.9	SS-index	1249.8	367.4

Query Processing. We experimentally compare the efficiency of the three algorithms against varying query size, and the results are shown in Fig. 6(2). The query processing of SS-search is up to three orders of magnitude faster than QuickSI, and five orders of magnitude faster than GADDI. The response time of SS-search drastically decreases as query size increases, whereas the response time of QuickSI only decrease slightly; on the contrary, the response time of GADDI increase slightly as query graph size increases. This phenomenon is more significant on large queries. We argue that as SS-search leverages the embeddings indexed by SS-index, the cost greatly depends on the selectivity of the chosen plan seeds. As query size increases, it is more likely to contain larger indexed templates subsuming some partial queries, which are more selective in general; however, this does not benefit QuickSI and GADDI.

Evaluating Constraints. The effect of varying constraint coverage is studied, as shown in Fig. 6(3). The x-axis is the percentage of query vertices with attribute constraints. We varied the percentage from 10% to 50%. With expectation, the results indicate GADDI and QuickSI are merely affected, since they do not take into consideration the selectivity of constraints. Contrarily, the response time of SS-search drastically reduces with the increasing percentage. The reason behind is that the more constraints there are in queries, the more selective the chosen indexed templates are. Consequently, it is likely we can obtain a small number of partial mappings at start-off, and effective mapping joins with indexed templates result in reductions on intermediate results. We also observe the performance of SS-search is slightly worse than QuickSI when only 10% of the queries have constraints. As the graphs in the default query set contains 16 edges on average, there are only approximately one or two constraints in each query. Therefore, it is less likely to find indexed templates with few mappings; however, SS-search is only slightly slower than QuickSI even in this extreme case. This also advises that existing subgraph search techniques do not handle substructure selections effectively.

We also study the effect of range size on response time, and report the results in Fig. 6(4). Range size is the average range of selections exerted by the range conditions. We varied the range size from 0.1 to 10. Intuitively, constraints with smaller range size are more selective. As expected, the response time decreases gradually as range size decreases. However, the impact of range size is comparatively insignificant under the given setting; i.e., similar value domains and small number of attributes. This implies the selectivity of a query is more dependent on the number

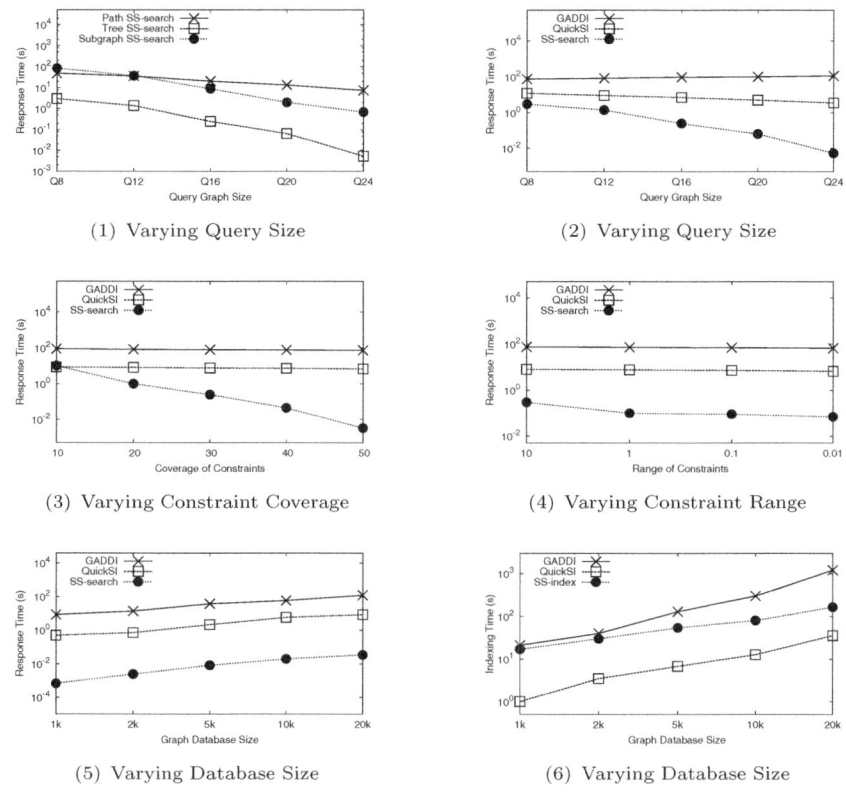

Fig. 6. Experiment Results-II

of constraints than the selectivity of each constraint, which can be used to further strengthen the index construction. From another angle, this also corresponds to the argument that substructure-based filtering is more effective than vertex/edge-based filtering techniques [11].

Evaluating Scalability. We evaluate the scalability of three techniques against varying graph database size. We sampled five graph databases by randomly choosing 1k, 2k, 5k, 10k and 20k graphs from the original dataset, so that the data distribution remains approximately the same. The results are plotted in Fig. 6(5). It suggests SS-search is expected to be more scalable than others, although all algorithms grows steadily towards larger database size. In addition, SS-search outperforms others with substantial gaps under all given database size settings.

The scalability of index construction is studied in Fig. 6(6). All three algorithms are scalable in terms of indexing time. SS-index showcases the smallest growth rate among the three, and the maximal frequent attribute sets based mining approach lends itself well to large graph databases. We note the performance of QuickSI is faster than the others. This is attributed to that QuickSI mines features from a portion of the database [8], which can be regarded as trading index quality for runtime performance. Similar concepts are applicable to SS-index for mining attribute sets and structural features. These are beyond the scope of this paper, and hence, left for future study.

7 Conclusion

In this paper, we have studied substructure selections in multi-attributed graph databases. We devise SS-index to provide effective support. On top of it, SS-search is proposed leveraging dynamical query execution plans. The effective index and efficient processing algorithms render our solution attractive in terms of performance and scalability.

References

1. Burdick, D., Calimlim, M., Flannick, J., Gehrke, J., Yiu, T.: Mafia: A maximal frequent itemset algorithm. IEEE Trans. Knowl. Data Eng. 17(11), 1490–1504 (2005)
2. Chen, C., Yan, X., Yu, P.S., Han, J., Zhang, D.-Q., Gu, X.: Towards graph containment search and indexing. In: VLDB, pp. 926–937 (2007)
3. Cheng, J., Ke, Y., Ng, W., Lu, A.: FG-index: towards verification-free query processing on graph databases. In: SIGMOD Conference, pp. 857–872 (2007)
4. Fan, W., Li, J., Ma, S., Tang, N., Wu, Y.: Adding regular expressions to graph reachability and pattern queries. In: ICDE, pp. 39–50 (2011)
5. Jiang, H., Wang, H., Yu, P.S., Zhou, S.: GString: A novel approach for efficient search in graph databases. In: ICDE, pp. 566–575 (2007)
6. Jin, C., Bhowmick, S.S., Xiao, X., Cheng, J., Choi, B.: GBLENDER: towards blending visual query formulation and query processing in graph databases. In: SIGMOD Conference, pp. 111–122 (2010)
7. Klein, K., Kriege, N., Mutzel, P.: CT-index: Fingerprint-based graph indexing combining cycles and trees. In: ICDE, pp. 1115–1126 (2011)
8. Shang, H., Zhang, Y., Lin, X., Yu, J.X.: Taming verification hardness: an efficient algorithm for testing subgraph isomorphism. PVLDB 1(1), 364–375 (2008)
9. Xie, Y., Yu, P.S.: CP-index: on the efficient indexing of large graphs. In: CIKM, pp. 1795–1804 (2011)
10. Yan, X., Han, J.: gSpan: Graph-based substructure pattern mining. In: ICDM, pp. 721–724 (2002)
11. Yan, X., Yu, P.S., Han, J.: Graph indexing: A frequent structure-based approach. In: SIGMOD Conference, pp. 335–346 (2004)
12. Yang, J., Zhang, S., Jin, W.: DELTA: indexing and querying multi-labeled graphs. In: CIKM, pp. 1765–1774 (2011)
13. Zhang, S., Li, S., Yang, J.: GADDI: distance index based subgraph matching in biological networks. In: EDBT, pp. 192–203 (2009)
14. Zhao, P., Han, J.: On graph query optimization in large networks. PVLDB 3(1), 340–351 (2010)
15. Zhu, K., Zhang, Y., Lin, X., Zhu, G., Wang, W.: NOVA: A novel and efficient framework for finding subgraph isomorphism mappings in large graphs. In: Kitagawa, H., Ishikawa, Y., Li, Q., Watanabe, C. (eds.) DASFAA 2010, Part I. LNCS, vol. 5981, pp. 140–154. Springer, Heidelberg (2010)
16. Zou, L., Chen, L., Yu, J.X., Lu, Y.: A novel spectral coding in a large graph database. In: EDBT, pp. 181–192 (2008)

Parallel Triangle Counting over Large Graphs

Wenan Wang, Yu Gu, Zhigang Wang, and Ge Yu

Northeastern University, China
{wangwenan6,wangzhigang_mail}@yahoo.cn, {guyu,yuge}@ise.neu.edu.cn

Abstract. Counting the number of triangles in a graph is significant for complex network analysis. However, with the rapid growth of graph size, the classical centralized algorithms can not process triangle counting efficiently. Though some researches have proposed parallel triangle counting implementations on Hadoop, the performance enhancement remains a challenging task. To efficiently solve the parallel triangle counting problem, we put forward a hybrid parallel triangle counting algorithm with efficient pruning methods. In addition, we propose a parallel sample algorithm which can avoid repeated edge sampling and produce high-precision results. We implement our patterns based on bulk synchronous parallel framework. Compared with the Hadoop-based implementation, 2 to 13 times gains can be obtained in terms of executing time.

1 Introduction

The triangle counting over various graph data is a basic problem to support many important high-level applications, which has attracted more and more attention in both academical and industrial communities, such as [1, 2].

With the rapid growth of graph data, counting and listing triangles in such large graphs will cause serious performance concerns. Faced with such massive data, parallelization and sampling become two potential solutions. Some researchers attempt to extend and implement triangle counting algorithms based on Hadoop platform [3, 4]. However, some important issues such as communication optimization have not been sufficiently addressed. Besides, Hadoop may suffer performance problems when multiple-step map-reduce execution processes are needed. In addition, as a most prominent alternative, effective sampling can remarkably reduce the data volume and consequently improve the evaluation efficiency. While, how to gain high-precision results becomes quite challenging. *Doulion* is a typical representative which can guarantee the precision [4]. Unfortunately, the available sampling algorithms can not be easily executed in parallel due to the "repeated edge sampling" problem.

Our major contributions are twofold. First, we step forward to explore some essential optimization techniques to improve the efficiency of parallel triangle counting and listing in terms of local computation and across-node communication costs. The proposed optimization methods can be easily implemented utilizing more fundamental frameworks such as bulk synchronous parallel to avoid the limitation of Hadoop-like systems. Second, we attempt to crack the nut of injecting sampling techniques into our parallel framework while guaranteeing quite high-precision analysis results, and hence further enhance the system capability in face of massive graph data. Specifically, (1) To tackle the problem of the overhead of communication and local computing, we propose a hybrid algorithm and a cut pruning technique. The hybrid algorithm combines the advantage of two available solutions namely *NodeIterator* and *EdgeIterator* [5]. And we propose the cut pruning to avoid repeated counting and reduce the message scale. (2) To solve the repeated edge sampling problem, we propose a partial-sampling method which can be embedded into our parallel framework.

The remaining sections are structured as follows. Section 2 reviews the related work. Section 3 proposes our optimization techniques and sampling algorithms. The experimental evaluation on various data sets is given in section 4 and we conclude in section 5.

2 Related Work

The centralized triangle counting and listing algorithms over graphs have been extensively studied. *NodeIterator* and *EdgeIterator* are two typical representatives [5]. *NodeIterator* is a vertex-centric algorithm which traverses every vertex and then checks the existence of an edge composed by any pair of the vertex's neighbors. While, *EdgeIterator* is an edge-centric algorithm, in which the source vertex and the destination vertex of every edge will be abstracted. Consequently, triangles can be found by searching common neighbors of these two vertices. In addition, some improved algorithms are proposed [6–8] which can gain better performance, but they are not suitable for parallel implementations as massive messages will be incurred. Some other works on graph data management can also indirectly offer the triangle counting function by issuing special queries. For example, R. Giugno et al. [9] propose a technique to count the three-node complete subgraph which composes a triangle actually. In [10], triangles can be counted as three-step-neighbors when the source vertex is assigned as the destination vertex.

With the rapid growth of graph data, some researchers are devoted to implementing classical centralized algorithms on parallel frameworks. S. Suri et al. [3] propose a parallel solution, *NodeIterator++*, by improving *NodeIterator*, and implement it on Hadoop. Although *NodeIterator++* counts the same triangle for several times repeatedly, the final result can be guaranteed to be correct due to designing different weights for edges.

Sampling techniques are regarded as feasible solutions on large data sets. Typically, C. E. Tsourakakis et al. [4] propose *Doulion* algorithm by using random sampling to process each edge, and *NodeIterator* to count triangles. Also, Rasmus Pagh et al. [11] introduce a new randomized algorithm for counting triangles in graphs. In the algorithm, one edge of a triangle is always sampled, if the other two have been sampled. However, these sampling algorithms can not be correctly executed in parallel because of the repeated edge sampling problem.

3 Optimization Policies and the Sampling Algorithm

3.1 SEN-Iterator

The definitions of symbols throughout the paper are given in Table 1.

Table 1. Symbols and Definitions

Sym	Definition	Sym	Definition
G	undirected graph(no self-edges)	V	vertex set of G
E	edge set of G	D_v	the degree of vertex v
$D(v)$	neighbor set of vertex v	$P(i)$	vertex set in $Node$ i
$Node$ i	a physical machine named i	N	the calculated number of triangles
M	the exact number of triangles		

Assume a *triangle* $\langle u, v, w \rangle$ exists and $u \in P(i)$, $v \in P(j)$, $w \in P(k)$, then triangles can be divided into three types: 1. Local-triangles, $i = j = k$. 2. Two-one-triangles, $i = j \neq k$ or $i = k \neq j$ or $k = j \neq i$. 3. Dis-triangles, $i \neq j \neq k$. Combining the partial-sampling algorithm (see section 3.2), *EdgeIterator* and *NodeIterator* [5], we propose a SEN-Iterator algorithm which has three phases. First, we generate a sampled graph G' by sampling edges which meet our policy with successful probability p (see section 3.2). Then local-triangles and two-one triangles in G' are counted by utilizing *EdgeIterator*, and we handle messages by using the concept of *NodeIterator* and cut pruning (see section 3.3). The third phase is to count the dis-triangles.

3.2 Partial-Sampling Algorithm

Assume an edge $\langle u, v \rangle \in E$, $u \in P(i)$, $v \in P(j)$, $i \neq j$. For existing sampling algorithms, in parallel environments, $\langle u, v \rangle$ will be processed on both $Node$ i and $Node$ j, which will be sampled twice. Therefore, for the parallel sampling process, *How to avoid sampling an edge repeatedly* is a critical problem. We propose a partial-sampling algorithm to overcome this issue. In the partial-sampling algorithm, the cross-$Node$ edges are sampled with the successful probability 1. While, the edges in the same $Node$ are sampled with the successful probability p.

Theorem 1. *The expected number of triangles in the sampled graph G' is equal to the actual number of triangles in G i.e. $E(N) = M$.*

Proof. For G', we assume that N_1 is the number of local-triangles, N_2 is the exact number of two-one-triangles and N_3 is the exact number of dis-triangles. For G, let M_1 denote the existing local triangles, M_2 be the number of two-one-triangles and M_3 be the number of dis-triangles. For each existing triangle with a specified ID i in G, ε_i is defined as a flag. Therefore, $\varepsilon_i = 0$ if triangle i does not exist in G', otherwise, $\varepsilon_i = 1$. According to the partial-sampling algorithm, the expected values of N_1, N_2, N_3 are:

$$E(N_1) = E\sum_{i=1}^{M_1}(\frac{1}{p^3} \times \varepsilon_i) = \frac{1}{p^3} \times \sum_{i=1}^{M_1} p^3 = M_1 \qquad (1)$$

$$E(N_2) = E\sum_{i=1}^{M_2}(\frac{1}{p} \times \varepsilon_i) = \frac{1}{p} \times \sum_{i=1}^{M_2} p = M_2 \qquad (2)$$

$$E(N_3) = E\sum_{i=1}^{M_3} \varepsilon_i = \sum_{i=1}^{M_3} E(1) = M_3 \qquad (3)$$

By above formulas, we can conclude that:

$$E(N) = E(N_1) + E(N_2) + E(N_3) = M_1 + M_2 + M_3 = M \qquad (4)$$

Furthermore, we analyze the variance of N in Theorem 2.

Theorem 2. *Let M be the exact number of triangles in graph G. The variance of N is:*

$$Var(N) = \frac{M \times (p^3 - p^6) + 2k \times (p^5 - p^6)}{p^6} + \Delta \times p \times (1-p) \qquad (5)$$

where, k is the number of triangles which share an edge with other triangles.

Proof. The deviation mainly comes from three parts: overall triangles' deviation, edge-shared triangles' deviation and local-triangles' deviation. In [4], the author gives the variance estimate of the overall triangles' deviation and edge-shared triangles' deviation as follows:

$$Var(N') = \frac{M \times (p^3 - p^6) + 2k \times (p^5 - p^6)}{p^6} \qquad (6)$$

Here, we will derive the deviation of the third case. First, we assume that Δ is the number of two-one-triangles. With the partial-sampling algorithm, only one edge will be sampled. The variance of the third case is:

$$Var(N'') = \Delta \times p \times (1-p) \qquad (7)$$

Finally, we get the variance estimate:

$$Var(N) = \frac{M \times (p^3 - p^6) + 2k \times (p^5 - p^6)}{p^6} + \Delta \times p \times (1-p) \qquad (8)$$

Using Theorem 2, we can get the following theorem to evaluate the stability of the expected number of triangles.

Theorem 3.

$$Pr(|X - M| \leqslant \varepsilon) \geq 1 - \frac{M \times (p^3 - p^6) + 2k \times (p^5 - p^6)}{P^6 \times \varepsilon^2} \\ - \frac{\Delta \times p \times (1-p)}{\varepsilon^2} \quad (9)$$

Proof. By using the Chebyshev's inequality, we have:

$$Pr(|X - M| \leqslant \varepsilon) \geq 1 - \frac{Var(X)}{\varepsilon^2} \quad (10)$$

and by substituting the Formula 8, we can analyze the bound.

This theorem gives an evaluation about the performance of our partial-sampling. The accuracy of the approximate value is affected by the number of triangles in the graph, the structure of the graph and the value of p. The larger the number of triangles in the graph, the more the probability to obtain a good approximate value is. Also, the fewer edge-shared triangles which exist in the graph, the better the approximate value is.

3.3 Cut Pruning for Messages

In parallel environments, messages are used to confirm the existence of dis-triangles. And in *EdgeIterator* or *NodeIterator* [5], the content of messages generated by vertex u is $D(u)$. If messages are generated on *Node i* and sent to the same *Node*, we define them as $self-messages$, while others are called $normal-messages$. For *EdgeIterator* or *NodeIterator* [5], to confirm the existence of a dis-triangle $\langle u, v, w \rangle$, where $u \in P(i)$, $v \in P(j)$, $w \in P(k)$ and $i < j < k$, there are six kinds of $normal-messages$: Node i to Node j, Node i to Node k, Node j to Node k, Node j to Node i, Node k to Node i and Node k to Node j. In fact, only one message is necessary to confirm the existence of dis-triangles. Therefore, we design an optimization policy to reduce the message scale in SEN-Iterator: First, Node i only generates messages whose destination Node ID is larger than i. Second, for a message sent to vertex u, its content only includes neighbors whose ID is larger than that of u. Then, only one message will be sent from *Node i* to *Node j* to confirm whether triangle $\langle u, v, w \rangle$ exists.

We analyze the effect of the policy. Assume $D_l(v, u) = \{w \mid w \in V, w \in D(v), w > u\}$ and the length of one message is measured by the number of neighbors included in its content. For $v, z \in V, D(v) = \{y_1, y_2, y_3...y_{k-1}, z, y_{k+1}...y_n\}$, where $y_1 < y_2 < y_3 < ...y_{k-1} < z < y_{k+1}... < y_n$, $v \in P(i)$, $z \in P(j)$, $i < j$, Node i will send a message to Node j according to our policy. Then we can compute the total length of messages sent by Node i as:

$$rLen(v) = (n-k+1)+(n-k)+(n-k-1)+...+1 = \frac{1}{2} \times (n-k+2) \times (n-k+1) \quad (11)$$

The length of messages based on *EdgeIterator* or *NodeIterator* is computed as:

$$Len(v) = n + n + n + ... + n = n^2 \quad (12)$$

Let $f(n,k) = Len(v) - rLen(v)$, we get:

$$f(n,k) = \frac{1}{2} \times n^2 + n \times k - \frac{1}{2} \times k^2 - \frac{3}{2} \times n + \frac{3}{2} \times k - 1 \qquad (13)$$

Then we evaluate the derivative functions of $f(n,j)$:

$$\frac{\partial f}{\partial n} = n + k - \frac{3}{2} \qquad (14)$$

$$\frac{\partial f}{\partial k} = n - k + \frac{3}{2} \qquad (15)$$

By analyzing the Formula 14 and Formula 15, we can infer the following properties: (1) A larger n will enhance the effect of the policy, in other words, it will have better performance for dense-graphs. (2) A larger k will enhance the effect of the policy. It means each vertex has fewer neighbors. Considering the number of edges is fixed, we can infer that this policy is more suitable for the scenario where the number of every vertex's neighbors is nearly equivalent.

4 Experiment

We implement the SEN-Iterator algorithm on the bulk synchronous parallel model and compare the performance with NI-Hadoop. NI-Hadoop is implemented on Hadoop by using the similar idea proposed by [3]. All of the datasets we used are publicly available [12] and described in Table 2. Self-loops and the direction of edges are removed. Our cluster is composed of 21 nodes. Every node contains 2 hyperthreaded 2.00GHz CPUs, 8GB RAM and a Hitachi disk drive with 500GB capacity and 7,200 RPM. All nodes are connected by gigabit Ethernet to an Ethernet switch.

Table 2. Characteristics of data sets

Data Set	Vertices	Edges	Triangles	Data Set	Vertices	Edges	Triangles
Ast	18,772	396,160	1,351,441	Web	875,713	5,105,039	13,391,903
Soc	131,828	841,372	4,910,076	Am2	262,111	1,234,877	717,719
Hep	12,008	237,010	3,358,449	Am5	410,236	3,356,824	3,951,063

4.1 Performance Analysis and Scalability of SEN-Iterator

We evaluate the SEN-Iterator algorithm over a large amount of real graphs. As shown in Fig 1(a), the overall gain of SEN-Iterator is tremendous. Exemplified by Web, the speedup of SEN-Iterator compared to NI-Hadoop is a factor of up to 13. Fig 1(b) demonstrates the scalability of SEN-Iterator. For the data set Am2, when the number of Nodes increases from 12 to 20, the running time reduces from 15s to 10s.

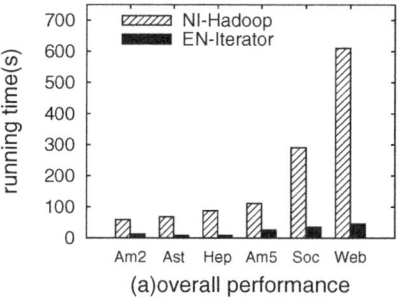 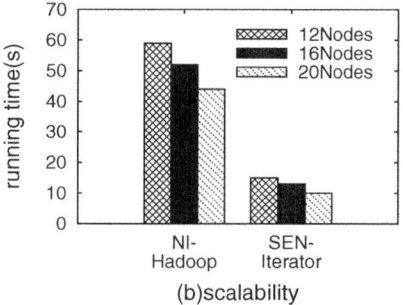

Fig. 1. Overall performance and scalability of SEN-Iterator

4.2 Analysis of Cut Pruning

The cut pruning policy improves the performance by reducing the number of messages. This suit of experiments is used to analyze the effect of the cut pruning policy by comparing SEN-Iterator with None-Iterator. The latter does not adopt the pruning policy. The message scale of None-Iterator is 2 times more than that of SEN-Iterator. For Web, SEN-Iterator only has 3845986 messages, while None-Iterator has 8644102 messages.

4.3 Accuracy Analysis for Partial-Sampling Algorithm

We run SEN-Iterator by five different values of p which ranges from 0.01 to 0.2. The examination is evaluated on real graphs. We define $Accuarcy = \frac{N}{M}$, where N is the calculated value of triangles and M is the exact value. Fig 2 shows the experimental results. We notice that the accuracy is always greater than 99%, when $p = 0.1$ or 0.15.

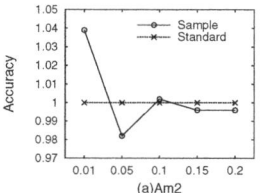 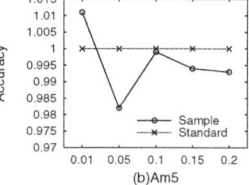 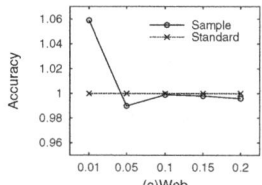

Fig. 2. The accuracy for different p in sampling

4.4 Performance Analysis of Partial-Sampling Algorithm

We analyze the gain of sampling by comparing SEN-Iterator and EN-Iterator without sampling. They are run on 12 Nodes with $p = 0.1$. We define the speedup as $Acc = \frac{R}{B}$, where R is the running time of EN-Iterator without cut pruning and B represents the running time of SEN-Iterator without cut pruning. The Acc is more than 200% for all data sets. What's more, the Acc of Web is 297%, which is the largest.

5 Conclusions

In this paper, we propose a new solution to efficiently solve the parallel triangle counting problem. A cut pruning policy is designed to optimize the overhead of communication, and we propose a partial-sampling method to avoid the repeated sampling in parallel environments. It can be embedded into our framework and improve the performance.

Acknowledgments. This research is supported by the National Natural Science Foundation of China (61272179, 61003058) and the Fundamental Research Funds for the Central Universities (N110404006, N100704001).

References

1. McPherson, M., Smith-Lovin, L., Cook, J.M.: Birds of a feather: Homophily in social networks. Annual Review of Sociology 27, 415–444 (2001)
2. Eckmann, J.-P., Moses, E.: Curvature of co-links uncovers hidden thematic layers in the World Wide Web. Proc. of the National Academy of Science, 5825–5829 (2002)
3. Suri, S., Vassilvitskii, S.: Counting triangles and the curse of the last reducer. In: Proc. of WWW, pp. 607–614 (2011)
4. Tsourakakis, C.E., Kang, U., Miller, G.L., et al.: DOULION: counting triangles in massive graphs with a coin. In: Proc. of KDD, pp. 837–846 (2009)
5. Alon, N., Yuster, R., Zwick, U.: Finding and counting given length cycles. Algorithmica 17(3), 209–223 (1997)
6. Schank, T., Wagner, D.: Finding, Counting and Listing All Triangles in Large Graphs, an Experimental Study. In: Nikoletseas, S.E. (ed.) WEA 2005. LNCS, vol. 3503, pp. 606–609. Springer, Heidelberg (2005)
7. Alon, N., Matias, Y., Szegedy, M.: The space complexity of approximating the frequency moments. In: Proc. of STOC, pp. 20–29 (1996)
8. Tsourakakis, C.E.: Counting triangles in real-world networks using projections. Knowl. Inf. Syst. 26(3), 501–520 (2011)
9. Giugno, R., Shasha, D.: Graphgrep: A fast and universal method for querying graphs. In: Proc. of ICPR, pp. 112–115 (2002)
10. Kang, U., Tong, H., Sun, J., et al.: Gbase: a scalable and general graph management system. In: Proc. of KDD, pp. 1091–1099 (2011)
11. Pagh, R., Tsourakakis, C.E.: Colorful triangle counting and a mapreduce implementation. Inf. Process. Lett. 112(7), 277–281 (2012)
12. SNAP, http://snap.stanford.edu/data/soc-LiveJournal1.html

Document Summarization via Self-Present Sentence Relevance Model

Xiaodong Li[1], Shanfeng Zhu[2,*], Haoran Xie[1], and Qing Li[1]

[1] Department of Computer Science, City University of Hong Kong, Hong Kong
xiaodonli2@student.cityu.edu.hk
[2] Shanghai Key Lab of Intelligent Information Processing and School of Computer Science, Fudan University, Shanghai 200433, China
zhusf@fudan.edu.cn

Abstract. Automatic document summarization is always attractive to computer science researchers. A novel approach is proposed to address this topic and mainly focuses on the summarization of plain documents. Conventional summarization methods do not fully use the inter-sentence relevance that is not preserved during the processing. In contrast, to tackle the problem and incorporate the latent relations among sentences, our approach constructs relevance structures at sentence-level for plain documents and each sentence is scored with a significance value. Accordingly, important sentences "present" themselves automatically, and the summary paragraph is then generated by selecting top-k scored sentences. Convergence of the algorithm is proved, and experiment, which is conducted on two data sets (DUC 2006 and DUC 2007), shows that the proposed model gives convincing results.

Keywords: Sentence relevance, Summarization.

1 Introduction

With the development of Internet, online textual information has been growing tremendously in recent years. Confronting with such a big volume of documents, people need a way of fast exploration and indexing. Despite summary quality, such as anaphoric references, grammar etc., computer-aided summarization has much faster speed. More and more computer-generated summaries have become beneficial for both users' exploration and search engines' indexing.

Computer-aided summarization approaches could be generally categorized as *abstract*-based and *extract*-based. *Abstract*-based approaches are more similar to rewriting of document(s) by human beings. *Extract*-based approaches, which mainly focus on the statistical information about terms' occurrences, extract sentences from the original document(s) by scoring the sentences. Approaches in this category could be originated from Luhn [9]. Vector Space Model (VSM), which is a commonly adopted model, has been playing an important role [16].

---

* Corresponding author.

However, while documents are divided into pieces of terms for analysis, sentence is no longer the smallest analyzing unit, and sentence relevance structure hidden in documents which presents latent relationship between sentences is not preserved during the processing. We think that the structure information is also critical to information processing and the relative importance order of sentences could be captured if the structure can be modeled and generated automatically.

Self-Present Sentence Relevance (SPSR) model proposed in this paper considers sentences as nodes and the relationship between sentences as the edges. Relevance values between sentences are calculated and taken as weights of the edges. Sentences' importance (significance) values are then scored interactively by measuring their contributions to the relevance values. Assumption here is that the more contributions one sentence makes, the more important the sentence is. Latent key sentences gain higher scores than the others and therefore "emphasize" themselves automatically.

Our approach lines up with the category of cluster-based and graph-based summarization. The main distinguishing points are: 1) SPSR does not need to determine the number of clusters. Cluster-based approaches implicitly or explicitly need to determine the number of clusters. Our method does not have an explicit cluster concept, where sentences are scored and summarization is generated from a sorted sentence list. The number of sentences in summary is determined by the requirement of summary length but not the clusters; 2) SPSR considers summarization's diversity in building algorithm and does not need to adopt Maximal Marginal Relevance (MMR) [1] for further processing, where MMR chooses sentence that is most relevant to document(s) but dissimilar to the sentences that have already been selected. Graph-based models, which takes sentences as individual entities and let them *mutual vote* each other, usually do not take into account summarization *diversity* during model building. They would employ MMR algorithm to refine sentence scores during sentence selection stage. In contrast, our approach uses *local-normalization* to make sentences' significance value be normalized in a local sub-graph, and local representative sentence will step out and compete with other local representatives for a position in the final summary.

The rest parts of this paper are organized as follows. A brief review of summarization approaches is provided in Section 2. In Section 3, SPSR model is formulated and the construction steps are illustrated by a running example. Details about experiment are reported in Section 4. In Section 5, conclusion is drawn.

2 Related Work

The main approaches of summarization could be visualized as a tree shown in Figure 1. As illustrated in Figure 1, summarization approaches could first be divided into two categories: *abstract*-based and *extract*-based. *Extract*-based approaches can be further categorized as supervised and unsupervised approaches.

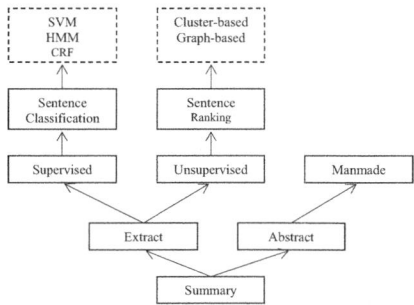

Fig. 1. Summarization approach categories

Supervised approaches take the summarization as a binary classification problem in which sentences are either classified as *in-summary* or *not-in-summary*. A fundamental assumption of these approaches is that sentences are independent from each other, however, this assumption ignores the relationship (or relevance) between sentences. Hidden Markov Model (HMM) [2] and Conditional Random Field (CRF) [3] based approaches are then proposed to relax that assumption.

Cluster-based approaches use clustering techniques to cluster sentences into several groups and select the most representative ones in each group to form the summary. Cluster-based approaches [22] usually have three steps: 1) partition sentences into n groups; 2) in each cluster, score sentences and sort them according to their scores; and 3) top scored sentences are selected as the summary. Radev et al. [15] think it better to leverage the cluster centroids. MEAD, a representative centroid-based approach, scores sentences based on the features of sentence, e.g. $tf \cdot idf$ value, cluster centroids, position, etc. Symmetric Non-negative Matrix Factorization (SNMF) is used in summarization by Wang et al. [19]. Latent Semantic Analysis (LSA), which helps add and analyze hidden topic features, is employed by Gong and Liu's work [5]. Graph-based model [11,4,17,18] is another category of unsupervised *extract*-based summarization approach. Wan and Yang [18] (ClusterHITS) improve Markov Random Walk model by integrating cluster information. Instead of using the score of centroids, Erkan and Radev (LexPageRank) [4] use PageRank-like method to let sentences "vote" and "recommend" each other. Yin et al. [21] consider the sentence ranking in a query-extraction scenario, where queries features provide extra information and improve the summarization quality when well incorporated.

The evaluation of summary is another important issue. Jones [7] start his survey paper by discussing many evaluation aspects. Also, Mani et al. [12] discuss this issue and thinks that the quality of the summary could be evaluated *intrinsic* and *extrinsic*. The main properties that people concerned about summary could be listed as follows:

- **Readability.** The sentences in the summary should be well-formed, well-connected and grammar-correct.
- **Relevance.** A good summary should contain sentences that are most relevant to the main concepts and topics of the original documents.

- **Diversity.** The sentences in the summary should be non-redundant.
- **Coverage.** A good summary should cover as many concepts and topics of the original documents as possible.
- **Distortion.** This is borrowed from information theory and first applied to summary evaluation by Ma and Wan [10]. They consider the summarization process as a data transmission system and think that minimizing the distortion should achieve better results.

ROUGE (Recall-Oriented Understudy for Gisting Evaluation) package [8], which is adopted as *"measures that count the number of overlapping units such as n-gram, word sequences, and word pairs between the computer-generated summary to be evaluated and the ideal summaries created by humans"*, is widely used together with DUC data sets nowadays. In our experiment, we also use ROUGE to evaluate summarization quality on DUC 2006 and DUC 2007 data sets.

3 Motivation and Formulation

The main purpose of SPSR model is to construct the latent relevance structure among sentences, and let key sentences gain higher scores for further summarization. Given n sentences, we treat each sentence as a node with the same significance value (for example, $1/n$) at beginning. Relevance between each pair of nodes is calculated as the weight of edges. We then divide each edge's weight value into two parts by measuring the contributions made by two nodes that the edge connects. Formulae of relevance and contribution measurement are defined as *abstract function* in our model, where concrete version could be user defined for specific cases. Based on the rule of "more contribution more gain", bigger parts of weight will be added to the significance value of sentence that contributes more to the relevance weight. After iterations, the significance values of important sentences will become higher than the others.

3.1 Model Formulation

Self-Present Sentence Relevance (SPSR) model is defined as a tuple of six elements $\langle T, V, F, E, W, S \rangle$:

- $T=\{t_i | t_i$ is a term in the corpus$\}$. T is the term space which contains all the terms (words) in the corpus without duplicates. Due to the existence of synonyms, documents may be written with different words but similar meanings. Therefore, comparing terms bit-by-bit could not fully represent the information of sentences. WordNet, built by Princeton, could measure the semantic distance between two words. Semantically close words are grouped together in one synonym set. Each set is indexed with a unique code. Without loss of generality, we use t_i to represent the set of synonyms.
- $V=\{v_j | v_j=\{t_k\}, t_k \in T\}$. Each sentence in the document is represented by a node v_j, which is a set of constituting terms in the sentence.

- $F = \{f_{ij} | f_{ij}$ is the number of occurrences of t_i in v_j, $t_i \in T$, $v_j \in V\}$. $\|v_j\|$ is defined as normalized sentence length, i.e. $\|v_j\| = \sqrt{\sum_{t_i \in v_j} f_{ij}^2}$.
- $E = \{e_{ij} | e_{ij}$ is an edge between nodes v_i and v_j, $v_i \in V$, $v_j \in V\}$. As it is an undirected graph, $e_{ij} = e_{ji}$.
- $W = \{w_{ij} | w_{ij}$ is the relevance weight of edge e_{ij}, $e_{ij} \in E\}$. Relevance weight, in another word *inverse distance*, could be measured in different ways, such as Manhattan distance and Euclidean distance etc. In our modeling, the measurement is declared as an *abstract function* which could be concretely defined by users. As commonly used in Information Retrieval, sentence relevance w_{ij} in vector space could be measured by cosine similarity, then

$$w_{ij} = \frac{v_i \cdot v_j}{\|v_i\| \cdot \|v_j\|} = \frac{\sum_{t_k \in v_i \cap v_j} f_{ki} \times f_{kj}}{\sqrt{\sum_{t_p \in v_i} f_{pi}^2} \cdot \sqrt{\sum_{t_q \in v_j} f_{qj}^2}} \quad (1)$$

Formula (1) defines one kind of similarity between v_i and v_j, where the maximum value is one, or zero if there is no common term between two nodes. As it is an undirected graph, $w_{ij} = w_{ji}$.
- $S = \{s_i | s_i$ is the significance value of node v_i, $v_i \in V\}$. Significance value is the sum of how much contribution the node v_i contributes to the weight of all the edges connected to it. The more weight v_i contributes, the higher value s_i will have, and therefore, sentence i is considered to be more important. As illustrated in Figure 2, significance value for node v_i is generated iteratively by: 1) splitting the weight of the edge e_{ij} which is connecting nodes v_i and v_j; 2) summing up all the fractions of weights gained from the edges connecting to v_i.

Between each pair of nodes, contribution c_i of node v_i is also declared as an *abstract function*. User could replace this function with what function they think is reasonable.

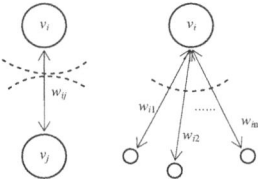

Fig. 2. Sentence v_i's significance: 1) split the weight w_{ij} and 2) sum up all the fractions of weights of w_{i*}.

The formula of calculating s_i iteratively is defined in Formula (2)[1].

$$s_i^{n+1} = s_i^n + c_i \quad (2)$$

---

[1] s_i^{n+1} is not normalized in Equation.

where n indicates the number of iteration and s_i^0 is initialized with $1/|V|$, which means all sentences are equally weighted at the beginning and have the same significance value. c_i is the contribution value of node v_i to all the edges connecting to it.

Equation 3 defines a trivial example of how we split edges' weights and generate sentences' contributions value.

$$c_i = \sum_{i \neq j} \left(\frac{\sum_{t_k \in v_i \cap v_j} f_{ki}}{\sum_{t_k \in v_i \cap v_j} (f_{ki} + f_{kj})} \cdot w_{ij} \right) \qquad (3)$$

If sentence v_i and v_j have common term t_k, which means that w_{ij} is not zero, we will divide w_{ij} and let $\frac{f_{ki}}{f_{ki}+f_{kj}} \cdot w_{ij}$ to v_i. The greater f_{ki} is, which means t_k appears more often in v_i, the bigger part of w_{ij} that v_i will get.

We also define $\sigma(s_i^n)$ in Equation (4), which is to normalize the significance value of v_i in each iteration. $\mathbf{1}\{w_{ij} \neq 0\}$ is used to ensure that the normalization is *localized*, which means s_i is normalized only by nodes that have non-zero-weight edge connected to it.

$$s_i^{n+1} = \sigma(s_i^n) = \frac{s_i^n + c_i}{s_i^n + c_i + \sum_{i \neq j}(s_j^n + c_j) \cdot \mathbf{1}\{w_{ij} \neq 0\}} \qquad (4)$$

Also, the normalization could be done *globally*, which is to remove the "connectivity" requirement in Equation (5). Thus, the formula for global normalization becomes:

$$s_i^{n+1} = \sigma(s_i^n) = \frac{s_i^n + c_i}{\sum_k (s_k^n + c_k)} \qquad (5)$$

where k is the index of all the nodes.

3.2 Algorithm

The detailed algorithm for model building is presented in Algorithm (1). The input of algorithm is the collection of document(s) which have been preprocessed (i.e. tokenized, stop word filtering, stemmed, synonyms grouped). The output is a vector that contains the significance values of each sentences accordingly.

The theoretical convergence of this algorithm is proved in Section 3.3, and iteration number N is determined specifically to different corpus. In our experiment, N is set to 50 which ensures that $|s_i^{n+1} - s_i^n| < 0.00001$.

After generating sentences' significance values, program could derive the final summary by selecting the top-k highest scored sentences, where k is the floor of the length requirement of summary paragraph divided by the average length of selected sentences. From this point of view, SPSR is more convenient than clustering based algorithm.

Algorithm 1. Algorithm of building SPSR model
1: **for all** f_{ij}, s_i, w_{ij} **do**
2:     $f_{ij} \leftarrow 0$
3:     $s_i^0 \leftarrow 1/|V|$
4:     $w_{ij} \leftarrow 0$
5: **end for**
6: **for all** v_j **do**
7:     **if** $t_i \in v_j$ **then**
8:         $f_{ij} \leftarrow f_{ij} + 1$
9:     **end if**
10: **end for**
11: **for all** e_{ij} **do**
12:     $w_{ij} \leftarrow relevance(v_i, v_j)$
13: **end for**
14: **for** $n = 1$ to N **do**
15:     $c_i \leftarrow contribution(v_i)$
16:     $s_i^{n+1} \leftarrow \sigma(s_i^n)$
17: **end for**

- Some clustering or centroid based approaches need people to decide the number of clusters (although sometimes heuristic helps decide cluster number, it is a still time-consuming work); SPSR does not need to know the cluster hierarchy of data set. Model quantifies the importance of the sentences and let higher ranked sentences "float to the front".
- After generating clusters, clustering based approach still need to decide the importance of sentences within each cluster. SPSR quantitatively measures sentence importance and order their significance scores in one shot.

3.3 Convergence

The meaning of convergence of SPSR building algorithm has two folds here: 1) each sentence's significance value converges; and 2) the order of all sentences' significance values converge.

Lemma 1. $\forall v_i \in V$, s_i^n converges.

Proof. Without loss of generality, assume v_i is connected with m other nodes. For the sake of simple notation, denote the m nodes as $v_1, v_2, \ldots, v_m$ and $i \notin [1, m]$.

To prove the converge of s_i^n is the same as to prove 1) s_i^n monotonically either increases or decreases and 2) s_i^n is upper or lower bounded, respectively.

$$s_i^{n+1} - s_i^n \\ = \frac{s_i^n + c_i}{s_i^n + c_i + \sum_j (s_j^n + c_j)} - s_i^n \\ = \frac{s_i^n + c_i}{\left(s_i^n + \sum_j s_j^n\right) + \left(c_i + \sum_j c_j\right)} - s_i^n \qquad (6)$$

Since s^n is normalized and summed to one, $s_i^n + \sum_j s_j^n = 1$. $\forall i$, c_i is constant and does not change along the iteration. Thus, $c_i + \sum_j c_j$ is a constant too, and equals the sum of weights of all edges, denoted as C. Therefore,

$$\begin{aligned} s_i^{n+1} - s_i^n &= \frac{s_i^n + c_i}{1+C} - s_i^n \\ &= \frac{s_i^n + c_i - s_i^n - C \cdot s_i^n}{1+C} \\ &= \frac{\frac{c_i}{C} - s_i^n}{1 + \frac{1}{C}} \end{aligned} \qquad (7)$$

Since s^0 is the same for all the nodes, for specific node v_i, the comparison between $\frac{c_i}{C}$ and s_i^n is only related to the comparison between $\frac{c_i}{C}$ and s_i^0. When node *gains* more from the weight than the average gain of the other nodes, its significance value will monotonically increase until $\frac{c_i}{C} = s_i^n$, and vice versa. From Equation (7), we could summarize the scenario as:

- if $\frac{c_i}{C} - s_i^0 > 0$, s_i^n is a sequence that monotonically increases until $\frac{c_i}{C} - s_i^n = 0$.
- if $\frac{c_i}{C} - s_i^0 < 0$, s_i^n is a sequence that monotonically decreases until $\frac{c_i}{C} - s_i^n = 0$.
- if $\frac{c_i}{C} = s_i^0$, s_i^n is a constant sequence that reaches equilibrium.

Also since

$$s_i^{n+1} = \sigma(s_i^n) = \frac{s_i^n + c_i}{1+C} \qquad (8)$$

where $\frac{s_i^n + c_i}{1+C}$ is obviously upper bounded by $\frac{1+c_i}{1+C}$ (for monotonically increase case), and lower bounded by $\frac{c_i}{1+C}$ (for monotonically decrease case), thus $\forall v_i \in V$, s_i^n converges.

Lemma 2. *The relative order $\langle s_1', s_2', \ldots, s_{m+1}' \rangle$ converges.*

Proof. To prove that the convergence of $\langle s_1', s_2', \ldots, s_{m+1}' \rangle$ is based on the convergence of each s_i. Since each s_i converges to s_i^* along with iteration, the final order which is sorted by sentences' significance will converge.

3.4 Example

1. [v_1]: Watching the new movie, "Imagine: John Lennon," was very painful for the late Beatles wife, Yoko Ono.
2. [v_2]: "The only reason why I did watch it to the end is because I'm responsible for it, even though somebody else made it," she said.
3. [v_3]: Cassettes, film footage and other elements of the acclaimed movie were collected by Ono.
4. [v_4]: She also took cassettes of interviews by Lennon, which were edited in such a way that he narrates the picture.
5. [v_5]: Andrew Solt ("This Is Elvis") directed, Solt and David L. Wolper produced and Solt and Sam Egan wrote it.
6. [v_6]: "I think this is really the definitive documentary of John Lennon's life," Ono said in an interview.

Paragraph in the above text frame is a running example used by Mihalcea and Tarau [13]. We use the same example to illustrate the construction of our SPSR model. After removing the stop words, we measure the distance of words in WordNet using the Path database and group synonyms (distance value is less than 3 which is empirically set) into the same set. Then, we assign each set with a unique ID to index that set of synonyms. For example, synonym set 0 of the sample paragraph contains words { "think", "reason", "why"} which are related to "reasoning", and set 4 contains {"ono", "andrew", "david", "lennon"} which are all human names.

Each sentence is translated into a vector which is the term frequency of terms in that sentence. We use Equation (1) to calculate the similarity between each pair of sentences to generate the weights of edges, which is

$$W = \begin{bmatrix} 0.00 & 0.15 & 0.33 & 0.30 & 0.34 & 0.58 \\ 0.15 & 0.00 & 0.00 & 0.27 & 0.10 & 0.22 \\ 0.33 & 0.00 & 0.00 & 0.20 & 0.14 & 0.19 \\ 0.30 & 0.27 & 0.20 & 0.00 & 0.13 & 0.35 \\ 0.34 & 0.10 & 0.14 & 0.13 & 0.00 & 0.25 \\ 0.58 & 0.22 & 0.19 & 0.35 & 0.25 & 0.00 \end{bmatrix}$$

Equation (3) is used to calculate the contribution of each sentence to the weights of connecting edges, which is

$$C = \begin{bmatrix} 0.00 & 0.17 & 0.71 & 0.71 & 0.67 & 0.60 \\ 0.83 & 0.00 & 0.00 & 0.75 & 0.75 & 0.83 \\ 0.29 & 0.00 & 0.00 & 0.50 & 0.33 & 0.33 \\ 0.29 & 0.25 & 0.50 & 0.00 & 0.33 & 0.43 \\ 0.33 & 0.25 & 0.67 & 0.67 & 0.00 & 0.50 \\ 0.40 & 0.17 & 0.67 & 0.57 & 0.50 & 0.00 \end{bmatrix}$$

Each node's significance value after one iteration can be then calculated using Equation (2). After normalization, the significance values are $s_1^1 = 0.2640$, $s_2^1 = 0.1908$, $s_3^1 = 0.1324$, $s_4^1 = 0.1329$, $s_5^1 = 0.1317$ and $s_6^1 = 0.1916$. We could illustrate changes of sentence significance in Figure 3. On the upper side is the initial status, all the nodes are treated with equal importance. While on the bottom part, paragraph structure is constructed.

It could be seen that initially all the nodes are with the same weights, and after one iteration nodes v_1 and v_6 gain higher scores than the other nodes. Comparing to the results of Mihalcea and Tarau (which is $s_1 = 1.34$, $s_2 = 0.70$, $s_3 = 0.74$, $s_4 = 0.52$, $s_5 = 0.91$ and $s_6 = 1.75$), both of us identify the important sentences 1 and 6 in the paragraph (the ranking results are listed in Table 1). However, the order of 1 and 6 are different: SPSR model gives sentence 1 the top score while benchmark approach tops sentence 6. It is hard to say which ordering is **correct**, but from position-wise assessment, the leading sentence of this news-like article would be more important.

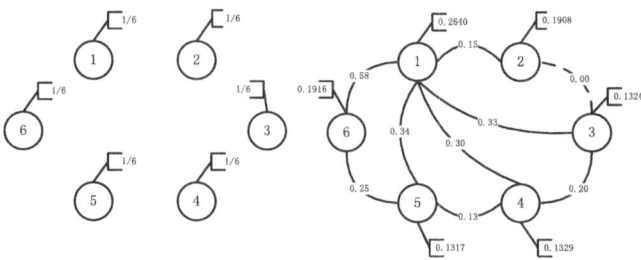

Fig. 3. Paragraph structure: 1) upper one is the initial status; 2) bottom one is the paragraph structure after one iteration. Circles in gray color indicate the higher ranked sentences. Not all the edges are drawn and dotted link means that the weight of edge is zero.

Table 1. The scoring result of benchmark approach and SPSR

Rank	Benchmark	SPSR
1	s_6=**1.7500**	s_1=**0.2640**
2	s_1=**1.3400**	s_6=**0.1916**
3	s_5=0.9100	s_2=0.1908
4	s_2=0.7000	s_5=0.1317
5	s_3=0.7400	s_3=0.1324
6	s_4=0.5200	s_4=0.1329

4 Experiment and Findings

4.1 Data Sets

Experiments are conducted on two data sets - DUC 2006 and DUC 2007 data sets[2], which are commonly used in text summarization evaluation. DUC 2006 has 50 topics and each topic has 25 documents. Ten NIST assessors write summaries for the 50 DUC 2006 topics. Each topic has 4 human summaries. Similar to DUC 2006, DUC 2007 has 45 topics and each topic has 25 documents. Ten NIST assessors wrote summaries for the 45 topics in the DUC 2007 main task. Each topic has 4 human summaries.

He *et al.* [6] implemented two basic benchmarks, which are

- **Random**: selects sentences randomly for each document set.
- **Lead** *(Wasson 1998): for each document set, orders the documents chronologically and takes the leading sentences one by one. [20]*.

and also three other state-of-art summarization algorithms, which are

- **LSA** *(Gong and Liu 2001): applies the singular value decomposition (SVD) on the terms by sentences matrix to select the highest ranked sentences. [5]*

---
[2] http://duc.nist.gov/

- **ClusterHITS** (Wan and Yang 2008): considers the topic clusters as hubs and the sentences as authorities, then ranks the sentences with the authorities scores. Finally, the highest ranked sentences are chosen to constitute the summary. [18]
- **SNMF** (Wang et al. 2008): uses symmetric non-negative matrix factorization(SNMF) to cluster sentences into groups and select sentences from each group for summarization. [19]

They also report algorithms' results on DUC 2006 and DUC 2007 in their AAAI'12 paper. We compare our model with those benchmark algorithms, and thus we have in total five benchmark models.

4.2 Evaluation Metric

For DUC 2006 and DUC 2007, ROUGE (Recall-Oriented Understudy for Gisting Evaluation) [8] is adopted in our experiment for evaluation. ROUGE-N (n-gram co-occurrence statistics) is defined as

$$\text{ROUGE-N} = \frac{\sum_{S \in RS} \sum_{gram_n \in S} Count_{match}(gram_n)}{\sum_{S \in RS} \sum_{gram_n \in S} Count(gram_n)} \quad (9)$$

where RS is the Reference Summaries. For multi-document summarization,

$$\text{ROUGE-N}_{multi} = \operatorname{argmax}_i \text{ROUGE-N}(r_i, s) \quad (10)$$

Stop words are filtered out and other words are stemmed using Porter's rules [14]. SPSR model is built on each topic, and summaries are then constructed within 250 words. Average F_1-measure results are calculated for ROUGE-1, ROUGE-2, ROUGE-3, ROUGE-4, ROUGE-L and ROUGE-W for both the benchmarks and SPSR model, where ROUGE-L is the Longest Common Subsequence and ROUGE-W is Weighted Longest Common Subsequence. We employ WordNet 2.1 in our programme, and as the aforementioned example, we use Path database and empirically set synonym threshold (distance upper-bound) as 3.

4.3 SPSR v.s. Baselines

The F_1 results over topics of DUC 2006 and DUC 2007 are listed in Table 2 and Table 3 respectively. The best score for one metric is marked in bold font, and the second best score for one metric is underlined. ROUGE-4 and ROUGE-W scores are not available for benchmarks, which are denoted with "N/A" in the tables.

From the results, we can see that SPSR (both local-normalized and global-normalized, denoted as L-SPSR and G-SPSR respectively) have better performance than benchmarks on F_1-measure, where in DUC 2006 SPSRs achieve six best scores and six second best scores and in DUC 2007 SPSRs achieve five best scores and five second best scores. Despite the average F_1-measure scores, we also follow the method of He *et al.* which plots the performance comparison of

Table 2. DUC 2006 results. SPSR v.s. Benchmarks

Algorithm	ROUGE-1	ROUGE-2	ROUGE-3	ROUGE-4	ROUGE-L	ROUGE-W
Random	0.28507	0.04291	0.01023	N/A	0.25926	N/A
Lead	0.27449	0.04721	0.01181	N/A	0.23225	N/A
LSA	0.25782	0.03707	0.00867	N/A	0.23264	N/A
ClusterHITS	0.28752	0.05167	0.01282	N/A	0.25715	N/A
SNMF	0.25453	0.03815	0.00815	N/A	0.22530	N/A
L-SPSR	**0.35383**	**0.05922**	0.01555	**0.00746**	**0.31354**	**0.11920**
G-SPSR	0.35376	0.05906	**0.01557**	0.00745	0.31307	0.11917

Table 3. DUC 2007 results. SPSR v.s. Benchmarks

Algorithm	ROUGE-1	ROUGE-2	ROUGE-3	ROUGE-4	ROUGE-L	ROUGE-W
Random	0.32028	0.05432	0.01310	N/A	0.29127	N/A
Lead	0.31446	0.06151	0.01830	N/A	0.26575	N/A
LSA	0.25947	0.03641	0.00854	N/A	0.22751	N/A
ClusterHITS	0.32873	0.06625	**0.01927**	N/A	0.29578	N/A
SNMF	0.28651	0.04232	0.00890	N/A	0.25502	N/A
L-SPSR	**0.37070**	**0.06716**	0.01844	**0.00845**	**0.32704**	**0.12348**
G-SPSR	0.36595	0.06541	0.01811	**0.00845**	0.32286	0.12202

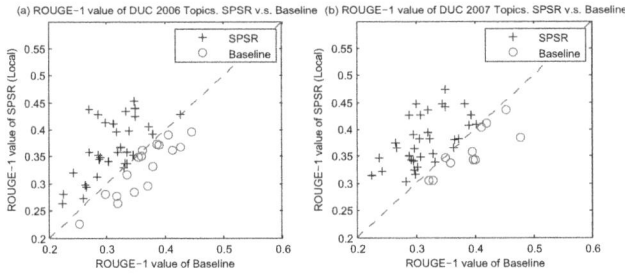

Fig. 4. Topic-to-topic ROUGE-1 value comparison of DUC 2006 and DUC 2007. The black '+' denotes local-normalized SPSR has a higher score while the red 'o' denotes otherwise.

each individual topic. In Figure 4 and Figure 5, we illustrate the topic-to-topic ROUGE-1 score comparison for L-SPSR and G-SPSR respectively. In penal (a), topics from DUC 2006 are depicted, and topics from DUC 2007 are depicted in penal (b). If SPSR outperforms baseline on that topic in terms of ROUGE-1, the point is above the blue dotted line and denoted with black '+'; on the other hand, point is below blue dotted line and denoted with red 'o'. It could be observed that '+'s are much more than 'o's, which means on topic-to-topic level, SPSR also outperforms baseline in majority of topics.

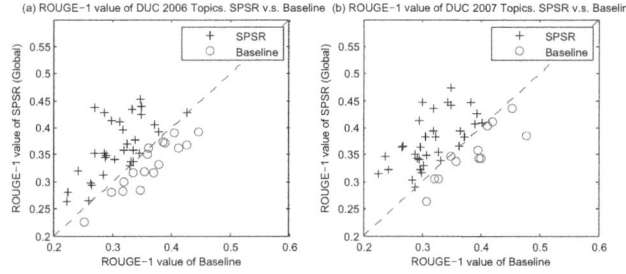

Fig. 5. Topic-to-topic ROUGE-1 value comparison of DUC 2006 and DUC 2007. The black '+' denotes global-normalized SPSR has a higher score while the red 'o' denotes otherwise.

5 Conclusion and Future Work

Summarization over textual documents, although it has a long history, still remains one of the hardest and yet not well solved problem in text mining area. In this paper, Sentence Self-Present Relevance model was proposed to address traditional document summarization problem by mining latent document implicit relevance structure. Different from reviewed approaches, SPSR constructs sentence-level structure on plain documents and iteratively generates the significance value for each sentence. SPSR building algorithm is proved to converge, and this approach tries to keep the generated summary as diverse as possible. Experiments are conducted on two data sets (DUC 2006 and DUC 2007 data sets). Comparing with other five benchmarks, SPSR gives convincing results. Different from G-SPSR, L-SPSR, which normalizes sentence's significance value in a local context, is proposed and analyzed. In the comparison between L-SPSR and G-SPSR, L-SPSR has better performance than G-SPSR.

For future exploration on this direction, we think of the following two points:

- Sentence grammar structure analysis. Grammar structure analysis could provide sentence microstructure information. While constructing SPSR, one key step is to setup and determine the relevance between sentences. In the proposed algorithm, it still considers words within one sentence as a "bag of words", and grammar structure within one sentence is broken. Bringing in sentence grammar structure may give SPSR a more accurate view of each sentence.
- Different concrete versions of similarity and split functions in the model. In our experiment, traditional cosine similarity and frequency based split function is used to generate the weights of edges and contribution values. For further exploration, different functions could be employed and analyzed.

Acknowledgement. This work is supported by the National Science Foundation of China (Grant Nos. 61173011 and 60903076).

References

1. Carbonell, J., Goldstein, J.: The use of MMR, diversity-based reranking for reordering documents and producing summaries. In: Proceedings of the 21st Annual International ACM SIGIR Conference on Research and Development in Information Retrieval, pp. 335–336. ACM (1998)
2. Conroy, J.M., O'leary, D.P.: Text summarization via hidden markov models. In: Proceedings of the 24th Annual International ACM SIGIR Conference on Research and Development in Information Retrieval, pp. 406–407. ACM (2001)
3. Dou, S., Sun, J.-T., Li, H., Yang, Q., Chen, Z.: Document summarization using conditional random fields. In: Proceedings of IJCAI, vol. 7, pp. 2862–2867 (2007)
4. Erkan, G., Radev, D.R.: LexRank: Graph-based lexical centrality as salience in text summarization. Journal of Artificial Intelligence Research 22(1), 457–479 (2004)
5. Gong, Y., Liu, X.: Generic text summarization using relevance measure and latent semantic analysis. In: Proceedings of the 24th Annual International ACM SIGIR Conference on Research and Development in Information Retrieval, pp. 19–25. ACM (2001)
6. He, Z., Chen, C., Bu, J., Wang, C., Zhang, L., Cai, D., He, X.: Document summarization based on data reconstruction. In: Twenty-Sixth AAAI Conference on Artificial Intelligence (2012)
7. Jones, K.S.: Automatic summarising: The state of the art. Information Processing & Management 43(6), 1449–1481 (2007)
8. Lin, C.-Y., Hovy, E.: Automatic evaluation of summaries using n-gram co-occurrence statistics. In: Proceedings of the 2003 Conference of the North American Chapter of the Association for Computational Linguistics on Human Language Technology, vol. 1, pp. 71–78. Association for Computational Linguistics (2003)
9. Luhn, H.P.: The automatic creation of literature abstracts. IBM Journal of Research and Development 2(2), 159–165 (1958)
10. Ma, T., Wan, X.: Multi-document Summarization Using Minimum Distortion. In: 2010 IEEE 10th International Conference on Data Mining, pp. 354–363. IEEE (2010)
11. Mani, I., Bloedorn, E.: Multi-document summarization by graph search and matching. In: AAAI 1997 (1997)
12. Mani, I., Klein, G., House, D., Hirschman, L., Firmin, T., Sundheim, B.: SUMMAC: a text summarization evaluation. Natural Language Engineering 8(01), 43–68 (2002)
13. Mihalcea, R., Tarau, P.: A language independent algorithm for single and multiple document summarization. In: Proceedings of IJCNLP, vol. 5 (2005)
14. Porter, M.F.: An algorithm for suffix stripping. Program: Electronic Library and Information Systems 14(3), 130–137 (1993)
15. Radev, D.R., Jing, H., Stys, M., Tam, D.: Centroid-based summarization of multiple documents. Information Processing & Management 40(6), 919–938 (2004)
16. Salton, G., McGill, M.J.: Introduction to modern information retrieval, vol. 1. McGraw-Hill (1983)
17. Wan, X., Yang, J.: Improved affinity graph based multi-document summarization. In: Proceedings of the Human Language Technology Conference of the NAACL, Companion Volume: Short Papers, pp. 181–184. Association for Computational Linguistics (2006)
18. Wan, X., Yang, J.: Multi-document summarization using cluster-based link analysis. In: Proceedings of the 31st Annual International ACM SIGIR Conference on Research and Development in Information Retrieval, pp. 299–306. ACM (2008)

19. Wang, D., Li, T., Zhu, S., Ding, C.: Multi-document summarization via sentence-level semantic analysis and symmetric matrix factorization. In: Proceedings of the 31st Annual International ACM SIGIR Conference on Research and Development in Information Retrieval, pp. 307–314. ACM (2008)
20. Wasson, M.: Using leading text for news summaries: Evaluation results and implications for commercial summarization applications. In: Proceedings of the 17th International Conference on Computational Linguistics, vol. 2, pp. 1364–1368. Association for Computational Linguistics (1998)
21. Yin, W., Pei, Y., Zhang, F., Huang, L.: Query-focused multi-document summarization based on query-sensitive feature space. In: Proceedings of the 21st ACM International Conference on Information and Knowledge Management, CIKM 2012, pp. 1652–1656. ACM, New York (2012)
22. Zha, H.: Generic summarization and keyphrase extraction using mutual reinforcement principle and sentence clustering. In: Proceedings of the 25th Annual International ACM SIGIR Conference on Research and Development in Information Retrieval, pp. 113–120. ACM (2002)

Active Semi-supervised Community Detection Algorithm with Label Propagation

Mingwei Leng, Yukai Yao, Jianjun Cheng, Weiming Lv, and Xiaoyun Chen*

School of Information Science and Engineering, Lanzhou University
Lanzhou, 730000, China
lengmw@163.com
chenxy@lzu.edu.cn

Abstract. Community detection is the fundamental problem in the analysis and understanding of complex networks, which has attracted a lot of attention in the last decade. Active learning aims to achieve high accuracy using as few labeled data as possible. However, so far as we know, active learning has not been applied to detect community to improve the performance of discovering community structure of complex networks. In this paper, we propose a community detection algorithm called active semi-supervised community detection algorithm with label propagation. Firstly, we transform a given complex network into a weighted network, select some informative nodes using the weighted shortest path method, and label those nodes for community detection. Secondly, we utilize the labeled nodes to expand the labeled nodes set by propagating the labels of the labeled nodes according to an adaptive threshold. Thirdly, we deal with the rest of unlabeled nodes. Finally, we demonstrate our community detection algorithm with three real networks and one synthetic network. Experimental results show that our active semi-supervised method achieves a better performance compared with some other community detection algorithms.

Keywords: Social Networks, community detection, active learning, semi-supervised learning, label propagation.

1 Introduction

A community in a network is a group of nodes that are similar to each other and dissimilar from the rest of the network[7]. It is also thought of as a group where nodes are densely inter-connected and sparsely connected to other parts of the network[1]. Community detection in complex networks is very important for us to understand the network structure and analyze the networks characters. Many community detection algorithms have been proposed, and they can be divided into four categories: divisive algorithm [1, 2], agglomerative algorithms [3, 4], optimisation algorithms [5, 6], and label propagation algorithms [7–12].

---
* Corresponding author.

Raghavan et al.[7] proposed a label propagation algorithm(LPA) for detecting network communities, which updates the labels of nodes by choosing the label that is the most frequent among its neighbors, and repeats the updating process till some terminate condition is reached. Several variations of LP algorithm have been proposed since 2007 [8–12]. Liu et al. updated all labels of the nodes simultaneously [8]. Barber and Liu et al modified the label updating rule so that modularity can be maximized [9, 10]. Xie et al. improved the computational efficiency by reducing the times of iteration and utilized the neighborhood strength to improve the quality of communities[11]. Šubelj updated the label of nodes by using two unique strategies of community formation, defensive preservation and offensive expansion of communities[12].

However, most of these label propagation algorithms are not stable, and sometimes the results of community detection are unsatisfied, especially for complex networks that the difference between community densities is large. Adding prior knowledge to the process of detecting community may be the most efficient method for improving the performance of community detection, and using the semi-supervised clustering method to guide the process of detecting should achieve better results. [13, 14] utilized the prior knowledge to improve the performances of the community detection algorithms. [13] proposed a $SNMF-based$ semi-supervised clustering algorithm for community detection based on pairwise constrains (cannot-link and must-link). [14] adapted the modularity method to the context of semi-supervised learning in the merging process based on the labeled nodes. Although [13, 14] introduced the semi-supervised method into the community detection, they did not mention how to get these prior knowledge.

Active learning technique aims to achieve high accuracy using as few labeled data as possible. It minimizes the cost of obtaining labeled data greatly without compromising the performance of community detection, and this is very attractive and valuable in real-world applications. Most of the existing active learning algorithms are pool-based [15, 16] or stream-based [17], and they are mainly applied in supervised learning. In recent years, active learning is introduced into clustering [18–24]. Different clustering algorithm exploits different active learning approaches. Nguyen et al. selected the most representative samples to avoid repeatedly labeling samples in the same cluster [18]. Vu et al. selected useful examples according to a Min-Max approach to determine the set of labeled data [19]. Zhao et al. selected informative document pairs for obtaining user feedback by using active learning approach, and incorporated instance-level constraints to guide the clustering process in DBSCAN [20]. Grira et al. defined an active mechanism for the selection of candidate constraints to minimize the amount of constraints required [21]. Wang et al. presented an active query strategy based on maximum expected error reduction and a constrained spectral clustering algorithm that can handle both hard and soft constraints [22]. Mallapragada et al. selected constraints through using a min-max criterion to improve the performance of semi-supervised clustering algorithms [23]. Huang et al. conducted a preliminary clustering process to estimate the true clustering assignments, and then chose informative document pairs by means of learning the intermediate cluster structure [24].

As far as we know, active learning has not been applied to community detection to improve the performance of the community detection algorithms, and we introduce active learning to community detection in this paper. Although most of the active learning algorithms select the node that is most uncertain to be labeled, the most uncertain node lies on the community boundary, and is not representative of other nodes in the same community. So knowing its label is unlikely to improve performance of the community detection as a whole. In this paper, we propose an active community detection algorithm, called active semi-supervised community detection with label propagation. Firstly, we calculate the density of each node and the weight of each edge based on the common neighbors of nodes, and find out all *core nodes*(the definition of *core nodes* is given in section 2). Secondly, we actively select a few *core nodes* based on the weighted shortest path method. Our algorithm tries to enable that the selected *core nodes* can cover as many communities as possible in a given complex network. These selected nodes are labeled by domain experts, and are to be viewed as the initial set of labeled nodes in the process of community detection. Thirdly, we expand the labeled nodes set by propagating label. The propagating process labels neighbors of the labeled nodes according to similarity threshold which is obtained automatically based on the characters of networks. Fourthly, the rest of unlabeled nodes according are assigned with the most frequent label among their neighbors. Our community detection algorithm has the following advantages.

- Our community detection algorithm translates a given unweighted network into a weighted network based on the similarities between nodes, then utilizes the weighted shortest path methods based on *core nodes* to find a few *core nodes* actively, and enables the selected *core nodes* can cover as many communities in a given complex network as possible.

- Our community detection algorithm finds out communities in complex networks by expanding the labeled nodes according to the similarities of nodes, and the expanding process gives priority to the nodes with maximum density.

The rest of the paper is organized as follows. Section 2 gives our community detection algorithm in detail. In section 3, we demonstrate our algorithm with standard network datasets, and compare it with some other community detection algorithm. We summarize our work in section 4.

2 Active Semi-supervised Community Detection with Label Propagation

In this section, we propose an active semi-supervised community detection algorithm with label propagation, which introduces active learning and semi-supervised learning into the label propagation algorithm for community detection. Degree of node is a very important character in networks, but it is not sufficient to measure the importance of a node only considering its degree. In this paper, we use density of node to measure the importance of it. Manual labeling nodes in complex networks is expensive, especially in large complex networks. Our method does not select nodes from the whole network, but from the important nodes set: *core nodes* set. We firstly delete nearly 25 percent of the nodes of lower density, and the rest of nodes is called *core nodes*. Some definitions is given below.

Definition 1 $density(i)$. Given one complex network G, the density of a node i is defined as following,

$$density(i) = \frac{\sum\limits_{j \in N(i)} n_{ij}}{2 * k_i} \quad (1)$$

where k_i is the degree of node i, and n_{ij} is defined as,

$$n_{ij} = |N(i) \cap N(j)| \quad (2)$$

$N(i)$ is the neighbors of node i, and let $density(G)$ denote the set of all densities of nodes in complex network G.

Definition 2 $25th\ percentile(S)$, given a set S, $25th\ percentile(S)$ is the value that there are 25 percent elements in S whose values are less than or equal to it, and there are 75 percent elements in S whose values are larger than it.

Definition 3 *core node*. Given one complex network G, i is a node in G, i is *core node* if and only if $density(i)$ is larger than or equal to $25th\ percentile$ $(density(G))$.

Definition 4 $sim(i,j)$. Given one complex network G, i and j are two nodes in G, the similarity between i and j is defined as,

$$sim(i,j) = \frac{n_{ij}}{(k_i + k_j)} \quad (3)$$

where the meaning of n_{ij} is the same as definition 1, k_i, k_j are the degree of node i and node j respectively.

Definition 5 $sim(i,S)$. Given one complex network G, i is a node in G, S is a nodes set and $S \subseteq G$, the similarity between i and S is defined as,

$$sim(i,S) = \frac{\sum\limits_{j \in N(i) \cap S} n_{ij}}{k_i} \quad (4)$$

where the meaning of n_{ij} is the same as definition 1, k_i is the degree of node i.

2.1 Active Nodes Selection

This subsection presents the idea of selecting nodes and and gives the details of algorithm for selecting nodes actively. Labeling nodes in complex network is very expensive, so in this paper, we want to select as few nodes as possible to achieve the best detecting results. We select nodes from the *core nodes* set using the weighted shortest path method. Selecting nodes from *core nodes* set is based on the following facts. Firstly, labeling a node in complex network is a difficult work, *core nodes* are more important ones in network, and they are the better representatives of communities, so *core nodes* are easy to be labeled compared with nodes lying in the boundary of the communities. Labeling *core nodes* can reduce efforts of domain experts, and we can get a higher quality of labeled nodes set. Secondly, *core nodes* can give more information, our semi-supervised community detection algorithm obtains a better community structure when giving small size of labeled *core nodes*. the shortest path method is used to selected *core nodes* can enables the selected nodes intersperse among as many communities as possible. The details of selecting *core nodes* are shown in algorithm 1.

Algorithm 1. SelectNodes(G,k)

1. let *NodesSet* denotes the nodes set in complex network G, and *NodesSet* = $\{1, 2, 3, \ldots, n\}$
2. calculate *density(G)*
3. calculate the *25th percentile(density(G))*
4. the nodes whose densities are not less than *25th percentile(density(G))* are viewed as *CoreNodes*
5. if there exists one edge e_{ij} between node i and node j, then the weight w_{ij} of e_{ij} is $(1 - sim(i,j))$.
6. $SelectedNodes = \phi$
7. $u, v \leftarrow arg \max\limits_{i,j \in CoreNodes} \{ShortestPathLength(i,j)\}$
8. find out the node u' with the max degree in nodes set $N(u) \cup \{u\}$, $SelectedNodes = SelectedNodes \bigcup \{u'\}$.
9. find out the node v' with the max degree in nodes set $N(v) \cup \{v\}$, $SelectedNodes = SelectedNodes \bigcup \{v'\}$.
10. $CoreNodes = CoreNodes \setminus \{u', v'\}$
11. **while** $|SelectedNodes| < k$
12. $\quad u \leftarrow arg \max\limits_{i \in CoreNodes} \min\limits_{j \in SelectedNodes} \{ShortestPathLength(i,j)\}$
13. $\quad$ find out the node u' with the max degree in nodes set $\{N(u) \cup \{u\}\} \setminus SelectedNodes$.
14. $\quad SelectedNodes = SelectedNodes \bigcup \{u'\}$.
15. $\quad CoreNodes = CoreNodes \setminus \{u'\}$
16. **end while**
17. **return** *SelectedNodes*

Algorithm 1 can be divided into two stages. The first stage of algorithm 1 is to find out *CoreNodes* based on the node density, and in the second stage, it selects k nodes from *CoreNodes*. Node density is proposed to measure the strongness of relation between its neighbors, the more edges between its neighbors, the larger density it has. In social networks, one people is presented as one node in the network, if two people are in contact with each other at least once, there exists one edge. If density of a node is large, its neighbors are in contact with each other frequently. In order to select more important nodes in complex network G, we select the nodes from *CoreNodes*. *core node* is proposed to represent the important nodes in complex networks.

The second stage is the core of algorithm 1, it starts at line 5 and ends at line 17. Since nodes labeling is time-consuming and costly, algorithm 1 aims to achieve better performance using as few labeled nodes as possible. In order to avoid selecting the nodes which lie in the boundary between communities, we select nodes from the *core nodes*. In order to measure the shortest path from one node to the other more effectively, we adopt the weighted shortest path. Since most of the complex networks are unweighted, we must translate a unweighted network into a weighted one. We assign a weight for each edge based on the dissimilarity between its two end nodes(line 5). Firstly, we select two nodes(u and v) with the max value of the shortest path, and then find out the node u' with max degree in the nodes set $N(u) \cup \{u\}$ as the first selected node, find out the node v' with max degree in the nodes set $N(v) \cup \{v\}$ as the second selected node. Both u' amd v' are added into *SelectedNodes*. Secondly, we select the node(u) from the *CoreNodes* which is most dissimilar with the selected nodes(line 12), then we find out the node u' with max degree in the nodes set $N(u) \cup \{u\}$, and add u' to *SelectedNodes*, we deploit the same method repeatly until the size of *SelectedNodes* is k . Finally, the k selected *core nodes* are viewed as the final result of algorithm 1.

2.2 Semi-supervised Community Detection with Label Propagation

In this subsection, the *core nodes* selected by algorithm 1 are labeled as the labeled *core nodes*. As the prior knowledge of the complex network, these labeled *core nodes* are used to detect community structure with label propagation. The details of community detection is depicted in algorithm 2.

Algorithm 2 can be divided into three stages: labeling the selected *core nodes* (lines 2 and 3), expanding the labeled *core nodes* with label propagation(lines from 4 to 13), and labeling the rest of the unlabeled nodes(lines from 14 to 20). The first stage actively obtains the labeled *core nodes*, and those labeled *core nodes* are viewed as the representatives of the initial communities of a complex network. If one or more communities have no node to be selected, the nodes in these communities will be assigned to other communities forcibly, and thus leads to worse performance of community detection. In this paper, we use the weighted shortest path method to select nodes from the *core nodes* set, and we try to enable the selected *core nodes* can cover as many communities in a given network as possible. The second stage propagates the label of the labeled node one by one based on a threshold which is determined by the characters of the network. If the similarity between a labeled node u and its neighbor is larger than or equal to the threshold, we assign the neighbor with the label of u, thus there would be some nodes which can not be assigned to any community when the second stage ends. The third stage deals with the rest of the unlabeled nodes according to the similarity between the unlabeled node and the communities. Suppose that

Algorithm 2. CommunityDetection(G)

1. let *NodesSet* denotes the nodes set in complex network G, and *NodesSet* = $\{1,2,3,\ldots,n\}$
2. *UnusedLabeledNodes=SelectNodes(G,k)*, and label *UnusedLabeledNodes* by domain experts.
3. suppose $u \in UnusedLabeledNodes$, let l_u denote the number of the community which u belongs to.
4. sort *UnusedSelectedNodes* in descending order according to their density.
5. $UsedLabeledNodes = \emptyset$
6. **while** *UnusedLabeledNodes* is not null
7.   take the first node from *UnusedLabeledNodes*, and let u denote this node.
8.   **for** v in $N(u)$
9.   **if** v is unlabeled and $n_{uv} >= (degree(v) - 1)/2$
10. $l_v \leftarrow l_u$, and insert v at the head of *UnusedLabeledNodes*.
11.   **end if**
12.   **end for**
13.   remove node u from *UnusedLabeledNodes*, and add it to *UsedLabeledNodes*.
14. **end while**
15. $UnlabeledNodes = NodesSet \backslash UsedLabeledNodes$.
16. **while** *UnlabeledNodes* is not null
17. $p \leftarrow \underset{v \in UnlabeledNodes}{\arg\max} \{\underset{l}{\max} \frac{\sum_{u \in N(v) \cap UsedLabeledNodes}[l_u==l]}{k_v}\}$.
18. $l_p \leftarrow \underset{l}{\arg\max} \frac{\sum_{u \in N(p) \cap UsedLabeledNodes}[l_u==l]}{k_p}$
19.   remove node p from *UnLabeledNodes*, and add it to *UsedLabeledNodes*.
20. **end while**
21. return *UsedLabeledNodes*

the number of the communities is *num*, we calculate $sim(v, C_i)(1 \leq i \leq num)$ for each unlabeled node v, where C_i is one of the existing communities, the node v is assigned to the community with the max value in $sim(v, C_i)(1 \leq i \leq num)$, and v and its label information will be added into *UsedLabeledNodes*. *UsedLabeledNodes* saves the community structure of the complex network G.

2.3 Time Complexity Analysis

Algorithm 1 needs $O(m+n)$ time in lines 1-6, where m is the total number of edges in network G. A shortest path using Dijkstra's algorithm needs $O(m + nlogn)$ time in the graph from a fixed node. So Line 7 needs $O(cm + cnlogn)$ time, where c is a constant means the number of *CoreNodes*. In order to avoid that each community has more than one core node, c is larger than $n/2$ in most case, so Line 7 needs $O(nm + n^2 logn)$ time. Lines 8-10 needs $O(\triangle)$ time, where $\triangle$ is the network average degree. Let k be the number of iterations and the number of nodes we want to select. It needs $O(kns(m + nlogn)/2 + k)$ time in lines 11-16, where s is the number of *SelectedNodes* which equals k in the worst case. Taken together with the complexity of Algorithm 1, the total worst-case complexity is $O(k^2nm) + O(k^2n^2logn)$. Algorithm 2 needs $O(k)$ time in lines 1-5, where k is the number of selected nodes. lines 6-14 need $O(k\triangle)$ time. Lines 15-21 needs $O(u\triangle)$ time, where u is the number of $UnlabeledNodes$. The total worst-case complexity of Algorithm 2 is $O(n)$. So the time complexity of our community detection method is $O(k^2nm) + O(k^2n^2logn)$ in total.

3 Experimental Results

In this section, in order to demonstrate our community detection algorithm with visual method better, we use three small networks and one larger artificial network to test it. Our proposed community detection algorithm is applied to several well-known networks, including three real-world networks and one Benchmark of Girvan and Newman [1]. In order to show the effectiveness of the our method, we compare our community detection algorithm with several existing community detection algorithms, EBC(hierarchical method)[2], LPAm(Label propagation method)[10] and semi-supervised learning method(SSLM) [14]. In the experiment, we suppose that the labeled nodes used in SSLM are obtained as follows [14]. In each complex network, we select one node as labeled node from each community randomly, and the rest of labeled nodes are selected from the whole network randomly. In this section, modularity and the number of nodes which are wrongly assigned to communities are use to demonstrate the performance of our method.

3.1 Evaluation with Number of Nodes Wrongly Assigned to Communities

In this subsection, we run EBC, LPAm, SSLM and our method on Karate network, Risk Map network, Collaboration network and one artificial network. Since the community detection results of EBC and LPAm do not change greatly, we give one running result for each of them. SSLM and our method are semi-supervised method, the results of them change with the increasing number of labeled nodes. So we do 10 experiments, each experiment use the same method to select different numbers of labeled nodes, and we run our method and SSML 10 times respectively in each experiment.

3.1.1 Zachary's Network of Karate Club Members

Zachary's network of karate club members [25] is a well-known graph, regularly used as a benchmark to test community detection algorithms (Section 15.1). It consists of 34 vertices, the members of a karate club in the United States, who were observed during a period of three years. Edges connect individuals who were observed to interact outside the activities of the club. At some point, a conflict between the club president and the instructor led to the fission of the club into two separate groups. Indeed, by looking at Fig.1, one can distinguish two aggregations, one around vertices 33 and 34 (34 is the president), the other

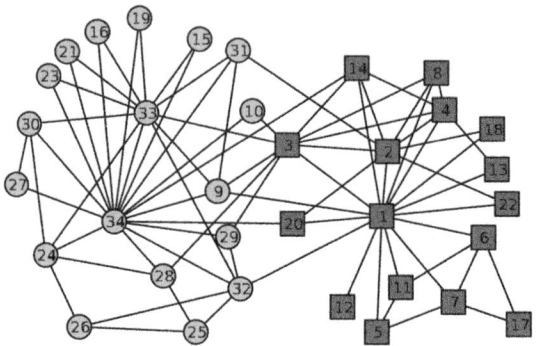

Fig. 1. Community structure in Karate Club network

around vertex 1 (the instructor). One can also identify several vertices lying between the two main structures, like 3, 9, 10; such vertices are often misclassified by community detection methods. Fig.1 shows community structure of the network.

EBC and LPAm wrongly assign the node '3' to a community, and the detected community results are the same in different running times. Since the method of selecting nodes is random, the experiments may be different in different running times on the same complex network. We adopt the intermediate result as the result of community detection for SSML on each experiment, and give the worst result on each experiment. Since we use the weighted shortest path to select the labeled nodes, the results of 10 times on each experiment are the same, and they are shown in the Fig.2.

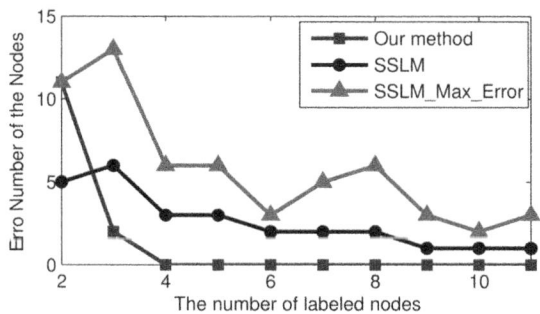

Fig. 2. Results of Our method and SSML on Karate Club network

Our method reaches a stable state when the number of labeled nodes is larger than 3, and it can rightly divide all the nodes into communities. Although all the nodes can be rightly assigned to communities in the best result of SSML in each experiment, the results of SSML is unstable. Even the numbers of nodes which

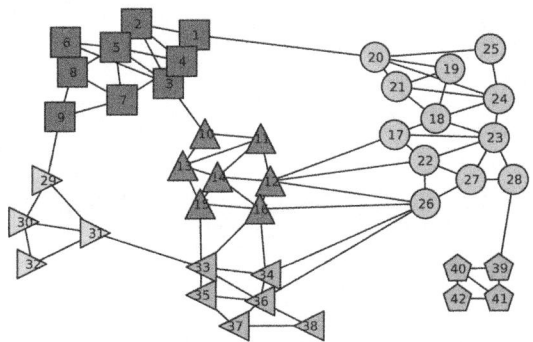

Fig. 3. Community structure in Risk Map network

are wrongly assigned to a community are the same, these nodes are different in different running times, and the performance is not improved with the increasing number of the labeled nodes. SSML_Max_Error in Fig.2 denotes the worst results of 10 running times on each experiment.

3.1.2 Risk Map Network

Risk Map Network was invented by French film director Albert Lamorisse, and was originally released in 1957 in France. Risk is a turn-based game for two to six players. The standard version is played on a board depicting a political map of the Earth, divided into forty-two territories, and these territories are grouped into six continents. The primary object of the game is "world domination," or "to occupy every territory on the board, and in so doing, to eliminate all other players." Players control armies with which they attempt to capture territories from other players, with results determined by dice rolls. The community structure is shown in Fig.3.

EBC has a better result on Risk Map network compared with LPAm, it wrongly assigns only the nodes '22', '26' to communities, but LPAm divides this network into 7 communities and wrongly assigns 14 nodes to communities. The experimental results of our method and SSML are shown in the Fig.4.

SSML can rightly assign each node to community in the best results when the number of labeled nodes is larger than 7. In this subsection, the intermediate result is viewed as the result of community detection for SSML on each experiment, and the worst results on the 10 experiments are denoted by SSML_Max_Error in Fig.4. Although the number of the nodes which are wrongly assigned to communities in the detected results of SSML is less than that of our method in some experiments, the worst results of SSML in 10 experiments are all worse than that of our method. At the same time, the community detection results of our method are all the same in 10 running times of each experiment, this shows that the experimental results of our method is more stable than that of SSML.

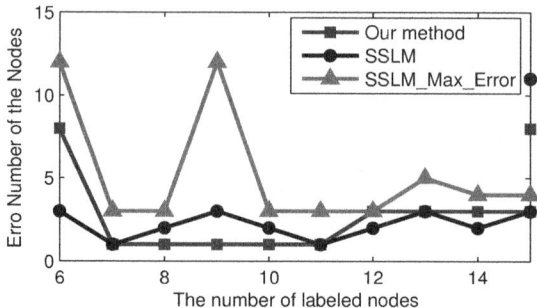

Fig. 4. Results of Our method and SSML on Risk Map Network

3.1.3 Collaboration Network

Collaboration network displays the largest connected component of a network depicting collaborations of scientists working at the Santa Fe Institute (SFI). There are 118 vertices, representing resident scientists at SFI and their collaborators. Edges associate with scientists that have published at least one paper together. The visualization layout allows to distinguish disciplinary groups. In this network one observes many cliques, as authors of the same paper are all linked to each other. There are but a few connections between most groups, collaboration network can be divided into 4 communities, and the community structure is shown in the Fig.5.

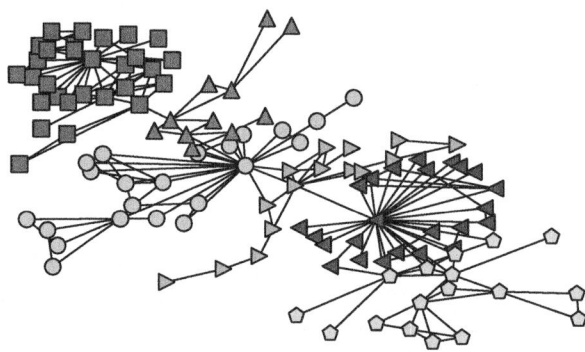

Fig. 5. Community structure in Collaboration network

EBC wrongly assigns 32 nodes to a community on collaboration network, this result can not be accepted. The community structure of collaboration network detected by LPAm is worse than that of EBC, LPAm divides collaboration network into 21 communities. The detecting results of our method and SSML are shown in Fig.6.

Our method can reach a stable state when the number of labeled nodes is larger than 19, and it wrongly assigns only one node to a community. Although

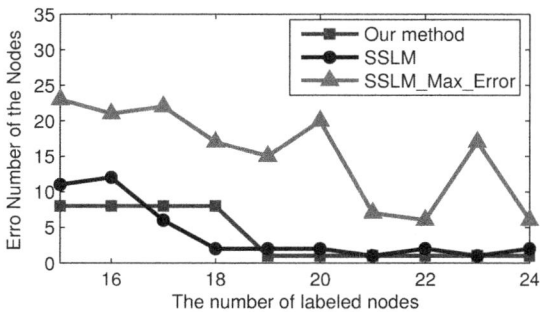

Fig. 6. Results of Our method and SSML on Collaboration network

the best results of SSML has no nodes be wrongly assigned to communities when the number of labeled nodes is larger than 17, the worst result in each experiment has too many nodes to be wrongly assigned to communities compared with our method. SSML wrongly assigns more than 15 nodes to communities in more than 6 experiments. In a large complex network, its information of community structure is unknown, we almost have no way to determine which community detection result is better than the rest of results. So in the real application, a stable community detection algorithm should be chosen to detecting the community structure of the large complex networks.

3.1.4 LFR Benchmark Network

LFR benchmark networks [27] are popular used in testing the performance of community detection algorithm. We generate a networks with 1000 nodes. The minimum community has 43 nodes, and the maximum community contains 214 nodes. We run our community detection algorithms, EBC, LPAm and SSML many times. Since the community structure of this network is clear, EBC and LPAm rightly assign all nodes to communities. Our method also can assigns all nodes to right communities when the number of selected nodes is larger than 10. Since SSML adopts random method to select nodes as labeled nodes, if some communities has no nodes to be selected, then the nodes in these communities will be assigned to other communities. Although SSML can assign all nodes to right communities in the best experimental results, the numbers of the nodes which are assigned to wrong communities in the worst result on each experiment are 113, 45, 180, 119, 77, 44, 48, 48, 108, 120 when the numbers of selected nodes are 30, 35, 40, 45, 50, 55, 60, 65,70, 75. These experiments show also show that SSML is an unstable algorithm.

3.2 Evaluation with Modularity

The modularity greedy algorithm, originally is described by Newman [27], ranges from 0 to 1. The modularity is viewed as a index to quantify how good a particular division of a network is, a larger value of modularity implies a better division of a complex network. Let e_{ij} be one-half of the fraction of edges in the network that connect vertices in group i to those in group j. e_{ii}, which are equal to the fraction of edges that fall within group i. The modularity is is described as,

$$Q = \sum_i (e_{ii} - a_i^2) \qquad (5)$$

where a_i is the fraction of all ends of edges that are attached to vertices in group i, and $a_i = \sum_j e_{ij}$. Since the performances of ECB and PLAm on Karate, Risk Map and Collaboration network are worse than that of SSML and our method, and ECB, LPAm and our algorithm rightly assigned all the nodes to communities on the artificial network, we only compare our method with SSML on Karate, Risk Map and Collaboration network. The modularity information of our method and SSML on 10 experiments are shown in table 1. nw1 is Zacharys Network of Karate Club Members, nw2 is Risk Map network, nw3 is Collaboration and A1 is our community detection algorithm in Table 1.

Table 1. The modularity information of our method and SSML on 10 experiments

data	Algor	exp1	exp2	exp3	exp4	exp5	exp6	exp7	exp8	exp9	exp10
nw1	A1	0.133	0.355	0.372	0.372	0.372	0.372	0.372	0.372	0.372	0.372
	SSML	0.294	0.303	0.298	0.352	**0.355**	**0.358**	**0.345**	0.360	0.372	0.358
nw2	A1	0.596	0.610	0.610	0.610	0.610	0.610	0.622	0.622	0.622	0.622
	SSML	0.628	0.631	**0.617**	0.587	**0.629**	0.608	**0.621**	0.621	**0.623**	0.610
nw3	A1	0.676	0.676	0.676	0.676	0.677	0.677	0.677	0.677	0.677	0.677
	SSML	0.666	0.678	0.678	0.677	0.677	0.677	0.677	0.677	0.677	0.677

Table 1 shows that the modularities of our method are larger than that of SSML on the three networks in most of the experiments, and the number of nodes which are wrongly assigned to communities by running our method is less than that of SSML in most of the experiments, and this is also shown in Fig 2. SSML has different modularities which are shown with bold in table 1 when the number of nodes wrongly assigned to communities is the same in different experiments, and this shows that the nodes wrongly assigned to communities are different even the number of them are the same. Although our method wrongly assigns '26', '33', '34' to communities since the seventh experiment on Risk Map network, the node '26' can be assigned to any one of three communities based on the ties in Fig 3, and thus the nodes '33' and '34' are wrongly assigned to communities because of the same reason, and this leads that the modularities are larger than that of the first six experiments. The modularity of our method will not change when our algorithm achieves its stable state, but the modularity of SSML has not any stable value.

4 Conclusion and Future Work

In this paper, we introduce active learning method into community detection, and present an algorithm, named active semi-supervised community detection algorithm with label propagation. We select nodes from the *core nodes* actively using the weighted shortest path and view them as labeled nodes, and experimential results show that the nodes selected by using this method can cover as many communities as possible, and thus we can obtain a stable community detection algorithm. Our community detection algorithm expands the labeled nodes by labeling the neighbors of labeled nodes with a threshold which is obtained automatically based on the characters of the networks. We demonstrate our algorithm with three real networks and one artificial network with well-known community structures. Althogh our algorithm has a better performance and more stable results compared with SSML, the time complexity is high, especially for the selecting method, algorithm 1. We will research on method of selecting nodes with lower time complexity in the future.

Acknowledgments. This paper is Supported by the Fundamental Research Funds for the Central Universities (lzujbky-2012-212).

References

1. Grvan, M., Newman, M.E.J.: Community Structure in Social and Biological Networks. Proc. Natl. Acad. Sci. USA 99, 7821–7826 (2002)
2. Brandes, U.: On Variants of Shortest-path Betweenness Centrality and Their Generic Computation. Social Networks 30, 136–145 (2008)
3. Newman, M.E.J., Girvan, M.: Finding and Evaluating Community Structure in Networks. Phys. Rev. E 69, 026113 (2004)
4. Comellas, F., Miralles, A.: A Fast and Efficient Algorithm to Identify Clusters in Networks. Appl. Math. Comput. 217, 2007–2014 (2010)
5. Brandes, U., Delling, D., Gaertler, M., Görke, R., Hoefer, M., Nikoloski, Z., Wagner, D.: On Finding Graph Clusterings with Maximum Modularity. In: Brandstädt, A., Kratsch, D., Müller, H. (eds.) WG 2007. LNCS, vol. 4769, pp. 121–132. Springer, Heidelberg (2007)
6. Chen, W., Liu, Z., Sun, X., Wang, Y.: A Game-theoretic Framework to Identify Overlapping Communities in Social Network. Data Min. Knowl. Discov. 21, 224–240 (2010)
7. Raghavan, U.N., Albert, R., Kumara, S.: Near Linear Time Algorithm to Detect Community Structures in Large-scale Networks. Phys. Rev. E 76, 036106 (2007)
8. Liu, X., Murata, T.: How Does Label Propagation Algorithm Work in Bipartite Networks? In: IEEE/WIC/ACM International Joint Conference on Web Intelligence and Intelligent Agent Technology, pp. 5–8. IEEE CPS Press, Piscataway (2009)
9. Barber, M.J., Clark, J.W.: Detecting Network Communities By Propagating Labels Under Constraints. Phys. Rev. E 80, 026129 (2009)
10. Liu, X., Murata, T.: Advanced Modularity-Specialized Label Propagation Algorithm for Detecting Communities in Networks. Physica A 389, 1493–1500 (2010)
11. Xie, J., Szymanski, B.K.: Community Detection Using a Neighborhood Strength Driven Label Propagation Algorithm. In: 1st International Network Science Workshop, pp. 188–195. IEEE CPS Press, Piscataway (2011)

12. Šubelj, L., Bajec, M.: Unfolding Network Communities by Combining Defensive and Offensive Label Propagation. In: International Workshop on the Analysis of Complex Networks, Catalonia, pp. 87–104 (2010)
13. Ma, X., Gao, L., Yong, X., Fu, L.: Semi-Supervised Clustering Algorithm for Community Structure Detection in Complex Networks. Physica A 389(1), 187–197 (2010)
14. Silva, T.C., Zhao, L.: Semi-Supervised Learning Guided by The Modularity Measure in Complex Networks. Neurocomputing 78(1), 30–37 (2012)
15. Scheffer, T., Wrobel, S.: Active Learning of Partially Hidden Markov Models. In: 12th ECML/PKDD Workshop on Instance Selection, Freiburg (2001)
16. Melville, P., Mooney, R.J.: Diverse Ensembles for Active Learning. In: 21st International Conference on Machine Learning, pp. 584–591. ACM Press, New York (2004)
17. Dasgupta, S., Hsu, D., Monteleoni, C.: A General Agnostic Active Learning Algorithm. In: Advances in Neural Information Processing Systems, pp. 353–360 (2008)
18. Nguyen, H.T., Smeulders, A.: Active Learning Using Pre-Clustering. In: Proceedings of the Twenty-First International Conference on Machine Learning, pp. 623–630 (2004)
19. Vu, V.V., Labroche, N., Meunier, B.B.: Active Learning for Semi-Supervised K-Means Clustering. In: 22nd International Conference on Tools with Artificial Intelligence, pp. 12–15. IEEE CPS Press, Piscataway (2010)
20. Zhao, W., He, Q., Ma, H., Shi, Z.: Effective Semi-Supervised Document Clustering Via Active Learning with Instance-Level Constraints. Knowl. Inf. Syst. 30(3), 569–587 (2012)
21. Grira, N., Crucianu, M., Boujemaa, N.: Active Semi-Supervised Fuzzy Clustering. Pattern Recogn. 41(5), 1834–1844 (2008)
22. Wang, X., Davidson, I.: Active Spectral Clustering. In: 2010 IEEE International Conference on Data Mining, pp. 561–568 (2010)
23. Mallapragada, P.K., Jin, R., Jain, A.K.: Active Query Selection for Semi-Supervised Clustering. In: 19th International Conference on Pattern Recognition, pp. 1–4. IEEE Press, Piscataway (2008)
24. Huang, R., Lam, W., Zhang, Z.: Active Learning of Constraints for Semi-Supervised Text Clustering. In: 7th SIAM International Conference on Data Mining, pp. 113–124. Society for Industrial and Applied Mathematics Publications Press, Philadelphia (2007)
25. Zachary, W.W.: An Information Flow Model for Conflict and Fission in Small Groups. J. Anthropol. Res. 33(4), 452–473 (1977)
26. http://en.wikipedia.org/wiki/Risk_(game)
27. Lancichinetti, L., Fortunato, F., Radicchi, R.: Benchmark graphs for testing community detection algorithms. Phys. Rev. E 78, 046110 (2008)
28. Newman, M.E.J.: Fast Algorithm for Detecting Community Structure in Networks. Phys. Rev. E 69, 66133 (2004)

Computing the Split Points for Learning Decision Tree in MapReduce

Mingdong Zhu, Derong Shen, Ge Yu, Yue Kou, and Tiezheng Nie

College of Information Science & Engineering, Northeastern University, China
dr.zhumd@gmail.com, {shenderong,yuge,kouyue,nietiezheng}@ise.neu.edu.cn

Abstract. The explosive growth of Data is bringing more and more challenges and opportunities to data mining. In data mining, learning decision tree is a common method, in which determining split points is the key problem. Existing methods of calculating split points in the distributed setting on large data either (1) cause high communication overhead or (2) are not universal for different levels of skewness of data distribution. In this paper, we study the properties of Gini impurity, which is a measure for determining split points, and design new algorithms for calculating split points in MapReduce. Empirical evaluation demonstrates that our method outperforms existing state-of-the-art techniques on communication cost and universality.

Keywords: Decision tree, split point, MapReduce.

1 Introduction

Large data have now become a trend, which has attracted lots of concerns from the academia and industry, and various models and systems are proposed [1–3], among which MapReduce [4, 5] is one of the most popular models because of its simplicity and scalability. Relevant developments for a wide range of data analysis are active, e.g., EARL [6], Cohadoop [7], MapReduce Online [8], Restore [9], Hive [12], Pig [10, 11] and others. Decision tree construction is an important method in data mining, and it has been extensively studied [13–16]. Nowadays the data are explosively growing, and they contain more useful information, from which the decision tree can be learned to facilitate a broad range of applications, such as predicting consumer preference, predicting topical events, classifying patients. However, it is a big challenge to efficiently learn decision tree from large data, which has received considerable attention [17–21]. In this work, we study how to efficiently compute optimal split points in MapReduce, which is the key problem in constructing decision tree. During computation, data are emitted with probability, that is, more useful data are more likely transmitted, so the communication cost becomes small while the performance hardly degrades. Note that when running only one analysis task, the communication cost may be not significant, but in a busy data cluster where thousands of tasks may be simultaneously running, the network bandwidth is relatively precious. We have implemented our ideas in hadoop cluster and demonstrated in hadoop cluster and the experimental results show traditional methods are not efficient for large data or not universal for different data distribution.

Contribution. Novel exact and approximation algorithms tailored to calculating optimal split points and the MapRedcue framework are proposed, which outperform existing methods by at least four fold in performance. Specifically, the contributions of this paper can be summarized below:

- An optimization for computing split points is presented and proved.
- A novel exact method to compute optimal split points, which can be efficiently instantiated in MapReduce, is proposed.
- We theoretically analyze the communication cost and standard deviation of a simple sample-based approximation method. And a new approximation method is showed and analyzed.
- Extensive experiments are conducted in a Hadoop cluster, the experimental results demonstrate efficiency and adaptability of our methods.

Note that the work focuses on Gini index and MapReduce model, however it is straightforward to extend the ideas to other impurity measures and master-slave models. Numerical attributes are considered in this work, and as for categorical attributes, conventional methods [17] can be used. The rest of the paper is organized as follows. Section 2 describes necessary background on CART(Classification and Regression Tree) and MapReduce. In Section 3, the new exact algorithm is showed. In Section 4, the novel approximation algorithm is presented. Section 5 describes our experiments and demonstrates the superiority of our methods. Section 6 reviews the related work and Section 7 concludes the paper.

2 Preliminaries

2.1 Classification and Regression Tree

Let $x_1, \ldots, x_m, c$ be random variables. $\text{Dom}(x_i)$ is the value domain of x_i, and similarly $\text{Dom}(c)$ is the value domain of random variable c, called class variable. We call $x_1, \ldots, x_m$ attribute variables, where m is the number of such attribute variables. We denote l as the number of the classes. A classification tree is essentially a function $f : \text{Dom}(X_1) \times \ldots \times \text{Dom}(x_m) \to \text{Dom}(c)$. As a kind of classification tree, Classification and Regression Tree(CART) is wildly used, because it is inherently non-parametric and relatively simple for non-experts to understand. Typically CART is built in two phases. In the first phase which is called growing phase, CART iteratively finds optimal split points. To compute optimal split points, each attribte variable should be combined with the class variable to form the attribute list [17]. In the attribute list, every tuple is the candidate split point. Hence, in this paper, if the context is clear, we use tuple to denote the candidate split point. For choosing optimal split points, CART seeks to maximize the average "purity" of the two children, and a number of "splitting functions" for measuring the "purity" can be chosen. One of the most common measures is Gini index. For a Set A, Gini impurity is $\text{Gini}(A)=1-\sum_{k=1}^{l} p_k^2$, where p_k is the probability of class k in A. Given a split point (value) v_{ij} of attribute variable x_i, and A is divided into two subset A_1 and A_2, setting p_1 and p_2 be percentage of A_1 and A_2 in A, N_1, N_2 the number of records in A_1, A_2, respectively. The reduction in impurity is: $\Delta Gini(A, x_{ij}) = Gini(A) - p_1 \cdot Gini(A_1) - p_2 \cdot Gini(A_2) = Gini(A) - \frac{N_1}{N} \cdot (1 - \sum_{k=1}^{l}(\frac{f_{c_k}}{N_1})^2) - \frac{N_2}{N} \cdot (1 - \sum_{k=1}^{l}(\frac{N_{c_k} - f_{c_k}}{N_2})^2)$, where $f_{c_k}(N_{c_k})$ is the number of records belonging to class c_k in $A_1(A)$, and N is the total number of records. CART iteratively conducts exhaustive searches to find optimal splits point in which each iteration makes $\Delta Gini$ take the maximum value.

Let

$$G(v_{ij}) = 1 - \frac{N_1}{N} \cdot (1 - \sum_{k=1}^{l}(\frac{f_{c_k}}{N_1})^2) - \frac{N_2}{N} \cdot (1 - \sum_{k=1}^{l}(\frac{N_{c_k} - f_{c_k}}{N_2})^2)$$
$$= \frac{N_1}{N} \cdot \sum_{k=1}^{l}(\frac{f_{c_k}}{N_1})^2 + \frac{N_2}{N} \cdot \sum_{k=1}^{l}(\frac{N_{c_k} - f_{c_k}}{N_2})^2 \,, \qquad (1)$$

then, $\Delta Gini(A, x_{ij}) = Gini(A) + G(v_{ij}) - 1$. $Gini(A)$ is always the same for any attribute x_j and any value v_{ij} of the attribute, so if $G(v_{ij})$ takes the maximum value, $\Delta Gini(A, x_{ij})$ takes the maximum value.

Hence the main task is to find the v_{ij} which make $G(v_{ij})$ take the maximum value. It is worth to note that each iteration in the phase are similar for every attribute variable, so we use one iteration and one attribte to illustrate our methods. The second phase is pruning phase. In this phase, the big tree is pruned to become simple and avoid "overfit" the information contained within the learning dataset, because the noise may be learned as well. Our paper focuses on the first phase, because it is much more complicated and time-consuming than the other.

2.2 MapReduce Framework

MapReduce [4] is a programming model that enables easy development of scalable parallel applications to process vast amounts of data on large clusters of commodity machines. A popular implementation of MapReduce is Hadoop [5]. Hadoop's default file system is HDFS, which MapReduce is based on. A HDFS cluster consists of a NameNode for maintaining all file meta-data, and multiple DataNodes that store the actual data. A file in HDFS is split into data blocks, 64M in size by default, which are allocated to DataNodes by the NameNode [22]. MapReduce primarily consists of a JobTracker task and many TaskTracker tasks. In Hadoop, typically Jobtracker and NameNode are configured to the same machine, called master, while TaskTrackers and DataNodes are deployed on other machines, called slaves.

Typically, one MapReduce job consists of three phases: Map, Shuffle-Sort, Reduce. Communication cost is mainly generated in Shuffle-Sort phase, and computation cost is primarily produced in Map and Reduce phases. Users can specify m, the number of Mappers and r, the number of Reducers. In Map phase, the JobTracker assigns each Mapper(TaskTracker) different portions of a file, and typically each Mapper is assigned locally stored data. Then every Mapper maps the key-value pairs$(k1, v1)$ from its data to intermediate key-value pairs $(k2, v2)$, which are sored on the local disk. When completing the task, the TaskTracker notifies the JobTracker. In Shuffle-Sort phase, each Reducer copies all value pairs that it is responsible for from DataNode, and sort them. In Reduce phase, the Reduce function processes pairs with the same key and produces final key-value pairs $(k3, v3)$.

3 Exact Computation

Before our algorithm is presented, a theorem for computing optimal split points is given.

Theorem 1. *The optimal split point can not be chosen among adjacent tuples belonging to the same class.*

Proof. Without loss of generalization, we assume there are two classes: H and L, and several continuous tuples belong to class H.

Equation 1 can be written as: $g = \frac{1}{N} \cdot [\frac{f_H^2 + f_L^2}{f_H + f_L} + \frac{(N_L - f_L)^2 + (N_H - f_H)^2}{N - f_H - f_L}]$, where $N_H(N_L)$ is the total record number of class H(L). We need prove that g can only have minimum value for varying f_H.

$$g'_{f_H} = -\frac{2}{N} \cdot [\frac{f_L^2}{(f_H + f_L)^2} - \frac{(N_L - f_L)^2}{(N_L + N_H - f_H - f_L)^2}]$$

Letting $g'_{f_H} = 0$, we have $f_H = \frac{f_L \cdot N_H}{N_L}$. When $f_H < \frac{f_L \cdot N_H}{N_L}, g'_{f_H} < 0$. When $f_H > \frac{f_L \cdot N_H}{N_L}, g'_{f_H} > 0$. So g can only have minimum value for varying f_H in any open interval and only two end points need to be checked to get the maximum. □

A simple example is shown in Figure 1.

Value	21	24	35	42	50	54	55	62
Class	H	L	L	H	L	L	L	H

Fig. 1. Candidate split points belonging to shaded region need not to be computed, because they can't be optimal

According to the theorem 1, only points between tuples belonging to different classes need be chosen as the candidate split points, which can save lots of computation resources. Hereafter, this optimization is used for computing optimal split points.

For the exact method [17], to obtain the optimal split point, firstly the data should be sorted globally, and then the values of Gini impurity for all candidate split points are calculated, at last the optimal split point is chosen. During the process, sorting incurs high communication cost and calculating for all candidate split points wastes computation resources. Motivated by this, we present a new distributed algorithm which can get the optimal split point but doesn't need to sort the data. Our algorithm requires three rounds.

Round 1: Estimate the quantiles for balancing the workload of analysis. Let r denote the number of Reducers in round 2. And $\frac{i}{r}$−quantile should be estimated, where $i \in (1, r-1)$. In this round, typically one Reducer(coordinator) is enough, and other nodes emit local histogram [27] to the coordinator. Remember that N is the number of records. It is well known to approximate a distributed quantile with a standard deviation of εN, the communication cost is $O(\frac{m}{\varepsilon})$, where m is the number of Mappers.

Round 2: Quantiles obtained in round 1 divide the data into r equal parts. In this round, each node(mapper) assigns the same key to each record belonging to the same part. The Reducers after receiving the records, compute the local optimal split point among the received records in the help of the meta information(containing the record number of each class, which is trivial to maintain. We assume it is ready to be used). The communication cost is $O(N)$.

Round 3 : All the local optimal split points computed in round 2 are collected to get the global optimal split point. The communication cost is $O(r)$.

Typically, $\frac{m}{\varepsilon} \ll N$ and $r \ll N$, so the main communication cost is $O(n)$ in round 2.

4 Approximate Computation

Obviously, the exact computation of split points is expensive. Although our algorithm avoids sorting the data before computation, it is still expensive due to the following: (1) the algorithm causes a lot of communication when it sends all data to Reducers; and (2) it has multiple rounds of MapReduce, which involves lots of overhead. Hence, an approximate method, which may not compute the optimal split point but can approximate it reasonably well, is needed.

It is difficult to directly estimate the value of Gini index for the candidate split point, so we approximate it by predicting f_{c_k} shown in equation 1. As f_{c_k} for different c_k are mutual independent, hereafter we focus on one class and denote f_{c_k} as f for convenience. To avoid the two problems mentioned above, only one Reducer(coordinator) can be used and at most one round of MapReduce are required. A natural attempt is *random sample*. More precisely, the Mappers sample the tuples with the probability $p = \frac{1}{\varepsilon^2 N}$ from the local data and send them to the Reducer. And then Reducer estimates the f. The communication cost is $p \cdot N = \frac{1}{\varepsilon^2}$, and the standard deviation is εN [22].

Next, we detail a new algorithm, *probability dispatch*(PD), which produces a unbiased estimator $\hat{f}$ of f with standard deviation $O(\varepsilon N)$, which is strictly better than random sample, improving communication cost to $(\frac{\sqrt{m}}{\varepsilon}) \cdot \log N$, as generally $\frac{1}{\varepsilon} < \sqrt{m} \cdot \log N$. The idea is to emit the tuples with high probability, which are more possible to be the optimal split point. And the Reducer estimates the f using the received data, and then computes the Gini impurity, at last selects the approximate split point. Specifically, in the Map phase, as shown in algorithm 1 , two boundary values are put into the dispatch set S, Gini impurity is locally computed, by which the tuples are sorted in descending order as a list L(line 1-3). In sequential order, each tuple l_i ($l_i \in L$) is emitted (put into the dispatch set) with probability

$$p = min(1, d_{ij} \cdot \frac{\sqrt{m}}{\varepsilon N}), \qquad (2)$$

where d_{ij} is the distance from the tuple l_i to the nearest tuples l_j in S. Note that if $p = 1$, we call the tuple l_i as *determinate tuple*, and put (l_i, f_{l_i}) in S, where f_{l_i} is the f corresponding to l_i; Otherwise the tuple is called *probability tuple*. If $p < 1$ and l_i is emitted, $(l_i, l_i \to l_j)$ is put in S.

Algorithm 1. Mapper
1: put the maximum and minimum value into the dispatch set S
2: locally compute the Gini impurity G_i for each tuple i
3: sort the tuples by G_i in ascending order as a List L
4: **for** $l_i \in L$ **do**
5:     t_j is the closet tuple in S to l_i
6:     use equation 2 to compute the sending probability p
7:     put the tuple into S with probability p
8: **end for**
9: emit all the tuples in S to Reducer

In the Reduce phase, an unbiased estimator $\hat{f}$ of f is constructed. Let f_i be the part of f in Mapper i and v be the candidate split point. For a value v, to estimate its $f_i(v)$, we find the nearest tuple t_n from Mapper i, if t_n is a *probability tuple*, we find the path of t_n to a determinate tuple t_d, and f for t_d is denoted as f_d. If the context is clear, we simplify $f_i(v)$ as f_i. For example, in Figure 2, for v, the nearest tuple is l_4, the path is $(v \to l_4)$, $(l_4 \to l_2)$, $(l_2 \to l_1)$. We denote C as the *length of the path*, which is calculated by $c_r - c_l$, where $c_r(c_l)$ is the step number of moving right(left). If t_n is a *determinate tuple*, $C = 0$. In Figure 2, $C = 2$ for v. Then f_i is estimated by

$$\hat{f}_i(v) = f_d + C \cdot \frac{\varepsilon N}{\sqrt{m}}. \qquad (3)$$

Then we estimate f as

$$\hat{f}(v) = \sum_{i=1}^{m} \hat{f}_i(v). \qquad (4)$$

The process is detailed in algorithm 2.

Algorithm 2. Reducer
1: receive tuples from Mappers
2: use equation 4 to estimate f for each tuple
3: compute the $\Delta Gini$ for each tuple using equation 1
4: choose the minimum as the approximate split point

An simple example is shown in Figure 2. Assuming $L=\{l_1, l_2, l_3, l_4\}$, first, l_1 is fetched, it is near to the minimal. By equation 2, $p = 1$, so l_1 is a *determinate tuple* and (l_1, f_{l_1}) is put in S. And then, l_2 is read, and it is near to l_1. after computation, $p < 1$, assuming it is emitted, so $(l_2, l_2 \to l_1)$ is put in S. It is similar to l_3 and l_4.

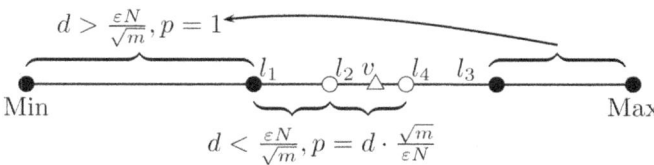

Fig. 2. An example of PD

Theorem 2. $\hat{f}$ *is an unbiased estimator of f with standard deviation at most* (εN).

Proof. Assume the step number is k for the candidate split point v, among which there are l leftward steps and r rightward steps, and the corresponding determinate tuple is t_d. Write C as $\sum_{i \in r} Y_i - \sum_{i \in l} Y_i$, where $Y_i = 1$ if the corresponding tuple is emitted in the path otherwise $Y_i = 0$. Each Y_i is an independent Bernoulli trail (In this setting, all tuples are ordered and each tuple is only checked once, assuming A is before B, then p(AB)=p(A(B|A)). And because p(B|A)p(A)=p(AB)=p(A(B|A), event (B|A) and event A are independent). Recall that d_{ij} is defined in equation 2. Hence, the expected value of Y_i is

$$\mathbf{E}[Y_i] = d_{ij} \cdot \frac{\sqrt{m}}{\varepsilon N},$$

and the variance is

$$\mathbf{Var}[Y_i] = p(1-p) \leq p = d_{ij} \cdot \frac{\sqrt{m}}{\varepsilon N}. \quad (5)$$

So, $\mathbf{E}[C] = d_d \cdot \frac{\sqrt{m}}{\varepsilon N}$, where d_d is the distance from v to t_d. Note that C is negative if v is less than t_d otherwise it is positive, and combining equation 3 and 4, we get $\hat{f} = f$, namely $\hat{f}$ is unbiased.

From (5), and because $p < 1$, $\sum d_{ij} < \frac{\varepsilon N}{\sqrt{m}}$, we have $\mathbf{Var}[C] < 1$. Hence from (3) we get

$$\mathbf{Var}[\hat{f_i}] < \frac{(\varepsilon N)^2}{m}.$$

Because $\hat{f_i}$ are mutual independent, $\mathrm{Var}(\hat{f}) = m \cdot \hat{f_i} \leq (\varepsilon N)^2$ and its standard deviation is at most εN. □

Theorem 3. *the expected communication cost of our algorithm is $O(\frac{\sqrt{m}}{\varepsilon} \cdot \log N)$.*

Proof. Because for $p = 1$, $d_{ij} \geq \frac{\varepsilon N}{\sqrt{m}}$, the number of determinate tuples is at most $\frac{\sqrt{m}}{\varepsilon}$. As for probabilistic tuple in each interval, the worst case is that the middle tuple is chosen, so the probabilistic tuples can be organized as a complete binary tree and each level has the equal expected number of emitted tuples, which is $\frac{1}{2}$. Because the number of levels of the tree is $\log(\frac{\varepsilon N}{\sqrt{m}})$, so the expected communication cost is $O(\frac{\sqrt{m}}{\varepsilon} \cdot \log(N))$. □

5 Experiment

In this section we demonstrate the performance of our algorithms in both communication cost and running time. For the exact methods, we denote our method as No-Sorting, which is compared with PLANET [19] in the experiments. For the approximation methods, We denote random sample method as RS, and our probability dispatch algorithm as PD. We use two state-of-the-art approximate distributed methods for constructing decision tree, COMBINE [21] and SPDT [25], to compare with our method. As COMBINE computes optimal split points locally, and then sends the result to the coordinator, finally the coordinator handles the conflicts. As a result, when the distribution is highly skewed, its performance is too bad. So we slightly changed it by sending candidate split point iteratively until there are no conflicts or no available candidate split points.

Setup and Datasets. All experiments are implemented in Hadoop v0.20.2. The Hadoop cluster consists of 20 machines with three different configurations : (1) 5 machines with Intel(R) Core(TM) i7 Quad 870 @2.93GHz CPU and 8 GB RAM, (2) 5 machines with Intel(R) Core(TM) i7-2600 CPU@3.40GHz CPU and 8 GB RAM, (3) 10 machines with Intel Core2 @2.8GHz and 2 GB RAM. A machine with configuration (2) is selected as the master, and another with the same configuration is selected as the (only) Reducer for PD algorithm.

In our experiments, a real-life dataset and several synthetic datasets are used. The real-life dataset is micro-blog dataset obtained from WISE 2012 Challenge, which is originally crawled from Sina Weibo [29]. Its total size is 74.7G and it contains two tables: user table and message table. We join the two tables together to form one big table, which approximately has 618 million records. Each record consists of 3 numeric attributes including time, following number and followed number(follower count), and 1 class attributes(event). Obviously the volume of the dataset is so large that it is difficult to be processed by centralized system.

The synthetic datasets built by the method proposed in [23] have 6 numeric attributes. Records of each synthetic dataset are divided into different classes by the classification functions. To model the behavior of a broad range of real datasets, the synthetic datasets are distributed with various degrees of skewness α and size n. Generally speaking, the more orderly the data blocks are, the higher the level of skewness is(the mean of each data block is more different from the population mean).

For all experiments, we vary one parameter while keeping the others fixed at their default values. The parameters and its values used in our experiments are listed in table 1. Unless otherwise specified, we use the default values.

Table 1. Experimental parameters

Parameter	Range	Default Value
ε	$10^{-6} \sim 10^{-2}$	10^{-4}
dataset size	$20G \sim 100G$	$60G$
α	0.5, 1.1, 1.4	0.5
split size	$64M \sim 512M$	$128M$

To ensure the quality of the approximation methods, we should compare its results with optimal split points calculated by exact method. We denote the number of records between the approximate value and the optimal value as absolute error(AE).

Effect of Parameter ε. We first study the impact of ε on our methods, by varying it from 10^{-6} to 10^{-2} in Figure 3, Figure 4 and Figure 5. In all cases, PD outperforms RS by four fold in accuracy. Both methods have larger AEs when ε increases. Figure 4 and 5 show that both methods have higher cost with ε decreasing. In all cases, PD has significantly lower communication cost as well as shorter running time compared to RS.

Comparing AE. As shown in Figure 6 and 7, we analyze the communication cost and running time of all the approximation methods with AE varying. Figure 6 indicates that the communication cost decreases as AE increases for all methods. PD communicates one order of magnitude less than RS and SPDT and two orders of magnitude less than COMBINE. Obviously PD gets the best communication cost to AE. Figure 7 shows that PD is the most efficient method in terms of running time.

In the following experiments, we omit the results on AEs, because the relative trends on AEs for all methods are similar to the results shown in Figure 6 and 7.

Effect of Split Size. The default split size is 64M. Figure 8 and 9 shows the impact of varying the split size from 64M to 512M. As the dataset size remains the same, the number of splits m grows with the split size increasing. Because the performance of PD and COMBINE is related to m, in Figure 8 their communication cost drops with a larger split size while RS's and SPDT's

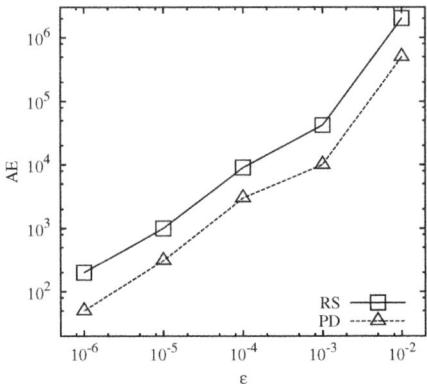

Fig. 3. AE: vary ε

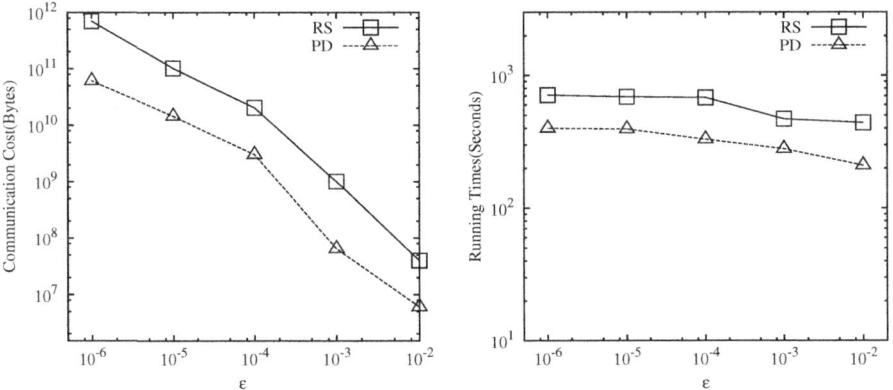

Fig. 4. Communication: vary ε **Fig. 5.** Running time: vary ε

performance hardly change. The running time for all methods reduce slightly for larger split size, but when the split size exceed 256M, the speedup are near to zero, justifying our default split size, because the concurrency capability reduces and the overhead of failure recovery increases.

Effect of Dataset Size. Figure 10 and 11 show the effect of varying dataset size. We can see that the performance for all methods degrade with the workload grows. No-Sorting outperforms PLANET by one order of magnitude in terms of communication cost. As analyzed above, the communication cost of PD is sublinearly proportional to N, which is strictly less than other methods. The running time mainly consists of access time and computation time, which linearly depend on dataset size. Overall, No-Sorting and PD are the best exact and approximate algorithm, respectively.

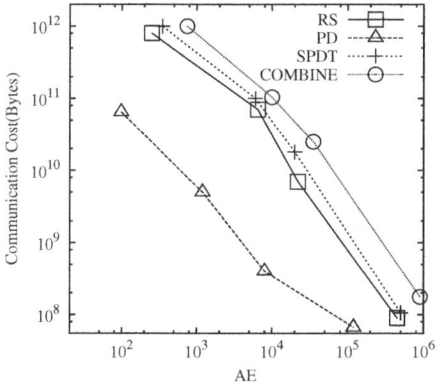

Fig. 6. Communication versus AE

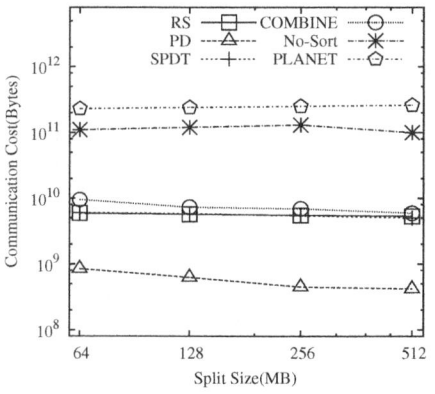

Fig. 7. Running time versus AE

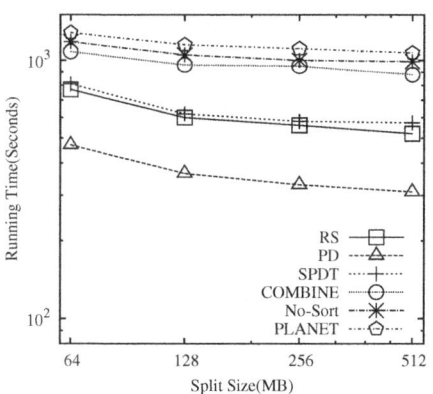

Fig. 8. Communication: vary split size   **Fig. 9.** Running time: vary split size

Effect of Degrees of Skewness. As shown in Figure 12 and 13, we study the effect of skewness α, with α as 0.5, 1.1 and 1.4. when data is highly skewed, the communication cost of COMBINE increases obviously, and results of PD and SPDT are slightly impacted. The running time of all methods have little changes. Overall, PD consistently performs the best.

Micro-blog Dataset. Finally, we study the performance of all methods on Micro-Blog with default values, and the results are shown in Figure 14 and 15. The performance of each method is similar to previous results, and PD outperforms other methods by at least one order of magnitude indicating our method is effective on large real-life datasets. Figure 16 shows that our approximate method improves the performance, not at the cost of quality.

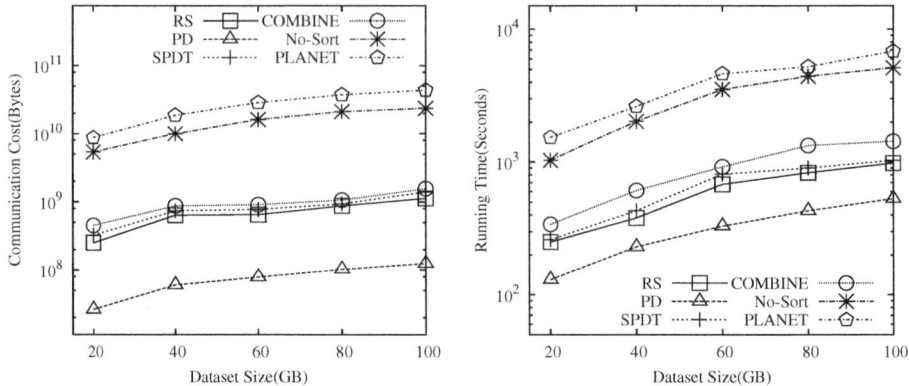

Fig. 10. Communication: vary dataset size **Fig. 11.** Running time: vary dataset size

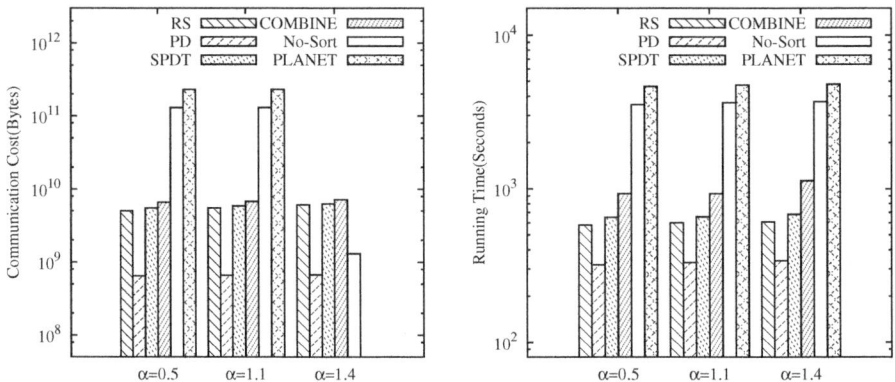

Fig. 12. Communication: vary skewness α **Fig. 13.** Running time: vary skewness α

6 Related Work

Since its introduction [13], CART has quickly emerged as a widely used tool in data mining. Lots of its variants are proposed, e.g., RainForest [16], SLIQ [15], Boat [14], and so on. However, they are centralized model and don't have scalability. As the management of large data is a pressing need, distributed and parallel methods emerge. SPRINT [17] is an exact method, and it eschews the need for any centralized, memory-resident data structures, memory-resident data structures, and can be easily and efficiently parallelizable. However, SPRINT needs to sort the data for construct the decision tree. SPIES [24], which is based on RainForest, needs to construct AVC group in memory and can not adapt to large data. As for approximation methods, VFDT is tailored to streaming data, and SPDT uses histograms to construct the decision tree. If the data distribution is not skewed, these methods can be sufficiently accurate.

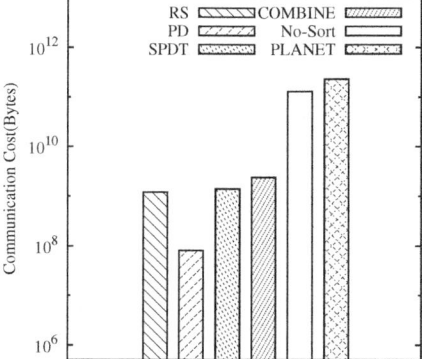
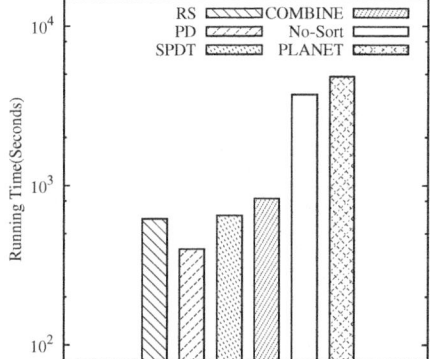

Fig. 14. Communication cost on Micro-Blog

Fig. 15. Running time on Micro-Blog

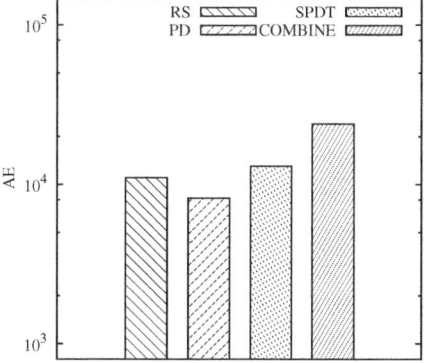

Fig. 16. AE on Micro-Blog

When MapReduce becomes the trend, several methods [19,26] that use MapReduce to construct the decision tree are proposed, however, they focus on parallel execution rather than communication cost which is of great importance in cluster.

To some extend, our work is related to quantile estimation. [27, 28] can efficiently track quantiles with relatively low communication cost. However, if they are straightforwardly applied to construct the decision tree, their performance is not so good as the performance of our algorithm, because they don't consider the characteristic of decision tree and transmit lots of unnecessary information.

The work that is most related to ours is PLANET [19]. It is the first to use MapReduce framework to construct the decision tree. However it is focus on the parallelism, our work mainly concentrates on communication cost.

7 Conclusion and Future Work

This paper studies how to efficiently compute optimal split points for constructing decision tree. We present both exact and approximation methods in MapReduce, which significantly outperform existing methods. To the best of our knowledge, our approximation method is the first method that use probability method to compute split points for constructing decision tree, and its communication cost is relatively low for the same accuracy and it is unaffected by high skewness of data distribution. And our approximation method has only one round, making it easy and appealing in practice.

Data mining have lots of important tools for data analysis. Decision tree is only one representative, and there are many other techniques we may consider in cloud, such as EM-algorithm, k-means clustering, support vector machine, outlier detection, and so on. Data mining and analysis on large data are still intellectually challenging problems, requiring lots of work on theory and practice.

Acknowledgments. This research was supported by the National Basic Research 973 Program of China under Grant No. 2012CB316201, the National Natural Science Foundation of China under Grant Nos. 60973021, 61033007, 61003060.

References

1. Chang, F., Dean, J., Ghemawat, S., Hsieh, W.C., Wallach, D.A., Burrows, M., Chandra, T., Fikes, A., Gruber, R.E.: Bigtable: A Distributed Storage System for Structured Data. In: Proc. of OSDI, pp. 205–218 (2006)
2. Cooper, B.F., Ramakrishnan, R., Srivastava, U., Silberstein, A., Bohannon, P., Jacobsen, H., Puz, N., Weaver, D., Yerneni, R.: Pnuts: Yahoo!'s Hosted Data Serving Platform. In: Proc. of VLDB, pp. 1277–1288 (2008)
3. DeCandia, G., Hastorun, D., Jampani, M., Kakulapati, G., Lakshman, A., Pilchin, A., Sivasubramanian, S., Vosshall, P., Vogels, W.: Dynamo: Amazons Highly Available Key-value Store. In: Proc. of SOSP, pp. 205–220 (2007)
4. Dean, J., Ghemawat, S.: MapReduce: Simplified Data Processing on Large Clusters. In: Proc. of OSDI, pp. 137–150 (2004)
5. Hadoop Project, http://hadoop.apache.org/
6. Laptev, N., Zeng, K., Zaniolo, C.: Early Accurate Results for Advanced Analytics on MapReduce. In: Proc. of VLDB, pp. 1028–1039 (2012)
7. Eltabakh, M.Y., Tian, Y., Özcan, F., Gemulla, R., Krettek, A., McPherson, J.: CoHadoop: Flexible Data Placement and Its Exploitation in Hadoop. PVLDB 4(9), 575–585 (2011)
8. Condie, T., Conway, N., Alvaro, P., Hellerstein, J.M., Gerth, J., Talbot, J., Elmeleegy, K., Sears, R.: Online Aggregation and Continuous Query Support in MapReduce. In: Proc. of SIGMOD, pp. 1115–1118 (2010)
9. Elghandour, I., Aboulnaga, A.: ReStore: Reusing Results of MapReduce Jobs. PVLDB 5(6), 586–597 (2012)
10. Gates, A., Natkovich, O., Chopra, S., Kamath, P., Narayanamurthy, S., Olston, C., Reed, B., Srinivasan, S., Srivastava, U.: Building a HighLevel Dataflow System on Top of MapReduce: the Pig Experience. PVLDB 2(2), 1414–1425 (2009)

11. Olston, C., Reed, B., Srivastava, U., Kumar, R., Tomkins, A.: Pig Latin: A Not-So-Foreign Language for Data Processing. In: Proc. of SIGMOD, pp. 1099–1110 (2008)
12. Thusoo, A., Sarma, J., Jain, N., Shao, Z., Chakka, P., Anthony, S., Liu, H., Wyckoff, P., Murthy, R.: Hive: A Warehousing Solution Over a Map-Reduce Framework. PVLDB 2(2), 1626–1629 (2009)
13. Breiman, L., Friedman, J., Olshen, R., Stone, C.: Classification and Regression Trees. Wadsworth (1984)
14. Gehrke, J., Ganti, V., Ramakrishnan, R., Loh, W.-Y.: BOAT-Optimistic Decision Tree Construction. In: Proc. of SIGMOD, pp. 169–180 (1999)
15. Mehta, M., Agrawal, R., Rissanen, J.: SLIQ: A fast scalable classifier for datamining. In: Proc. of EDBT, pp. 18–32 (1996)
16. Gehrke, J., Ramakrishnan, R., Ganti, V.: RainForest - A Framework for Fast Decision Tree Construction of Large Datasets. In: Proc. of VLDB, pp. 416–427 (1998)
17. Shafer, J., Agrawal, R., Mehta, M.: SPRINT: A Scalable Parallel Classifier for Data Mining. In: Proc. of VLDB, pp. 544–555 (1996)
18. Domingos, P., Hulten, G.: Mining High-Speed Data Streams. In: Proc. of KDD, pp. 71–80 (2000)
19. Panda, B., Herbach, J., Basu, S., Bayardo, R.: PLANET: Massively Parallel Learning of Tree Ensembles with MapReduce. PVLDB 2(2), 1426–1437 (2009)
20. Ye, J., Chow, J., Chen, J., Zheng, Z.: Stochastic Gradient Boosted Distributed Decision Trees. In: Proc. of CIKM, pp. 2061–2064 (2009)
21. Hall, L., Chawla, N., Bowyer, K.W.: Decision tree learning on Very Large Data Dets. In: Proc. of SMC, vol. 3, pp. 2579–2584 (1998)
22. Jestes, J., Yi, K., Li, F.: Building Wavelet Histograms on Large Data in MapReduce. PVLDB 5(2), 109–120 (2011)
23. Agrawal, R., Ghosh, S., Imielinski, T., Iyer, B., Swami, A.: An Interval Classifier for Database Mining Appliation. In: Proc. of VLDB, pp. 560–573 (1992)
24. Jin, R., Agrawal, G.: Communication and Memory Efficient Parallel Decision Tree Construction. In: Proc. of SDM, pp. 119–129 (2003)
25. Ben-Haim, Y., Tom-Tov, E.: A Streaming Parallel Decision Tree Algorithm. Journal of Machine Learning Research (JMLR) 11, 849–872 (2010)
26. He, Q., Zhuang, F., Li, J., Shi, Z.: Parallel implementation of classification algorithms based on mapReduce. In: Yu, J., Greco, S., Lingras, P., Wang, G., Skowron, A. (eds.) RSKT 2010. LNCS, vol. 6401, pp. 655–662. Springer, Heidelberg (2010)
27. Yi, K., Zhang, Q.: Optimal Tracking of Distributed Heavy Hitters and Quantiles. In: Proc. of PODS, pp. 167–174 (2009)
28. Huang, Z., Wang, L., Yi, K., Liu, Y.: Sampling Based Algorithms for Quantile Computation in Sensor Networks. In: Proc. of SIGMOD, pp. 745–756 (2011)
29. Sina Weibo, http://www.weibo.com/

FP-Rank: An Effective Ranking Approach Based on Frequent Pattern Analysis

Yuanfeng Song, Kenneth Leung, Qiong Fang, and Wilfred Ng

Department of Computer Science and Engineering
The Hong Kong University of Science and Technology, Hong Kong, China
{songyf,kwtleung,fang,wilfred}@cse.ust.hk

Abstract. Ranking documents in terms of their relevance to a given query is fundamental to many real-life applications such as document retrieval and recommendation systems. Extensive studies in this area have focused on developing efficient ranking models. While ranking models are usually trained based on given training datasets, besides model training algorithms, the quality of the document features selected for model training also plays a very important aspect on the model performance. The main objective of this paper is to present an approach to discover "significant" document features for learning to rank (LTR) problem. We conduct a systematic exploration of frequent pattern-based ranking. First, we formally analyze the effectiveness of frequent patterns for ranking. Combined features, which constitute a large portion of frequent patterns, perform better than single features in terms of capturing rich underlying semantics of the documents and hence provide good feature candidates for ranking. Based on our analysis, we propose a new ranking approach called *FP-Rank*. Essentially, *FP-Rank* adopts frequent pattern mining algorithms to mine frequent patterns, and then a new pattern selection algorithm is adopted to select a set of patterns with high overall significance and low redundancy. Our experiments on the real datasets confirm that, by incorporating effective frequent patterns to train a ranking model, such as RankSVM, the performance of the ranking model can be substantially improved.

Keywords: Learning to rank, frequent pattern, combined features, feature selection, ranking performance.

1 Introduction

Ranking is a well-recognized problem in the areas of knowledge management and information retrieval, since it is an integral part of many data-intensive applications such as advertising, documents retrieval, recommender systems, and many others. For example, given a query, in a document retrieval system, an effective ranking algorithm is essential to estimate the relevance of each document with respect to this query, so that users can easily find the most relevant documents.

A high-quality ranking method is vital to guarantee the retrieval qualities. The problem of finding effective ranking (or the ranking problem) has attracted a lot of researchers' attention in recent years. Many empirical ranking models, like the boolean model, the vector space model, and the probabilistic model, were then adopted to solve the ranking problem [2]. However these methods usually suffer high cost for parameter tuning. Later, machine learning approaches, such as RankSVM[15], RankNet[3], SoftRank[25], CRR[24], etc. have been derived to automatically learn ranking functions, and they are collectively regarded as the *learning to rank (LTR)* methods. By representing the documents with a large amount of features and making use of advanced machine learning techniques, most existing LTR methods give rise to very effective ranking functions.

While the majority of the research focuses on the design of more effective ranking models, limited studies are carried out to improve the quality of the document features used in LTR approach. In fact, besides the ranking model training algorithms, the performance of a ranking model is also highly related to the choice of the features used for ranking. In this paper, we systematically investigate the possibility of frequent pattern-based ranking approach, where a ranking model is built in terms of single features as well as significant frequent patterns. We propose a new ranking approach, *FP-Rank*, which optimizes the set of features used in LTR to improve the accuracy of ranking methods.

Combined features, which constitute a large portion of frequent patterns, are proved to be effective to capture underlying semantics of datasets [6] [7]. For ranking problem, a good example is that in order to extract features to represent documents, compared to single words (single feature), phrases (combined features) can better deliver the semantics of the documents. In this paper, we first formally analyze the ranking effectiveness of frequent patterns. In particular, we adopt a well-acknowledged criterion called *pattern significance* to measure the ranking capability of a pattern. Then, we show combined patterns, which consist a large portion of frequent patterns tend to have higher significance than single patterns. Furthermore, we prove the significance of low frequency patterns is limited due to their small coverage in the dataset. This work provides us a theoretical support to use frequent patterns as feature candidates for ranking problem and to filter the infrequent patterns when mining frequent patterns.

Our important observation is that not every frequent pattern is equally helpful for ranking. A good example is stop words which appear frequently in the documents but tends to be useless in differentiating the documents. In addition, due to large amount of possible frequent patterns, including all patterns in the extended feature space not only increases model training time, but also deteriorates the ranking accuracy due to problem of over-fitting the model. These conclusions provide us the necessity to do further feature selection on frequent pattern set after mining frequent patterns. Therefore, we propose a new algorithm to further select a pattern subset with high overall significance and low redundancy after frequent pattern mining.

We now highlight all the components of our ranking approach called *FP-Rank*, as shown in Figure 1(b), which consists of the following three phases: (1) frequent pattern mining, (2) pattern selection, and (3) model training. In this paper, we employ FP-Close [13] as the frequent pattern mining method, which is shown

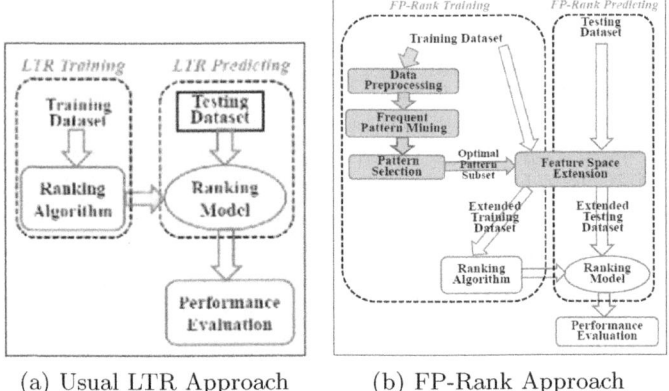

Fig. 1. Traditional LTR Approach vs. Our Proposed FP-Rank Approach

to be effective to mine closed frequent itemsets. Then, by adopting the pattern significance criterion, our proposed pattern selection method does the further pattern selection. Finally, the selected patterns are used to extend the original feature space of training dataset, and the extended dataset is used to train the ranking model.

In summary, the major contributions are fourfold:

- We formally justify that frequent patterns are important in ranking. By incorporating frequent patterns, the quality of training datasets can be improved, and eventually the performance of ranking methods can be boosted.
- We propose a novel pattern selection algorithm to select a pattern set with high overall significance and low redundancy. The pattern set is proved to be effective for ranking.
- We present a new ranking approach called *FP-Rank*. In our proposed approach, the ranking models are built in terms of single features as well as significant frequent patterns.
- We provide experimental evaluation of our proposed algorithm on real datasets. By incorporating the selected patterns as new features for ranking, the ranking performance of current widely-used LTR model such as RankSVM has been greatly improved.

The rest of this paper is organized as follows. The notations and basic concepts are introduced in Section 2. The related work is discussed in Section 3. In Section 4, the details of the frequent pattern-based ranking approach and the *FP-Rank* approach are presented. Extensive experiments on real datasets have been conducted in Section 5. Finally Section 6 concludes the paper.

2 Preliminaries

We introduce notations and basic concepts that are used throughout the paper.

Documents and the Training Dataset. We denote by A the set of m attributes that are used to represent documents, and the domain of each attribute $a_i \in A$ is either a range $[l_i, u_i]$ or a discrete value set R_i. The training dataset is denoted by $\mathcal{D}$, and each record in $\mathcal{D}$ is in the form of $\langle q, \boldsymbol{d}, y \rangle$, where q is a query, $\boldsymbol{d}$ is a document, and y is the relevance score of the document $\boldsymbol{d}$ with respect to the query q. A document $\boldsymbol{d}$ is a set of attribute-value pairs, denoted as $\boldsymbol{d} = \{\langle a_1, v_1 \rangle, \ldots, \langle a_m, v_m \rangle\}$, where v_i is the value of attribute a_i for $1 \leq i \leq m$. The relevance score y of a document is a value in the range $[0, K]$, where 0 means no relevance between the query and the document and the (maximum) value K means a "perfect" relevance.

Patterns (Features), single patterns and combined patterns. A *pattern* is a set of attribute-value pairs, and we denote it as $\alpha = \{\langle a_{i_1}, v_{i_1} \rangle, \ldots, \langle a_{i_k}, v_{i_k} \rangle\}$. We call the set of attributes contained in a pattern α the *associated attribute set* of α and denote it as A^α. A^α is a subset of A, i.e., $A^\alpha \subseteq A$. Given a pattern, if the size of its associated attribute set is 1, we call this pattern a *single pattern*; if the size of its associated attribute set is larger than 1, we call it a *combined pattern*. Since the patterns are used as features in *FP-Rank*, we use interchangeably the concepts patterns and features, single patterns and single features, combined patterns and combined features, when no ambiguity arises.

Frequent patterns. Given a pattern α, we denote by $\mathcal{D}_\alpha$ the set of records $\langle q_i, \boldsymbol{d_i}, y_i \rangle$ in $\mathcal{D}$ such that, $\boldsymbol{d_i}$ contains pattern α. For example, suppose we have a record $\langle q, \{\langle a_1, v_1 \rangle, \langle a_2, v_2 \rangle, \langle a_3, v_3 \rangle\}, y \rangle$, the record is said to belong to $\mathcal{D}_\alpha$ with the pattern $\alpha = \{\langle a_1, v_1 \rangle, \langle a_3, v_3 \rangle\}$. Given a threshold θ_0, a pattern α is said to be a *frequent pattern* if $P(\alpha) = \frac{|D_\alpha|}{|D|} \geq \theta_0$. We use F to denote a set of frequent patterns.

Learning to rank problem. The LTR approach solves the ranking problem in the following way. First, it takes a training dataset $\mathcal{D}$ as the input, and a ranking model is then constructed on $\mathcal{D}$. The testing dataset $\mathcal{T}$ contains the records in the form of $\langle q, \boldsymbol{d}, \bar{y} \rangle$ and $\bar{y}$ is the relevance score to be estimated. Then, the ranking model is applied on $\mathcal{T}$ to estimate $\bar{y}$ of each record in it. Finally, records in $\mathcal{T}$ are given in the form of a list sorted in term of their estimated relevance scores. The LTR approach is shown on Figure 1(a).

MAP and NDCG. MAP and NDCG are two criteria to evaluate the performance of the ranking model. The details can be found in [12].

3 Related Works

Frequent Pattern Mining Based Classification. Frequent pattern mining has been a focused theme in data mining research, which gives rise to a large number of scalable methods. A comprehensive survey can be found in [14]. Besides traditional techniques of deterministic frequent pattern mining, mining frequent itemsets over uncertain databases has also attracted much attention recently. For example, Tong et al. [26] [27] compare eight representative approaches of uncertain frequent itemset mining and develop a comparable software platform.

The frequent pattern-based classification is inherently related to associative classification. In associative classification, a classifier is built upon high quality rules, such as the ones with high-confidence and high-support. The association between frequent patterns and class labels is then used for prediction. The work related to this area includes: CBA[19], CMAR[18], CPAR[34] etc. These methods differ in their rule selection criteria (confidence, support, etc), number of rules they select (dataset coverage, top N, etc), and prediction result combination methodology. Cheng [6] provides a theoretically analysis about why frequent patterns are helpful for classification and bridges the gap between pattern's support with its information gain. Recent work in this area focuses on how to mine the discriminative pattern efficiently. For example, Cheng [6] provides a pattern selection method MMRFS to select frequent patterns from the candidate pattern set. HARMONY [32] adopts an instance-centric rule generation approach and achieves high accuracy and efficiency. DDP-Mine [7] provides a more effective pruning technique and directly mines out informative patterns for classification.

Learning to Rank. Ranking is a fundamental problem in many application areas such as recommendation systems, document retrieval and advertising etc. Previous work such as boolean models, vector models and probabilistic models [2] usually suffers high cost of parameter tuning since we usually consider a large number of relevant features for documents and queries.

Machine learning techniques provide many feasible solutions, since they can automatically learn parameters and make use of a large part of features in the model learning process, and this approach is referred *Learning to rank* (LTR) approach. According to [4], [5], current LTR methods can be classified into three categories: (i) Pointwise approach, (ii) Pairwise approach and (iii) Listwise approaches. In pointwise approach, each training example is treated as an independent instance and a model is trained to map each document's features to its relevance score which could be based on regression [9] or classification [20] [17]. The pairwise approach train ranking function to minimize a loss function which is based on pair-wise preferences. The ranking problem is then transformed into binary classification problem. Typical examples of such models includes RankSVM [15], RankNet [3], FRank [28], MHR [23], RankBoost[11], and CRR[24]. etc. In listwise approach, the models consider the whole document list instead of document pairs by either directly optimizing the IR measures, or indirectly optimizing the IR measures by employing a loss function correlated to IR measures. Directly optimizing the IR measures is difficult since they depend on the rank and are not differentiable. Example methods include [8], SVM^{map} [35], AdaRank [33], Boltzrank [31], NDCG-Boost [29], and [16]. Indirectly optimizing the IR measures includes RankCosine [21], and ListNet [5].

Beside the above approaches, association rules have also been applied to solve the LTR problem by Veloso [30]. When predicting the orders, several high confidence rules are used and the final relevance score is computed by weighted combination of the relevance score of all these selected rules. Our approach is inspired by the success of existing frequent pattern based classification approaches, however, we differ from these approaches in the following three aspects: (1) We use frequent patterns to extend the feature space instead of only using association rules [30]. (2) Rather than only considering confidence or support of patterns or association rules, we consider the characteristic of ranking problem and provide pattern selection method to select high significance, low redundancy pattern set for effective ranking. (3) Our approach is compatible with most of current LTR algorithms and it demonstrates significant ranking improvement.

4 Frequent Pattern-Based Ranking Approach

In this section, we present the frequent pattern based-ranking approach *FP-Rank*, which carries out ranking by the following phases: (1) frequent pattern mining, (2) pattern selection, and (3) model training. We first prove the effectiveness of frequent patterns for ranking, and adopt the frequent pattern mining methods such as FP-Close [13] to mine frequent patterns. By adopting the pattern significance criterion, a greedy method is developed to select the pattern set with high overall significance and low redundancy. Finally, the selected patterns are used to extend the original feature space of training dataset, and the extended dataset is used to train the ranking model.

4.1 The Effectiveness of Frequent Pattern for Ranking

Frequent patterns have two essential properties: *combined patterns* and *high frequency*. We analyze how these properties contribute to the ranking problem.

The Significance of Combined Patterns. A large portion of frequent patterns are combined patterns. Compared with single patterns, combined patterns are better at capturing the underlying semantics of the documents, and thus they can be more effective for producing more accurate ranking.

In order to formally analyze ranking capability of frequent patterns, we adopt a well-acknowledged criterion called *pattern significance* for ranking.

Pattern significance. Given a pattern α, pattern significance $S(\alpha)$ measures the correlation between α *w.r.t* relevance score. For the ranking problem, MAP and $NDCG$ are used to evaluate the effectiveness of a feature, which are proved to be helpful in [12]. Here we adopt the same methodology and define pattern significance be a pattern's MAP or $NDCG$, denoted as $MAP(\alpha)$ and $NDCG(\alpha)$, and they can be computed by MAP and $NDCG$ of the ranking model trained solely based on this pattern (using RankSVM, RankNet, etc.).

We utilize the Microsoft LETOR MQ2008 and OHSUMED datasets [22], and plot the MAP and $NDCG$ of single patterns as well as combined patterns. We can see that the combined patterns tend to have higher significance, e.g. Figures 2(a) and 2(b).

Pattern Significance vs. Pattern Frequency. We now study the relationship between the significance of a frequent pattern and its frequency, and demonstrate that the significance of patterns with low frequency is limited. In addition, patterns with low frequency may lower the ranking accuracy due to model overfitting. We provide the following lemma for detailed illustration.

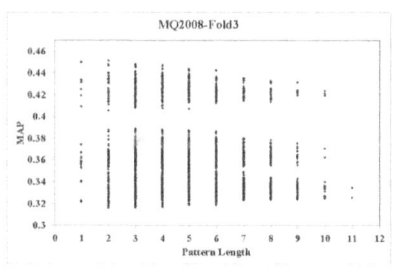
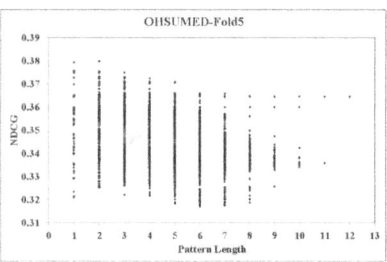

(a) MQ2008 Fold3 MAP (b) OHSUMED Fold5 NDCG

Fig. 2. Pattern Significance vs. Pattern Length on LETOR dataset

Lemma 1. *Given dataset D and pattern α, suppose pattern frequency $P(\alpha) = \frac{|D_\alpha|}{|D|} = \theta$. To simplify our analysis, we further assume relevance score $y \in \{0, 1\}$, the percentage of relevant documents $P(y = 1) = \frac{|D_{y=1}|}{|D|} = p$, the possible significance upper bound of α, denoted as $S(\alpha)_{ub}$, is monotonically increasing with θ, when θ is small. i.e., $0 \leq \theta < \min\{1 - p, p\}$.*

In order to prove Lemma 1, we now cast ranking problem into a multiple classification problem by treating relevance scores as class labels, since perfect classifications lead to perfect DCG scores according to the definition of DCG in Section 2. This view connects two intrinsically different problems of ranking and classification. In addition, Li et al.[17] further proved that a model's DCG error is bounded by the converted classification error by Lemma 2.

Lemma 2. *Suppose there are n documents $\{d_1, d_2, \ldots, d_n\}$. Given a query q, the ground truth ranked list of documents is G, which is produced by ranking documents in terms of their true relevance scores. Suppose a classifier estimates the relevance score $\bar{y}_i$ of document d_i to be an integer in $[0, K]$, for $1 \leq i \leq n$. Then the documents are sorted in terms of their estimated relevance scores to produce the estimated ranked list R. The corresponding DCG error of R with respect to G is bounded by the square root of the classification error, that is,*

$$DCG_G - DCG_R \leq \left(2^K - 1\right) \left(\sum_{i=1}^n c_{[i]}^2 - n \prod_{i=1}^n c_{[i]}^{2/n}\right)^{1/2} \left(\sum_{i=1}^n 1_{y_i \neq \bar{y}_i}\right)^{1/2}. \quad (1)$$

Based on Lemma 2, we now prove that the significance of patterns with low frequency is limited. To simplify our analysis, we further assume relevance score $y \in \{0, 1\}$. Given a dataset D, let $P(\alpha) = \frac{|D_\alpha|}{|D|} = \theta$, $P(y = 1) = \frac{|D_{y=1}|}{|D|} = p$, where $D_{y=1}$ is the set of documents with relevance score $y = 1$, and $P(y = 1|\alpha) = \frac{|D_\alpha \cap D_{y=1}|}{|D_\alpha|} = q$. Then

$$1 - S(\alpha) = 1 - \frac{DCG(\alpha)}{DCG_G} \leq \lambda * \left(\frac{\sum_{i=1}^n 1_{y_i \neq \bar{y}_i}}{|D|}\right)^{1/2}, \quad (2)$$

where $\frac{\sum_{i=1}^{n} 1_{y_i \neq \bar{y}_i}}{|D|}$ is the relevant classification error of the classifier built solely on α, denoted as $\mathcal{E}(\alpha)$, and

$$\lambda = \frac{\left(2^K - 1\right)\left(\sum_{i=1}^{n} c_{[i]}^2 - n \prod_{i=1}^{n} c_{[i]}^{2/n}\right)^{1/2} * |D|^{1/2}}{DCG_G} \qquad (3)$$

is a constant for a given dataset.

From the above assumption, we deduce that $P(\alpha, y = 1) = q\theta$, $P(\bar{\alpha}, y = 1) = p - q\theta$, $P(\alpha, y = 0) = \theta - q\theta$ and $P(\bar{\alpha}, y = 0) = 1 - p - \theta + q\theta$. So the error of the classifier built on α is given by:

$$\mathcal{E}(\alpha) = min\left\{\theta + p - 2\theta q, 1 - (\theta + p - 2\theta q)\right\}. \qquad (4)$$

For fixed p and θ, $\mathcal{E}(\alpha)$ varies with q, and reaches the lower bound at the following conditions. When $p \leq 0.5$,

$$\mathcal{E}(\alpha)_{lb} = \begin{cases} p - \theta, & \text{for } q = 1, 0 \leq \theta < p \\ \theta - p, & \text{for } q = \frac{p}{\theta}, p \leq \theta < 0.5 \\ 1 - \theta - p, & \text{for } q = 0, 0.5 \leq \theta < 1 - p \\ \theta + p - 1, & \text{for } q = 1 - \frac{1-p}{\theta}, 1 - p \leq \theta \leq 1 \end{cases}, \qquad (5)$$

and when $p > 0.5$,

$$\mathcal{E}(\alpha)_{lb} = \begin{cases} 1 - \theta - p, & \text{for } q = 0, 0 \leq \theta < 1 - p \\ \theta + p - 1, & \text{for } q = 1 - \frac{1-p}{\theta}, 1 - p \leq \theta < 0.5 \\ p - \theta, & \text{for } q = 1, 0.5 \leq \theta < p \\ \theta - p, & \text{for } q = \frac{p}{\theta}, p \leq \theta \leq 1 \end{cases}. \qquad (6)$$

We take one case of $\mathcal{E}(\alpha)_{lb}$ as an example, i.e., $p \leq 0.5$ and $0 \leq \theta < p$. $\mathcal{E}(\alpha)$ gets its lower bound when $q = 1$. The partial derivative of $\mathcal{E}(\alpha)_{lb|q=1}$ w.r.t. θ is

$$\frac{\partial \mathcal{E}(\alpha)_{lb|q=1}}{\partial \theta} = -1 < 0. \qquad (7)$$

The above analysis demonstrates that when $p \leq 0.5$, $\mathcal{E}(\alpha)_{lb}$ is a function of the pattern frequency θ. When θ is small, i.e., $0 \leq \theta < p$, $\mathcal{E}(\alpha)_{lb|q=1}$ is monotonically decreasing with θ, i.e., the smaller θ is, the larger $\mathcal{E}(\alpha)_{lb|q=1}$ is, and according pattern significance $S(\alpha)$ is likely to be smaller as well. The conclusion is the same for the cases with $p > 0.5$. When θ is small, i.e., $0 \leq \theta < 1 - p$, $\mathcal{E}(\alpha)_{lb|q=0}$ is monotonically decreasing with θ. Therefore, the significance of patterns with low frequency is bounded by a small value.

We have discussed the effectiveness of combined patterns for ranking in Section 4.1. One possible way to generate combined patterns from the original dataset is to enumerate all the combinations of the single features. This naive method suffers from the high cost due to large number of combinations ($O(2^n)$). The formal analysis in this section indicates that we can use frequent patterns with frequency large than some threshold min_sup instead of all the single pattern combinations without suffering too much performance loss, since significance of patterns with low frequency is limited.

4.2 Pattern Selection

Although frequent patterns are useful for improving accuracy of ranking, it does not mean that every frequent pattern is equally helpful. A good example is stop words which appear quite a lot in most of the documents, but almost useless

Algorithm 1. FP-Rank Feature Selection
Input: Frequent pattern set F; Training dataset D; Pattern Number N;
Output: Pattern set F_s;
 1: $F_s = \Phi$;
 2: **while** ($|F_s| < N$) **do**
 3:    $\alpha = arg\max_{\alpha \in F - F_s} \Phi(\alpha)$;
 4:    **if** α can correctly cover at lease one instance in D **then**
 5:       $F_s = F_s \cup \{\alpha\}$;
 6:    **end if**
 7:    $F = F - \{\alpha\}$;
 8:    **if** $F = \Phi$ **then**
 9:       break;
10:    **end if**
11: **end while**
12: **return** F_s

in differentiating documents. Since frequent patterns are generated by only considering frequency, the mined frequent patterns may contain a large portion of insignificant patterns. Including insignificant patterns for model training does not only increase the model training time, but also leads to the reduction of the ranking performance due to model overfitting. The objective of pattern selection is to find a pattern set from all the mined frequent patterns, such that the overall pattern significance is high, while the redundancy among the patterns in the set is low. This problem is known to be NP-hard [12]. Since the number of mined frequent patterns is usually extremely large, we therefore need to devise an efficient pattern selection method, which searches for the pattern set in a greedy way. We have defined pattern significance in Section 4.1, and the redundancy criterion is defined as follows.

Redundancy between two patterns. Given two patterns α and β, redundancy $R(\alpha, \beta)$ measures the correlation between these two patterns. Particularly, we consider the redundancy between two patterns based on the prediction results given by the models solely built on each of them. Many methods have been proposed to measure the distance between two ranked lists, such as Sperman's footrule, Kendall's tau distance, etc [12]. We choose the Kendall's tau distance, which has been proved to be effective in measuring distance of ranked lists [12], and thus the $R(\alpha, \beta)$ is defined as follows:

$$R(\alpha, \beta) = \tau(\alpha, \beta) \times min(S(\alpha), S(\beta)), \text{ with } \tau(\alpha, \beta) = \frac{\sum_{q \in Q} \tau_q(\alpha, \beta)}{|Q|}. \quad (8)$$

$\tau_q(\alpha, \beta)$ is the Kendall's tau value between two rankings respectively generated based on two patterns for query q, which is defined as follows:

$$\tau_q(\alpha,\beta) = \frac{|\{(d_i,d_j) \in D_q\}|d_i \prec_\alpha d_j \ and\ d_i \prec_\beta d_j|}{|\{(d_i,d_j) \in D_q\}|}, \qquad (9)$$

where D_q denotes the set of documents given by query q. $\tau(\alpha,\beta)$ is the average Kendall's tau value over all the queries in set Q.

We define a score for a pattern α, denoted as $\Phi(\alpha)$, as follows:

$$\Phi(\alpha) = S(\alpha) - max_{\beta \in F_s} R(\alpha,\beta). \qquad (10)$$

The greedy pattern selection algorithm is presented in Algorithm 1. It searches over all the mined frequent patterns in F and find the one with maximal Φ value (Line 3), and if this pattern can correctly cover at least one instance in the training dataset, we include it to the selected pattern set F_s (Lines 4-5). We keep searching the mined frequent pattern set until N patterns are found (Line 2) or set F is empty (Lines 8-10).

4.3 FP-Rank Approach

We present the two algorithms in our *FP-Rank* Approach: *FP-Rank* Training (Algorithm 2) and *FP-Rank* Predicting (Algorithm 3). In the training part, after we preprocess the dataset (Line 1), the frequent pattern mining algorithm, such as FP-Close [13], is adopted for mining frequent patterns (Line 2). Our proposed pattern selection algorithm 1 is used to select a set of patterns F_s (Line 3). The selected patterns are used to extend the original feature space of the dataset (Line 4), and extended dataset is used to train a ranking model M, using RankSVM, RankNet, and etc (Line 5). In the prediction part, we use the pattern set F_s to extend the feature space of the testing instances (Line 1), and then ranking model M is used to predict the relevance scores of testing instances (Line 2).

Algorithm 2. FP-Rank Training
Input: Training dataset D;
Output: Ranking model M. Pattern set F_s
1: D' =Preprocessing(D). //data discretization etc.
2: F =FP-Close(D'). //closed frequent pattern mining.
3: F_s =FeatureSelection(F). //pattern selection (Algorithm 1).
4: D'' =FeatureSpaceExtension(F_s, D). //feature space extension using F_s and D.
5: M =ModelTraning(D'') //model training based on extended dataset.
6: **return** F_s and M

Algorithm 3. FP-Rank Predicting
Input: Pattern set F_s, Ranking model M, Testing instance t
Output: Predicted relevance score y for t
1: t' =FeatureSpaceExtension(F_s, t) //feature space extension for t using F_s.
2: y =Prediction(M, t') //relevance score prediction for t' using model M
3: **return** y

5 Experiments

In this section, we evaluate the effectiveness of *FP-Rank* framework. We introduce the datasets and the relevant setup algorithms used in the experiments in Section 5.1. Then, we evaluate the ranking performance in Section 5.2.

5.1 Experimental Setup

Dataset. In our experiments, the Microsoft's LETOR benchmark [22] is used. LETOR is a benchmark for research on LTR, which composes of several data subsets, evaluation tools, and baseline evaluation results (such as RankSVM, RankBoost, etc) for ranking performance evaluation. Each data subset contains a set of queries, a set of features for query document pairs, and a set of corresponding relevance scores for the evaluation. We choose the LETOR4.0 MQ2008 dataset, the statistics of which is listed in Table 1. For each fold, the training set is first used to learn a ranking model. The validation set is used for model parameters tuning, and the ranking model is then used on testing set. The estimated relevance scores on the testing set are employed to derive the standard $NDCG@n$, $P@n$, and MAP measures in the ranking evaluation.

Table 1. Statistics of the MQ2008 dataset

No. of Features	No. of Queries	No. of Query-Document	No. of Document
46	784	15211	14384

Ranking Model. In our experiments, *RankSVM* is employed to derive the ranking model. It utilizes instance pairs and their preference labels in the training. The optimization formulation of RankSVM is given by:

$$\min \frac{1}{2} w^T w + C \sum_{i,j,q} \varepsilon_{i,j,q}$$
$$s.t. \forall (d_i, d_j) \in r_q^* : \omega\phi(q, d_i) \geq \omega\phi(q, d_j) + 1 - \varepsilon_{i,j,q}.$$

We employ RankSVM^{Struct} [15] in the *FP-Rank* framework. RankSVM^{Struct} is the most up-to-data implementation with optimized speed and performance, and previous studies [15] have already shown the effectiveness of RankSVM^{Struct}.

Data Preprocessing. Most pattern mining algorithms, such as Apriori [1], FP-Close [13], can only handle discrete attributes. However, since the attributes of most of the ranking datasets (e.g, Microsoft's LETOR datasets, Yahoo's LTR competition[1] datasets) are continuous, data discretization should be performed before frequent pattern mining. Naive discretization methods such as binary discretization or n-equal-width bins discretization suffers from two major problems: 1) information loss, which decreases the significance of frequent patterns, and 2)

---
[1] http://learningtorankchallenge.yahoo.com/datasets.php

useless patterns, which are patterns that have limited effect for ranking but make mining and pattern selection more expensive. Since if the discretization is not fine enough, it assigns many different values into the same bins, and thus generating useless patterns with information loss. In our experiment we compared several discretization methods, and we use MDL methods [10], which gives the best results due to the minimal information loss.

Frequent Pattern Mining Algorithm. Frequent pattern mining is a well-studied theme with various available algorithms and software tools. Based on the redundancy definition in section 4.2, instead of frequent patterns, we use closed frequent patterns as features in our framework, since a closed pattern is a concise representation of all its redundant non-closed sub-patterns. We choose FP-Close [13] to mine closed frequent patterns in our experiment. To maximize the number of significant patterns, we divide each dataset into several partitions according to the relevance scores. We first mine the frequent patterns in each partition. The mined patterns are merged together, and pattern selection is then applied on the merged pattern set to find the pattern subset.

To compare different pattern selection criteria, we also adopt information gain, which is a widely-used feature quality measurement for classification, to measure significance of a pattern, and adopt an extension based on Jaccard distance for measuring the redundancy. This criterion is effective for classification according to [6].

5.2 Ranking

Accuracy. The ranking results in terms of MAP and $NDCG@n$ for the MQ2008 dataset are presented in Figures 3 and Table 2. From the results, we observe that the newly added frequent patterns can significantly improve the ranking performance. Both the two feature selection criteria (i.e.,IG+Jaccard, MAP+KenTau) achieve much better results compared to the baseline method (RankSVM with no pattern added). This aligns with our claim in Section 4.1 that ranking performance can be improved by including selected frequent patterns subset.

Table 2. Summary of Ranking Improvement on MQ2008 dataset

Fold	MAP			NDCG		
	Baseline	FP-Rank	Improv.	Baseline	FP-Rank	Improv.
F1	0.4502	0.4672	3.78%	0.4577	0.4784	4.52%
F2	0.4213	0.4377	3.89%	0.4296	0.4378	1.91%
F3	0.4529	0.4529	0%	0.4686	0.4686	0%
F4	0.5284	0.5472	3.56%	0.5442	0.5604	2.98%
F5	0.495	0.5059	2.20%	0.5159	0.5232	1.42%
Ave.	0.46956	0.48172	2.69%	0.4832	0.4931	2.17%

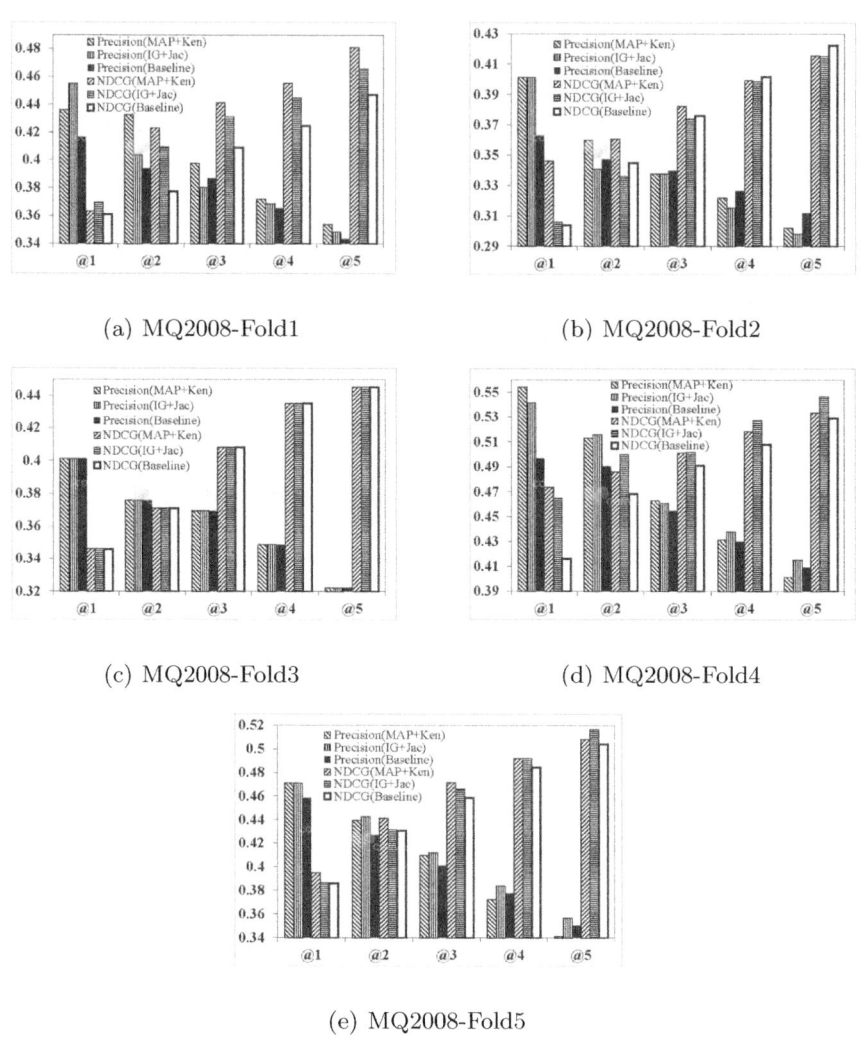

Fig. 3. Detailed Ranking Improvement on LETOR MQ2008 dataset

We find that our proposed MAP significance with Kendall tau redundancy criterion in FP-$Rank$ achieve better results compared to IG with Jaccrad methods, showing that our proposed ranking pattern selection method is more effective comparing to methods (e.g., IG and Jaccrad) for classification (Figure 3). We observe that our method significantly improves the ranking performance (Maximum: 4.52% and Average: 2.17% in terms of $NDCG@n$; Maximum 3.89% and Average: 2.69% in terms of MAP) compared to the baseline RankSVM[Struct] method (Table 2).

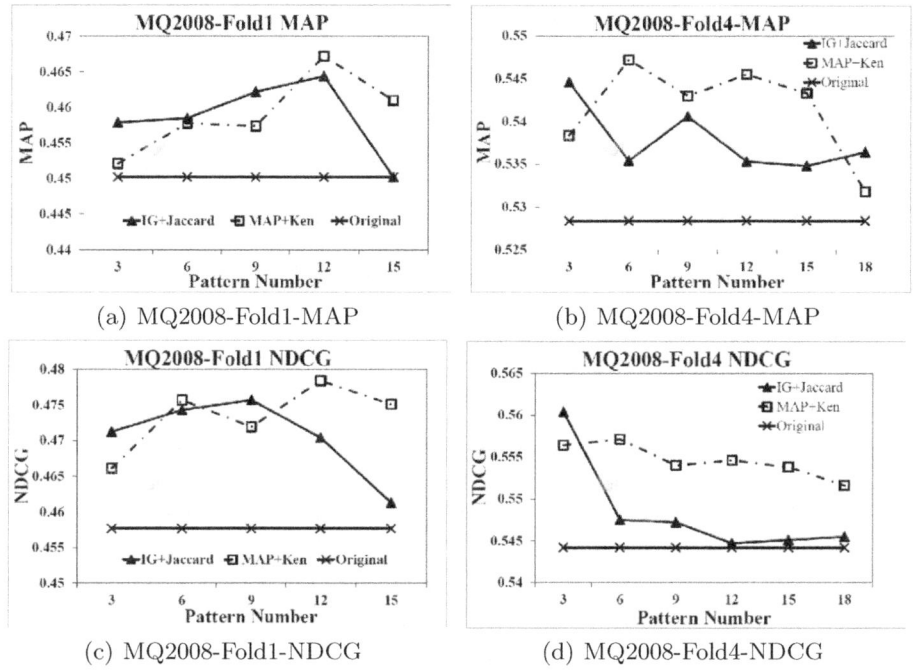

Fig. 4. Ranking Performance Improvement vs. Pattern Number N

The Effect of Pattern Set Size N. In our pattern selection algorithm, parameter N denotes the subset size of the selected pattern. In our experiment, we try different N to train the model with training set, and the models with the best performance on the validation set are used. As N varies, the ranking results in terms of MAP and $NDCG@n$ for the MQ2008 dataset are presented in Figure 4. Besides confirming the effectiveness of the new added patterns and our pattern selection algorithm, we conclude that the subset size N of the new added pattern is small (less than 20), which makes the model training time similar as the baseline RankSVM^{Struct} method.

6 Conclusions

In this paper, we propose a new approach *FP-Rank* that aims to achieve a more effective learning to rank approach by using frequent patterns. Our study confirms that frequent patterns offer high quality features that can be used to improve the performance of a ranking model. Compared with commonly used feature selection approaches, our ranking feature selection method is able to find a pattern subset that is specific for a ranking problem. The improvement is clearly evidenced by the ranking accuracy measured by MAP and $NDCG$ in *FP-Rank* in a spectrum of experiments.

Acknowledgments. This work is partially supported by GRF under grant numbers HKUST 617610 and 618509. We also wish to thank the anonymous reviewers for their comments.

References

1. Agrawal, R., Srikant, R.: Fast algorithms for mining association rules in large databases. In: VLDB 1994, pp. 487–499 (1994)
2. Baeza-Yates, R., Ribeiro-Neto, B.: Modern Information Retrieval. Addison Wesley (1999)
3. Burges, C., Shaked, T., Renshaw, E., Lazier, A., Deeds, M., Hamilton, N., Hullender, G.: Learning to rank using gradient descent. In: ICML 2005, pp. 89–96 (2005)
4. Cao, Y., Xu, J., Liu, T.-Y., Li, H., Huang, Y., Hon, H.-W.: Adapting ranking svm to document retrieval. In: SIGIR 2006, pp. 186–193 (2006)
5. Cao, Z., Qin, T., Liu, T.-Y., Tsai, M.-F., Li, H.: Learning to rank: from pairwise approach to listwise approach. In: ICML 2007, pp. 129–136 (2007)
6. Cheng, H., Yan, X., Han, J., Hsu, C.-W.: Discriminative frequent pattern analysis for effective classification. In: ICDE 2007, pp. 169–178 (2007)
7. Cheng, H., Yan, X., Han, J., Yu, P.S.: Direct discriminative pattern mining for effective classification. In: ICDE 2008, pp. 169–178 (2008)
8. Burges, C.J., Ragno, R., Le, Q.V.: Learning to rank with nonsmooth cost functions. In: NIPS 2006, pp. 193–200 (2006)
9. Cossock, D., Zhang, T.: Subset ranking using regression. In: Lugosi, G., Simon, H.U. (eds.) COLT 2006. LNCS (LNAI), vol. 4005, pp. 605–619. Springer, Heidelberg (2006)
10. Fayyad, I.: Multi-interval discretization of continuous-valued attributes for classification learning. In: UAI 1993, pp. 1022–1027 (1993)
11. Freund, Y., Iyer, R., Schapire, R.E., Singer, Y.: An efficient boosting algorithm for combining preferences. J. Mach. Learn. Res. 4, 933–969 (2003)
12. Geng, X., Liu, T.-Y., Qin, T., Li, H.: Feature selection for ranking. In: SIGIR 2007, pp. 407–414 (2007)
13. Grahne, G., Zhu, J.: Efficiently using prefix-trees in mining frequent itemsets. In: FIMI 2003 (2003)
14. Han, J., Cheng, H., Xin, D., Yan, X.: Frequent pattern mining: current status and future directions. Data Min. Knowl. Discov. 15(1), 55–86 (2007)
15. Joachims, T.: Training linear svms in linear time. In: KDD 2006, pp. 217–226 (2006)
16. Karimzadehgan, M., Li, W., Zhang, R., Mao, J.: A stochastic learning-to-rank algorithm and its application to contextual advertising. In: WWW 2011, pp. 377–386 (2011)
17. Li, P., Burges, C.J.C., Wu, Q.: Mcrank: Learning to rank using multiple classification and gradient boosting. In: NIPS 2007, pp. 845–852 (2007)
18. Li, W., Han, J., Pei, J.: Cmar: Accurate and efficient classification based on multiple class-association rules. In: ICDM 2001, p. 369 (2001)
19. Liu, B., Hsu, W., Ma, Y.: Integrating classification and association rule mining. In: KDD 1998, pp. 80–86 (1998)
20. Nallapati, R.: Discriminative models for information retrieval. In: SIGIR 2004, pp. 64–71 (2004)

21. Qin, T.: yan Liu, T., feng Tsai, M., dong Zhang, X., Li, H.: Learning to search web pages with query-level loss functions. Tech. rep. (2006)
22. Qin, T., Liu, T.-Y., Xu, J., Li, H.: Letor: A benchmark collection for research on learning to rank for information retrieval. Information Retrieval 13, 346–374 (2010)
23. Qin, T., Zhang, X.-D., Wang, D.-S., Liu, T.-Y., Lai, W., Li, H.: Ranking with multiple hyperplanes. In: SIGIR 2007, pp. 279–286 (2007)
24. Sculley, D.: Combined regression and ranking. In: KDD 2010, pp. 979–988. ACM, New York (2010)
25. Taylor, M., Guiver, J., Robertson, S., Minka, T.: Softrank: optimizing non-smooth rank metrics. In: WSDM 2008, pp. 77–86 (2008)
26. Tong, Y., Chen, L., Cheng, Y., Yu, P.S.: Mining frequent itemsets over uncertain databases. In: PVLDB 2012, vol. 5(11), pp. 1650–1661 (2012)
27. Tong, Y., Chen, L., Ding, B.: Discovering threshold-based frequent closed itemsets over probabilistic data. In: ICDE 2012, pp. 270–281 (2012)
28. Tsai, M.-F., Liu, T.-Y., Qin, T., Chen, H.-H., Ma, W.-Y.: Frank: a ranking method with fidelity loss. In: SIGIR 2007, pp. 383–390 (2007)
29. Valizadegan, H., Jin, R., Zhang, R., Mao, J.: Learning to rank by optimizing ndcg measure. In: NIPS 2009 (2009)
30. Veloso, A.A., Almeida, H.M., Gonçalves, M.A., Meira Jr., W.: Learning to rank at query-time using association rules. In: SIGIR 2008, pp. 267–274 (2008)
31. Volkovs, M.N., Zemel, R.S.: Boltzrank: learning to maximize expected ranking gain. In: ICML 2009, pp. 1089–1096 (2009)
32. Wang, J., Karypis, G.: On mining instance-centric classification rules. IEEE Trans. on Knowl. and Data Eng. 18, 1497–1511 (2006)
33. Xu, J., Li, H.: Adarank: a boosting algorithm for information retrieval. In: SIGIR 2007, pp. 391–398 (2007)
34. Yin, X., Han, J.: Cpar: Classification based on predictive association rules. In: SDM 2003 (2003)
35. Yue, Y., Finley, T., Radlinski, F., Joachims, T.: A support vector method for optimizing average precision. In: SIGIR 2007, pp. 271–278 (2007)

A Hybrid Framework for Product Normalization in Online Shopping

Li Wang[1], Rong Zhang[1], Chaofeng Sha[2], Xiaofeng He[1], and Aoying Zhou[1]

[1] East China Normal University, Shanghai, China
[2] Fudan University, Shanghai, China

Abstract. The explosive growth of products in both variety and quantity is an obvious evidence for the booming of C2C (Customer-to-Customer) E-commerce. Product normalization, which determines whether products are referring to the same underlying entity, is a fundamental task of data management in C2C market. However, product normalization in C2C market is challenging because the data is noisy and lacks a uniform schema. In this paper, we propose a hybrid framework, which achieves product normalization by the schema integration and data cleaning. In the framework, a graph-based method was proposed to integrate the schema. The missing data was filled and the incorrect data was repaired by using the evidence extracted from surrounding information, such as the title and textual description. We distinguish products by clustering on the product similarity matrix which is learned through logistic regression. We conduct experiments on the real-world data and the experimental results confirm the effectiveness of our design by comparing with the existing methods.

1 Introduction

Online retailing has been rapidly developed in the past decades, especially on the C2C (Customer-to-Customer) sites. At Bing Shopping (www.bing.com/shopping), there are over 5 million products offered by 10 thousand sellers. At Taobao site (www.taobao.com), the biggest C2C site in China, there are over 800 million products and over 370 million registered users.

However, the huge amount of products in both variety and quantity on C2C sites poses great challenge to effective and efficient data management mechanisms. One important task in data management is to determine whether products are referring to the same underlying entity, namely *Product normalization*. Product normalization is important in improving user experience due to the following reasons.

- **Easing Browsing.** Customers usually feel bored of too many instances of the same product in the search result list. Search diversification technique that leverages product similarity based on product normalization rather than string similarity will have stronger robustness to the data description noise so as to help to improve the user experience.

- **Facilitating Comparison.** Customer usually wants to see a list of the same to make comparison. However the search results contain too many different items to check, which is caused by the noisy information in item description So product normalization is on demand to facilitate the comparison.
- **Improving Recommendation.** Product recommendation takes customer-product purchase matrix as input for collaborative filtering. Usually this matrix is very sparse, which leads to inaccurate model prediction. With the help of product normalization, the purchase matrix can be nicely compressed by merging the similar products.

On the other hand, product normalization is a challenging job in that

- **Lack of a uniform schema.** Generally, each product has a structured attribute table which consists of attribute-value pairs as shown in Tab.1. The table provides detailed and rich information and is critical to product normalization. But the lack of a uniform schema (e.g., attribute *"product model"* was named as *"model"* or *"product number"* in different product description) has severely reduced its contribution to product normalization.
- **Noisy data.** As shown in Tab.1, product description is very noisy (For example, for the title of the second product model identity *"TL-WR703N"* are missing and *"iPad3"* and *"iPhone5"* is misleading. In attribute table field, the value of *"Brand"* in the first product is missing and the value of *"Product Model"* in the third product is incorrect). The noise causes great troubles in product normalization.

In this paper, we present a general hybrid framework for product normalization, which accomplishes product normalization by schema integration and data cleaning. Schema Integration aims to provide a uniform and meaningful representation of the products. Then we conduct data cleaning to provide a precise and comprehensive description for each product, which includes missing value filling, incorrect value detection and value confirmation. In our work, schema integration and data cleaning reinforce each other and are the critical process for the remaining work. Finally, we employ a logistic regression model to train the similarity between products and cluster products based on the trained similarity to achieve product normalization.

The rest of the paper is organized as follows. Section 2 discusses related work. The motivation of this work was presented in Section 3. Section 4 shows the framework of our approach. Section 5 describes data preprocessing. Section 6 introduces product normalization. Section 7 presents the experimental results and analysis. We conclude our work in Section 8.

2 Related Work

Product normalization is a variant of Entity Resolution which tries to identify instances of the same real-world entity. Entity resolution was first proposed in [13].

The traditional approach to entity resolution considers similarity of text. There has been extensive work on approximate string similarity measures [11][12], used in unsupervised entity resolution. Machine learning approaches are introduced to learn string similarity measures from labeled data [3][15][1]. There have been some techniques that enhance the traditional techniques by utilizing certain types of context entity reference to improve the quality [1][6].

The groundwork for posing entity resolution as a probabilistic classification problem was done by Fellegi and Sunter [7]. There are some following works, such as [14][17]. Some recent works leverage knowledge acquired from external sources, such as Wikipedia and WorldNet, for domain-independent entity resolution [5][4]. Some works employ negative evidence for entity resolution [6][16][9], which do not need that every entities have corresponding external resource. In the e-commerce domain, [2] and [8] proposed methods for clustering merchant offers, which is similar to our work. However, their work didn't take noise into consideration and the similarity measures they used are sensitive to noise. As a result, the accuracy and recall of this method is low when the data is noisy.

3 Motivation

To reduce the side effort of too many same products, C2C sites developed various methods for product normalization, which are either based on Universal Product Codes (UPCs) or keywords. However, the data in C2C markets is very noisy and UPC is usually not available, so these methods are either inaccurate or has extremely low recall.

Having carefully checked the dataset, we have the following observations which inspire the idea of our work.

- Identical attribute expressions usually have many same values (e.g. attribute *"Product Model"* and *"Product Number"* will have many same values). So we can integrate schema by value statistics. The details see Section 5.1.
- The missing value and incorrect value in attribute table can be replaced by the correct one predicted from surrounding information and other products (e.g., for the second product in Tab.1, the value of *"Product Model"* is missing. We found that that *"TL-WR703N"* is a candidate value for the

Table 1. Example of products' descriptions

Title	Price	Attribute table	Textual description
3G TAX FREE 150M portable Wi-Fi Wireless Router TP-LINK TL-WR703N	$29.99	Brand: null Model number:TL-WR703N Type: Wireless router Ports number: 4 ports	With a compact form factor, the TP-Link TL-WR703N 150 Mbps Wi-Fi router provides 3G wireless...
TP-LINK 150M Mini wireless router 300M speed iPad3 iPhone5	$31.99	Brand: TP-Link Max.Rate:150Mbps Product Model: null	TL-WR703N is small enough to put into your pocket and...
TL-WR703N Mini Wi-Fi wireless router 300M speed	$31.99	Max.Rate:150Mbps Product Model: TL-MR3220	TP-link TL-WR703N is a truly plug and play wireless router...

attribute in the first product, and *"TL-WR703N"* appears in the second product's textual description, so we can deduce that the missing value is *"TL-WR703N"*). See Section 5.2 for the details.
- Attributes, such as title, product number in Tab.1, have different discrimination ability (e.g., the attribute *"Product Model"* is more discriminative than title and the attribute of *"Brand"*). In our method, we automatically weight the features according to their discriminative ability, the detail will be given in Section 6.2.

4 Framework Overview

To resolve problems mentioned above, we propose a hybrid framework for product normalization as shown in Fig.1. The process consists of two parts:

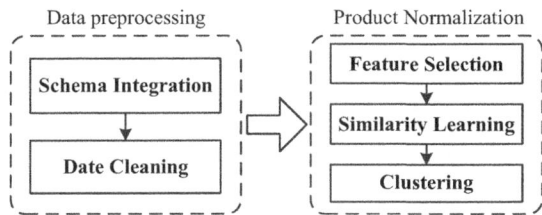

Fig. 1. Framework of our approach

- **Data Preprocessing:** This part first integrates schema by merging different schemas to a global schema. Then it conducts schema fusion to resolve the heterogeneity problems such as synonyms, abbreviation and so on. After that, we address the missing and incorrect data issue. See Section 5 for the details.
- **Product Normalization:** After data preprocessing, we pick features for the products and employ logistic regression to learn the product similarity. Finally, products are clustered based on the learned similarity to achieve product normalization. The details will be given in Section 6.

5 Data Preprocessing

5.1 Schema Integration

The usefulness of the attribute table to product normalization will be negatively affected by lacking of uniform schema. So we first integrate the schema, e.g., unifying synonyms and abbreviation. The naive strategy is to merge the strings with high string similarity. This method is not effective, because identical strings may be quite different in the form(e.g., *"RAM"* and *"Random Access Memory"*) while strings that are not identical may be quite similar(e.g., *"Cores*

Number" and *"Model Number"*). Our method needs to consider not only string similarity but also the neighbor information.

To facilitate the remaining work, we first combine attribute tables from all products into a global schema graph $G = <A, V, E>$ as shown in Fig.2(a), where A is the node set of attributes, V is the node set of values, and E is edge set. For nodes $a \in A$ and $v \in V$, edge (a,v) with weight k exists in G if and only if attribute-value pair (a,v) appears in k products' attribute tables. The weight of edge (a, v) is represented by $w(a, v)$.

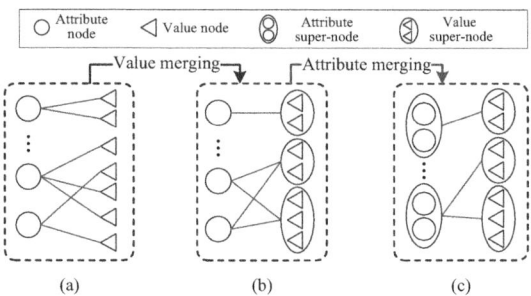

Fig. 2. Process of Schema Integration

Before grouping the similar attribute, we need to merge the identical values(e.g., merge *"TL-WR703N"* and *"WR703N"*). Either Edit Distance [10] or N-gram can be used as the similarity/dissimilarity measures. We denote the string similarity as $Sim_{Str}(\cdot)$. If two value nodes (value super-nodes) have a higher similarity than a predefined threshold $\delta (0 \leq \delta \leq 1)$, then we combine them into a super-node. The similarity between a value node a and a super-node v^{sup} is the average similarity between a and nodes in v^{sup}. After merging, G is converted into $G' = <A, V^{sup}, E'>$ as shown in Fig.2(b), where V^{sup} represents the set of the generated super-nodes and E' represents the set of super-edges. The weight of a super-edge between an attribute a and a value super-node v^{sup} is defined as: $w(a, v^{sup}) = \sum_{v \in v^{sup}} w(a, v)$.

For any value $v \in V$, $sup(v)$ denotes the super-node in V^{sup} which contains v. From any super-node $v^{sup} \in V^{sup}$, we choose the node the most frequently in the data as the representative, denoted by $v^{sup}.rcp$. denotes the representative node in v^{sup}, which is the value in v^{sup} that appears more frequently in the data than any other value in v^{sup}.

Now we start to merge the identical attributes. As discussed previously, attribute similarity is measured by the string similarity and the neighbor similarity. The similarity of two attributes $a_i, a_j \in A$ is defined as:

$$S_{att}(a_i, a_j) = \lambda Sim_{Str}(a_i, a_j) + (1 - \lambda) S_{neighbor}(a_i, a_j)$$

$$S_{neighbor}(a_i, a_j) = \frac{\sum_{v \in V(a_i) \cap V(a_j)} Min(w(a_i, v), w(a_j, v))}{\sum_{v \in V(a_i) \cup V(a_j)} Max(w(a_i, v), w(a_j, v))}$$

$S_{att}(\cdot)$ is the weighted sum of string similarity and neighbor similarity. In our experiments, we set $\lambda = 0.3$. For any $a \in A$, $V(a)$ denotes the set of sup-nodes in V^{sup} which are adjacent to a. Neighbor similarity is the global ratio of common values the two attribute a_i and a_j have. Using $S_{att}(\cdot)$ and a given threshold θ, we merge the attributes according the same strategy as merging values. After the attribute merging, we get $G^* =< A^{sup}, V^{sup}, E^* >$ as shown in Fig.2(c). For any sup-node $a^{sup} \in A^{sup}$, $a^{sup}.rep$ denotes the representative node in a^{sup}, which is the attribute in a^{sup} that appears the most frequently in the data.

Now, the integrated schema is stored in G^*, where different attributes (values) within a super-node are considered identical. Then we can use graph G^* to convert all products into a uniform data schema. The converting process is below: for each attribute-value pair (a,v) of product p, there will be two sup-nodes $a^{sup} \in A^{sup}$ and $v^{sup} \in V^{sup}$ in graph G^* containing a and v respectively. We convert (a,v) to $(a^{sup}.rep, v^{sup}.rep)$. After the conversion, all the products are in a unified data schema, so a more accurate comparison based on attribute table can be achieved.

5.2 Data Filling and Cleaning

As described in Section 3, the missing and incorrect values greatly deteriorate the performance for product normalization. This subsection introduces the details in data filling and cleaning.

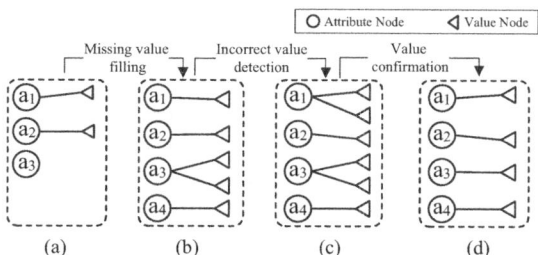

Fig. 3. Process of Data Cleaning

As shown in Fig.3(a), we use a bipartite graph $g^{(p)} =< A^{(p)}, V^{(p)}, E^{(p)} >$ to model the attribute table of product p, where $A^{(p)}$ is the set of attributes, $V^{(p)}$ is the set of values, an edge in $E^{(p)}$ represents an attribute-value pair in p (the original weight for each edge is 1). And we use $tit(p)$ and $des(p)$ to denote title and textual description for product p, respectively.

Missing Value Filling. There are two types of missing values:(i) *Value-level missing*: In value-level missing, an attribute's corresponding value is *"null"*. E.g., in Fig.3(a) a_3 is the value-level missing;(ii)*Schema-level missing*: Schema-level missing occurs when a product do not have an attribute it should have. E.g. the third product in Tab.1 does not have the attribute *"brand"* owned by other similar products.

Algorithm 1. Missing Data Filling
Input: $G^* =< A^{sup}, V^{sup}, E^* >$, $tit(p)$, $des(p)$,
$\quad\quad g^{(p)} =< A^{(p)}, V^{(p)}, E^{(p)} >$
Output: $g^{'(p)}$
1: $V_{miss} = \emptyset, g^{'(p)} = g^{(p)}$
2: **for each** $a_i \in A^{sup}$ **do**
3: $\quad$ **if** $a_i.rep \notin A^{(p)}$ or $(a_i.rep, null) \in E^{(p)}$ **then**
4: $\quad\quad A_{miss} \leftarrow a_i.rep$
5: $\quad$ **end if**
6: **end for**
7: **for each** $a_i \in A_{miss}$ **do**
8: $\quad$ **for each** $v \in \bigcup_{a \in a_i} V(a)$ **do**
9: $\quad\quad$ **if** v appears in $tit(p)$ or $des(p)$ **then**
10: $\quad\quad\quad$ **if** $(a_i, sup(v).rep) \in E^{(p)}$ **then**
11: $\quad\quad\quad\quad w(a_i, sup(v).rep) = w(a_i, sup(v).rep) + 1$
12: $\quad\quad\quad$ **else**
13: $\quad\quad\quad\quad E^{'(p)} \leftarrow (a_i, sup(v).rep)$
14: $\quad\quad\quad\quad w(a_i, sup(v).rep) = 1$
15: $\quad\quad\quad$ **end if**
16: $\quad\quad$ **end if**
17: $\quad$ **end for**
18: **end for**
19: **return** $g^{'(p)}$

For any product p, the process of filling the missing value is described in Algorithm 1. The first step is to locate those attributes that are either value-level missing or schema-level missing, and store them into A_{miss}. Note that only the representative attribute are stored into A_{miss} to avoid the redundancy. For each attribute a_i in A_{miss}, we scan $tit(p)$ and $des(p)$ for any sub-string which is equal to any element $v \in \bigcup_{a \in a_i} V(a)$. If any sub-string is equal to v, then it is considered as the candidate value for a_i. So we create a new edge (a_i, v) into $g^{'(p)}$ with weight 1, if the edge already exists, we plus 1 to the weight. Based on the result of algorithm 1, we get a new bipartite graph $g^{'(p)} =< A^{'(p)}, V^{'(p)}, E^{'(p)} >$, which has some new nodes and new edges compared with $g^{(p)}$, as shown in Fig.3(b). New attribute nodes in $A^{'(g)}$ are evidences found for schema-level missing, and new value nodes in $V^{'(p)}$ are candidates found for value-level missing. The weights of new edges in $E^{'(p)}$ are the support degree of evidences found from $tit(p)$ and $des(p)$. For any new attribute nodes in $g^{'(p)}$ which connect to two or more nodes such as a_3 in Fig.3(b), we need to determine which value node is the right one. The determining process is described in section 5.2.

Incorrect Value Detection. For product p, the incorrect value can be deduced from $tit(p)$ or $des(p)$. E.g. for the third product in Tab.1, the title and textual description imply that the true value for *"Product Model"* should be *"TL-WR703N"*.

Algorithm 2. Incorrect Value Detection
Input: $G^* =< A^{sup}, V^{sup}, E^* >, tit(p), des(p),$
$\quad\quad g^{'(p)} =< A^{'(p)}, V^{'(p)}, E^{'(p)} >, A^{(p)}$
Output: $g^{'(p)}$
1: **for each** $a_i \in A^{(p)}$ **do**
2:    **for each** $v \in \cup_{a \in a_i} V(a)$ **do**
3:       **if** v appears in $tit(p)$ or $des(p)$ **then**
4:          **if** $< a_i, sup(v).rep > \in E^{'(p)}$ **then**
5:             $w(a_i, sup(v).rep) = w(a_i, sup(v).rep) + 1$
6:          **else**
7:             $g^{'(p)} \leftarrow (a_i, sup(v).rep)$
8:             $w(a_i, sup(v).rep) = 1$
9:          **end if**
10:       **end if**
11:    **end for**
12: **end for**
13: **return** $g^{'(p)}$

Algorithm 2 shows the process for incorrect value detection. For each attribute $a_i \in A^{(p)}$, we check whether there is any sub-string in $tit(p)$ or $des(p)$ which suggests that the current value for a_i may be improper. If there is such a substring, we add a new edge with weight 1 into $g^{'(p)}$. And if the edge is already in $g^{'(p)}$, we increase the weight. After the process, evidences which support that the current value may be incorrect, will have a corresponding edge with some weight in $g^{'}$, as shown in Fig.3(c).

Value Confirmation. After missing value filling and incorrect value detection, the graph $g^{'(p)}$ for product p may contains some multi-valued attributes which connect to more than one values, such as a_1 and a_3 in Fig.3(c). We should decide which value is the right one.

For any $a_i \in A^{'(p)}$, we denote $V^{'(p)}(a_i)$ to be the set of value nodes that connect to a_i.

For each attribute node $a_i \in A^{'(p)}$ that connects to two or more value nodes, we choose one from all the value nodes connected to a_i according to the following two criteria:

1. $v = \arg\max_{v \in V^{'(p)}(a)} w(a_i, v)$
2. $v = \begin{cases} original\ value & if \frac{\max w(a_i, v)}{\sum_{v \in V^{'(p)}(a)} w(a_i, v)} < \gamma \\ \arg\max_{v \in V^{'(p)}(a)} w(a_i, v) & otherwise \end{cases}$

The first criterion is straightforward, which selects the value with the maximum weight. The second criterion is more passive. It chooses the value with the maximum weight only if the maximum weight exceeds a percentage(a threshold γ) of total weights. Otherwise we keep the original value to avoid incorrect "repair" of the right value. γ is between [0,1]. The greater γ is, the more easily the original value was changed.

After removing the contradicted values in $g'(p)$, we obtain a new graph as shown in Fig.3(d), where the missing data is filled and the incorrect value is repaired.

6 Product Normalization

After the data preprocessing was done by unifying attribute table from the view of schema and value, we now use the integrated data for product normalization.

6.1 Product Feature Selection

We pick the features and define the similarity measurement for each feature.

Attribute table contains the strongest features for product distinguishing because it is detailed, especially after schema integration and data cleaning. We use every attribute node in G^* as a feature, and we get $|A^{sup}|$ features. The similarity between two product on the ith feature is defined as:

$$s_i(value_k, value_j) = \begin{cases} 1 & if\ value_k = value_j \\ 0 & otherwise \end{cases}$$

Title is another important feature. Sellers often add popular but irrelevant keywords to titles as discussed in Section 3. We use word segmentation tools to partition t_i and t_j into two sets of words $w(t_i)$ and $w(t_j)$, and use $tf\text{-}idf$ to give a less weight to these irrelative words, then the similarity between two titles is:

$$Sim_{title}(t_i, t_j) = \frac{\sum_{w \in w(t_i) \cap w(t_j)} tf(w) \times idf(w)}{\sum_{w \in w(t_i) \cup w(t_j)} tf(w) \times idf(w)}$$

Price is important feature for product normalization, because a considerable differ in price between two products may give evidence that they refer to two different entities. The price similarity is defined as:

$$Sim_{price}(price_i, price_j) = 1 - \frac{|price_i - price_j|}{\max(price_i, price_j)}$$

Online Review is also an important feature. We observe that reviews for the similar products are always involved in many common aspects.

To extract feature from reviews, we use word segmentation and POS tagging tool [18] to partition sentences in reviews into words and pos tags, and extract nouns as aspect words. Many aspect words such as *"quality"* and *"shipping"* are not discriminative since they appear frequently in many products. So we use $tf\text{-}idf$ to score every aspect word and pick the top k aspect words. We use $Asp(p) = \{w_1, ..., w_k\}$ to denote the top k aspect words of product p. The similarity of two products on the feature of reviews is:

$$Sim_{reviews}(p_i, p_j) = \frac{|Asp(p_i) \cap Asp(p_j)|}{|Asp(p_i) \cup Asp(p_j)|}$$

6.2 Model Training

Now for any product, we have $k = |A^{sup}+3|$ features ($|A^{sup}|$ features in attribute table, and 3 features in title, price and reviews). Using the predefined similarity functions, we can obtain a similarity vector $\mathbf{s}(p_i, p_j) = <s_1, ..., s_k>$ for products p_i and p_j. We convert the problem of whether two products are matching, to be two-class classification problem, and use linear logistic regression model for classification. The reason of using two-classification instead of multi-class classification is twofold: (i)The number of parameters for two-classification is much smaller than multi-class classification, especially when the entity number is large.(ii)In multi-class classification, the training set must covers all classes, which is difficult because entity number is usually either too large or not known.

We set C_0 to be matching and C_1 to be mismatching. The posterior probability of class C_0 can be modeled as logistic sigmoid acting on a linear function of the feature vector $\mathbf{s}$ so that:

$$p(C_0|s) = y(\mathbf{s}) = \frac{1}{1+e^{-\mathbf{w}^T \mathbf{s}}} = \frac{1}{1+e^{-(w^T s + w_0)}}$$

In $P(C_0|s)$, $\mathbf{w} = [w_0, \mathbf{w}]$, where w_0 is a bias and $\mathbf{w}$ is the weight vector for features, and $\mathbf{s} = [1, \underline{s}]$. We use training data to train $\mathbf{w}$ so that $\mathbf{w}^T \underline{s} > 0$ is for matching and $\mathbf{w}^T \underline{s} < 0$ is for mismatching. After the training, the k-th value in $\underline{w}$ indicates the importance of the k-th feature in the discriminative function.

6.3 Product Normalization via Clustering

After we trained a model for the probability of two products' matching. We need to convert pair-wise matching into partitions so that products in each partition refer to a unique underlying entity.

The naive way is to generate a graph where nodes represent products and there is an edge between two nodes if and only if the probability of two products' matching is more than 0.5. However, this method will cause low precision since an incorrect prediction of matching will mistakenly merge two partitions.

Our solution is to treat the probability of two products' matching as similarity, then apply clustering algorithm to partition. This solution is effective since clustering makes global decision rather than local decision. For n products, we can get a similarity matrix $\mathbf{M}_{n*n}$ where m_{ij} is the similarity (estimated probability of matching) between product i and j. We use existing cluster algorithm such as the hierarchical Aggregation Cluster (HAC) or k-means to partition products into clusters. The number of clusters is determined in the process of clustering according to the purity or diameter. When n is large, the cost for both storage and computation is very large. Our solution is to use taxonomy information to divide the n products into several disjoin subsets so that each subset is a separate category with smaller number of products.

7 Experiment and Evaluation

7.1 Dataset

The dataset is crawled from Taobao, the largest e-commerce site. The data set covers 168 categories, over 1,400,000 products, 500,000 sellers, and 78,000,000 reviews. We choose 12 representative categories, and randomly sample 15% products for experiments as shown in Tab.2.

Table 2. The categories of our dataset

Categories	#.of products	#.of entities	Categories	#.of products	#.of entities
Phone	5,345	98	Wallet & purse	2,090	108
Camera	8,980	89	Jacket	3,334	79
Notebook	5,879	168	perfume	1,090	67
Network devices	12,324	127	Shampoo	2,073	103
T-shirt	8,977	333	Women's shoes	3,909	159
Jeans	9,006	206	Sport shoes	6,348	298

7.2 Data Noise Statistics

Noise in Titles. We denote the noise by the percentage of irrelevant words in titles. Word that cannot indicate what entity the product refers to is called a irrelevant word(e.g., advertising words such as *"excellent"*, *"free Tax"*,etc. and misleading words such as *"iPad3"* for a router).

Irrelevant words are noise for product normalization. We randomly choose 1,000 products from our dataset and manually label irrelevant words in their titles. Fig.4 shows the percentage of irrelevant words (items are ranked in descending order of the percentage). It's surprising that over 30% titles have at least 30% irrelevant words, and over 80% titles have at least 15% irrelevant words. Due to the noise, the performance of product normalization that only uses title is extremely low as shown in Fig.6

Noise in Attribute Table. To evaluate the noise in attribute table, we randomly sample 1,000 products and manually label missing and incorrect values, then calculate the number of null values and incorrect values respectively for each product. The result is shown in Fig.5 (Products are ranked in descend order according to the percentage respectively). The surprising result shows that 90% products have data quality problems in their attribute tables, and data quality is severely low for 50% products.

7.3 Evaluation on Data Cleaning

We use our data cleaning algorithm to fill the missing values and repair the incorrect values on the 1,000 products we have labeled with the prudent parameter γ to be 0, 0.2, and 0.4 respectively, then evaluate the accuracy and recall. The

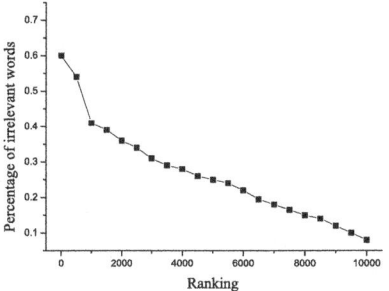
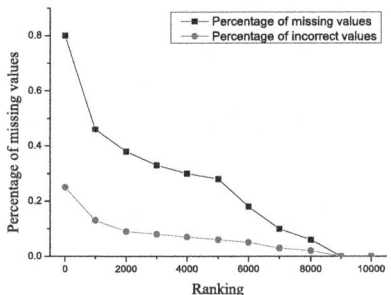

Fig. 4. Percentage of non-relative words in titles

Fig. 5. Percentage of missing and incorrect values in attribute table

results are shown in Tab.3. The results show that our data cleaning algorithm wins a high precision and recall, especially when $\gamma = 0.2$ (the rest of our experiments use this setting). Note that the precision is more important than the recall in our scenario, because a mistaken value filling(repair) is serious while leaving some missing(incorrect) values unfilled(unrepaired) is acceptable.

Table 3. The Precision and Recall of data cleaning(the best results are in bold)

	Missing values filling		Incorrect values repair	
	Precision	Recall	Precision	Recall
$\gamma = 0$	0.85	0.78	0.72	0.75
$\gamma = 0.2$	**0.95**	**0.75**	**0.92**	**0.65**
$\gamma = 0.4$	0.98	0.43	0.93	0.25

7.4 Evaluation on Product Normalization

This subsection will compare the effectiveness of our framework with three Baselines:

- **Baseline#1** (String similarity between titles) This method just uses String similarity on titles. Two products are considered to be matching if the similarity is over The threshold η.
- **Baseline#2** (Weighted String similarity between titles) This method is a melioration for Baseline#1 by giving a tf-idf weight for each word.
- **Baseline#3** (Adaptive Product Normalization) This method is based on [2], which uses title, price and textual description for features, and employs averaged perceptron to train the weight for each feature. In our experiment, we take attribute table and reviews as additional features for Baseline#3

Evaluation Metrics. We use Precision, Recall, and F-measure to evaluate the performance of product normalization:

$$Prec = \frac{|TP|}{|TP|+|FP|}, Rec = \frac{|TP|}{|TP|+|FN|}, F = \frac{Prec \times Rec \times 2}{Prec + Rec}$$

Where $|TP|$, $|FP|$, and $|FN|$ are the number of true positive, false positive and false negative respectively.

Results and Discussion. We ran the three Baseline approaches and our approach on the 12 categories respectively. For Baseline#1 and Baseline#2, there is no model needed to train. The threshold η is the tradeoff parameter for Precision and Recall. For Baseline#3 and our approach, we use 5-cross validation and the balance between precision and recall can be controlled by the bias w_0.

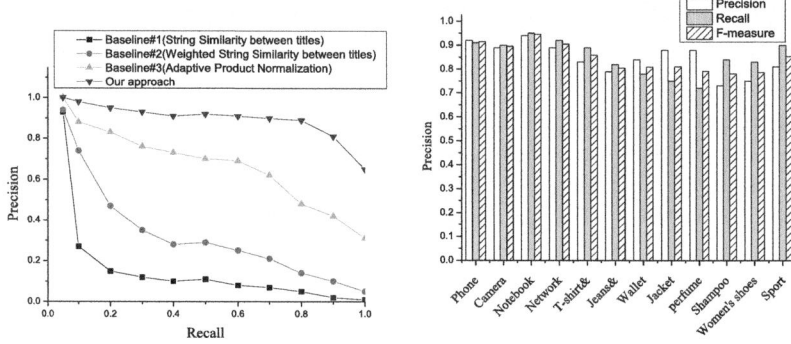

Fig. 6. Performance comparison between the baselines and ours

Fig. 7. The performance of the three variant approaches

Fig.6 shows the average Precision, Recall and F-measure score of the 12 categories. It's clear that the performance of Baseline#1 and Baseline#2 are very low. We found that the Recall value is almost 0 when the Precision is high. That's because when we set the threshold η too high, Baseline#1 and Baseline#2 will only predict products with the same titles to be matching. Note that the Precision drops significantly when we increase the recall requirement. It is because the titles of many different products are similar due to the high rate of irrelevant words. Baseline#2 is better than Baseline#1, because we give a less weight to irrelevant words. However, due to the existence of noise in titles, the performance of these approaches are very low.

Baseline#3 is much better than the two approaches. That's because Baseline#3 uses more features and employ a machine learning method to train the weight for each feature. Compared with Baseline#3, our approach is better. The reason is twofold: (i)We integrate schema so that the attribute table can be used much more efficiently, since the identical attributes with different expressions can be traded as the same one. (ii)We fill the missing values and repair the incorrect values.

To show the importance of our schema integration and data cleaning for product normalization, we defined the following variants of our approach:

1. Complete approach
2. Without data cleaning
3. Neither schema integration nor data cleaning

We ran experiments on these three kinds of settings in each category respectively. Fig.7 shows the F-measure scores. It shows that with the help of schema integration, the performance is improved in all categories, which validates the role of Schema Integration. After data cleaning, the F-measure is improved significantly in all categories, which validates the need of data cleaning for product normalization in C2C sites and the effectiveness of our data cleaning method.

8 Conclusions

With the development of C2C e-commerce, the large scale of products increases the demand for product normalization. However, product normalization is difficult because of the serious noise in product description. In this paper, we proposed a hybrid framework, which realized product normalization by data preprocessing. The experimental results on a real-world data validate the effectiveness of our approach. There are some parameters to tune in the process of Schema integration and Data cleaning, which is left for the future work.

Acknowledgments. The work is partially supported by the Key Program of National Natural Science Foundation of China (Grant No.61232002), National High Technology Research and Development Program 863 (Grant No.2012AA011003), National Natural Science Foundation of China (Grant No.61103039, No.61021004, No.60903014).

References

1. Ananthakrishna, R., Chaudhuri, S., Ganti, V.: Eliminating fuzzy duplicates in data warehouses. In: VLDB, pp. 586–597. Morgan Kaufmann (2002)
2. Bilenko, M., Basu, S., Sahami, M.: Adaptive product normalization: Using online learning for record linkage in comparison shopping. In: ICDM, pp. 58–65. IEEE Computer Society (2005)
3. Bilenko, M., Mooney, R.J.: Adaptive duplicate detection using learnable string similarity measures. In: Getoor, L., Senator, T.E., Domingos, P., Faloutsos, C. (eds.) KDD, pp. 39–48. ACM (2003)
4. Bunescu, R.C., Pasca, M.: Using encyclopedic knowledge for named entity disambiguation. In: EACL. The Association for Computer Linguistics (2006)
5. Cucerzan, S.: Large-scale named entity disambiguation based on wikipedia data. In: EMNLP-CoNLL, pp. 708–716. ACL (2007)
6. Dong, X., Halevy, A., Madhavan, J.: Reference reconciliation in complex information spaces. In: SIGMOD 2005, pp. 85–96. ACM Press, New York (2005)
7. Fellegi, I.P., Sunter, A.B.: A theory for record linkage. Journal of the American Statistical Association 64(328), 1183–1210 (1969)

8. Kannan, A., Givoni, I.E., Agrawal, R., Fuxman, A.: Matching unstructured product offers to structured product specifications. In: Apté, C., Ghosh, J., Smyth, P. (eds.) KDD, pp. 404–412. ACM (2011)
9. Lee, T., Wang, Z., Wang, H., Hwang, S.W.: Web scale taxonomy cleansing. PVLDB 4(12), 1295–1306 (2011)
10. Levenshtein, V.: Binary codes capable of correcting deletions, insertions, and reversals. Soviet Physics Doklady 10, 707–710 (1966)
11. Monge, A.E., Elkan, C.: The field matching problem: Algorithms and applications. In: KDD, pp. 267–270 (1996)
12. Navarro, G.: A guided tour to approximate string matching. ACM Computing Surveys 33(1), 31–88 (2001)
13. Newcombe, H.B., Kennedy, J.M., Axford, S., James, A.: Automatic linkage of vital records. Science 130(3381), 954–959 (1959)
14. Ravikumar, P.D., Cohen, W.W.: A hierarchical graphical model for record linkage. In: Chickering, D.M., Halpern, J.Y. (eds.) UAI, pp. 454–461. AUAI Press (2004)
15. Ristad, E.S., Yianilos, P.N.: Learning string edit distance. CoRR, cmp-lg/9610005 (1996)
16. Whang, S.E., Benjelloun, O., Garcia-Molina, H.: Generic entity resolution with negative rules. VLDB J 18(6), 1261–1277 (2009)
17. Winkler, W.E.: Methods for record linkage and bayesian networks. Technical Report Statistical Research Report Series RRS2002/05, U.S. Bureau of the Census, Washington, D.C. (2002)
18. Zhang, H., Yu, H., Xiong, D., Liu, Q.: Hhmm-based chinese lexical analyzer ictclas. In: Proceedings of the Second SIGHAN Workshop on Chinese Language Processing, vol. 17, pp. 184–187 (2003)

Staffing Open Collaborative Projects Based on the Degree of Acquaintance

Mohammad Y. Allaho[1], Wang-Chien Lee[1], and De-Nian Yang[2]

[1] The Pennsylvania State University, University Park, PA, USA
{mya111,wlee}@cse.psu.edu
[2] Academia Sinica, Taipei, Taiwan
dnyang@iis.sinica.edu.tw

Abstract. We consider the team formation problem in open collaborative projects existing in large community setting such as the *Open Source Software* (OSS) community. Given a query specifying a set of required skills for an open project and an upper bound of team size, the goal is to find a team that maximizes the *Degree of Acquaintance* (DoA) and covers all the required skills in the query. We define the DoA in terms of the team graph connectivity and edge weights, corresponding to the local *Clustering Coefficient* for each team member and the strength of social ties between the team members, respectively. We perform a statistical analysis on historical data to show the importance of the connectivity and social tie strength to the overall productivity of the teams in open projects. We show that the problem defined is NP-hard and present three algorithms, namely, PSTA, STA and NFA, to solve the problem. We experiment the algorithms on a dataset from the OSS community. The results show the effectiveness of the proposed algorithms to find a well acquainted teams satisfying a given query.

1 Introduction

Motivation. Large collaborative online communities have become a phenomena in the presence of Web 2.0 technology, witnessed by the massive success of Open Source Software (OSS) projects such as the Apache projects and GNU/Linux. As members of OSS projects are usually volunteers [12], they usually work out of personal goals/interests, e.g., practicing existing skills and gaining experience, following fellow peers, networking with the OSS community members, or simply supporting free open software projects [7]. Consequently, the amount of participation and commitment by the volunteering developers are crucial factors in the cause of OSS projects success [13].

Research on software engineering reveals a number of factors that assist in increasing the developers participation in OSS projects, including the computer language required, the operating system used, or the type of license for open software [14,4]. For example, in [14], it is observed that OSS projects requiring popular computer languages (e.g., Java and C variants) attract more participants since many developers are experienced in these languages. This finding suggests that developers tend to participate in a project if they possess the skill(s) required for that project. Recently, the importance of social factors, such as for a participant to join fellow peers or build a professional network with fellow developers in the community, for a project to ensure the rapport between group members, are noticed. Backstrom et al. [1] study the group formation and the membership growth in large open communities and find that the probability of an individual joining a community group increases as the number of friends inside the community and the internal connectedness of friends in the community increase. Hinds et al. [8] suggest that, when forming a new team, people tend to join others whom they have established work ties before. Moreover, Hahn et al. [6] point out that existing ties and relationships in OSS communities affect the formation of new project teams. It has been pointed out that prior collaborative ties among developers increases the probability of developers to join a new project where prior ties with the initiators exist. These studies suggest that existing ties (acquaintances) between individuals in open collaborative communities are crucial for those individuals to connect and join a new emerging project to form a team with rapport. These findings give a new insight to the effective factors in forming a successful team in open collaborative projects, which is our motivation to conduct this work. In this paper, we study the problem of team formation in the volunteer-based community of open collaborative projects.

Related Works. There exist several works that address the team formation problem in literature. However, they are not suitable for the open source projects that are based on volunteers. An early work by Barreto et al. [2] defines the staffing of software projects as a constraints satisfaction prob-

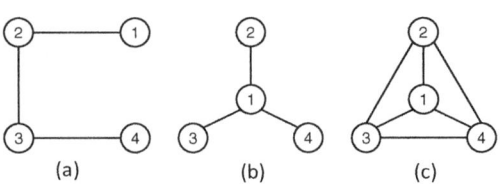

Fig. 1. Different graph shapes: (a) line, (b) star, (c) full graph

lem. The work considers only skills matching but does not consider the social ties between members. Other recent works optimize the team utility in terms of the team communication cost, which is mostly measured by different graph distance measures in a connected graph. The first work involving the communication cost is the work of Lappas et al. [10], where they use the diameter and the spanning tree of the team graph to measure the team communication cost. Li et al. [11] enhance the Steiner Tree algorithm used in [10] to solve a generalized team formation problem that assigns different number of experts for each required skill as a constraint. Both [10] and [11] tend to add mediator members, i.e., members that do not possess the skills required, to the team in order to minimize the communication cost. Nevertheless, it is worth noting that mediator members, in most cases, are not involved in OSS projects because these mediators will not feel the obligation to participate in open projects if they do not possess the required skills [14]. Another issue is that OSS projects usually consist of volunteers that sometimes form several subgraphs or even individual members that work on their own [3]. Therefore, it is difficult to optimize the communication cost in OSS projects. Recently, Gajewar et al. [5] introduce a measure of team communication quality based on a graph density which is claimed to be more robust than the diameter and the spanning tree measure, and thus more suitable for modular graph structures. However, the density measure ignores the aspect of graph structure and connectivity between the team members. Recall that a graph density is the total edges weights divided by the total number of vertices in a graph. In Figure 1, assuming all edge weights in graphs (a), (b) and (c) equal one. The density of graphs (a) and (b) is 0.75, regardless of their different graph structures. However, in terms of social ties, each vertex has different tie structure, e.g., vertex 1 is connected to one vertex in graph (a), however, it is connected to all the vertices in graph (b) which indicates a different social influence for vertex 1 in the two graphs, (a) and (b). Moreover, when considering the diameter measure, both graphs (b) and (c) have diameter 2, while it is obvious that graph (c) is more socially tight than graph (b). In Section 3, we show a statistical evidence that each team member's connectivity and tie strength are important in increasing the team productivity. Therefore, we need a more fine-grained measure to find a well socially tight team with high participation and productivity outcome.

Our Contributions. In light of the above observations, this paper proposes a novel concept, which is called the Degree of Acquaintance (DoA), which seamless integrates the connectivity, measured using the local *Clustering Coefficient* for each member, and the strength of ties specified by the frequency of co-participation. Thus, given an emerging OSS project that requires members with certain skills, our goal is to create a team that covers all the required skills and maximizes the two social factors which are defined by the DoA of the team. In this paper, we have made a number of contributions in achieving our goal:

- Our key contribution is to account for the *connectivity* (local clustering coefficient) and *ties strength* for users (in a social network) in the team formation problem. This is fundamentally different than the other approaches that uses aggregated graph metrics such as graph Diameter or graph Density .
- Through a statistical analysis, we demonstrate that the two social factors (connectivity and ties strength) play crucial role in the contributors' productivity and commitment in open projects (refer to Section 2.2). Therefore, we define the Degree of Acquaintance (DoA) based on these two factors.
- We formulate the problem of team formation based on DoA and prove that it is NP-hard and hard to approximate problem (refer to Section 3).
- We propose three new algorithms to solve the DoA team formation problem. The first one, *Partial Selective Tree search Algorithm* (PSTA), and the second one, *Selective Tree search Algorithm* (STA), are based on a BFS tree search, where PSTA produces the optimal solution but is less scalable while STA is more scalable but does not guarantee optimality. The third algorithm, *Neighbor-First search Algorithm* (NFA), is an efficient and scalable greedy approach (refer to Section 4).

– We evaluate the scalability and performance of the proposed algorithms and two existing approaches (i.e., Diameter and Density based) using real data from *ohloh.net* that includes hundreds of thousands of developers and over a million relationships. Experimental results show that the proposed algorithms can find the teams with high DoA in much higher magnitude than the existing approaches (refer to Section 5).

The rest of the paper is organized as follows. The following section defines the DoA and gives statistical analysis on the DoA factors. Section 3 defines the DoA based team formation problem and discusses the DoA properties used in our algorithms. Section 4 discusses in details the proposed algorithms, and Section 5 evaluates the proposed algorithms. Finally, Section 6 concludes the paper.

2 The Degree of Acquaintance

The social network of an OSS community is modeled as an undirected graph $G(X, E)$, where $X = \{x_1, ..., x_n\}$ is the set of n vertices that represent all the active developers in the community, and E is the set of weighted edges that represent the relationships between developers. We use an $n \times n$ matrix M to present the social network. Also, let w_{ij} in M denotes the edge weight between individuals x_i and x_j.[1]

Let $S = \{s_1, ..., s_m\}$ be a universe of m skills. We define an $n \times m$ developer-skill matrix A, where the rows consist of n developers and the columns consist of m skills. Each element of A, denoted as $a_{i,j}$, is a binary value, where $a_{i,j} = 1$ indicates that developer x_i possesses skill s_j; and $a_{i,j} = 0$, otherwise. Each row i in A, denoted by $\mathbf{x}_i$, is a vector of skills possessed by developer x_i. Also, each column j in A, denoted by $\mathbf{s}_j$, is a vector of developers who possess skill s_j. We refer to $\mathbf{x}_i$ as the *developer profile* of x_i and $\mathbf{s}_j$ the *skill profile* of s_j.

Next, we define the Degree of Acquaintance (DoA) and then demonstrate, by statistical analysis, that the DoA factors are crucial for team overall productivity in OSS projects.

2.1 Definition of DoA

Let G be a collaborative community. The *Degree of Acquaintance (DoA)* for an individual x_i in a team T, where $T \subseteq G$, consists of two factors: (i) the total weights of edges incident to x_i, and (ii) the connectivity among x_i's neighbors. While the first factor is easy to understand, we exploit the *local Clustering Coefficient (CC)* of vertex x_i in graph T, defined in Eq. (1), to capture the second factor.

$$CC_T(x_i) = \frac{2(k_{N_i})}{|N_i|(|N_i| - 1)} \quad (1)$$

where N_i is the set of x_i's neighbors in T, and k_{N_i} is the number of edges connecting the vertices in N_i. Formally, let $w_{i,j}$ denote the edge weight between vertex

---

[1] In this paper, w_{ij} is the collaboration counts between individuals x_i and x_j normalized by dividing by the maximum weight in G.

x_i and x_j, the DoA for an individual x_i in a team T is the linear combination of the total weights and the CC factors as defined in Eq. (2).

$$DoA_T(x_i) = \alpha \left(\sum_{\forall j \in N_i} w_{ij} \right) + (1 - \alpha) CC(x_i) \qquad (2)$$

where $\alpha = [0, 1]$ is a control parameter to balance the two factors. A proper value for α would depend on the nature of the team desired in terms of connectedness or ties strength. In our experiment, we set $\alpha = 0.5$. The DoA is a team structure-dependent metric which does change when adding or removing vertices from T.

The DoA of a team T is defined as the summation of $DoA_T(x_i)$ for every vertex $x_i \in T$ as in Eq. (3).

$$DoA(T) = \sum_{\forall x_i \in T} DoA_T(x_i) \qquad (3)$$

To eliminate the naming confusion between Eq. (2) and Eq. (3), we refer to Eq. (2) as the *Individual DoA* ($IDoA$), and Eq. (3) as the *Team DoA* ($TDoA$). In the next section, we present a statistical evidence to demonstrate the importance of the two DoA factors on the contribution and commitment exerted by developers in the OSS projects.

2.2 Statistical Analysis on the DoA Factors

We conducted a statistical analysis on more than 1300 OSS projects of multiple team sizes and topics (see Section 5.1 for detail description of the dataset). Our main goal is to investigate the correlation between the two DoA social factors (connectivity and ties strength) and *the amount of contribution* and *commitment* observed in OSS projects. The amount of contribution for a developer in a project is measured by the average number of commits made per month (denoted as commits/month). In OSS projects, a commit is an update that a developer submits to a project. The commitment by a developer to a project is measured by the number of active months where a developer submitted at least one commit (denoted by months-work). The amount of contribution shows the average amount of work produced by a developer per month whereas the commitment shows the period of time that a developer stayed committed and active in a project. Moreover, to observe the effects of social factors on various sizes of projects, we group projects in our dataset into the following team size ranges: [5,10), [10,25), [25,50), [50,75), [75,100) and [100 and above].

Effect of Connectivity. First, we study the effect of connectivity on the amount of contribution and commitment, under different team size ranges. To proceed, we classify contributors as (i) Highly Connected (HC), (ii) Lowly Connected (LC); and (iii) Non-Connected (NC). The HC contributors are those with $CC > 0$, forming one or several cliques pattern. The LC contributors are those with $CC = 0$, forming a star, line or circle shape patterns. Finally, the NC contributors are those with no acquaintants. Figure 2(a) shows that the average commits per month is larger for HC contributors than those LC ones in most team sizes. Similarly, Figure 2(b) shows that the average months-work is larger for HC contributors than those LC ones in most team sizes. Moreover, both figures show that the NC contributors make the lowest contribution and commitment. These results show a strong statistical evidence that, in general, HC contributors contribute more and are more committed to open projects than LC contributors.

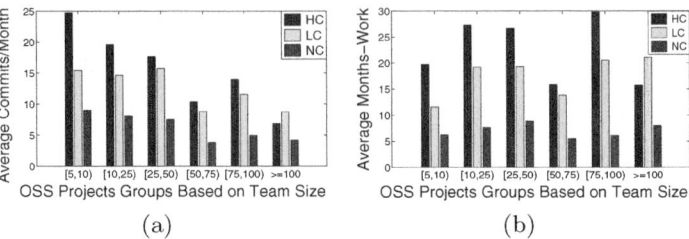

Fig. 2. Effect of connectivity on (a) Ave. Commits/Month and (b) Ave. Months-Work

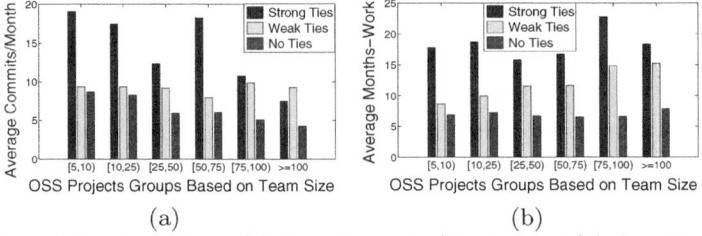

Fig. 3. Effect of tie strength on (a) Ave. Commits/Month and (b) Ave. Months-Work

Effect of Ties Strength. Next, we study the effect of tie strength on the amount of contribution and commitment, under different team size ranges. We measure the strength of a tie between two contributors by the frequency of their co-participations in projects (i.e., one co-participation is counted as one unit of link weight between two developers). Accordingly, we consider a contributor belonging to the Weak Tie set if she has no more than two units of link weights. Contributors with more than two units of link weights are in the Strong Tie set. Finally, contributors with no acquaintances are in the No Tie set. To proceed, we select the projects that show all the three types of tie weights (Strong, Weak and No Ties) in order to conduct the comparison. Figure 3(a) shows that the average commits per month is larger for the Strong Ties contributors than those Weak Ties ones in most team sizes. Also, Figure 3(b) shows that the average months-work is larger for the Strong Ties contributors than those Weak Ties ones in all team sizes. These results show a strong statistical evidence that, in general, contributors connected with Strong Ties make more contribution and commitment to open projects than the Weak Ties and No Ties contributors.

In both Figure 2 and Figure 3, very large teams (those with one hundred and above members) demonstrate different trends. We believe that, in these large teams, contributors with strong social ties may have more administrative and controlling roles. On the other hand, contributors with weaker social ties may commit more time to contribute to the project since they may join big projects to gain experience and form connections with reputable developers.

In summary, these statistical results show an obvious effect of connectivity and ties strength between developers on the amount of contribution and commitment in OSS projects. Therefore, taking into account these two factors in the *Team DoA* would improve the teams' productivity. Accordingly, we formulate the team formation problem and define a query in the next section.

3 The DoA Based Team Formation Problem

In this section, we formulate the team formation problem based on the notion of DoA, and present several properties useful to our proposed algorithms.

3.1 Problem Formulation

Definition. Degree of Acquaintance Based Team Formation (DoA-TF): *Given a social graph $G(X, E)$ and a developer-skill matrix A, a DoA based team formation query $Q = \{S_q, \tau\}$, where S_q is a set of skills required ($|S_q| = l$) and τ is the maximum number of developers allowed in a team, finds the set of developers to form a team T that covers all the skills required by Q such that $|T| \leq \tau$ and that $DoA(T)$ is maximized.*

Proposition 1. *The DoA-TF problem is NP-complete.*

Proof. We consider a special case of the DoA-TF problem, where $\alpha = 0$ and every vertex in G_F covers all required skills. In this case, only the Clustering Coefficient is considered. Then we prove the proposition by a reduction from the k-clique problem, a well known NP-complete problem. An instance of the k-clique problem consists of a graph $\hat{G}(\hat{X}, \hat{E})$ and k, where $\hat{X}$ is the set of vertices in $\hat{G}$, $\hat{E}$ is the set of edges in $\hat{G}$, and k is a positive integer. A decision problem version of the k-clique asks whether there exists a clique of size k in $\hat{G}$ or not.

We transform an instance of the k-clique problem to an instance of the DoA-TF special case problem by a direct mapping from $\hat{G}(\hat{X}, \hat{E})$ to $G_F(X_F, E_F)$. Having $\tau = k$, the solution for DoA-TF is a clique of size τ in G_F. Therefore, solving the DoA-TF special case problem instance can obtain the solution to the k-clique instance. The proposition follows. □

Note that finding the maximum clique problem is both NP-hard and hard to approximate (not approximable within $|X|^{(1-\epsilon)}$ for any $\epsilon > 0$) [9]. Consequently, the general DoA-TF problem is NP-hard and hard to approximate, which makes the problem very challenging.

3.2 DoA Properties

We aim to have, the *Team DoA* serving as the objective function for team formation. One may think that adding more members to a team would increase the objective function. However, Eq. (2) is not monotonic because the CC is not monotonic. As illustrated in Figure 4(a), $CC(x_3) = 1$. By adding x_4 and $e_{3,4}$ (see Figure 4(b)), CC for x_3 decreases since not all of its neighbors are connected.

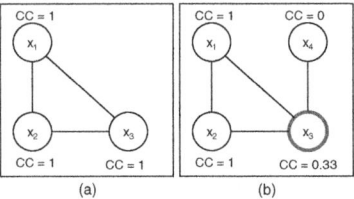

Fig. 4. Example of CC property

Nevertheless, the first term in Eq. (2), $\sum_{\forall j \in N_i} w_{ij}$, is monotonic. Therefore, if the second term in Eq. (2), $CC(x_i)$, does not decrease when adding candidate members to T during query processing, we can ensure the monotonicity of Eq. (2). Consequently, Eq. (3) becomes monotonic as well.

As monotonicity is important to assure the optimization of the objective function in any optimization algorithm, we introduce two cases where adding acquainted members does not change the clustering coefficient of the members in the team. In Case (1), assume all members in some team know each other, i.e., their CC equals 1. If a new member joins the team, where everyone in the team knows the new member, their clustering coefficient remains 1.

In Case (2), assume each member in some team knows at least one member in the team, but there is no mutual acquaintance among the members (i.e., their CC equals 0). If a new member joins the team, and the new member is acquainted to only one member in the team, then the CC remains 0 for each member in the team.

Our strategy to process the DoA-TF query, in Section 4, is to carefully examine the structure of the community graph in order to select individuals and construct a team subgraph in consecutive steps. As a result, when we add a vertex x to T, where x is part of a full-clustering structure (Case (1)), the objective function increases. Likewise, when we add a vertex x to T, where x is part of a zero-clustering structure (Case (2)), the objective function increases as well.

For a graph $G(X, E)$ with undirected weighted edges and a subgraph $T(\hat{X}, \hat{E})$, where $T \subseteq G$, such that $\hat{X} \subseteq X$ and $\hat{E} \subseteq E$, suppose we want to add a vertex $x_i \in X$ to T with k existing vertices in T that are neighbors to x_i, where $k \leq |N_i|$ (N_i is the set of neighboring vertices to x_i in G). When adding x_i to T, edges $e_{i,j_1}, e_{i,j_2}, \cdots, e_{i,j_k}$ are included in $\hat{E}$, then the objective function for T must be increased according to the following two lemmas:

Lemma 1. *If $CC_G(x_i) = 1$ and $CC_G(x_j) = 1, \forall x_j \in N_i$, then adding x_i to T increases the Individual DoA values for x_i and its k neighboring vertices in T, and as a result increases the objective function value.*

Proof. Since $CC = 1$ for x_i and $\forall x_j \in N_i$, then edges $e_{i,j_1,\ldots,k} \in \hat{E}$ are part of a full-clustering subgraph. Therefore, when adding x_i to T, only the edge weights dominate the objective function monotonically. □

Lemma 2. *If $CC_G(x_i) = 0$ and $CC_G(x_j) = 0, \forall x_j \in N_i$, then adding x_i to T increases the Individual DoA values for x_i and its k neighboring vertices in T, and as a result increases the objective function value.*

Proof. Since $CC = 0$ for x_i and $\forall x_j \in N_i$, then edges $e_{i,j_1,...,k} \in \hat{E}$ are part of a zero-clustering subgraph. Therefore, when adding x_i to T, only the edge weights dominate the objective function monotonically. □

4 Team Formation Algorithms

A straightforward approach to solve the DoA-TF problem is to find every team following the constraints and select the one with the maximum $TDoA$ value. Nevertheless, this approach is computation intensive and requires $O(2^n)$ time to find the optimal solution. With a vast OSS community, with over a million developer, this straightforward approach is not efficient. To address this issue, we propose three algorithms for DoA-TF.

4.1 Partial-Selective Tree Search Algorithm

The *Partial-Selective Tree search Algorithm* (PSTA) takes a tree-search approach to find the optimal team. The algorithm starts with developers with a required skill as seeds to grow search trees of team solutions. Therefore, the whole search space have multiple trees. Notice that a tree grows into lower level branches by adding a candidate team member one at a time. Thus, each node on a tree contains a partial team solution, denoted by $T_{d,b} \subseteq G_F$, where d is the *depth* level of the node, b is the *branch* count in level d and $G_F \subseteq G$ is the feasible graph containing every developer possessing the required skills in Q.[2] As mentioned, PSTA adds one vertex to each node (i.e., team) in the current level to create the child nodes in the next level. Therefore, the number of vertices in a node at level d equals d, $|T_{d,b}| = d$. In other words, the maximum level of each search tree is τ. PSTA algorithm is shown in Alg. 1.

First, lines 1-2 find G_F and calculates the $IDoA$ value for each vertex in G_F. Then, line 3 finds the rarest skill profile, $\mathbf{s}_{rare}$, in S_q, i.e., the rarest skill has the lowest number of developers possessing that skill. We choose the seeds from $\mathbf{s}_{rare}$ to start the tree search in order to minimize the number of search trees.

At each node, there exist a set of candidate developers/vertices, X_c. If the vertices in $T_{d,b}$ do not cover the query skills, i.e., $T_{d,b}$ is not a solution, then X_c contains the vertices possessing the uncovered skills. The crux of PSTA is to divide the vertices in X_c into monotonic and non-monotonic candidate sets. In details, X_c is divided into N and H sets, where N is the set of neighboring vertices to $T_{d,b}$ and H is the set of non-neighboring vertices. Moreover, N is divided into N_1, N_0 and N_p, where $CC_{G_F}(x_i) = 1$ for $\forall x_i \in N_1$, $CC_{G_F}(x_i) = 0$ for $\forall x_i \in N_0$, and $0 < CC_{G_F}(x_i) < 1$ for $\forall x_i \in N_p$. Vertices in N_1 follow Case (1) and vertices in N_0 follow Case (2) and thus are monotonic, while vertices in N_p

---
[2] In this paper, a node is related to a tree, while a vertex is related to a network graph.

Algorithm 1. The *Partial-Selective Tree Search* Algorithm

Input : $G(X, E)$; matrix A; query Q; α.
Output: $T^* \subseteq G$; $DoA(T^*)$.

1. *Init.*: $G_F(X_F, E_F) \leftarrow \bigcup_{\forall j \in S_q} \mathbf{s}_j$; $T^* = \Phi$;
2. $\forall_{x_i \in G_F} DoA_{G_F}(x_i)$;   // Calculate $IDoA$
3. $\mathbf{s}_{rare} \leftarrow \arg\min_{j \in S_q} |\mathbf{s}_j|$;
4. **seeds**$_0$, **seeds**$_1$ and **seeds**$_p$ $\leftarrow \mathbf{s}_{rare}$;
5. $T_{d,b++} \leftarrow$ **seeds**$_0$, **seeds**$_1$, **seeds**$_p$;
6. **for** $d = 1; d \leq \tau; d++$ **do**
7.   **for** $b = 1; b \leq BreadthSize; b++$ **do**
8.   Get T^*, $DoA(T^*)$ and X_c for $T_{d,b}$;
9. N_1, N_0, N_p, $H \leftarrow X_c$;
10.   **if** $N_1 \neq \Phi$ **then**
11. $x_{selected} \leftarrow \arg\max_{\forall i \in N_1, \forall j \in T_{d,b}} w_{i,j}$;
12. $T_{d+1, b_c++} \leftarrow T_{d,b} \cup x_{selected}$;
13.   **if** $N_0 \neq \Phi$ **then**
14. $x_{selected} \leftarrow \arg\max_{\forall i \in N_0, \forall j \in T_{d,b}} w_{i,j}$;
15. $T_{d+1, b_c++} \leftarrow T_{d,b} \cup x_{selected}$;
16.   **if** $H \neq \Phi$ **then**
17.   **seeds**$_0$, **seeds**$_1$ and **seeds**$_p$ $\leftarrow H$;
18. $T_{d+1, b_c++} \leftarrow$ **seeds**$_0$, **seeds**$_1$, **seeds**$_p$;
19.   **if** $N_p \neq \Phi$ **then**
20.   **foreach** $x_i \in N_p$ **do** $T_{d+1, b_c++} \leftarrow T_{d,b} \cup x_i$;

are not. Therefore, selecting vertices from N_1 and N_0 monotonically guarantees to optimize the solution and otherwise for N_p. PSTA selects the vertex with the highest total link wight from N_1 and N_0 (lines 11 and 14). This *forward-pruning* process reduces the search space tremendously. On the other hand, a selection from N_p does not guarantee to optimize the next level solution, therefore, PSTA creates a child solution from each vertex in N_p (line 20).[3]

We actually adopt the same idea in seeds selection, i.e., $\mathbf{s}_{rare}$ is divided into three sets (line 4), i.e., **seeds**$_1$, **seeds**$_0$ and **seeds**$_p$, corresponding to N_1, N_0 and N_p respectively. However, PSTA selects the vertex with the highest $IDoA$ in G_F from **seeds**$_1$ and **seeds**$_0$ if $\tau \geq |N_{x_{selected}}|$ to guarantee that $IDoA$ is obtainable after τ steps. Otherwise, each vertex in **seeds**$_1$ and **seeds**$_0$ grows as a separate search tree. Finally, vertices in set H are treated as seed sets since they are not connected to the current solution. The algorithm stops after processing all the nodes at level τ and the solution would be the team that satisfies the constraints and having the maximum $TDoA$.

---

[3] In lines 11, 14 and 20, the index b_c represents the branch count of the child nodes.

Proposition 2. *The PSTA algorithm finds the optimal solution for a given query Q.*

Proof. According to Lemma 1, we conclude that any selected vertex $x \in N_1$ added to solution T does optimize the objective function of $T \cup x$. Also, according to Lemma 2, we conclude that any selected vertex from set $x \in N_0$ added to solution T does optimize the objective function of $T \cup x$. Since the selection process is monotonic and each vertex from set N_p is added to solution T, the PSTA algorithm assures finding the optimal solution. □

In the above proof, the set H is ignored in the argument because it creates a subgraph that follows the same procedure of creating solutions for the whole graph. The time complexity of PSTA is $O(n^\tau)$. The complexity comes close to the upper bound if at each node the candidate vertices are in set N_p. However, in reality the time complexity is much smaller than the upper bound. Yet, PSTA is not scalable to large graphs. Therefore, we developed the complete *Selective Tree search Algorithm* (STA) to mitigate the scalability issue and bring it to a practical level.

4.2 Selective Tree Search Algorithm

The *Selective Tree search Algorithm* (STA) is the same as the PSTA algorithm except that it uses the *forward-pruning* on each selection set at each node. The seed selection is as follows. For sets **seeds**$_0$, **seeds**$_1$ and **seeds**$_p$, STA always selects the vertex with the maximum $IDoA$ from each set. Hence, the maximum fan-out of the root is three. At each node, from level one to level $\tau - 1$, STA selects the vertex with the maximum $IDoA$ from sets N_p and H. The selection heuristic for N_0 and N_1 is the same as in PSTA. Thus, the tree fan-out is reduced tremendously. STA is scalable but does not guarantee an optimal solution. STA may still not scale to large graphs and queries with large team size upper bounds. To further address the scalability issue, we introduce a polynomial-time algorithm in the following section.

4.3 Neighbor-First Search Algorithm

We developed a greedy algorithm, called *Neighbor-First search Algorithm* (NFA), which uses $IDoA$ as a heuristic for vertex selection. The NFA algorithm, shown in Algorithm 2, starts by finding the feasible graph G_F. Then it selects a seed vertex from G_F with the maximum $IDoA$ and adds the seed vertex into the team solution set T_t (lines 2-3), where T_t is the team set at step t. With T_1 containing the seed vertex, the algorithm proceeds by entering a while loop. At the start of each iteration (line 5), NFA finds the candidates set X_{c_t}, which contains the developers possessing the uncovered skills and $G_F - T_t$ if all skills are covered by T_t. Moreover, it calculates the objective function if T_t is valid.

Having X_{c_t}, NFA proceeds by finding the set of vertices neighboring to the vertices in T_t (line 6), denoted by $N_t \subseteq X_{c_t}$. Afterwards, if N_t is not empty, NFA selects the vertex with maximum $IDoA$ from N_t to the current solution

Algorithm 2. The *Neighbor-First* Algorithm

Input : $G(X, E)$; matrix A; query Q; α.
Output: $T^* \subseteq G$; $DoA(T^*)$.

1 *Init.*: $G_F(X_F, E_F) \leftarrow \bigcup_{\forall j \in S_q} \mathbf{s}_j$; $T_1 \leftarrow \Phi$; $t = 1$;
2 $x_{seed} \leftarrow \arg\max_{i \in X_F} DoA_{G_F}(i)$;
3 $T_1 \leftarrow x_{seed}$;
4 **while** $|T_t| \leq \tau$ **do**
5   Get T^*, $DoA(T^*)$ and X_{c_t} for T_t ;
6     $N_t \leftarrow \{i | i \in X_{c_t} \land e_{i,j \in T_t} \neq \Phi\}$;
7   **if** $N_t \neq \Phi$ **then**
8         $x_{selected} \leftarrow \arg\max_{i \in N_t} DoA_{G_F}(i)$;
9   **else**
10        $x_{selected} \leftarrow \arg\max_{i \in X_{c_t}} DoA_{G_F}(i)$;
11    $T_{t+1} \leftarrow T_t \cup x_{selected}$;

T_t. On the other hand, if there are no vertices neighboring to T_t, then the algorithm selects the vertex with the maximum $IDoA$ from X_{c_t} to T_t (lines 7-11). The algorithm stops when the team size reaches τ, and the result is the team formation T^* with the maximum $TDoA$ obtained. Again the team solution may not be necessary of τ members as τ is only the upper bound of team size. The time complexity of NFA is $O(\tau(\frac{n^2}{4}) + nl)$ or simply $O(n^2)$.

5 Performance Evaluation

In this section we evaluate the proposed algorithms NFA, PSTA, and STA in addition to the Brute-Force Approach (BFA). Also, we compare the proposed algorithms with two existing approaches that form teams based on graph *Diameter* [10] and *Density* [5].

5.1 Dataset

The real dataset, in this paper, is collected from *Oholoh.net*, a fast growing OSS social site. It is an online community web service that provides a platform for developers and users to interact. There are over 600,000 developers in *Ohloh.net* with over 1,096,000 relationships and 83 different skills, mainly related to computer languages. *Ohloh.net* is unique because, first, it hosts OSS projects from multiple version control repositories (e.g. Subversion, CVS, Git, Mercurial, etc.) which increases the number of OSS projects hosted and includes various software topics. Second, and more importantly, it provides social ties information for the developers, such as the recognition and approval network, where developers are allowed to explicitly express approval and recognition to each other based on previous collaboration. In the dataset, we only consider the developers who have announced the skills they possess and contributed to OSS projects. Finally, we realize that many previous work use the DPLB dataset but we opted not to use it since it does not represent the sentiment of open projects environment which is the focus of this paper.

5.2 Scalability and Accuracy Evaluation

Scalability Evaluation. First, we evaluate the scalability of the proposed algorithms in terms of the number of Execution Runs (ER) and the Execution Time (ET) with different τ and G_F sizes, i.e., $\tau = [3, 20]$ and $|G_F| = [26, 7748]$, respectively. The size of G_F depends on the number of skills specified in a query, thus the increment of the graph size is not equally distanced in the experiment. The ER represents the number of iterations for BFA and NFA, while it is the number of nodes in the search tree for the PSTA and STA. Figure 5(a) shows the ER scalability as G_F increases. BFA and PSTA ER increase exponentially, however, PSTA ER is smaller than BFA. NFA ER is constant since it, always, iterates τ times. STA ER is higher than NFA but does not increase exponentially. Figure 5(b) shows the ET as G_F increases. Again the BFA and PSTA ET increase exponentially, while the STA and NFA ET scale well on large graphs.

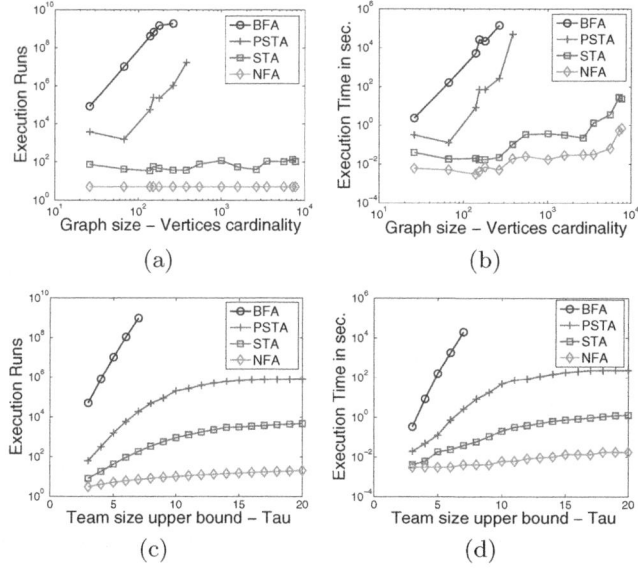

Fig. 5. Scalability Performance: (a) Execution Runs vs. Graph Size. (b) Execution Time vs. Graph Size. (c) Execution Runs vs. τ. (d) Execution Time vs. τ. (log scale).

Furthermore, we evaluates the scalability when τ increases. Figure 5(c) shows how the ER scales as τ increases. PSTA and STA converge as τ increases because as the tree search explores more levels, the choices become limited and fewer nodes (runs) are created. BFA ER increases exponentially, and NFA ER increases linearly. Figure 5(d) shows how the ET scales as τ increases. Again it follows the same trend as the ER.

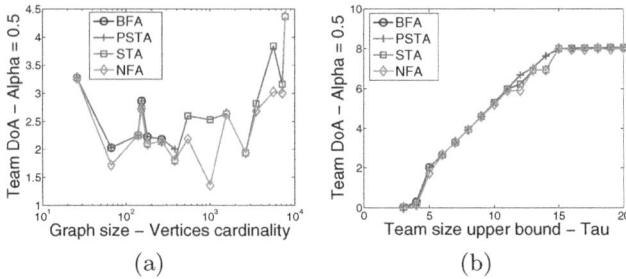

Fig. 6. Accuracy: (a) TDoA vs. Graph Size (log scale on x-axis). (d) TDoA vs. τ.

Accuracy Evaluation. Figure 6 evaluates the accuracy, which is the difference between the *Team DoA* value and the optimal one for different graph sizes and τ. Figure 6(a) plots the $TDoA$ value for each algorithm with different G_F sizes. BFA and PSTA $TDoA$ results are identical since they output the optimal solution. The first seven runs compare the STA and NFA results to the optimal solution; beyond run seven the computation becomes excessive for BFA and PSTA while the graph size increases. Meanwhile, the other runs only compare the difference between STA and NFA, where higher $TDoA$ value is preferred. Figure 6(b) shows the $TDoA$ value for each algorithm as τ increases. From Figure 6(b), the BFA and PSTA $TDoA$ results are identical since they output the optimal solution. We calculate the Mean Absolute Error (MAE) for both STA and NFA, where STA algorithm's MAE equals 0.0797, and NFA algorithm's MAE equals 0.1506.

5.3 Comparison with the Graph Diameter and Density Approaches

We compare the proposed algorithms in terms of *Team DoA* with two prominent team formation approaches of different objectives. The first approach finds the team with the smallest graph diameter in an effort to minimize the communication cost in a team. This approach is implemented by the *RarestFirst* algorithm in [10] denoted by *Diameter*. The second approach finds the team with the maximum density in [5] and denoted by *Density*.

Experiment Setup. The *Diameter* approach is a minimization problem, where it treats edge weights as distances, i.e., the higher the weight is the farther the distance is, and vise versa. In order to apply the *Diameter* approach on our dataset, we take the reciprocal of each edge weight, thus, a high edge weight indicates a closer distance, and vise versa. Also, we assign a high weight between not connected vertices as a penalty. On the other hand, the *Density* approach is a maximization problem, thus, we can apply it directly on our dataset.

Since the *RarestFirst* algorithm iterates every skill set and selects one member from each set that minimizes the diameter, the algorithm often results in a team cardinality identical to the number of required skills. In contrast, our proposed algorithms have an upper bound for the team size. Therefore, if $\tau > l$, the solution of the *RarestFirst* algorithm will always have fewer members than our algorithms. Also, if $\tau < l$, the solution of the *RarestFirst* algorithm will always have more members than our algorithms. Therefore, on each experiment we assign $\tau = l$ as in Figure 7(a). On the other hand, the *Density* approach does not have this restriction.

Results. Figure 7(a) compares *TDoA* results for STA, NFA and *Diameter* as l and τ increase. It shows that *Diameter* outperforms NFA in small teams with $\tau \leq 4$ but its performance starts degrading tremendously on finding larger teams. Similarly, Figure 7(b) shows *TDoA* results for STA, NFA and *Density* as τ increases for each l. It shows that *Density* performs similar to NFA in small teams ($\tau = 5$), but its performance starts degrading tremendously on finding larger teams. Moreover, Figure 7(c) shows *TDoA* results for STA, NFA and *Density* as l increases for each τ. Figure 7(c) shows that STA and NFA, consistently, outperform *Density*. These results show that our proposed approaches outperform the *Diameter* and *Density* approaches in finding a well acquainted team members with high connectivity and tie weights.

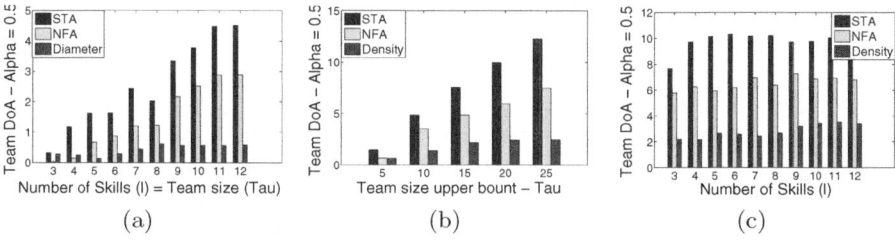

Fig. 7. The *Team DoA* comparison for STA, NFA, *Diameter* in (a), *Density* with changing τ in (b), and *Density* with changing l in (c)

6 Conclusion

In this paper, we defined a new DoA based team formation problem and proved that it is NP-hard. We proposed three algorithms, namely PSTA, STA and NFA, to address the problem. We evaluated the proposed algorithms on a real dataset collected from OSS community that consist of over 600,000 developer and over 1,096,000 relationships. The PSTA is proved to find the optimal solution, and STA and NFA demonstrated scalable performance. Also, the proposed algorithms outperform the *Density* and *Diameter* approaches in maximizing the *Team DoA*. Furthermore, we presented a statistical analysis to demonstrate the influence of the DoA factors on OSS teams' overall productivity and found a strong influence.

References

1. Backstrom, L., Huttenlocher, D., Kleinberg, J., Lan, X.: Group formation in large social networks: membership, growth, and evolution. In: Proceedings of the 12th ACM SIGKDD, KDD 2006, New York, NY, USA, pp. 44–54 (2006)
2. Barreto, A., De O Barros, M., Werner, C.M.L.: Staffing a software project: A constraint satisfaction and optimization-based approach. Computers & Operations Research 35(10), 3073–3089 (2008)

3. Bird, C., Pattison, D., D'Souza, R., Filkov, V., Devanbu, P.: Latent social structure in open source projects. In: Proceedings of the 16th ACM SIGSOFT, SIGSOFT 2008/FSE-16, New York, NY, USA, pp. 24–35 (2008)
4. Casaló, L.V., Cisneros, J., Flavián, C., Guinaliu, M.: Determinants of success in open source software networks. Industrial Management and Data Systems 109(4), 532–549 (2009)
5. Gajewar, A., Sarma, A.D.: Multi-skill collaborative teams based on densest subgraphs. CoRR, abs/1102.3340 (2011)
6. Hahn, J., Moon, J.Y., Zhang, C.: Emergence of new project teams from open source software developer networks: Impact of prior collaboration ties. Information Systems Research 19(3), 369–391 (2008)
7. Hertel, G., Niedner, S., Herrmann, S.: Motivation of software developers in open source projects: an internet-based survey of contributors to the linux kernel. Research Policy 32(7), 1159–1177 (2003)
8. Hinds, P.J., Carley, K.M., Krackhardt, D., Wholey, D.: Choosing work group members: Balancing similarity, competence, and familiarity. Organizational Behavior and Human Decision Processes 81(2), 226–251 (2000)
9. Håstad, J.: Clique is hard to approximate within $n^{(1-e)}$. Acta Mathematica, 627–636 (1996)
10. Lappas, T., Liu, K., Terzi, E.: Finding a team of experts in social networks. In: Proceedings of the 15th ACM SIGKDD, KDD 2009, New York, NY, USA, pp. 467–476 (2009)
11. Li, C.-T., Shan, M.-K.: Team formation for generalized tasks in expertise social networks. In: Proceedings of the 2010 IEEE Second International Conference on Social Computing, SOCIALCOM 2010, Washington, DC, USA, pp. 9–16 (2010)
12. Michlmayr, M., Hill, B.M.: Quality and the reliance on individuals in free software projects. In: Proceedings of the 3rd Workshop on Open Source Software Engineering, Portland, OR, USA, pp. 105–109 (2003)
13. Robles, G., Gonzalez-Barahona, J.M., Michlmayr, M.: Evolution of volunteer participation in libre software projects: Evidence from Debian. In: Proceedings of the First International Conference on OSS, Genova, Italy, pp. 100–107 (2005)
14. Subramaniam, C., Sen, R., Nelson, M.L.: Determinants of open source software project success: A longitudinal study. Decis. Support Syst. 46(2), 576–585 (2009)

Who Will Follow Your Shop? Exploiting Multiple Information Sources in Finding Followers

Liang Wu[2], Alvin Chin[1], Guandong Xu[3], Liang Du[5],
Xia Wang[4], Kangjian Meng[4], Yonggang Guo[4], and Yuanchun Zhou[2]

[1] Xpress Internet Services, Nokia, Beijing, China
[2] Computer Network Information Center, Chinese Academy of Sciences
[3] Advanced Analytics Institute, University of Technology Sydney, Australia
[4] Beijing NoyaXe Technologies Co. Ltd., China
[5] Institute of Software, Chinese Academy of Sciences, China
alvin.chin@nokia.com, {wuliang,zyc}@cnic.cn
guandong.xu@uts.edu.au, xia_s_wang@yahoo.com, duliang@ios.ac.cn

Abstract. WuXianGouXiang is an O2O(offline to online and vice versa)-based mobile application that recommends the nearby coupons and deals for users, by which users can also follow the shops they are interested in. If the potential followers of a shop can be discovered, the merchant's targeted advertising can be more effective and the recommendations for users will also be improved. In this paper, we propose to predict the link relations between users and shops based on the following behavior. In order to better model the characteristics of the shops, we first adopt Topic Modeling to analyze the semantics of their descriptions and then propose a novel approach, named INtent Induced Topic Search (INITS) to update the hidden topics of the shops with and without a description. In addition, we leverage the user logs and search engine results to get the similarity between users and shops. Then we adopt the latent factor model to calculate the similarity between users and shops, in which we use the multiple information sources to regularize the factorization. The experimental results demonstrate that the proposed approach is effective for detecting followers of the shops and the INITS model is useful for shop topic inference.

Keywords: User Behavior, Location Based Services, Matrix Factorization.

1 Introduction

The growth of intelligent mobile devices like smart phones and tablets have contributed to the popularity of location-based services. The convenience of mobile devices with GPS and wireless technologies has now enabled the linking between offline physical location and online access. A hot business model for mobile service providers to create profit is O2O(offline to online and vice versa) commerce, which helps the shops to find new customers by guiding the online users to the real stores in the physical world.

In this paper, we aim to predict the follow relationship between users and shops, which can be formalized as a relational prediction task. However, in real applications there exists many difficulties for modeling the relationships due to the following reasons: 1)*Heavily-tailed distribution:* The distribution of the relationships between users and shops are heavy-tailed. 2)*Sparsity*: The relationships between users and shops are sparse. 3)*Incompleteness*: The information of the merchants is incomplete. The descriptions and even the names for many shops are lost. A basic intuition to tackle these problems is to use some complementary data. Thus, we propose to exploit some useful information sources from multiple heterogeneous domains. In addition, based on the link analysis algorithm Hyperlink-Induced Topic Search (HITS) [4], we propose a novel approach called INtent Induced Topic Search (INITS), which exploits the user intent to predict the hidden topics of a shop by analyzing the contextual information. The contributions of this paper lie in four aspects: 1)Based on our mobile application WuXianGouXiang, we put forward a new problem for discovering the followers of a shop, by which we can offer the better recommendations for users and help merchants improve the effectiveness of targeted advertising. 2)We design a novel approach called INITS to predict the hidden semantics of the shops. By analyzing the contextual information of user behaviors, the correlations between the topics and each of the shops are discovered to tackle the incompleteness of shop description and reveal the in-depth topic distribution of shops. 3)We exploit several useful auxiliary data to tackle the problems of sparsity and cold-start and matrix factorization is adopted to combine the heterogeneous information sources in a unified manner to solve the optimization problem in global scale. 4)We evaluate our method using real-world data collected by WuXianGouXiang.

The remaining sections of the paper are organized as follows: In Section 2, we present some related work. Section 3 describes the application, the motivations and the architecture of our proposed approach. The shop topic modeling and the INITS algorithm are discussed in Section 4 and Section 5. In Section 6, we present the proposed algorithm to predict the followers of a shop. The experiments are shown in Section 7. Section 8 concludes the paper.

2 Related Work

In this section, we will describe some related work on the location-based recommendation, the link prediction in social networks and the collective link prediction.

Location-Based Recommendation: As the mobile applications become popular among users, more information like the location, time and the user behaviors can be collected from the intelligent devices. The recommender system on mobile devices can now provide help based on the contextual information. In [16,17], Zheng et al. propose novel approaches to model the user trajectories and the locations, they recommend the locations and the travel package based on the users' GPS logs. Park et al. propose to recommend the restaurants by taking the user preferences and the location contexts[19]. Bayesian learning is incorporated to compute a score for each shop thus providing recommendations.

Link Prediction in Social Networks: When a user follows a shop, the user can be regarded as a follower or a fan of the shop, which is very similar to the social network like Facebook and Twitter. So there exist many related work which also focus on link prediction in social networks. In [1], Backstrom et al. leverage the training data and the random walk model to predict the relational links. The polarity of the relational link is also considered in [5], where a logistic regression model is used to predict the sign of the edge based on the social sciences. More approaches can be found in [13,8,10] and a survey [7] reviews some of them. Different from these methods, our model is designed for recommending the shop following behavior. Since our problem is obtained from a real world application, the information is very sparse and incomplete, so we leverage other data sources to tackle this while the methods mentioned above model the relationship between the user and the item individually.

Collective Link Prediction: In this paper, we leverage several useful information sources to solve the problem of sparsity, cold-start and data incompleteness. Thus, our work is similar to the multi-relational learning problem. In [11], Singh et al. propose a framework to collectively factorize the matrices, where the same entities in different relationships share the same coordinates in latent spaces. In [14], Xu et al. extend such collective matrix factorization models to a Gaussian processes-based nonparametric Bayesian framework. In [15,18], the tensor is composed to provide both the location recommendation and the activity recommendation simultaneously. Unlike the approaches which tackle multiple tasks at the same time, our approach factors only one matrix, the user-shop matrix, i.e., we only predict the links between users and shops. Another similar work is [6] where Li et al. propose to improve One-Class Collaborative Filtering by exploiting the rich user information. However, the difference in our approach is that besides the user information, we also use the shop information from the search engine, the topic model and the INITS model.

3 Overview

In this section, we present a brief introduction of our mobile application WuXianGouXiang and the architecture of the proposed approach.

3.1 WuXianGouXiang

Nokia Research Center developed a mobile application called WuXianGouXiang in April 2011, which is an O2O-based mobile application, with over 20,000 registered users as of April, 2012. The application can guide the online users to the real shops by offering users the deals and coupons of nearby merchants. The users can download the coupons they like and follow the shops to know their latest deals. WuXianGouXiang has several versions including Java, Symbian and Android, and can be downloaded from http://www.gouxiang.com.

3.2 System Architecture

Figure 1 illustrates the framework of our proposed approach. In order to solve the cold-start and sparsity problems, we propose to leverage heterogeneous information sources from other domains. Firstly, we extract the data of the most frequent behaviors, including downloading coupons and clicking products, from the user logs archived in the application server to measure the user-user similarity. Secondly, we used the shop names as queries and use the aggregated search results to compute the shop-shop similarity. Thirdly, we adopt the topic modeling technique to mine the topics of the shops. To alleviate the data incompleteness, we propose the INITS algorithm to estimate the topic distribution of the shops. The proposed model incorporates the user's intents into the HITS link analysis algorithm, and infers the topics by analyzing the contextual information. Based on the useful information sources, Matrix Factorization is used to decompose them and predict the relational link between the shops and the users. All the above steps will be introduced in the following sections.

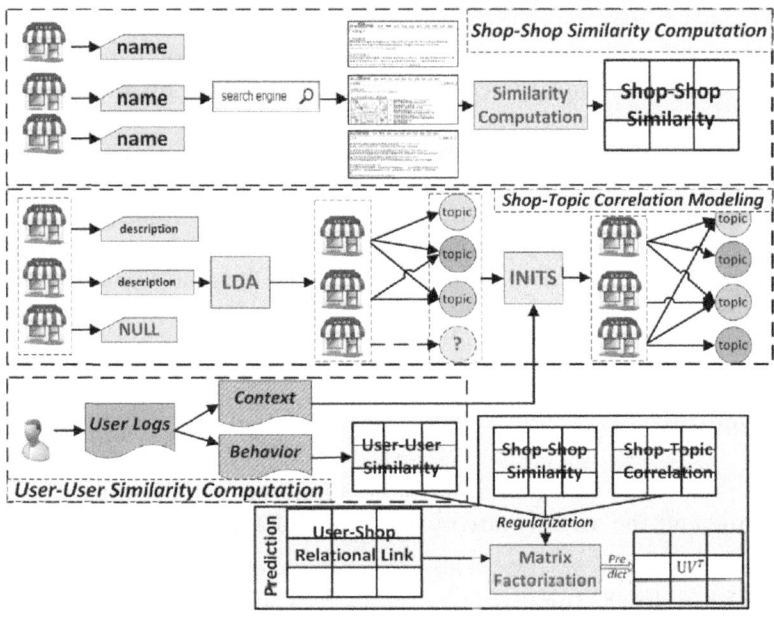

Fig. 1. The framework of the proposed approach on relational link prediction

4 Shop Topic Modeling

When using WuXianGouXiang, the merchants can write a description for their shops. The descriptions contain rich information about the services they provide. The features of the shops are characterized by the words of the descriptions and also by the hidden topics underlying the bag of words. Thus, we adopt the Latent Dirichlet Allocation (LDA)[3] to discover the hidden topics of the descriptions to better model the shops, which is a probabilistic topic model that has been proven to be useful for extracting the latent semantics of documents. Table 1 depicts the high weighted words of some topics. Here we set the number of latent topics k as 10 based on experiments, since some redundant and meaningless topics may be produced given a larger k and some important topics may be neglected when k becomes smaller.

Table 1. The examples of topic-word distribution of experimental results from shop descriptions

Topic 1	ice cream	queen	dairy	instant
Topic 2	italian	spa	luxury	vip
Topic 3	mcdonald	subway	american	hamburger

5 Intent Induced Topic Search

As discussed in Section 4, LDA can discover the topics of the shops by analyzing their descriptions. Since the descriptions are generated by the merchants, there exist many difficulties when analyzing the user generated contents. A major difficulty is that we cannot judge the correctness and quality of the descriptions, many of the merchants even refuse to write one for their shops. If the contents are directly used for the shops' topic inference, noises may be brought in and the features of a shop will be characterized by the shop owner, rather than the customers. Therefore, to circumvent this problem, we propose a novel approach, INtent Induced Topic Search (INITS) to estimate the topic distributions of shops.

Since simply relying on the merchants' descriptions alone will bring in unavoidable bias, we aim to exploit the user ratings to alleviate it. Though there are no explicit ratings between users and shops, the user behaviors can be regarded as implicit feedbacks. In particular, if a shop is visited by many experienced users online, it is highly possible to be a good shop; on the other hand, if a user visits many good shops, she is very likely to be a shopping expert, i.e., there is a mutual enforcement relation between the users and shops. This indicates that the Web page ranking algorithm (HITS) [4] is applicable here. To better study the relationship between users and shops, we construct a bipartite graph, where the users and shops are the two sets of vertices and a directed link is built if a user visits the deals of a shop. As all the edges are pointing from the users to the shops, every shop will get an authority score after applying the HITS algorithm.

Based on the authority scores, we can get the overall popularity of each shop. To further discover the authority scores of each topic of a shop, we propose a novel approach, named INtent Induced Topic Search (INITS) algorithm based on the user contexts. The basic intuitions are as follows: When a user visits a shop online (e.g., downloading the coupons of a shop), she may be interested in a hidden topic the shop bears and the timestamp when the user visits the shop can represent the user's intent, for instance, if a user viewed McDonalds at 11:00 and visited KFC at 11:20, the intents of the user for viewing them are probably the same, which most likely is the user wants to find something to eat. If McDonalds has a latent topic which is "fast food", but KFC does not, then we can possibly infer that KFC is also fast food based on the view and visit.

Theoretically, each shop has k hidden topics. The INITS model assigns the shop's authority score to the hidden topics based on the user's intent. In the HITS algorithm, a shop's authority score is the sum of the hub scores of all the users which have visited it. Given that a user visited a shop h times, then we say the user contributes $\frac{1}{h}$ of its hub score to the shop for each visit. The INITS algorithm assigns the authority score of the shop to each of the shop's latent topic according to the following intuitions: 1): Given a user's log data, when a user visits from one shop to another, if the $\Delta time$ is within a certain threshold, we say that the intents of the user do not change, and the similar topics of the two shops gain a larger share of the $\frac{1}{h}$ of the hub score, e.g., if a user visits McDonalds and KFC successively, and the topic value of "fast food" is similar for both the shops, then the topic of "fast food" of the KFC store will get a higher authority score. 2): If the intents of the user change, i.e., the user may seek for different topics, therefore, the different topics should get a larger share. 3): The topic with a richer value should get a higher authority score, e.g., if a store is famous for the fast food, the user should be more interested in the fast food of it.

We show an example of the INITS model below:

Given that a user first visits shop $shop_i$ and then visits shop $shop_j$, the example shows how much authority score a hidden topic $p \in \{1, \ldots, k\}$ of shop $shop_j$ gains for this visit.

Notations:

k: The number of latent topics.

S^p: The share of the authority score that topic p gets. If the topic is to get a larger share of the authority score, it should have a larger S^p. $\frac{S_p}{\sum_{q=1}^{|k|} S_q}$ is the ratio of the authority score the topic gets.

c_i^p: The correlations between $shop_i$ and the hidden topic p. For the shops which have no descriptions, each topic will be assigned with a default value of $\frac{1}{k}$.

T_i: The time when the user visits shop $shop_i$.

T_{thrd}: The time threshold. If the time between the two visits exceeds the threshold, we say the user's intent changes.

$T_{i \cdot hour}$: The hour of T_i, e.g., if T_i is 13:00, $T_{i \cdot hour}$ equals 13.

hub: The hub score of the user.

$AuthorityGain_j^p$: The gain of the authority score of $shop_j$'s topic p.

N_j: The number of times that the user visits $shop_j$.

Equation 1 computes the dissimilarity of the authority score of topic p between the two shops. For some shops which have a same value of a topic, the dissimilarity will be 0, which will cause the division by 0 in equation 4. Thus, μ is used to avoid this. On the other hand, the S_p cannot be larger than 1, so the μ should be small enough. In this paper, we fix it as 0.01 for simplicity.

$$S_p = |c_i^p - c_j^p| + \mu \tag{1}$$

Equation 2 calculates the time interval between the two visits.

$$\Delta T = T_j - T_i \tag{2}$$

$$ind = \begin{cases} 1, \Delta T > T_{thrd} & AND \quad T_{i \cdot Hour} \neq T_{j \cdot Hour} \\ -1, \Delta T < T_{thrd} & OR \quad T_{i \cdot Hour} = T_{j \cdot Hour} \end{cases} \tag{3}$$

In Equation 3, the indicator ind indicates whether the user intent has changed, that is, if the time span between two visits is within the time threshold, it is highly probable that the user intents are similar. For some users who do not use the application frequently, they may visit one shop and then another shop after one or several days, though the time span exceeds the threshold, their intents may hold. Thus, if the hour of day of the two visits are the same, we say the intents do not change.

$$S_p' = (S_p)^{ind} \times c_j^p \tag{4}$$

$$AuthorityGain_j^p = \frac{1}{N_j} \times hub \times \frac{S_p'}{\sum_{q=1}^{|k|} S_q'} \tag{5}$$

Equation 5 computes the authority score that is assigned to the topic p, where S_p' is the ratio the topic occupies and $\sum_{q=1}^{|k|} S_q'$ is the sum of all the topics' share. Equation 4 computes the share of the topic p. ind is used for implementing the second intuition: if the user's intent changes, ind will make the share of the more different topics bigger, and vice versa. c_j^p is used to implement Intuition III, which makes the share of the topics with a larger value bigger.

Based on the INITS model, the correlations between the hidden topics and the shops which have no descriptions and low-quality descriptions can also be estimated. In addition, the topics of the shops with descriptions will also be updated based on user's intents. Thus, after updated by INITS, the problems of sparsity and incompleteness are relieved.

6 Followers Discovery

In this section, we will introduce the proposed approach for predicting the followers of shops, which leverages several auxiliary information sources to help the prediction.

6.1 User Information Extraction

As introduced in Section 3.1, the users can download and use the digital coupons in offline shops, view the deals and the shops when using WuXianGouXiang. The behaviors of the users can reflect their habits and preferences. If the users have similar preferences, then they tend to follow similar shops. We extract several common online actions to model the user behaviors, including downloading deals, coupons, get the latest deals and expiring deals. The downloaded information like the price, discount, description, the unique ID of the products, deals and the corresponding shops, are obtained from the logs.

Since we have thousands of deals and coupons in WuXianGouXiang, the Cartesian product of the actions and the deals is very huge, which makes the action vectors of users sparse. Thus, we replace the deals and coupons with their shops. Then we build a vector for each of the users as follows:

$$\boldsymbol{user_i} = <\#action_1(user_i), \ldots, \#action_n(user_i)> \qquad (6)$$

where $\#action_j(user_i)$ is the number of the $action_j$ performed by the $user_i$. An action contains a behavior and the object(shop) of the behavior. The similarity of any two users is calculated by cosine measure:

$$CosSim(\boldsymbol{user_i}, \boldsymbol{user_j}) = \frac{\boldsymbol{user_i} \cdot \boldsymbol{user_j}}{||\boldsymbol{user_i}|| \cdot ||\boldsymbol{user_j}||} \qquad (7)$$

6.2 Shop Information Extraction

The users will follow the shops which can meet their needs and preferences. So the correlations between the shops are useful for predicting the missing values based on the training data. A direct way to measure the similarity between the shops is to use the description information. The descriptions, however, are sparse and incomplete as discussed above. One possible solution is to exploit the external data source. Fortunately, the information from the World Wide Web can be used to measure the similarity, which is similar to the approach in [18].

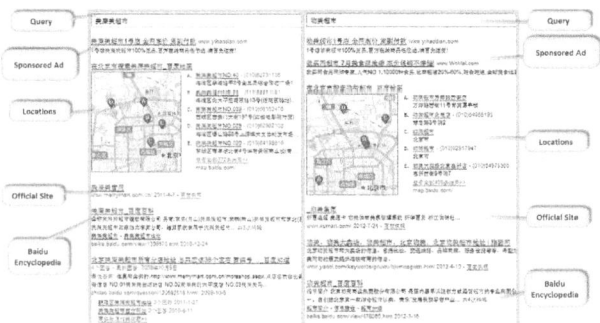

Fig. 2. The results returned by Baidu by using MeiLianMei supermarket and WuMei supermarket as queries

We use the name of the shops as queries and the Chinese search engine Baidu is adopted in this work. The results returned by the search engine are very useful for measuring the similarity. Figure 2 on the previous page illustrates the results when we use the names of two supermarkets as queries. As displayed in the figure, the search results are semi-structured for similar searched entities. That is, for similar kinds of merchants, the search engine has a semi-structured template to display the returned items. The semi-structured results are not proper for modeling the semantics of the shops, but are useful to compute the similarity between shops.

We extract the words of the search results on the first page of each query. The words are then used to describe the shops in a vector space as follows, where $\#word_j(shop_i)$ denotes the word count of $\#word_j$ that appears in $(shop_i)$'s search results.

$$shop_i = < \#word_1(shop_i), \ldots, \#word_n(shop_i) > \quad (8)$$

To filter out the noises from the sponsored advertising, the weighting scheme TF-IDF [2] is adopted to generate a weight for the words of each of the shops. The weighting scheme can avoid the computation to be dominated by the common words.

$$IDF(word_j) = log\frac{D}{|d \in D : word_j \in d|}$$
$$Weight_{i,j} = \#word_j(shop_i) \times IDF(word_j) \quad (9)$$

Equation 9 calculates the weight of $shop_i$'s $word_j$ and we get the weight vector of the shops.

$$shop'_i = < \#weight_1(shop_i), \ldots, \#weight_n(shop_i) > \quad (10)$$

Cosine similarity is adopted here and the similarity between two shops is calculated as follows:

$$CosSim(shop'_i, shop'_j) = \frac{shop'_i \cdot shop'_j}{||shop'_i|| \cdot ||shop'_j||} \quad (11)$$

The shop-shop similarity and user-user similarity information are then leveraged to help the prediction task, which will be introduced in Section 6.3.

6.3 Link Relation Prediction

In order to solve the problems of sparsity, cold-start and data incompleteness, we propose to leverage the useful complementary information sources. As these external sources are all heterogeneous, which cannot be exploited directly, we use matrix factorization to borrow the knowledge by using them as the regularizer.

Definitions:

$F_{m \times n}$: The relationship matrix of the users and the shops, where each entry represents whether the shop is followed by the user. m is the number of users and n is the number of shops.

$U_{m \times k}$: The low rank factor of the users, where k is the number of hidden topics and $k < n$.

$V_{n \times k}$: The low rank factor of the shops.

$C_{m \times m}$: The user-user similarity matrix, which is obtained from the user logs and described in Section 6.1.

I_C: The indicator matrix of $C_{m \times m}$, $I_{C,ij} = 1$ if $C_{i,j}$ is not null.

$M_{n \times n}$: The shop-shop similarity matrix, which is based on the search results and described in Section 6.2.

I_M: The indicator matrix of $M_{m \times m}$, $I_{M,ij} = 1$ if $C_{i,j}$ is not null.

$T_{n \times k}$: The shop-topic matrix based on the descriptions and the INITS model.

$\lambda_1, \lambda_2, \lambda_3, \lambda_4$: The first three parameters are used to control the influence of the complementary information sources and the last controls the regularization over the factorized matrices so as to avoid over-fitting.

The basic intuition of the proposed model is, the users with similar behaviors tend to follow similar shops and the shops with similar topics tend to be followed by similar users. As illustrated in Figure 1, given the relational links $F_{m \times n}$, we decompose it as a product of $U_{m \times k}$ and $V_{n \times k}$. The factorization leverages the auxiliary data sources by sharing the user-topic matrix $U_{m \times k}$ with the user-user matrix $C_{m \times m}$, the shop-topic matrix $V_{n \times k}$ with the shop-shop similarity $M_{n \times n}$ and the shop-topic correlation $T_{n \times k}$. Hence, the objective function is:

$$L(U,V) = \frac{1}{2}\|F - UV\|_F^2 + \frac{\lambda_1}{2}\|I_C \circ (C - UU^T)\|_F^2 + \\ \frac{\lambda_2}{2}\|I_M \circ (M - VV^T)\|_F^2 + \frac{\lambda_3}{2}\|V - T\|_F^2 + \\ \frac{\lambda_4}{2}(\|U\|_F^2 + \|V\|_F^2) \quad (12)$$

where $\|\cdot\|_F$ is the Frobenius norm and the operator $\circ$ denotes the entry-wise product. The objective function is a non-convex optimization problem. Therefore, we use stochastic gradient descent(SGD) to get the local optimal solution. The gradients (denoted as ∇) for U and V are as follows:

$$\nabla_U L = (UV^T - F)V + 2\lambda_1[I_C \circ (UU^T - C)]U + \lambda_4 U \\ \nabla_V L = (UV^T - F)^T U + 2\lambda_2[I_M \circ (VV^T - M)]V + \quad (13) \\ \lambda_3(V - T) + \lambda_4 V$$

7 Experiments

7.1 Dataset

We conducted experiments with the user logs of WuXianGouXiang from September 2011 to March 2012. We obtained a dataset from the server which contains 998 shops and 681 users.

7.2 Evaluation

In order to measure the accuracy of the prediction, we use two methods to evaluate the recommendation performances. The first one is Mean Absolute Error (MAE):

$$MAE = \frac{\sum_{i=1}^{m}\sum_{j=1}^{n}|f_{i,j} - f_{i,j}^{p}|}{m \times n} \qquad (14)$$

where m and n are the number of the users and the shops, $f_{i,j}$ is the ground truth of whether user i follows shop j and $f_{i,j}^{p}$ is the predicted result. Noticeably, for the predicted results we transform the values which are larger than one as one and transform the values which are less than zero as zero, since the correlations between the users and the shops cannot be negative and will be one at most. Thus, a smaller MAE score means better prediction performance.

Another measure method we employ is the normalized discounted cumulative gain(nDCG)[9]. This measure is useful for computing the quality of search engines as it considers both the returned contents and the rank of the results. To evaluate the quality of ranking list, we rank the shops for each of the users based on the online visit behavior to get the ground truth. That is, given a shop and a user, if the shop is followed by the user, the shop is relevant for the user. The more times the user visits the shop online, the higher the shop ranks for the user. When testing our proposed approach, we generate the recommended list for each user based on the correlations between the user and the shops in UV^T. A higher $nDCG$ represents a better ranking result.

7.3 Settings

In order to investigate the effectiveness of the auxiliary information sources, we experiment on the following methods:

1) *IM*: The proposed integration method which is based on the INITS model.

2) *WU, WS, WT*: The methods that use all the information sources except the user-user similarity(WU), the shop-shop similarity(WS) or the shop-topic information(WT).

3) *CF*: The Collaborative Filtering method which only uses the training data to predict the missing values. Low-rank matrix factorization [12] is adopted to act as the baseline.

To test the impact of the INITS model, we perform another experiment using the following methods:

1) *INITS*: The integrating method which uses the INITS model to generate the shop-topic information

2) *HITS*: The integrating method which uses the HITS model to generate the shop-topic information, where the HITS model computes an authority score for each of the shops. For the shops without a proper description, the authority score is divided equally.

For each experiment, we repeat five rounds by randomly choosing different entries of matrix F as the training data and the rest as the testing data. The average value of MAE and nDCG are used to measure the performance.

7.4 Experimental Results

Figure 3 illustrates the experimental results of the different methods introduced above based on Mean Absolute Error(MAE). It can be observed that the best performance is achieved by our proposed approach. Thus, we can say that the model which combines the useful auxiliary information sources performs better. When we use the WS model which ignores the shop-shop similarity, the experimental result is closest to the best performance, which means the shop-shop similarity contributes the least to the prediction task. This may be caused by the poor quality of the search results: 1) The search results often contain advertisements of the merchants' competitors, which may bring in some noise when compared with other shops and 2) for the shops which are not so famous, the search engine returns very few results. This makes the feature vector of the shop very sparse and the shops which are not famous and are quite different will be judged to be similar. Though some noises may be taken from the search results, it is still proven to be helpful (about 0.04% improvement). When we use the WU model which ignores user-user similarity and the WT model which ignores shop-topic information, the Mean Absolute Error increases significantly. Based on this we can say that: 1) The users who view the similar deals, download the

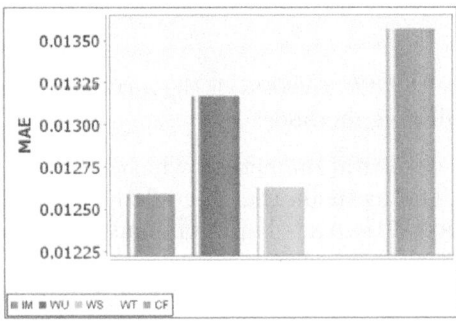

Fig. 3. The overall experimental results on the complementary information sources based on Mean Absolute Error

similar coupons, and will follow similar shops online, and 2) the shops which own similar topics and are viewed at similar times will be followed by the similar group of users, as the INITS model is based on the contextual information of the users.

Table 2. The overall experimental results on the complementary information sources based on normalized discounted cumulative gain

	nDCG[5]	Ratio
IM	0.6261	0.0%
WU	0.5828	7.43%
WS	0.5957	5.10%
WT	0.5607	11.66%
CF	0.5484	14.17%

Table 2 illustrates the experimental results of different methods based on nDCG[5], i.e. we measure the system performance based on the top five recommended shops. The third column denotes the improvement ratio of the proposed method. The integrating method that uses all the information sources achieves the best result. Similarly, the second best result is achieved when ignoring the shop-shop similarity. An interesting difference between the experimental results of MAE and nDCG[5] is that the shop-topic information is most important for the ranking quality among all the auxiliary data. The result indicates that the users will visit the shops more frequently if the topics of the shops can match the users' needs.

Table 3. The influence of the INITS model on different shops

	no description	description
HITS	0.01297	0.01336
INITS	0.01240	0.01304
Ratio	4.40%	2.40%

Table 3 shows the experimental results in terms of MAE on the shops which have a description and the shops without a description. The first model (HITS) is the integration model which uses the HITS algorithm to update the user-topic information, and the second model uses the INITS algorithm. The ratio is the improvement ratio. We can observe that on both datasets, the INITS model outperforms the baseline. Another observation is that, the INITS model has a higher ratio of improvement on the shops without descriptions than the shops with a description, which proves that the INITS model is useful for inferring the hidden topics and is effective for assigning the score to the sub-topics of an authority. Notice that the shops without descriptions have a lower error rate on average, for they are followed by less users and the relationships are more sparse, which lead to less prediction errors but a low recall.

Figure 4 illustrates the performances of our approach varying the parameters, where λ_1 controls the influence of user-user similarity, λ_2 controls the contribution of the shop-shop similarity to the objective function, λ_3 controls the information source of shop-topic, λ_4 is used to avoid over-fitting. Mean Absolute Error is adopted to measure the error rate. The four parameters are tested individually. When testing one of the parameters, the other three are fixed to be 0.1. The results show that the error rate increases when the parameters are either too large or small.

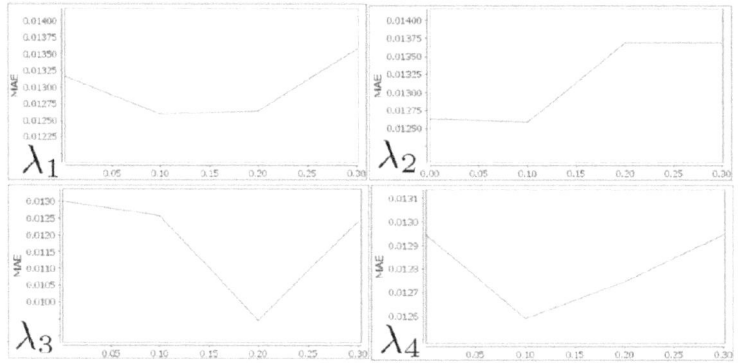

Fig. 4. The impact on Mean Absolute Error of different parameters

8 Conclusion and Future Work

In this paper, we propose a novel approach to predict the link relations between users and shops based on a real world application. The contributions of our work are the following. By surveying the application and analyzing the user logs, we put forward a new problem of discovering the potential followers of a shop. In order to better model the characteristics of the shops, LDA is adopted to process the descriptions offered by the merchants to recover the latent topics underlying the texts. We propose to use several useful auxiliary data sources to tackle the sparsity problem. A novel approach, namely INtent Induced Topic Search (INITS) is introduced to revise the coordinates of the merchants in the latent semantic space. In the future, we will validate our method with other datasets to see and improve the effectiveness of our approach.

Acknowledgments. We are very grateful to the WuXianGouXiang project team, Hao Yang, Minggang Wang, Shenghua Wang, Ke Zhang and Olivia Li. They designed and developed the service from scratch and allowed us to get access to the data and conduct this research work. Alvin Chin and Yuanchun Zhou are the corresponding authors. The work is supported by the Natural Science Foundation of China(NSFC) under Grant No. 61003138.

References

1. Backstrom, L., Leskovec, J.: Supervised random walks: predicting and recommending links in social networks. In: ACM International Conference on Web Search and Data Mining, pp. 635–644 (2011)
2. Baeza-yates, R.A., Ribeiro-neto, B.A.: Modern Information Retrieval (1999)
3. Blei, D.M., Ng, A.Y., Jordan, M.I.: Latent dirichlet allocation. Journal of Machine Learning Research 3, 993–1022 (2003)
4. Kleinberg, J.M.: Authoritative sources in a hyperlinked environment. Journal of the ACM 46, 604–632 (1999)
5. Leskovec, J., Huttenlocher, D.P., Kleinberg, J.M.: Predicting positive and negative links in online social networks. Computing Research Repository abs/1003.2:641–650 (2010)
6. Li, Y., Hu, J., Zhai, C., Chen, Y.: Improving one-class collaborative filtering by incorporating rich user information. In: International Conference on Information and Knowledge Management, pp. 959–968 (2010)
7. Liben-Nowell, D., Kleinberg, J.M.: The link prediction problem for social networks. In: International Conference on Information and Knowledge Management, pp. 556–559 (2003)
8. Lichtenwalter, R.N., Lussier, J.T., Chawla, N.V.: New perspectives and methods in link prediction. In: Knowledge Discovery and Data Mining (2010)
9. Manning, C.D., Raghavan, P., Schtze, H.: Introduction to information retrieval (2008)
10. Newman, M.E.J.: Clustering and preferential attachment in growing networks. Physical Review E 64 (2001)
11. Singh, A.P., Gordon, G.J.: Relational learning via collective matrix factorization. In: Knowledge Discovery and Data Mining, pp. 650–658 (2008)
12. Srebro, N., Jaakkola, T.: Weighted Low-Rank Approximations. In: International Conference on Machine Learning, pp. 720–727 (2003)
13. Wang, C., Satuluri, V., Parthasarathy, S.: Local probabilistic models for link prediction. In: International Conference on Data Mining, pp. 322–331 (2007)
14. Xu, Z., Kersting, K., Tresp, V.: Multi-Relational Learning with Gaussian Processes. In: International Joint Conference on Artificial Intelligence, pp. 1309–1314 (2009)
15. Zheng, V.W., Cao, B., Zheng, Y., Xie, X., Yang, Q.: Collaborative Filtering Meets Mobile Recommendation: A User-Centered Approach. In: National Conference on Artificial Intelligence (2010)
16. Yoon, H., Zheng, Y., Xie, X., Woo, W.: Smart Itinerary Recommendation Based on User-Generated GPS Trajectories (2010)
17. Zheng, Y., Xie, X.: Learning Travel Recommendations from User-Generated GPS Traces 2, 1–29 (2011)
18. Zheng, V.W., Zheng, Y., Xie, X., Yang, Q.: Collaborative location and activity recommendations with GPS history data. In: World Wide Web (2010)
19. Park, M.-H., Hong, J.-H., Cho, S.-B.: Location-Based Recommendation System Using Bayesian User's Preference Model in Mobile Devices. In: Indulska, J., Ma, J., Yang, L.T., Ungerer, T., Cao, J. (eds.) UIC 2007. LNCS, vol. 4611, pp. 1130–1139. Springer, Heidelberg (2007)

Performance of Serializable Snapshot Isolation on Multicore Servers

Hyungsoo Jung[1], Hyuck Han[3,*], Alan Fekete[1], Uwe Röhm[1], and Heon Y. Yeom[2]

[1] The University of Sydney
{firstname.lastname}@sydney.edu.au
[2] Seoul National University
yeom@dcslab.snu.ac.kr
[3] Samsung Electronics
hyuck.han@samsung.com

Abstract. Snapshot isolation (SI) is a widely studied concurrency control approach, with great impact in practice within platforms such as Oracle or SQL Server. Berenson *et al.* showed though that SI does not guarantee serializable execution; in certain situations, data consistency can be violated through concurrency between correct applications. Recently, variants of SI have been proposed, that keep the key properties such as (often) allowing concurrency between reads and updates, and that also guarantee that every execution will be serializable. We have had the opportunity to use three implementations of two different algorithms of this type, all based on the InnoDB open source infrastructure. We measure the performance attained by these implementations, on high-end hardware with a substantial number of cores. We explore the impact of the differences in algorithm, and also of the low-level implementation decisions.

1 Introduction

At the core of a database engine is the concurrency control component that provides isolation of concurrently running transactions (the 'I' in ACID). For many years, strict two-phase locking with refinements for indices and counters was seen as a good all-round choice, and concurrency control was often considered a solved problem, one of those where further work would just be "polishing the round ball" [1].

In 1995, a new approach was described [2], and deployed in Oracle DB [3], and thereafter in other platforms like PostgreSQL and Microsoft SQL Server. Called Snapshot Isolation (SI), this multiversion concurrency control mechanism has attractive features. Among these are that "readers never block" and that each transaction observes the database in a transaction-consistent state (it sees all or none of the changes of each other transaction). But for some transactions, SI allows executions that do not meet the definition of serializability. For example, a "write skew" can occur [2], and in production code, this has led to data corruption where the final state of the data does not satisfy an integrity property that would be maintained by each program running alone [4].

Recently, several new mechanisms were published that are minor variants of SI (and thus allow concurrent reading and writing in many cases), with the extra property that

---
* Work done while at Seoul National University.

they do enforce that every execution will be serializable. Cahill *et al.* introduced this idea in SIGMOD'08 [5], and the definitive account of Serializable Snapshot Isolation (SSI) was published in TODS [6]. The key feature is that one uses the normal SI mechanism, but also tracks information about read-write conflicts (without blocking any of these). When a particular pattern of conflicts occurs, one of the transactions involved must abort. The theory of [7] implies that any non-serializable execution will display a conflict fitting this pattern; thus SSI prevents any non-serializable execution. In ICDE'11, Revilak *et al.* [8] gave another proposal called Precise Serializable Snapshot Isolation (PSSI), which aims to eliminate any cases of unnecessary aborts (the "false positive" cases described in [6]). In essence, PSSI does serialization graph testing [9] on top of SI.

In this paper, we are interested in the performance and scalability of these algorithms on modern multicore servers, and how much overhead they introduce as compared to a standard SI approach. In doing so, we seek to gain an understanding of the impact of essential algorithmic design decisions on the performance on multicore servers. In particular, we are interested in which factors are intrinsic to multi version concurrency control, and which are not.

Our experiments are conducted on a high-end server with 24 cores across 4 chips. This type of multicore environment is becoming increasingly common, though it is known that many software systems do not perform well in such a setting [10,11]. Recent research has proposed some special techniques to improve locking-based DBMS operation on multicore hardware [12,13]; similar work for SI-based systems is beyond the scope of this paper, where we take the code used in published papers, and measure how well each performs "as is".

2 SSI Implementations

We consider three implementations of serializable snapshot isolation: SSI is the implementation by Cahill [6]; ESSI is Revilak's implementation of the same algorithmic concept; and PSSI is Revilak's proposal that detects cycles in a dependency graph.

Cahill's original SSI was implemented by Cahill as an experimental prototype as modifications to Berkeley DB and InnoDB, and it has been deployed recently in PostgreSQL [14]. Revilak implemented PSSI as an experimental prototype modifying InnoDB, and for comparison, he also implemented SSI; for clarity we follow [8] and refer to this implementation as ESSI. In this paper, we take advantage of the presence of three implementations that all are done in the context of InnoDB: Cahill's SSI, Revilak's ESSI, and Revilak's PSSI. As a baseline, we use an implementation of the standard, not-always-serializable SI algorithm, done by Cahill for [6], that is very straightforward given the existing code of InnoDB for version and lock management. We measure the performance of these four systems, which share a common code structure.

The open-source InnoDB system keeps multiple versions of each record, and a version is tagged with the id of the transaction that produced it. These versions are not used by transactions running InnoDB's SERIALIZABLE isolation level, but the code exists to read a version with a given timestamp (by converting from transaction ids to the commit time of the transaction) - this code is used in InnoDB for lower isolation levels, and has been called by the modified functions in SI, SSI, ESSI and PSSI. InnoDB also has a fairly standard lock manager (supporting a form of range locking to prevent phantoms). SI, SSI, ESSI and PSSI all use this lock manager. All the systems take write locks when updating, to implement the "First Updater Wins" principle of SI, preventing concurrent transactions from modifying the same item. SSI, ESSI and PSSI also take read-like locks that do not cause blocking, but instead allow one to detect that read-write conflicts have occurred during execution. Using these observed conflicts, one can form the serialization graph for an execution as it happens.

In ESSI and PSSI, the needed subset of the serialization graph is actually built as a data structure, called the CTG (Cycle Testing Graph); in SSI it is implicit, with the essential information stored in two flags associated with each transaction (and kept in the transaction list data structure). In SSI, a variable T.in records whether there is some edge of the graph leading to T, and T.out records whether some edge starts at T. Furthermore, when there is only one edge in (or out) we also note as part of the transaction information the identity of the transaction at the other end; but when there are multiple incoming (or outgoing edges) SSI does not track them individually, but just notes the existence of more than one. In SSI, the information about conflict edges is updated in each read or write operation; in ESSI and PSSI it is during commit processing for transaction T that the CTG is updated by adding edges incident on T. ESSI and PSSI also remove edges involving T if T aborts; SSI in contrast may keep flags set in other transactions, showing conflict edges that involved T and hence are no longer valid.

In PSSI, the main algorithmic idea of the concurrency control is that when a transaction T tries to commit, the system will add the appropriate edges to the CTG, do a search from T to find if the new node lies in a cycle; if so, T is aborted instead.

In both SSI and ESSI, the essential step is to abort a transaction if one finds what is called a "dangerous structure": both an incoming edge and an outgoing edge involving concurrent transactions, occurring at a single transaction (which is often but not always the one that is aborted). Certain extra conditions are applied, involving the ordering between commit time of the transactions involved among the two edges; these extra conditions reduce the frequency of unnecessary aborts (called "false positives" in [6]). In fact, aborting the cases that are done in SSI and ESSI is enough to guarantee that all executions will be serializable; this is shown in [6] using the proof details of the main theorem from [7].

In InnoDB, all the internal data structures of the system itself are protected from race conditions by a single, shared kernel mutex. In all its modifications that we consider, this same mutex is used to cover any access to any shared data structures introduced in the modification, such as the serialization graph.

3 Experimental Environment

We have conducted an extensive performance evaluation of four SI-based concurrency control implementations on a state-of-the-art multicore environment: SI and SSI from Cahill *et al.* [6], and ESSI and PSSI from Revilak *et al.* [8]. With the exception of SI, all three SSI variants ensure serializable execution.

System Setup. We have deployed all four implementations into a MySQL 5.1.31 instance. The experiments are conducted on a 24-core Intel Xeon server under Linux 3.1.5 that has a total of four 1.86 GHz Intel Xeon MP Processor 8000 series dies (or chips) (each die with 6 cores). Each core has access to 48 KB private L1 cache and 256 KB private L2 cache; all 6 cores on each die share one 18 MB L3 cache. To make the database run on the actual cores, we disabled Intel Hyper-Threading throughout all experiments. The server is further equipped with a 500 GB 7200 RPM SATA II hard disk and 512 GB of RAM.

In all experiments, there is a single MySQL instance running on the database server, whereas a varying number of clients is emulated on a separate client computer (1.9 GHz Opteron, 128 GB of RAM, running Linux 3.0.0). Client and server machines are connected with a 1 Gbps Ethernet network.

The Benchmark. Our benchmark uses three tables called `ssibench-{1, 2, 3}` with two non-null integer and ten variable sized character columns (`b_value-{1, 2, ..., 10}`); one of integer value columns (`b_int_key`) is a primary key. Each table is populated with randomly chosen 100K items. For this study, we use two types of transactions: query transactions (read-only) and update-after-read transactions (read-update).

The read-only transaction consists of a single Select-From-Where query:
```
SELECT sum(b_int_value) FROM ssibench-i
    WHERE b_int_key > :id and b_int_key <= :id+100
```
Note that through this query the DBMS scans `100` rows in the table and aggregates their integer column, so that the final result is small (minimized network cost).

The read-update transaction is designed to create conflict cycles: it first reads `100` rows from `ssibench-i` and updates `20` rows from `ssibench-((i+1)%3)`. The reading part of this transaction uses the same range query as the read-only transaction, and the update part consists of just a single SQL statement (to minimize network cost):
```
UPDATE ssibench-((i+1)%3) SET b_value-k = :rand_str
    WHERE b_int_key = :id1
       OR b_int_key = :id2
       OR ... b_int_key = :id20
```
In our experiments, we vary the multiprogramming level (MPL) from 1 to 30, with all clients trying to execute transactions as fast as possible without think time in between. Each experiment was repeated five times, with each run consisting of 1 minute ramp-up period and 1 minute measurement period. All plotted points in the figures of Section 4 are the average of these 5 test runs, and confidence intervals are shown.

To evaluate the impact of various CPU configurations, we use the CPU hotplug feature of modern CPU architectures to make CPUs (or cores) available (/unavailable) to the Linux kernel, be setting the `/sys/devices/system/cpu/cpuX/online` values to 0 or 1, where X is the corresponding core number to enable, respectively disable. We used OProfile to profile various system activities, a system-wide, statistical, continuous profiler for Linux systems. For brevity, we classify all profiled system functions into three categories: 'Kernel' for Linux kernel functions, 'MySQL' for non-mutex related MySQL functions, and 'Mutex' for all mutex related functions in MySQL.

4 Evaluation Results

4.1 SI Performance on a Multicore Server

In our first experiment, we are interested in the general efficiency of the different SI-based concurrency control algorithms when running on a state-of-the-art multicore server. To this end, we measured the throughput of plain SI and the three variants of the SSI approach with different multiprogramming levels and all 24 cores of our server being enabled. The workload consisted of 75% of read-only transactions (RO) and 25% read-updater transactions (RU). The data accesses are uniformly distributed across the tables. The performance results and runtime analysis are shown in Figures 1 to 3.

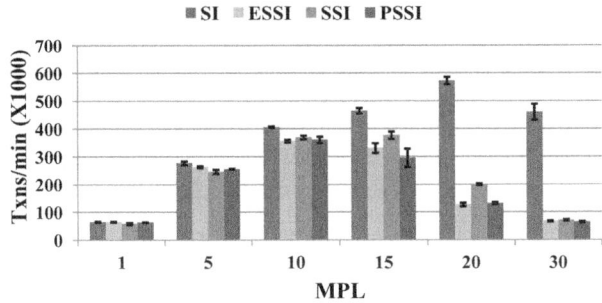

Fig. 1. Throughput of SI, ESSI, SSI, and PSSI on 24 cores, under 75%RO-25%RU workload

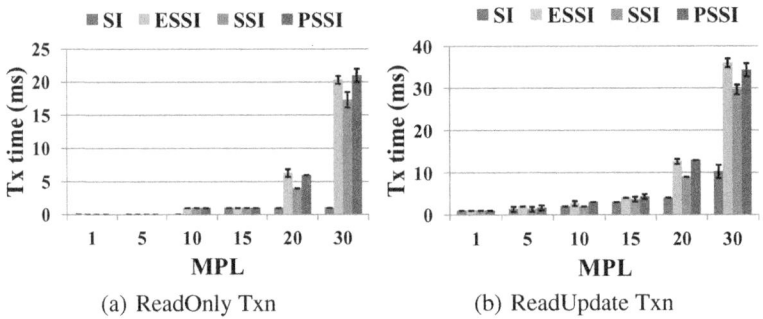

Fig. 2. Response times of SI, ESSI, SSI, and PSSI on 24 cores, under 75%RO-25%RU workload

In general, all four SI and SSI implementations show the same overall performance curve: initially, the throughput increases with increasing MPL up-to a peak performance point after which the throughput starts degrading (cf. Figure 1). The major difference between plain SI and the three SSI variants is when and how this performance degradation occurs. In this initial setting, SI's throughput scales well up to MPL 20, reaching a peak throughput of over 550,000 tpm, before its throughput starts gradually decreasing. In contrast, the performance of the three SSI implementations peaks already between MPL 10 and 15, and then degrades drastically. At MPL 30, the throughput of all SSI variants has collapsed to basically the same level as with that of MPL 1. Among the

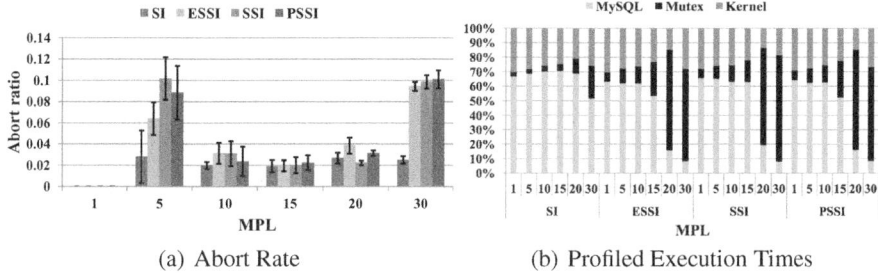

Fig. 3. Abort ratio and profiled execution times on 24 cores, under 75%RO-25%RU workload

three SSI implementations, Cahill's SSI shows the highest peak performance with just under 400,000 tpm, and a slightly better resistance to the performance degradation, though the overall throughput collapse is experienced too. What is the cause of these major scalability problems of all current SSI implementations?

In Figure 3(a), we compare the abort ratios (the ratio of the number of aborted transactions over the total number of committed transactions) of the implementations. Plain SI shows a stable abort ratio at around 2% of committed transactions for all MPL. The three SSI variants show similar abort ratios around 2% for most MPL levels, with the exception of two peaks of higher abort ratios at MPL 5 and MPL 30. However, these peaks for aborted transactions under the SSI variants do not align with the measured performance behaviour, which indicates that the poor scalability of the SSI variants is not due to an increase in concurrency conflicts.

Next, we have a look at the transaction runtimes. As shown in Figure 2, the higher the parallelism is (level of MPL), the longer the runtimes of both read-only and read-update transactions become. Especially update transactions experience massive delays with high MPL, and for the SSI variants even read-only transactions run much longer. This clearly contradicts the usual rule-of-thumb that *"readers don't wait"* under SI.

To further explore this phenomenon, we executed the workload again with profiling of all system activities, including the Linux kernel, switched on. Figure 3(b) gives the breakdown of the profiled execution times for all approaches with increasing MPL. These results show a very clear and sudden increase of the 'Mutex' portion (black bars) in all SSI implementations after MPL passes a certain point. Apparently, as MPL grows a significant portion of time is spent on spin waiting for the crucial kernel mutex. This in turn causes the transaction runtimes to steeply increase with higher MPL, which in consequence leads to a collapse of the transaction throughput under all SSI variants.

The excessive waiting for the shared kernel mutex to synchronise access to the internal data structures leads to a reduction of useful work done in the database system, which is similar to the performance effect of *lock thrashing*. Conventional wisdom says that a lock thrashing problem usually arises in locking-based concurrency control algorithms and under update-heavy workloads. Our result here has a particular importance because this type of performance collapse happened even with a read-mostly workload and under all SSI implementations. When MPL is over 20, SSI spends 70% of its execution time spin waiting whereas ESSI (and PSSI) spend 65% on mutex waiting. This mutex waiting increases transaction runtimes as compared to MPL 1 by factor 30.

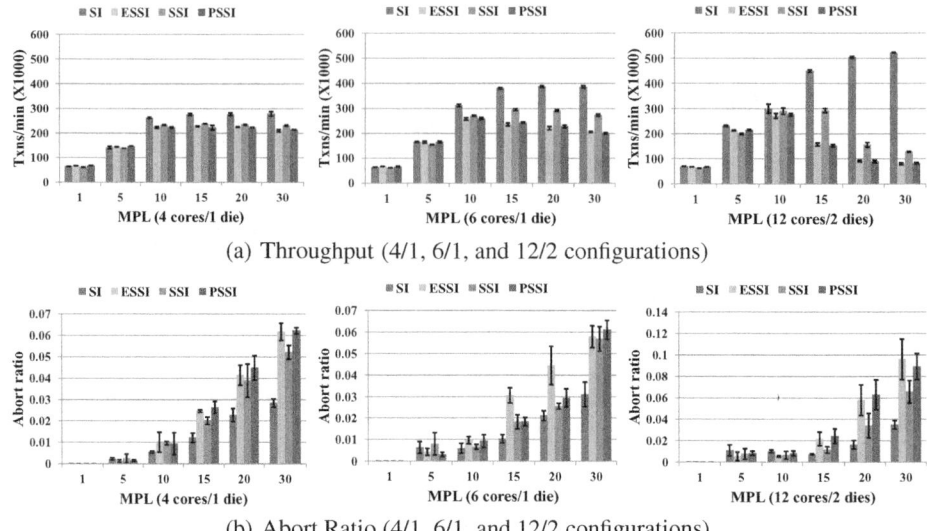

Fig. 4. Multicore experiment: throughput and abort rates of SI, ESSI, SSI, and PSSI with varying number of cores, under a 75%RO(Read-Only)-25%RU(Read-Update) workload

4.2 Varying the CPU Configuration

Next, we strive to gain a better understanding of the influence of the number of cores on the performance behaviours of the different concurrency control implementations. To do so, we varied the number of cores visible to the Linux operating system by using the aforementioned CPU hotplug feature of the Linux kernel. This runtime configuration gave us control of making a particular set of cores visible to the Linux kernel, so that the operating system (and thus the DBMS) use only a well-defined set of cores. We describe a configuration as n/m representing n physical cores arranged among m chips. Thus the full hardware is 24/4; we also report on measurements at 4/1, 6/1 and 12/1.

Scalability with Varying Number of Cores. Figure 4 shows the performance results of all implementations with various core settings, from 4 core to 12 cores. In a single core configuration (not shown in the figure), the throughput with all implementations is very similar, reaching about 80,000 tpm very early at MPL 5, and all concurrency control approaches can also sustain their peak performance as MPL increases. There is no overhead noticeable for SSI in a single core environment.

As we increase the number of cores to 4, the peak throughput of all implementations improves: The peak throughput of SI on 4 cores is $3.5\times$ higher, the peak values of the SSI implementations are about $3\times$ higher than that on a single core. On this 4 core configuration, we start to see the overhead incurred by all SSI implementations; in particular, the abort ratio of SSI implementations increases more steeply than with SI, resulting in less throughput. However, we can hardly observe any performance collapse even up to our maximum of MPL 30.

With a 6 core configuration, the peak throughputs increase further, while at the same time performance gap among the different implementations widens. The overhead of

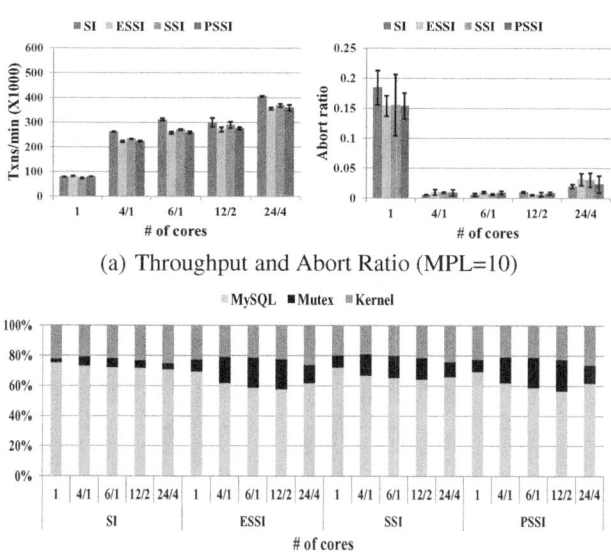

(a) Throughput and Abort Ratio (MPL=10)

(b) Breakdown of Profiled Execution Times (MPL=10)

Fig. 5. Throughput and abort ratios of SI, ESSI, SSI, and PSSI with varying number of cores under a 75%RO-25%RU workload at MPL 10

SSI implementations in this setting becomes more prominent than before: ESSI (and PSSI) show now early performance drop-offs from MPL=10, whereas SSI drops off later from MPL=30. When we look at the abort ratio of ESSI, its abort ratio abruptly increases from MPL 10 on, while SSI matches it only at MPL 30.

Finally with a 12 core configuration, we see quite different performance behaviours from each implementation. SI keeps increasing its peak throughput up to 1.37X higher than that on 6 cores, and it does not show any performance drop-off. In contrast, all SSI implementations show severe performance drop-offs after their peak throughput, even though the abort ratio of all three is only slightly increased than before. As we have seen in the previous setting with 6 cores, the performance drop-off of SSI is a bit later point than ESSI (or PSSI), due to the different data structures and algorithms used.

Profiling MPL 10 vs MPL 30. For a more detailed analysis of the performance situation with different CPU configurations, we ran another set of varying-core experiments with system-level profiling turned on. We focus on two interesting levels of MPL: Firstly we fix to MPL 10 as a proxy for "peak throughput" – because in basically all measurements, the different algorithms achieved close to their highest throughput at this level. Secondly, we also chose MPL 30 as the measurement points with the most extreme system load. In the following, we hence fix the workload to one of these two MPL values, and then vary the number of cores from a single core up-to the full 24 cores of our server.

The results are summarised in Figure 5 (MPL 10) and Figure 6 (MPL 30). In the (a) subgraphs of each of these figures, we show the performance and abort ratios for each implementation with varying number of cores; this is followed in the subgraphs (b) by an analysis of the profiling information for the varying CPU configurations.

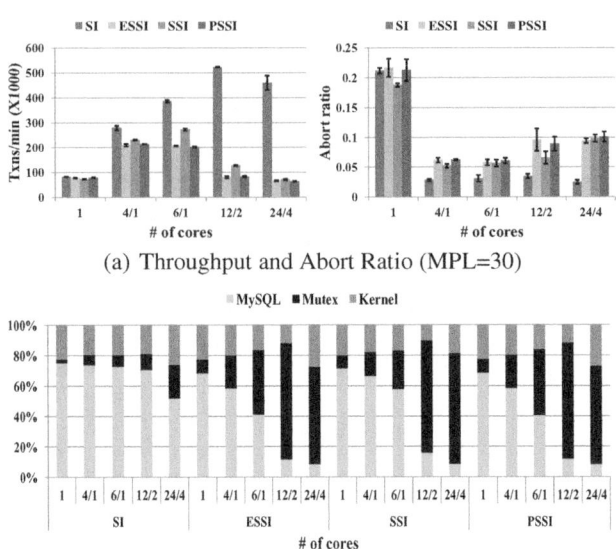

(a) Throughput and Abort Ratio (MPL=30)

(b) Breakdown of Profiled Execution Times (MPL=30)

Fig. 6. Throughput and abort ratio of SI, ESSI, SSI, and PSSI with varying number of cores under a 75%RO-25%RU workload at MPL 30

With MPL 10, the throughput for all implementations basically increases linearly with the number of cores, without showing a steep increase of the abort ratio (cf. Figure 5(a)). All schemes achieve their highest throughput on a 24 core setting. At MPL 10 the mutex contention is still low, so that the Mutex regions (black bars) in Figure 5(b) do not show any sudden increase. It shows that all SSI implementations have a general low runtime overhead above that of SI. Indeed, Figure 5(a) is a typical performance graph as it was found in prior SSI studies. These performance results seem to be promising, as they would enable all schemes to perform well with increasing number of cores.

With MPL 30 the situation looks quite different (Figure 6). When measuring the workload at MPL 30, the synchronisation cost between concurrent transactions becomes more expensive and all SSI implementations experience the performance collapse shown before. In the system community, this type of performance collapse is referred to as the *scalability collapse* problem, which has been addressed well in prior studies [10,11] for Linux operating systems. The large mutex waiting portion elongates transaction lifetimes enormously and increases the abort ratio (up to 10X higher), leading to a performance collapse. In particular, the throughput of all three SSI implementations on 24 cores is lower than on just a single core. We do not claim that SI is free from any performance collapse, but at least in this experiment its overall performance only gradually degrades after its peak.

Effect of Separate L3 Caches. The final experiment in this section is to measure the effect of an enlarged L3 cache on the performance. We investigate two scenarios: 4 cores on 4 separate dies (4/4) as compare to all four on just one die, and 12 cores on 4 dies (12/4; each die shared by three cores) compared to be on just 2 dies (6 cores

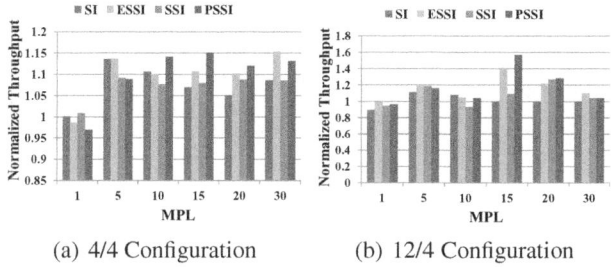

Fig. 7. Throughput with different core/die configurations under 75%RO-25%RU workload

sharing same die). Spreading cores across multiple dies has the advantage of expanding the amount of available L3 cache. The potential disadvantage is a decrease in cache locality among cores due to the distant cache locations. Figure 7 shows the normalized throughput – we show the ratio of values for 4/4 to the corresponding values for 4/1 configuration, and we plot the ratio of 12/4 to the previously shown 12/2. Between the two configurations, the 4/4 setting gives a better increase effect than the 12/4 setting: While the 4/4 setting has a 4X larger L3 cache than under a 4/1 setting, the 12/4 setting augments only a twice as large cache than under its comparison point 12/2. As shown in Figure 7(a), SI and SSI always benefit from an enlarged L3 cache; the throughput of SI increases up to 13%, while SSI gains up to 8%. However, the throughput of ESSI and PSSI initially degrades by 3% when at MPL=1, before they both enhance up to 15%.

In Figure 7(b), the performance of all schemes shows larger variations, compared to Figure 7(a). Although we increase L3 cache twice as much than before, the combined effect of an enlarged cache while maintaining reasonable locality between cores sometimes leads to greater impact on performance, either positively or negatively. For example, PSSI increases its throughput up to 55% with MPL=15 than that with a 12/2 setting, while its throughput decreases by 5% at MPL=1. The unexpected improvement at MPL=15 may be because the cycle testing graph (CTG) structure could have fitted better on the enlarged L3 cache, and 4 cores on each die exploit cache locality much better than with other MPL conditions. Measuring such fine-grained cache hits/misses, induced penalty cycles, and delays is however beyond the scope of this paper.

4.3 Varying the Read/Update Ratio

The final set of experiments investigates the effect of varying the' ratio between read-only transactions and updater transactions. In Figures 8 to 10, we compare the performance of each implementation with different reader/updater ratios, from an update-only workload (0% read-only / 100% read-update) over an update heavy workload (25% read-only / 75% read-update) and a mixed workload (50% read-only / 50% read-update) to finally the read-only case (100% read-only transactions). Note that the vertical scales on the graphs are not the same. In all cases our server has all available 24 cores enabled.

In general, the performance graphs show the same overall shape as we initially measured in Section 4.1. The notable exception is when the update rate is very high. With 75% of transactions being updaters (Figure 8(b)), we see the throughput collapse with growing MPL for SSI a bit delayed, though throughput has clearly dropped at MPL=20,

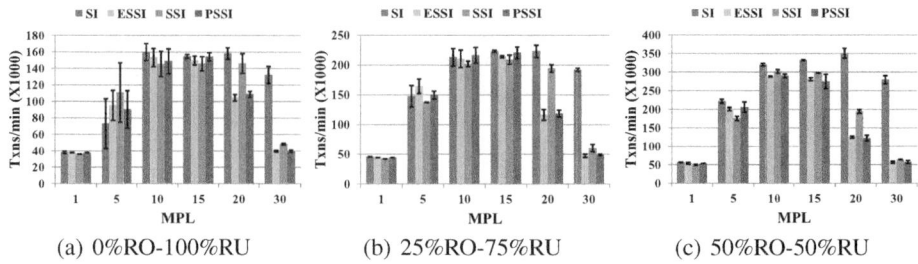

Fig. 8. Throughput on 24 cores, with varying the portion of RO (Read-Only) transactions

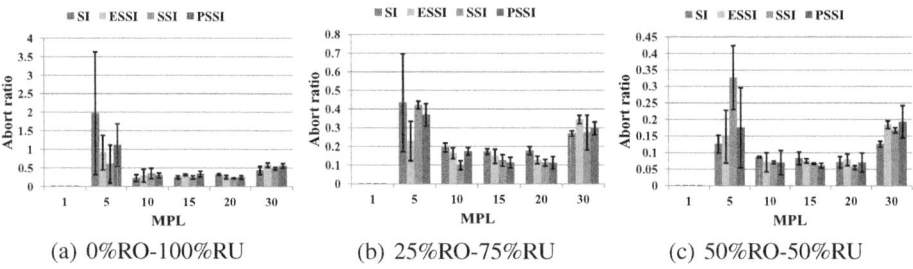

Fig. 9. Abort rates on 24 cores, with varying the portion of RO (Read-Only) transactions

and collapsed by MPL=30. In all cases, there exists a crossover that ESSI initially performs better than SSI, but becomes inferior to SSI later, or vice versa. This suggests that different data structures and associated algorithms used in these SSI variants show their merits and demerits in different workload conditions; there seems to be no universal solution to implement serializability that always performs better.

Figure 8(a) shows the update-only scenario with 100% read-update transactions. At MPL=5, the average throughput of the SSI schemes seems to be even higher than SI, but please note that the throughput deviation is quite large and that the confidence intervals overlap. Hence our results out of just 5 runs are not be stable enough to draw this conclusion. Once MPL increases beyond 5, SI always achieves higher throughputs than all others. In this update-intensive setting, the abort ratio of SI reaches 2 (i.e., aborting twice as many transactions than committed ones) at MPL=5. Other SSI variants also have high abort ratio (1.1). The irregular behavior of having high abort ratio at MPL=5 continues in all scenarios, except the 100% read-only workload. In Figures 8(b) and 8(c), the pattern is similar in that the throughput of SI drops off gradually, whereas SSI shows its performance collapse between MPL 20 and 30.

The most interesting phenomenon shows in the read-only case (Figure 10): We now run 100% read-only transactions, each of which just read 100 rows, and so SI alone gives serializable executions — there are no rw conflict edges, and the SSI variants should need to abort nothing. However, the throughput of the SSI variants still collapses with higher MPL: In Figure 10, the throughput of SI is well maintained while the throughput with the remaining three approaches collapses to about the level as with MPL=1. At MPL 30, SI achieves a ten-times higher throughput than SSI (or ESSI). This phenomenon has not previously been reported, and it contradicts our intuitive expectations

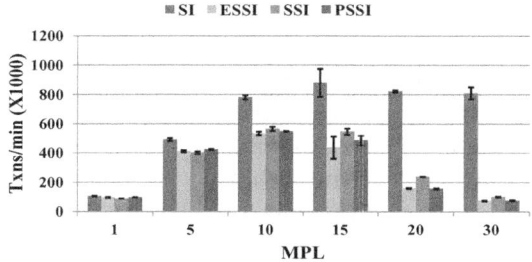

Fig. 10. Throughput of SI, ESSI, SSI, and PSSI on 24 cores, with 100% Read-Only transactions

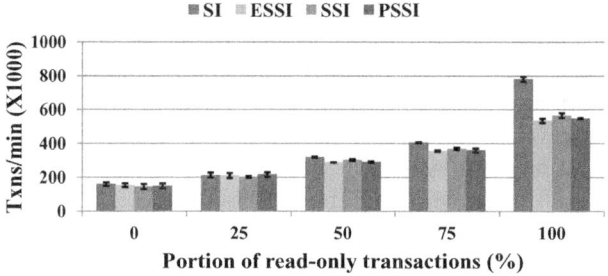

Fig. 11. Throughput on 24 cores and MPL 10, with varying the portion of Read-Only transactions

that SSI implementations would behave well at least under read-mostly workloads, regardless of MPL or the number of cores. We discuses the possible causes in Section 5.

In terms of peak throughput, we can say that all schemes perform well at lower MPL. This is shown in Figure 11, which compares the throughputs at fixed MPL 10 and plotted against the ratio of read-only transactions. But at our highest MPL of 30, the situation looks much more dire, with the throughput values of the SSI variants degrading to almost the level of MPL 1 for read-only workloads. To summarize, even though we vary the reader/updater ratio, the performance collapse can still be observed throughout the experiments. The only difference from the previous experiments is the higher abort ratio (or lower throughput) due to the increased contention with an higher update ratio.

5 Discussion

5.1 False Positive Aborts

The design objective of PSSI is to be precise: aborting a transaction only when the abort is necessary to prevent non-serializable execution. ESSI and SSI, in contrast, use a simpler test, that does not require tracing out whole cycles, but rather looks just at two adjacent edges in the serialization graph. The simplicity of the test means that ESSI and SSI will sometimes abort a transaction that could have been allowed to commit, because the edges do not lie within a cycle. Between ESSI and SSI there are also some differences in precision, because ESSI uses the CTG, a graph where some edges have been pruned away, such as those passing through aborted transactions. SSI may keep the impact of those "dangling" edges in transaction flags, because it does not spend the effort to clear flags that reflected an edge involving a transaction that subsequently aborted. Thus one might expect the lowest abort rate in SI (which aborts transactions only for *ww*-conflicts, that also lead to aborts in the other systems too); followed by PSSI, ESSI and then SSI should have the highest abort rate.

In the experiments, we did not find a reliable pattern though. The abort rates and throughputs of SSI, ESSI and PSSI remained quite comparable in all the experiments where we tested the different hardware configurations. The tradeoff against precision comes from the added overhead of carefully tracking the edges, and pruning them aggressively. When there are many conflict edges for each transaction, this becomes important. We did hence run some additional experiments where we varied the size of the transactional hotspots and the transactions lengths [15]. For space reasons, we cannot include a detailed discussion here; but we can report that in experiments with 500 or 5000 rows selected per transaction (rather than the 200 in our evaluation in the previous Section), we see that SSI gets better throughput than ESSI or PSSI, as long as MPL is low enough to avoid the performance collapse from competition over the kernel mutex. For more details please see our corresponding technical report [15].

We conclude that precision, whether from careful cycle detection or from cleaning up dangling edges in the serialization graph, is worth the effort when contention in concurrency control is dominant; in many conditions its impact is unclear.

5.2 Shared System Data Structures

The implementations of PSSI and ESSI track conflicts between transactions in a data structure, the CTG, that is shared between all transactions, and thus between all threads in the server. In contrast, SSI tries to keep additional data structures local to each transaction and thread, so there is no need to synchronize the access to these data structures. SSI adds two bookkeeping variables to each transaction block.

One might expect the reduction in shared structures to make SSI much more scalable than PSSI or ESSI, and while we do see that SSI throughput does not degrade quite as drastically with slightly increased MPL, it still shows a performance collapse under high MPL, and once MPL is 30, all three implementations are performing as badly as when single threaded, under a broad range of workloads. To explain this, we look in more detail at the code paths of the three systems.

In ESSI and PSSI, the work is concentrated in the commit step; reading or writing involves no more than setting a lock and accessing the appropriate version. The commit step is slow. As shown in Figure 12(a), ESSI first holds a `kernel_mutex` before it proceeds to update a CTG, then it starts checking concurrent *rw*-edges by traversing its read records (or updated records). This is done inside the `trx_do_cycle_test()`, and this cycle testing function has a branch towards either precise cycle testing (for PSSI) or checking pairs of edges (for ESSI). The bad situation is when transactions have a large number of incident edges. A higher degree of concurrency induced by many cores can increase the chance of having incident *rw*-edges. This explains the early throughput collapse of ESSI or PSSI in many experiments.

In SSI, the data variables that record conflicts are transaction-specific, trying to reduce the coverage of the kernel mutex. The code updates them in every read and write step as soon as a conflict first occurs. Although this frequent checking makes certification at commit very easy, we find that it incurs remarkable overhead as parallelism

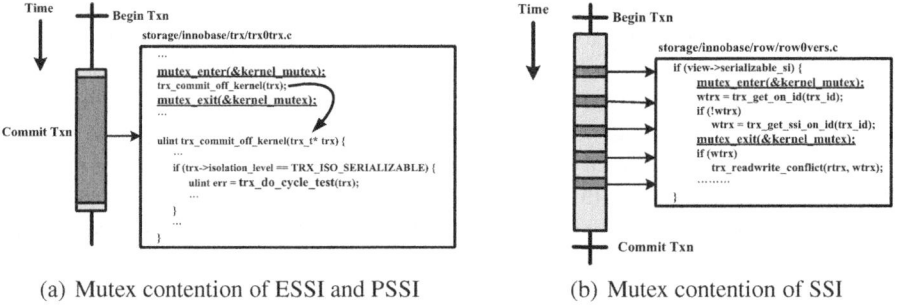

(a) Mutex contention of ESSI and PSSI (b) Mutex contention of SSI

Fig. 12. Mutex contentions with the serialisable SI implementations

grows. See Figure 12(b). The routine is to check presence of a *rw*-edge when it detects concurrent access to the same version of data by an updater and a reader. When retrieving data structures for a reader and an updater (by `trx_get_on_id()`), SSI scans a globally maintained list structure, each node of which points to metadata for an active (running) transaction. This necessitates the use of a kernel mutex to guarantee mutual exclusion to avoid data race on the list; and accessing the list by `trx_get_on_id()` is protected by a big kernel mutex. Since this routine is executed for every record a transaction reads (i.e., a small dark gray region in Figure 12(b)), the routine could increase mutex contention enormously as the degree of concurrency increases. For instance, if we employ more cores with higher MPL, the length of the shared transaction list is substantially increased; and the time to find a target transaction from this list also grows linearly; then mutex contentions and associated spin-waiting would become worse.

As we have seen from many figures showing profiled executions with a large portion of time spent in the Mutex region, employing more cores increases the degree of concurrency; then it leads to higher contention on the mutex, and longer waiting. If we increase the number of read records, this also introduces more chances of conflicting accesses to `kernel_mutex`. In SSI, the critical region is always executed for all records accessed by a transaction, so that read-only transactions can not escape from this *lock trashing* in the worst case. This is how we could observe the undesirable throughput collapse phenomenon in 100% read-only workloads in SSI.

6 Conclusions

To the best of our knowledge, this is the first extensive performance evaluation of snapshot isolation concurrency control on multicore servers. Our results show that while forms of serializable snapshot isolation (SSI) can be implemented with a low runtime overhead for the actual algorithm, the synchronisation overhead for some central shared data structures starts dominating the costs for highly parallel configurations. We evaluated three different existing implementations of these ideas in InnoDB, for a variety of effects from varying hardware and workload configurations.

We found the performance of all three SSI variants to be very similar, with the original SSI from [6] showing a slightly more robust performance in some settings, and with PSSI from [8] showing benefits of precision in some cases when the concurrency control itself becomes the bottleneck. But compared to plain SI, all three SSI approaches showed a rapid performance degradation on a 24 core machine once MPL reached around level 20 when implemented in InnoDB. Even worse, this effect occurs for pure read-only workloads, contradicting the common belief that under SI variants 'readers never wait'. When analysing the underlying cause, we identified a few synchronisation points in these algorithms, where some internal shared data structures of their serializability check are protected by a shared kernel mutex against race conditions. This finding mirrors similar findings from the systems community. The elimination of these synchronisation points will be important future work leveraging the benefits of SSI into the world of modern multicore servers.

Acknowledgments. This research was supported by the Australian Research Council grant DP0987800.

References

1. Stonebraker, M.: Are We Polishing a Round Ball? (Panel). In: Proceedings of the 1993 IEEE International Conference on Data Engineering (ICDE), p. 606 (1993)
2. Berenson, H., Bernstein, P., Gray, J., Melton, J., O'Neil, E., O'Neil, P.: A Critique of ANSI SQL Isolation Levels. In: Proceedings of SIGMOD 1995, pp. 1–10 (1995)
3. Jacobs, K.: Concurrency Control: Transaction Isolation and Serializability in SQL92 and Oracle7. Technical Report A33745 (White Paper), Oracle Corporation (1995)
4. Jorwekar, S., Fekete, A., Ramamritham, K., Sudarshan, S.: Automating the Detection of Snapshot Isolation Anomalies. In: Proceedings of VLDB 2007, pp. 1263–1274 (2007)
5. Cahill, M.J., Röhm, U., Fekete, A.D.: Serializable Isolation for Snapshot Databases. In: Proceedings of SIGMOD 2008, pp. 729–738 (2008)
6. Cahill, M.J., Röhm, U., Fekete, A.: Serializable Isolation for Snapshot Databases. ACM Transactions on Database Systems 34, 1–42 (2009)
7. Fekete, A., Liarokapis, D., O'Neil, E., O'Neil, P., Shasha, D.: Making Snapshot Isolation Serializable. ACM Transactions on Database Systems 30, 492–528 (2005)
8. Revilak, S., O'Neil, P., O'Neil, E.: Precisely Serializable Snapshot Isolation (PSSI). In: Proceedings of ICDE 2011, pp. 482–493 (2011)
9. Casanova, M., Bernstein, P.: General purpose schedulers for database systems. Acta Informatica 14, 195–220 (1980)
10. Boyd-Wickizer, S., Clements, A.T., Mao, Y., Pesterev, A., Kaashoek, M.F., Morris, R., Zeldovich, N.: An analysis of Linux scalability to many cores. In: OSDI 2010, pp. 1–8 (2010)
11. Clements, A.T., Kaashoek, F., Zeldovich, N.: Scalable Address Spaces Using RCU Balanced Trees. In: Proceedings of ASPLOS 2012 (2012)
12. Johnson, R., Pandis, I., Hardavellas, N., Ailamaki, A., Falsafi, B.: Shore-MT: a scalable storage manager for the multicore era. In: Proceedings of EDBT 2009, pp. 24–35 (2009)
13. Pandis, I., Johnson, R., Hardavellas, N., Ailamaki, A.: Data-Oriented Transaction Execution. In: Proceedings of VLDB 2010, pp. 928–939 (2010)
14. Ports, D.R.K., Grittner, K.: Serializable Snapshot Isolation in PostgreSQL. In: Proceedings of VLDB 2012, pp. 1850–1861 (2012)
15. Jung, H., Han, J.H., Fekete, A., Röhm, U., Yeom, H.Y.: Performance of serializable snapshot isolation on multicore servers. Technical Report TR693, School of Information Technologies, The University of Sydney (December 2012)

A Hybrid Approach for Relational Similarity Measurement

Zhao Lu[1] and Zhixian Yan[2]

[1] Department of Computer Science and Technology, East China Normal University, 200241, Shanghai, China
zlu@cs.ecnu.edu.cn
[2] Samsung Research America, Silicon Valley, USA
zhixian.yan@samsung.com

Abstract. Relational similarity measurement between word-pairs is important in many natural language processing tasks such as information extraction and information retrieval. The paper proposes a hybrid approach for relational similarity measurement based on various aspects including *term co-occurrence, lexicon-syntactic patterns*, as well as their *combinations*. In this approach, we first extract two relation-term sets from sentences of Wikipedia documents in which two words coincide, and compute the semantic relatedness score of each word-pair in the two relation-term sets. Second, we model the semantic relatedness value of two words together with their frequencies as a point in the three-dimensional space. Afterward, we apply DBSCAN - the classic density-based spatial clustering algorithm to group these 3D points. We finally calculate the similarity based on the clusters. We evaluate this hybrid approach using the well-known 374 SAT analogy questions. The experimental results show that our approach can significantly reduce computational time for measuring relational similarity with a relatively higher score of 52.9% compared to the state-of-the-art.

Keywords: Relational similarity, Semantic relatedness, Density-based clustering algorithm, WordNet.

1 Introduction

Relational similarity measurement can be applied widely in numerous natural language processing tasks such as detecting word analogies, classifying semantic relations, information extraction, and information retrieval [1,7]. Latent Relation Search (LRS) is recently proposed as a query-by-example technique for solving queries in which the user specifies a triplet of terms $q=(A,B,C)$ and seeks from a search engine a fourth term D such that the (C,D) relation is analogous to the (A,B) relation [16,17]. For example, we know *Obama* is the president of *U.S.*, if we want to know who is the president of *France*, we can query $q=(Obama,U.S.,France)$ to seek the term *Hollande* as the answer. This requires the LRS search engine to correctly measure the similarities between two relations hold by two words in two word-pairs, (*Obama,U.S.*) and (*Hollande,France*).

In general, there are two kinds of similarity, i.e., *attributional similarity* and *relational similarity*. Attributional similarity is a correspondence between attributes of two words. A measure of attributional similarity (sim_a) is a function that maps two words, w_a and w_b to a real number. The higher correspondence there is between the properties of w_a and w_b, the greater value their attributional similarity has. For example, *dog* and *wolf* have a relatively high degree of attributional similarity.

Relational similarity refers to the similarity of semantic relations between pairs of words [1]. The relational similarity between two word-pairs, (w_a,w_b) and (w_c,w_d), is defined as the correspondence between semantic relations that exist between the two words in each word-pair. For instance, the two word-pairs, (*ostrich,bird*) and (*lion,cat*), are considered relationally similar because the relation X *is larger than* Y holds between the two words X and Y, in each of two word-pairs. Here we use X to replace *ostrich* and *lion*, while use Y to replace *bird* and *cat*. A measure of relational similarity sim_r is the function that maps two word-pairs, (w_a,w_b) and (w_c,w_d), to a real number. The more correspondence there is between the relations of (w_a,w_b) and that of (w_c,w_d), the greater their relational similarity is.

Many algorithms have been proposed for measuring attributional similarity between two words [9], while measurements of relational similarity are not well studied [1]. Existing relational similarity measurements can be broadly divided into two categories: *lexicon-based* methods and *corpus-based* methods.

The *lexicon-based* methods are built on the basis of a hypothesis, the amount of relational similarity between two word-pairs, (w_a,w_b) and (w_c,w_d), depends on the degree of correspondence between the relations between w_a and w_b and the relations between w_c and w_d [1]. Veale [2] proposed a relational similarity measure algorithm based on WordNet. This approach was tested using a set of 374 SAT (Scholastic Aptitude Test) analogy questions, achieving a score of 47%.

Existing *lexical-based* approaches have two major limits: (1) They depend on the relations between two concepts described in the lexical resources, where there are limited relations supported by a lexical resource. (2) The part of speech in a semantic dictionary is also limited, e.g., Wordnet only contains nouns, verbs, adjectives and adverbs, it does not supply some proper nouns (e.g. *XML*), prepositions (e.g. *for* and *into*), personal pronouns (e.g. *him*, *her* and *himself*) and combinations of words. These characteristics restrict the use of these approaches. For these reasons, the *corpus-based* methods attract more focus.

The *corpus-based* approaches are built on the basis of a hypothesis, two words have a certain degree of semantic similarity if and only if they appear in the same context. This method views the probability distribution of context information as the reference of lexical semantics. The basic idea of these approaches is to calculate the relational similarity between word-pairs using the co-occurrence context information of a word-pair from large corpus (e.g., Wikipedia or by a search engine like Google) [15,16].

The VSM (Vector Space Model) approach is introduced which adopts 128 manually patterns and counts the hit frequency [1]. Later, Turney [18] provides the LRA (Latent Relational Analysis) method. LRA automatically derives the patterns from the corpus and uses synonyms to explore reformulations of word-pairs. However, LRA is computationally intensive. It cost 8 days to process the 374 analogy questions. Bollegala et al. [4,5] proposed another relational similarity algorithm which adopts web text snippets. Later, Bollegala et al. [11] describe eight different types of relational symmetries that are frequently observed in proportional analogies and use those symmetries to robustly and accurately estimate the relational similarity between two given word-pairs. The existing approaches are mostly based on extracting semantic features as feature matrixes from large-scale corpus. The extracted semantic features are loosely distributed, which cause the sparseness of feature matrixes. MTLRel compresses the feature matrix into a feature vector using a multi-task lasso method, then measures relational similarity between two word-pairs by the cosine of the angle between two feature vectors [12].

These *corpus-based* methods build the matrixes that heavily depend on statistical calculations. They calculate the relational similarities by the cosine or the Mahalaobis distance. For the VSM method, users need to manually create the 128 patterns, which is a huge workload. In LRA, the matrix is high dimensional and sparse, which typically brings time-consuming process. These methods use lexicon pattern frequency as the feature vector for a word-pair. They mainly depend on statistical data and cannot mine the implicit semantic relations. Other approaches [11,12] use snippets. However snippets just summarize the main points of the text and the extracted patterns from them are not enough to represent the semantic relations hidden in word-pairs.

To tackle the problem of mining the *semantic relations* among words for *corpus-based* methods, we proposed an approach by combining Wordnet3.0 and the Wikipedia [3]. Experimental evaluation based on the same 374 SAT analogy questions, the score of the approach is 43.9%, which is higher than the approach suggested by Bollegala [4]. And the computation time is 3.x days which is less than that of LRA (8 days). During our experiments, we noticed that these semantic relatedness are scattered, which is one main reason of a relatively lower score compare with that of LRA.

In this paper, we propose an extended approach from the following three aspects: (1) We design a method to represent relations of a word-pair using relation-terms extracted from the corpus (e.g., *Wikipedia*). Both relation-terms and their corresponding frequencies are viewed as attributions of relations. The relation-terms can be nouns, verbs, adjectives and prepositions. (2) The semantic relatedness score of two words extracted from the relation-term sets respectively is computed by Gloss Vector. A three-dimension model is created using two frequencies of the two words and their semantic relatedness score. Furthermore to cluster these points, a Density-based clustering algorithm (DBSCAN) is employed. (3) The relational similarity between two word-pairs is conducted by these clustered semantic relatedness scores. Experimental results conducted on the 374 SAT analogy questions show that the proposed method attains a relatively higher accuracy score of 52.9% with relatively lower time cost.

In summary, the main contributions of this paper include:

- We propose a novel hybrid model that integrates three components to measure relational similarity, i.e., term concurrence, syntactic patterns, and a clustering algorithm.
- We extensively validate the approach using the well-known SAT 374 analogy questions dataset.

The rest of this paper is organized as follows: Problem statement is introduced in Section 2. The details of the suggested approach are presented in Section 3. Section 4 shows the evaluation and its performance. Section 5 describes related work. Finally, we conclude this paper in Section 6.

2 Problem Statement

Our problem can be summarized as follows: Let R_1 and R_2 be the semantic relations between a word-pair (w_a, w_b) and another word-pair (w_c, w_d). Our aim is to measure the similarity between the two relations R_1 and R_2.

To completely describe the latent semantic relations between a word-pair, in this paper, we represent the relations between a word-pair using bag-of-words model. That is, $R_1(w_a, w_b)$ and $R_2(w_c, w_d)$ are represented by two sets of terms $T_1(w_a, w_b) = \{ t_1, t_2, ..., t_i, ... t_m \}$ and $T_2(w_c, w_d) = \{ t_1', t_2', ..., t_j', ..., t_n' \}$ respectively. Two terms t_i and t_j' are viewed as relation-terms of $R_1(w_a, w_b)$ and $R_2(w_c, w_d)$ respectively. The frequency of the term t_i is f_i and the frequency of t_j' is f_j'. There are $F_1(w_a, w_b) = \{f_1, f_2, ..., f_i, ..., f_m\}$ and $F_2(w_c, w_d) = \{f_1', f_2', ..., f_j', ..., f_n'\}$.

For two relations, $R_1(w_a, w_b)$ and $R_2(w_c, w_d)$, we view the two relation-term sets $T_1(w_a, w_b)$ and $T_2(w_c, w_d)$, together with their frequencies $F_1(w_a, w_b)$ and $F_2(w_c, w_d)$ as the attributes of the two relations respectively. Originate from the definition of attributional similarity, that is, the attributional similarity between two words is a correspondence between attributes of the two words, we give two hypotheses about the relational similarity between two word-pairs in this paper:

Hypothesis 1: for two word-pairs, their *relational similarity* is a correspondence between the attributes of two relations.

In this paper, we model the features of a relation between two words using a relation-term set and the frequency of each term. Thus we can give another hypothesis.

Hypothesis 2: the *relational similarity* between two word-pairs depends on the *attributional similarity* of two terms extracted from two relation-term sets respectively together with their frequencies.

For two word-pairs, (w_a, w_b) and (w_c, w_d), existing related works reduce *relational similarity* to *attributional similarity* based on a hypothesis, that there is a high degree of relational similarity between (w_a, w_b) and (w_c, w_d), if there is also a high degree of attributional similarity between (w_a, w_c), and between (w_b, w_d). The algorithm proposed by Veale [2] uses the depth of the relation between w_a and w_c, the depth of the relation between w_b and w_d, and the common adjective words in the two glosses of each two words (w_a and w_c, w_b and w_d) in Wordnet. The quality measure was based on the similarity between the (w_a,w_c) paths and the (w_b,w_d) paths.

Different from Veale's approach, we measure relational similarity by measuring attributional similarity between two relations R_1 and R_2. Since we model the two relations using relation-terms and their frequencies, we measure the relational similarity of two word-pairs by measuring their relation-terms. Symbols used in the following sections are summarized in Table 1.

Table 1. Table of symbols

Symbol	$Meaning$
w	A word in a word-pair
X, Y	A word used to replace the first word or the second word in a word-pair
R	The relations two words in a word-pair hold
t, t'	A relation-term a word-pair holds
T	The relation-term set of a word-pair
f, f'	The frequency of a relation-term
F	The frequency set of a word-pair
Doc	Collected web documents
S_k	A sentence
$\mathbf{S}$	A set of sentences
sim_a	Attributional similarity
sim_r	Relational similarity
rel	Semantic relatedness value
clu	A cluster
$weight$	The weight of a cluster
$\overline{rel}$	The arithmetic mean of all semantic relatedness values in a cluster
$MinPts$	The minimum number of objects
p, q, o	A data object
ω	The subsequence threshold
ε	A given radius
Ph	The set of all subsequences

3 Our Approach

For two word-pairs (w_a,w_b) and (w_c,w_d), since two relations $R_1(w_a,w_b)$ and $R_2(w_c,w_d)$ are not given, our task is (1) inferring these hidden (latent) relations and expressing the relations using bag-of-words model, (2) computing semantic relatedness values of any two relation-terms from two relation-term sets, and (3) computing the relational similarity value for the two relations. For the convenience of discussion, we take two word-pairs (*museum,exhibit*) and (*theater,performance*) as an example. The outline of the suggested approach is shown in Fig.1.

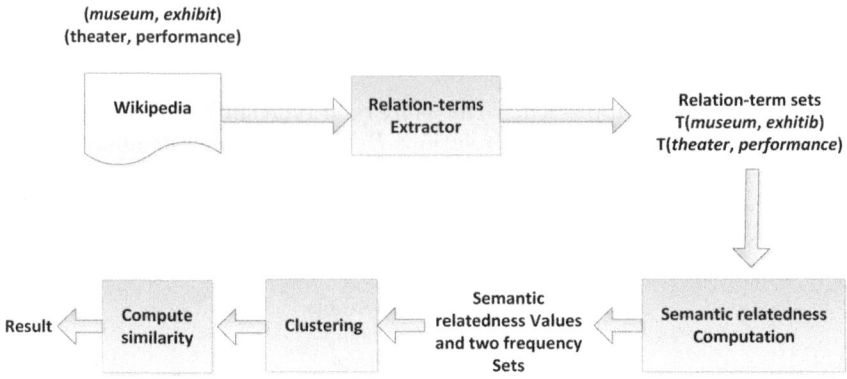

Fig. 1. Outline of the suggested approach

3.1 Extracting Relation-Terms

To identify the implicit relations for two word-pairs (w_a, w_b) and (w_c, w_d), both two word-pairs are input into the search engine of Wikipedia. In this paper, we choose Wikipedia as the corpus, since it is currently the largest knowledge repository on the Web with over 3.4 million articles in English now. Web pages that contain both terms w_a and w_b or both terms w_c and w_d are gathered as the text corpus for the two word-pairs. We represent all collected pages as $\text{Doc}(w_a \wedge w_b)$ and $\text{Doc}(w_c \wedge w_d)$ respectively. Note that, we do not know the relations in advance. We extract all pairs that might hold some relations and then use several filters to obtain informative pairs as describe late.

All text documents in both $\text{Doc}(w_a \wedge w_b)$ and $\text{Doc}(w_c \wedge w_d)$ are fed to the Extractor, which extracts subsequences containing both w_a and w_b or both w_c and w_d in the same sentences. The lengths of the extracted subsequences are not longer than a subsequence threshold ω (We will discuss the determination of ω in Section 4). For example, given a word-pair (w_a, w_b), we extract subsequences that might represent semantic relations between two words w_a and w_b in a sentence S_k. S is the set of sentences containing both w_a and w_b, there is $S_k \in S$. We consider the gaps between w_a and w_b (or w_b and w_a) of the sentence S_k.

We then generate all subsequences from the above sentences. These subsequences start with the first word w_a (and its other formats) and end at the second word w_b (and its other formats), or in turn.

For all subsequences, we change them into the form $X v_1 \ldots v_i \ldots v_k Y$. The set of all subsequences is denoted as $Ph(w_a, w_b)$ in which the word-pair (w_a, w_b) appear. For example, for the word-pair $(museum, exhibit)$, the extractor would extract all subsequences that represent the semantic relations between museum and exhibit or their other forms. We replace museum and its other formats with X, while replace exhibit and its other formats (such as *exhibits, exhibited, exhibiting*) with Y. We also count the frequency of each subsequence.

Choosing Relation-Terms
To eliminate the differences between inflected forms of a word, all words in all subsequences between X and Y are stemmed and POS (Part-of-Speech) Tagging. In this paper, we use the Part-of-Speech Tagger developed by Natural Language Processing Research Group of Stanford University [14]. We consider *nouns, verbs, adjectives* and *prepositions* for the following reasons: (1) Article and interjection have less effect on expressing semantic relations between two words; (2) Prepositions can be used to express time, position and reason; (3) Wordnet is a lexical dictionary which contains nouns, verbs, adjectives and adverbs, where adverbs generally are used to decorate adjectives and verbs.

For the relation-terms extracted in the previous step, all nouns, verbs, adjectives and prepositions in subsequences are remained. These terms are viewed as the relation-terms. Through this way, for the word-pair (w_a, w_b), we can get its relation-terms $T_1(w_a, w_b) = \{t_1, t_2, ..., t_i, ..., t_m\}$. We denote the frequencies of the relation-terms as $F_1(w_a, w_b) = \{f_1, f_2, ..., f_i, ..., f_m\}$. Similarly, for another word-pair (w_c, w_d), we can get the relation-terms $T_2(w_c, w_d) = \{t'_1, t'_2, ..., t'_j, ..., t'_n\}$ and their frequencies for $F_2(w_c, w_d) = \{f'_1, f'_2, ..., f'_j, ..., f'_n\}$.

Compared to the relation presentation algorithms in previous researches, our approach is adapted to improve the recall of relation presentation between two words: (1) We eliminate the differences between inflected forms of a word by using various forms of two words, and by stemming the subsequences between two words. (2) We use X and Y to replace the two words respectively. For example, consider the two sentences: *Obama is the 45th and current president of the U.S* and *Hollande is the current president of France*. We generate the terms between *Obama* and *U.S* as *X president of Y* then we have a common terms between two word-pairs (*Obama,U.S*) and (*Hollande,France*).

3.2 Semantic Relatedness Computation

Two words are semantically related if they have any kind of semantic relation [8]. They are semantically related to the degree that they share attributes. Examples are synonyms (*bank* and *trust company*), meronyms (*car* and *wheel*), antonyms (*hot* and *cold*), and words that are functionally related or frequently associated (*pencil* and *paper*) [1]. Attributional similarity in cognitive science corresponds to the term semantic relatedness in computational linguistics [1]. We do not distinguish semantic relatedness and attributional similarity in this paper.

Gloss Vectors is a popular algorithm based on WordNet to measure semantic relatedness between two words by combining the structure and content of WordNet with co-occurrence information derived from raw text [9]. In our approach, we use Gloss Vectors to measure semantic relatedness value (*rel*) between any two terms extracted from two relation-term sets. For the relation-term set $T_1(w_a, w_b)$ of the word-pair (w_a, w_b) and the relation-term set $T_2(w_c, w_d)$ of the word-pair (w_c, w_d), we compute the semantic relatedness value rel_{ij} between a word t_i in $T_1(w_a, w_b)$ and a word t'_j in $T_2(w_c, w_d)$. The number of all semantic relatedness values is $m * n$. For example, for two word-pairs, (*museum,exhibit*) and (*theater,performance*), the samples of $T(museum, exhibit)$ and

Table 2. Semantic relatedness measurement between two terms

t_i	t_j'	Semantic relatedness values
be	be	1
build	outdoor	0.466874489
display	tour	0.798024459
feature	group	0.800974744
with	and	0
contain	have	0.933292187
house	school	0.811479648
several	many	0.898463388
provide	host	0.883077082
...	...	...

$T(theater, performance)$ are shown in the first two columns of Table 2. The samples of semantic relatedness values between one term of $T(museum, exhibit)$ and one term of $T(theater, performance)$ are shown in the third column of Table 2.

3.3 Semantic Relatedness Clustering

As mentioned above, frequencies of terms in a relation-term set are attributions of a word-pair also. We organize the frequency f_i of the term t_i, the frequency f_j' of the term t_j' and their semantic relatedness value rel_{ij} as a point with a coordinate (f_i, f_j', rel_{ij}) in the 3D space. For instance, the coordinates of two word-pairs, $(museum, exhibit)$ and $(theater, performance)$, are shown in Fig.2. In Fig.2, x–axis and y–axis represents the frequencies of the terms in $T(museum, exhibit)$ or T($theater, performance$) respectively, z-axis represents the semantic relatedness values between any two terms in two relation-term sets.

From Fig.2, we observe that, most of the points are intensive, while some of the points are discrete without fixed shape. The concentrated points indicate that their relations are similar, while individual points mean that they are noise or their relations are different. Considering these features of points, it is better to use a density-based clustering method to cluster these points. The main ideas of the density-based clustering are: looking for the higher-density areas separated by the lower-density areas from the data set, and view each individual higher-density area as a cluster. DBSCAN (Density Based Spatial Clustering of Applications with Noise) is one of the most popular clustering algorithm [6]. DBSCAN defines that: for each object in a cluster, there is at least the minimum number ($MinPts$) of objects which are the neighbors of the object within a given radius (ξ).

It is necessary to organize the data set before we use the DBSCAN algorithm. Our method is summarized here: the frequency f_i of the term t_i is marked as the x-dimension, the frequency f_j' of the term t_j' is marked as the y-dimension, and the semantic relatedness value rel_{ij} between two terms $t_i(i = 1, 2, ..., m)$ and $t_j'(j = 1, 2, ..., n)$ is marked as the z-dimension. For any two terms of the two relation-term sets, all these points are represented as $\{p_1, p_2, ..., p_k\}$. We use the DBSCAN algorithm to cluster these points.

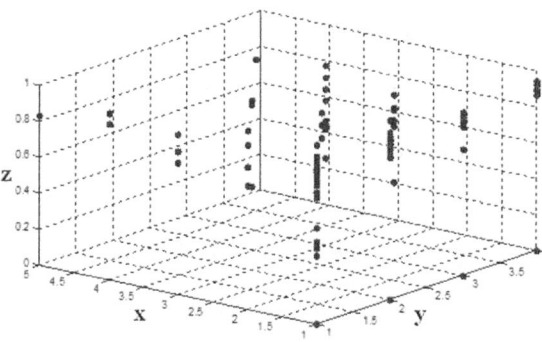

Fig. 2. The points in 3D for two word-pairs (museum,exhibit) and (theater,performance)

The DBSCAN algorithm begins from a starting object o which is not accessed. The algorithm searches all ξ-*neighborhoods* of the object o with the distance ξ. If there is $\mid N_\xi(p) \mid \geq$ *MinPts*, thus a cluster is created (which means the point o is a core object). After that, the algorithm clusters all objects which are density-reachable from the core object. Then, a new never accessed point is used to explore a fresh cluster. Both ξ and *MinPts* are global variables. Their determinations will be discussed in Section 4.

After using the DBSCAN algorithm, all points are clustered into (k) clusters (clu), here $clu_i = \{rel_{i1},...,rel_{ij},...,rel_{ip}\}$. The number of points contained in i-th cluster (clu_i) is viewed as the size of the cluster. Furthermore, we view the size of a cluster as the weight ($weight_i$) of the cluster.

To indicate the central tendency of the collection of all semantic relatedness values in a cluster, we compute the arithmetic mean ($\overline{rel}$) of the cluster. That is, for a cluster (clu_i), its semantic relatedness score is,

$$\overline{rel}_i = \frac{1}{weight_i} \sum_{j=1}^{weight_i} rel_{i,j} \qquad (1)$$

3.4 Relational Similarity Computing

For the two word-pairs, (w_a,w_b) and (w_c,w_d), we have extracted their relation-terms above. Considering that some relation-terms contribute more than others, we use the weighted mean of these clustered relatedness values to compute the relational similarity (sim_r) between the two word-pairs. The relational similarity (sim_r) is the weighted mean of a data set, $\{\overline{rel}_1,...,\overline{rel}_i,...,\overline{rel}_k\}$, with their weights, $\{weight_1,...,weight_i,...,weight_k\}$, that is,

$$sim_r(w_a, w_b :: w_c, w_d) = \frac{\sum_{i=1}^{k} weight_i \overline{rel}_i}{\sum_{i=1}^{k} weight_i} \qquad (2)$$

Therefore the cluster with a higher weight contributes more to the weighted mean than that of the element with a lower weight.

4 Experimental Evaluations

We evaluate the proposed approach on the 374 SAT analogy questions. The task of the 374 SAT analogy questions is to find out the solution from the five choices which is the most analogous to the stem. The performance of our approach is measured by Precision (Accuracy or Score), Recall and F-measure (F). The correct choice is called the solution, and the incorrect choices are distracters [1]. Table 3 shows one of 374 analogy questions, along with the relational similarity values between the stem and each choice, using the proposed approach. The choice with the highest relational similarity values is viewed as the answer.

Table 3. A sample of SAT analogy question with their relational similarity values

		Relational similarity values
Stem	ostrich, bird	
Choices	(a) cub, bear	0.074
	(b) ewe, sheep	0.063
	(c) goose, flock	0
	(d) lion, cat	0.12
	(e) primate, monkey	0.083

4.1 Thresholds Determination

There are three thresholds in the proposed approach, the subsequence threshold ω, the distance threshold ξ and the minimum number threshold $MinPts$. In our experiments, we determine the values of the three thresholds as follows: (1) For the subsequence threshold ω, we set $\omega=5$ for the reason that LRA uses the value 5 which is discussed in [18]. (2) The other two thresholds ξ and $MinPts$, are related to the DBSCAN algorithm. There have been a lot of works on the two thresholds. However certain rules are not getting yet. In our approach, the two threshold are empirically determined. Our empirical results show that the ranges of ξ and $MinPts$ are $0 < \xi \leq 4$ and $3 \leq MinPts \leq 7$ respectively. We show the corresponding experimental results in Fig.3.

In Fig.3, the accuracies represent the scores of the 374 SAT questions using the proposed approach. We notice that our approach gets the highest accuracy rate 52.9% while $\xi=3.25$ and $MinPts=5$.

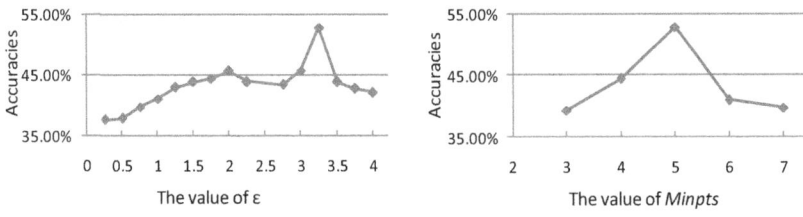

Fig. 3. The accuracies of the 374 SAT questions under various ξ and $MinPts$

4.2 Experiment Analysis

We show the various comparisons between our approach with other existing methods in Table 4. All values of existing methods (VSM-AV, VSM-WMTS and LRA) are reported in [18]. The time costs of three main steps of the proposed approach are shown in Table 5.

Table 4. Comparison among various relational similarity measurements

	VSM-AV	VSM-WMTS	LRA	Our approach
Correct questions	176	144	210	190
Wrong questions	193	196	160	169
Skipped questions	5	34	4	15
Precision	47.7	42.4	56.8	52.9
Recall	47.1	38.5	56.1	50.8
F	47.4	40.3	56.5	51.8
Time (Day:Hour:Min)	17:00:00	1:00:00	8:17:49	2:05:18

Table 4 shows that the precision rate 52.9% of the proposed approach is slightly lower than 56% of LRA, while the time cost 2.x days of our approach is far less than 8 days of LRA. Two main factors of these experimental results are: (1) the size of documents which are entered into Extractor, and (2) 15 skipped questions are skipped.

Table 5. Three main steps used in the proposed approach

Main steps	Time cost (Hour:Min:Sec)	Hardware
Extracting relation-terms	4:25:08	1CPU
Computing semantic relatedness values	48:50:00	1CPU
Clustering and computing relational similarity	0:03:00	1CPU
All time cost	53:18:08	

For the first factor, the size of documents, in our experiments, we summarize the relation between the size of documents and the accuracy, and the relation of time cost and the accuracy in Fig.4 and Table 6 respectively. The larger the number of documents is, the higher the accuracy is, while the time cost is ever greater. In this paper, considering the tradeoff between accuracy and time cost, we limit the number of documents to be 100. There are more corpora in LRA (nearly all pages returned) than that in the proposed approach (the top 100 pages returned), while the time cost of LRA is much higher.

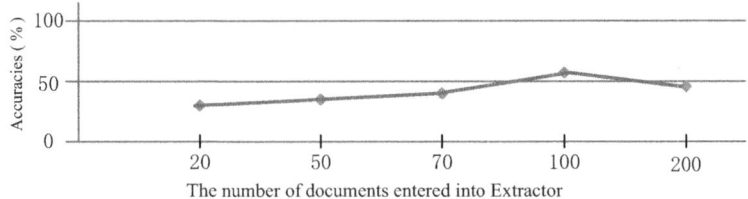

Fig. 4. Accuricies with the number of documents entered into Extractor

For the second factor, the 15 questions are skipped for the following two reasons: (1) There are no terms left after POS tagging. For example, all middle words of the stem (*amplifier,ear*) are removed for there are not nouns, verbs, adjectives and prepositions. (2) If all semantic relatedness values are zero for two word-pairs, thus the question is skipped.

Table 6. Time cost with the number of documents entered into extractor

Number of documents	20	50	70	100	200
Time cost(Day:Hour:Min)	0:05:08	1:01:00	1:15:14	2:04:05	5:02:16

One main feature of the proposed approach is the usage of the DBSCAN algorithm to cluster points. Without clustering, all points are used in the relational similarity measurement. After clustering, the points are clustered and the noise points are removed. We show the experimental results of employing DBSCAN and that of excluding DBSCAN in Table 7.

Table 7. Comparison between Excluding clustering and employing clustering

	Excluding DBSCAN	Employing DBSCAN
Time cost(Day:Hour:Min)	2:04:05	2:05:18
Correct questions	137	190
Wrong questions	161	169
Skipped questions	76	15
Accuracy	46%	52.90%

As shown in Table 7, by employing the DBSCAN clustering, the number of correct questions increases significantly, from 137 to 190, the number of skipped questions reduces greatly also, from 76 to 15, and the number of wrong questions increased slightly. The experimental accuracy rate of employing clustering is larger than that of excluding clustering around 7 percent. The main reason is related to the 15 skipped questions. The points of the word-pairs corresponding to these 15 questions are more concentrated and there are no noise. Consequently, the clustering procedure has no effect.

5 Related Work

For relational similarity measurement, characterizing semantic relations between two words are important. Generally, there are two ways to characterize semantic relations between two words, *lexical patterns* and *bag-of-words* model.

Lexical patterns are used to represent semantic relations between two words of a word-pair that appear in the same sentence [1,11]. The relation between two words in a word-pair is therefore represented by a vector of lexical pattern frequencies.

In the VSM approach, a vector is first created for a word-pair (w_a, w_b) by counting the frequencies of various lexical patterns in which slot markers X and Y are substituted respectively by w_a and w_b[1]. They used 128 manually created patterns such as X of Y, Y of X, X to Y, and Y to X in their experiments. The numbers of hits for respective queries are used as elements in a vector the represent the word pair. The VSM approach achieves a score of 47% on the 374 SAT questions. LRA extends the VSM approach in three ways [1]: (1) The connecting patterns are derived automatically from the corpus, instead of using a fixed set of patterns. (2) The Singular Value Decomposition (SVD) is used to smooth the frequency data. (3) LRA considers transformations of the word-pair, generated by replacing one of the words by synonyms. LRA achieves a score of 56% on the same dataset. LRA is time consuming for it adopts synonymous variants of the given word-pair and needs a large number of search engine queries. The formed matrix in the algorithm is so large and sparse that the cost is heavy to reduce the noise. LRA takes over 8 days to process the 374 SAT analogy questions, which can be problematic in many Web mining related tasks.

Bollegala *et al.* [5] proposed a method which adopted web text snippets to automatically extract lexical patterns. The characteristics of the method are: (1) Support Vector Machine is trained to recognize word pairs with similar semantic relations. (2) It applies the Mahalanobis distance to calculate the vector similarity. Bollegala's method achieves the SAT score of 51.1%. The reason of a lower score is that the method use snippets as their corpus: (1) Which cause some wrong relations extracted by their method. And (2) the information carried by snippets is lower than the information carried by web documents returned by a web search engine.

Another way to characterize semantic relations is retrieving terms in specific relation and model these terms as *bag-of-words* model. For relational similarity of two word-pairs, in bag-of-words model, a relation is represented as a collection of words. Church and Hanks [9] measured semantic relatedness between two words with mutual information. Turney [1], Baroni and Bisi [13] proposed methods that calculate the level of synonyms for two words by using the number of Web documents returned by search engines. Their methods use the concurrences of words and mutual information. Kato *et al.* [11] represent the relations between two words by *bag-of-words*, that is, the extracted words from the sentences containing two words. Their assumption is that, the relations between two words are better expressed by the sentences which frequently appear in the documents containing both the two words.

In our approach, we model semantic relations between two words by retrieving terms in specific relation which satisfy special lexical patterns, and we model the relations between a word-pair using bag-of-words.

6 Conclusion and Future Work

Relational similarity measurement attracts great interest in natural language processing. In the paper, we proposed a hybrid approach for relational similarity measurement, which integrates WordNet and Wikipedia to measure relational similarity between two word-pairs. The main features of the proposed approach are: (1) We use relation-terms to represent relations, and both relation-terms and their frequencies are viewed as attributions for measuring the two word-pairs, in order to represent latent relations between two word-pairs. Furthermore, we measure relational similarity between two word-pairs using *attributional similarity* measurement. (2) In this approach, Gloss Vectors are used to measure semantic relatedness between word-pairs. (3) The DBSCAN algorithm is adopted to cluster the points. Experiments show that the similarity measuring accuracy is improved using the clustering algorithm.

In the future, the following aspects will be further conducted: (1) Gloss Vectors has show their higher performance in semantic relatedness measurement between two words. However it is computationally intensive. To reduce time cost, we will introduce other semantic relatedness measurements with higher accuracy with less time cost. (2) Other clustering algorithms will be evaluated in our approach to validate the performance. (3) We further evaluate our approach in other works such as Latent Relational Search and web entity similarity measurement.

Acknowledgement. This work is sponsored by the grant from the Shanghai Science and Technology Foundation (No. 11511504000 and No.11511502203).

References

1. Turney, P., Pantel, P.: From frequency to meaning: Vector space models of semantics. Journal of Artificial Intelligence Research 37, 141–188 (2010)
2. Veale, T.: WordNet sits the sat: A knowledge-based approach to lexical analogy. In: 16th European Conference on Artificial Intelligence, pp. 606–612. IOS Press (2004)
3. Cao, Y., Lu, Z., Cai, S.: Relational Similarity Measure: An Approach Combining Wik-ipedia and WordNet. Applied Mechanics and Materials 55-57, 955–960 (2011)
4. Bollegala, D., Matsuo, Y., Ishizuka, M.: Www sits the sat: Measuring relational similarity on the web. In: 18th European Conference on Artificial Intelligience, pp. 333–337. IOS Press (2008)
5. Bollegala, D., Matsuo, Y., Ishizuka, M.: Measuring the similarity between implicit semantic relations from the web. In: 18th Int. World Wide Web Conference, pp. 651–660. ACM (2009)

6. Ester, M., Kriegel, H., Sander, J., Xu, X.: A density-based algorithm for discovering clus-ters in large spatial databases with noise. In: Second International Conference on Knowledge Discovery and Data Mining, pp. 226–231. AAAI Press (1996)
7. Bollegala, D., Matsuo, Y., Ishizuka, M.: Measuring the Degree of Synonymy between Words Using Relational Similarity between Word Pairs as a Proxy. In: IEICE Transaction on Information and System, vol. E95D, pp. 2116–2123 (2012)
8. Budanitsky, A., Hirst, G.: Semantic distance in WordNet: An experimental, application-oriented evaluation of five measures. In: Workshop on WordNet and Other Lexical Resources, Second Meeting of the North American Chapter of the Association for Computational Linguistics, pp. 29–24 (2001)
9. Patwardhan, S., Pedersen, T.: Using WordNet-based context vectors to estimate the se-mantic relatedness of concepts. In: EACL 2006 Workshop, Making Sense of Sense: Bringing Computational Linguistics and Psycholinguistics Together, pp. 1–8. IOS Press (2006)
10. Kato, M., Ohshima, H., Oyama, S., Tanaka, K.: Query by analogical example: relational search using web search engine indices. In: 18th ACM Conference on Information and Knowledge Management, pp. 27–36. ACM Press (2009)
11. Bollegala, D., et al.: Improving relational similarity measurement using symmetries in proportional word analogies. In: Information Processing and Management (2012)
12. Yan, D., Lu, Z.: Relational Similarity Measurement Between Word-pairs using Multi-Task Lasso. In: International Conference on Cloud and Service Computing (2012)
13. Baroni, M., Bisi, S.: Using cooccurrence statistics and the web to discover synonyms in a technical language. In: Fourth International Conference on Language Resources and Evaluation, pp. 1725–1728 (2004)
14. http://nlp.stanford.edu/software/tagger.shtml (2012)
15. Nakov, P., Hearst, M.: Solving relational similarity problems using the web as a corpus. In: ACL 2008-HLT, pp. 452–460 (2008)
16. Duc, N., et al.: Using relational similarity between word pairs for latent relational search on the web. In: Intl. Conf. on Web Intelligence, pp. 196–199 (2010)
17. Liang, C., Lu, Z.: Chinese Latent Relational Search Based on Relational Similarity. In: Xiang, Y., Pathan, M., Tao, X., Wang, H. (eds.) ICDKE 2012. LNCS, vol. 7696, pp. 115–127. Springer, Heidelberg (2012)
18. Turney, P.: Measuring semantic similarity by latent relational analysis. In: Proceedings of the Nineteenth International Joint Conference on Artificial Intelligence, pp, pp. 1136–1141 (2005)

Subspace MOA: Subspace Stream Clustering Evaluation Using the MOA Framework

Marwan Hassani, Yunsu Kim, and Thomas Seidl

Data Management and Data Exploration Group
RWTH Aachen University, Germany
{hassani,kim,seidl}@cs.rwth-aachen.de

Abstract. Most available static data are becoming more and more high-dimensional. Therefore, subspace clustering, which aims at finding clusters not only within the full dimension but also within subgroups of dimensions, has gained a significant importance. Recently, *OpenSubspace* framework was proposed to evaluate and explore subspace clustering algorithms in WEKA with a rich body of most state of the art subspace clustering algorithms and measures. Parallel to it, MOA (**M**assive **O**nline **A**nalysis) framework was developed also above WEKA to provide algorithms and evaluation methods for mining tasks on evolving data streams over the full space only.

Similar to static data, most streaming data sources are becoming high-dimensional, and tracking their evolving clusters is also becoming important and challenging. In this demonstrator, we present, to the best of our knowledge, the first subspace clustering evaluation framework over data streams called *Subspace MOA*. Our demonstrator follows the online-offline model which is used in most data stream clustering algorithms. In the online phase, users have the possibility to select one of three most famous summarization techniques to form the microclusters. In the offline phase, one of five subspace clustering algorithms can be selected. The framework is supported with a subspace stream generator, a visualization interface to present the evolving clusters over different subspaces, and various subspace clustering evaluation measures.

1 Introduction

Clustering on the high-dimensional data becomes more and more important as modern databases tend to be huge. Due to the curse of dimensionality, excessive number of attributes makes data points unique, and the distances between the points become more alike as the dimensionality grows in high-dimensional space [3]. For such kinds of data with higher dimensions, distances grow more and more alike (cf. the toy example in Figure 1(a)). Applying traditional clustering algorithms (called in this context: *full-space* clustering algorithms) over such data objects will lead to useless clustering results. In Figure 1(a), the majority of the black objects will be grouped in a single-object cluster (outliers) when using a full-space clustering algorithm, since they are all dissimilar, but apparently they are not as dissimilar as the gray objects. The latter fact motivated the

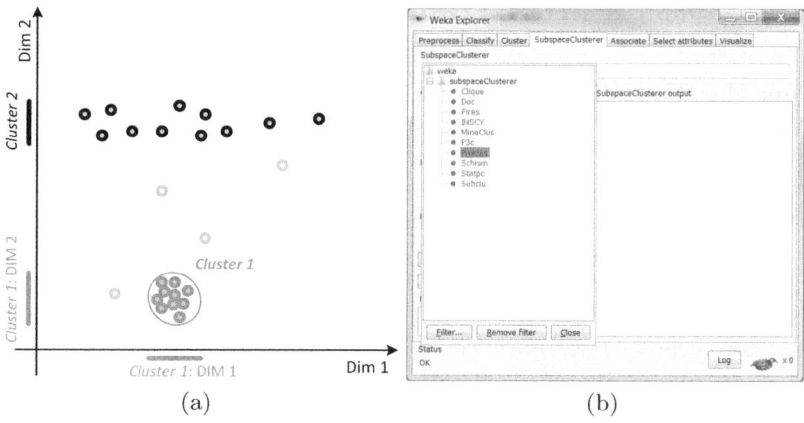

Fig. 1. (a) A Toy Example of a Subspace Clustering Output, (b) A Screen Shot of *OpenSubspace* Framework

research in the domain of *subspace* and *projected clustering* in the last decade which resulted in an established research area for static data.

OpenSubspace framework [8] was proposed to evaluate and explore subspace clustering algorithms in WEKA with a rich body of most state of the art subspace/projected clustering algorithms and measures (cf. Figure 1(b)).

In this research, these algorithms are applied to the streaming cases. Other than static data that do not vary over time, stream data are given in different rate and pattern changing dynamically, which makes it challenging to analyze its evolving structure and behavior. In streaming scenarios, we also often face limitations on processing time and storage, since a vast amount of continuous data are coming rapidly.

Data stream mining has been an emerging research topic in the previous decade and a rich body of stream mining algorithms has been created. MOA (Massive Online Analysis) framework [7] was built on experience with both WEKA and VFML (Very Fast Machine Learning) toolkit [6] to support the research in the stream mining area with generators, visualization methods, and interesting evaluation measures. Similar to static data, evolving data streams are also becoming naturally high-dimensional with their existence in multiple applications with many attributes. However, different to subspace clustering algorithms over static data, only few subspace stream clustering algorithms has been developed recently (HPStream [2] and PreDeConStream [5]. Such kinds of algorithms are a bit tricky since they have to track the all changes of evolving clusters over the streams (splitting, merging, appearance, decaying, moving, ... etc.), by considering the fact the these clusters might exist in all possible subspaces and not only in the full-space. In *Subspace MOA*, users can select any of ten subspace clustering algorithms to be the offline part of subspace clustering algorithm, where one of seven summarization methods for the online part can be also selected.

2 The *Subspace MOA* Framework

1. Under the Setup Tab: (cf. Figure 2(a)), the selection of the **data stream input**, Subspace MOA offers the possibility of reading external ARFF files, a synthetic random RBF generator, and a synthetic random RBF subspace generator with the possibility of varying the subspace event. The online-offline model is followed by most stream clustering algorithms (cf. [1], [4], [5]). In the **online phase**, a summarization of the data stream points is performed and the resulting microclusters is given by sets of cluster features $CF_i = (N, LS_i, SS_i)$ which represent the number of points within that microcluster, their linear sum and their squared sum, respectively. Subspace MOA offers three algorithms to form these microclusters and continuously maintain them. These are the main ones supported by MOA: ClusterGenerator, CluStream, and DenStream . In the **offline phase**, the clustering features are used to reconstruct an approximation to the original N points using Gaussian functions to reconstruct spherical microclusters centered at $c_i = \frac{LS_i}{N}$ with a radius: $r = \sqrt{\frac{SS}{N} - (\frac{LS}{N})^2}$ ($SS = \frac{1}{d}\sum_{i=1}^{d} SS_i$ and $LS = \frac{1}{d}\sum_{i=1}^{d} LS_i$). The generated N points are forwarded to one of the five most famous subspace clustering algorithms that are supported by OpenSubspace: SubClu, ProClus, P3C, FIRES and CLIQUE. Up to eight **evaluation measures** (such as CE, CMM, SubCMM, Entropy, F1, RNIA) (cf. Figure 2(a)) can be used to reflect the quality of the clustering directly after processing each horizon. These values are printed gradually in the output panel under the Setup tab as the stream evolves.

2. Under the Visualization Tab: (cf. Figure 2(b)), the evolving of the final clustering of the selected subspace clustering algorithms as well as the evolving of the ground truth stream is visualized in a two dimensional representation. Users can select any pair of dimensions to visualize the evolving ground truth

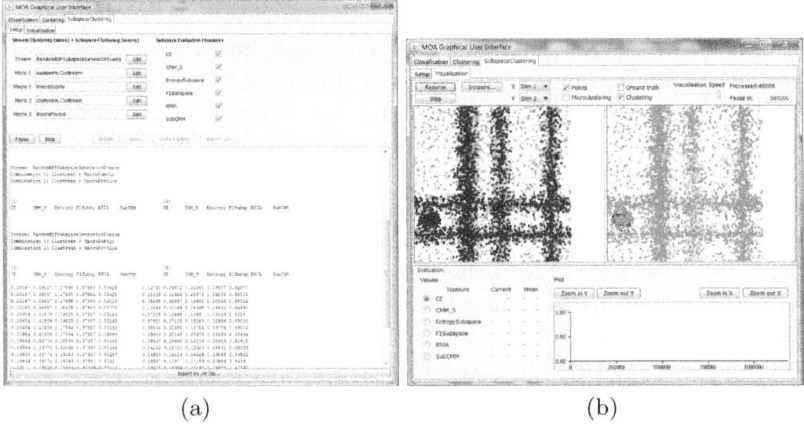

Fig. 2. Subspace MOA Screen Shots of (a) The Setup Tab, (b) The Visualization Tab

as well as the resulted clustering. Different to MOA, Subspace MOA is able to visualize and get the quality measures of arbitrarily shaped clusters.

3 Website, Demo Plan and Conclusion

Subspace MOA can be found at http://dme.rwth-aachen.de/en/subspacemoa. In the demonstrator, we want to explain the main idea of two subspace clustering algorithms as well the online-offline model, with the motivation for getting the final subspace stream clustering algorithms. The framework will offer researchers the possibilities to detect weak and strong points of different subspace clustering algorithms when applied in the streaming scenario, as well as the suitable online/offline combination for a certain dataset. This is all done in a user friendly interface that is in line with the MOA framework style.

Acknowledgments. This work has been supported by the UMIC Research Centre, RWTH Aachen University, Germany.

References

1. Aggarwal, C.C., Han, J., Wang, J., Yu, P.S.: A framework for clustering evolving data streams. In: Proc. of the 29th Int. Conf. on Very Large Data Bases, VLDB 2003, vol. 29, pp. 81–92 (2003)
2. Aggarwal, C.C., Han, J., Wang, J., Yu, P.S.: A framework for projected clustering of high dimensional data streams. In: Proc. of the 30th Int. Conf. on Very Large Data Bases, VLDB 2004, vol. 30, pp. 852–863 (2004)
3. Beyer, K., Goldstein, J., Ramakrishnan, R., Shaft, U.: When is "nearest neighbor" meaningful? In: Int. Conf. on Database Theory, pp. 217–235 (1999)
4. Cao, F., Ester, M., Qian, W., Zhou, A.: Density-based clustering over an evolving data stream with noise. In: 2006 SIAM Conference on Data Mining, pp. 328–339 (2006)
5. Hassani, M., Spaus, P., Gaber, M.M., Seidl, T.: Density-based projected clustering of data streams. In: Hüllermeier, E., Link, S., Fober, T., Seeger, B. (eds.) SUM 2012. LNCS, vol. 7520, pp. 311–324. Springer, Heidelberg (2012)
6. Hulten, G., Domingos, P.: VFML – a toolkit for mining high-speed time-changing data streams (2003)
7. Kranen, P., Kremer, H., Jansen, T., Seidl, T., Bifet, A., Holmes, G., Pfahringer, B., Read, J.: Stream data mining using the moa framework. In: Lee, S.-g., Peng, Z., Zhou, X., Moon, Y.-S., Unland, R., Yoo, J. (eds.) DASFAA 2012, Part II. LNCS, vol. 7239, pp. 309–313. Springer, Heidelberg (2012)
8. Müller, E., Assent, I., Günnemann, S., Jansen, T., Seidl, T.: Opensubspace: An open source framework for evaluation and exploration of subspace clustering algorithms in weka. In: Open Source in Data Mining Workshop at PAKDD, pp. 2–13 (2009)

Symbolic Trajectories in SECONDO: Pattern Matching and Rewriting

Fabio Valdés[1], Maria Luisa Damiani[2], and Ralf Hartmut Güting[1]

[1] FernUniversität Hagen, Germany
[2] Università degli Studi di Milano, Italy
{fabio.valdes,rhg}@fernuni-hagen.de
mdamiani@dico.unimi.it

Abstract. In this paper, we introduce a novel data model for representing symbolic trajectories along with a pattern language enabling both the matching and the rewriting of trajectories. We illustrate in particular the trajectory data type and two operations for querying symbolic trajectories inside the database system SECONDO. As an important application of our theory, the classification and depiction of a set of real trajectories according to several criteria is demonstrated.

1 Introduction

Recently, pattern matching with movement history has been studied extensively [1,2]. Since raw GPS records are inconvenient for most matching applications, [3,4,5] and especially [6] focus on trajectories containing semantic information.

A symbolic trajectory is a sequence of temporally annotated labels each of which is a semantically meaningful description, e.g., a street or city name, an activity, or a means of transportation. Inside our database system SECONDO [7,8], a symbolic trajectory is called a moving label, since similar data types (moving point, moving real, etc.) are supported. SECONDO converts GPS data into a moving label by matching the segments onto a map and storing the according street names from OpenStreetMap [9]. For ease of exposition, we provide a short movement history of a person inside the city of Dortmund in nested list syntax:

```
((("2012-03-31-13:17:01" "2012-03-31-13:18:21" TRUE FALSE) "Hansastr.")
 (("2012-03-31-13:18:21" "2012-03-31-13:20:37" TRUE FALSE) "Kampstr.")
 (("2012-03-31-13:20:37" "2012-03-31-13:21:07" TRUE FALSE) "Brueckstr.")
 (("2012-03-31-13:21:07" "2012-03-31-13:21:47" TRUE FALSE) "Hohe Luft")
 (("2012-03-31-13:21:47" "2012-03-31-13:21:57" TRUE FALSE) "Bissenkamp"))
```

Each line of the quoted moving label corresponds to a so-called unit label, i.e., a combination of a time interval and a description. The boolean expressions indicate whether the start and end instant belong to the interval, respectively.

We created and implemented the operators `matches` and `rewrite`, the former for matching and the latter for rewriting a symbolic trajectory, both available in SECONDO. For using the first one, the user has to provide a mobility pattern, arbitrarily many additional conditions, and a symbolic trajectory. The result is true if and only if the specified pattern matches the trajectory, thus the operator yields a boolean value and may be applied as a selection criterion for filtering trajectories fulfilling a specific pattern – and related conditions, if specified – from a large database relation. For example, from all the trips of a person during the last 12 months, we may obtain exactly the paths from home to work.

The second operator requires the same input as `matches`, extended by a rewrite part consisting of result variables and optional assignments. Again, a matching decision is computed, but `rewrite` focuses on the fact that there may be numerous matching possibilities (e.g., if a place is visited repeatedly inside a trajectory). Its result is a set of moving labels, more exactly, according to the selection of the user, a specific part of the trajectory is returned, and due to the possible matching ambiguity, there may be several results. In addition, these results can be rewritten in order to enrich them with further knowledge.

The main contributions of our research, compared to the abovementioned related work, are the expressiveness of our pattern language allowing the use of SECONDO database queries for filtering moving labels as well as the ability to rewrite parts of the moving label. The latter enables the user to classify large sets of symbolic trajectories concerning business trips to particular customers, migration behaviors of different bird species, positions of components for an automatic manufacturing process, or private journeys to special destinations. A comprehensive example is presented below.

In Section 2, we introduce the pattern language for matching and rewriting a symbolic trajectory. The demonstration is reported in Section 3.

2 Pattern Language

This section is dedicated to the language for specifying a `matches` or `rewrite` query in SECONDO. As mentioned above, we distinguish four parts of the input. This and the next section refer to the following example:

```
X (_ "Alte Teichstr.") Y * Z (_ "Alte Teichstr.") // get_duration(X.time)
+ get_duration(Y.time) + get_duration(Z.time) < (duration (0 1200000))
=> A // A.label := "short walk", A.start := X.start, A.end := Z.end
```

Unit Patterns. We consider the first query line until // containing a sequence of unit patterns, each of which may be assigned a variable by prepending it. For both operators, this part is crucial for the matching decision and thus mandatory. Each unit pattern has one of the forms $(t\ l)$, $((t\ l))$, +, or *, where t is a time interval either in semantic or in numerical form (e.g., `afternoon`, 2012-05-12~2012-05-13-23:45:00) and l is a label.[1] A simple pattern $(t\ l)$ matches a unit label $(t_u\ l_u)$ if and only if $t_u \subset t$ and $l = l_u$. An underscore acts as a wildcard, i.e., guarantees a time or label match, respectively. The second form has the same matching criteria but refers to one or more unit labels, as long as a matching is possible. The two remaining alternatives match any sequence of unit labels and differ insofar as * may also match no unit label at all.

The values of the assigned variables are accessed in the following parts.

---
[1] The user may specify sets of time intervals resp. labels instead of single ones.

Conditions. Subsequent to the unit patterns, the user may append conditions (until =>) obeying the following two rules: Each condition has to be evaluable into a boolean result by SECONDO and to contain at least one expression of the form *v.attr*, where *v* is one of the variables from the previous part, and *attr* is one of the attributes *label, time, start, end, card* (the latter denotes the number of unit labels matched by the respective unit pattern) of this variable determined by the matching. Notably, several unit patterns may be linked in one condition.

Results. In this required part, ranging from => to //, the user controls the `rewrite` operator's result by specifying the output variables, either known from the unit patterns section or new. The latter must be enriched with information in the following part. For every matching possibility (if any), the unit labels assigned to the output variables are merged into a moving label which is returned.

Assignments. Finally (fourth line after //), the user may rewrite the result by assigning new values (either by referring to values from other variables or by typing them in SECONDO syntax) to certain result unit labels. Left from the assignment symbol :=, a *v.attr* expression (as explained in the conditions part, although the use of *card* is not allowed here) is required. The assigned value has to be a SECONDO-evaluable expression of the respective data type.

3 Demonstration

In the following, we demonstrate the functionality of the operators. The used relation contains 169 moving labels of different lengths from one person, covered by car, train, bike, or foot, including trips from home to work or leisure walks. Our objective is to extract all walks starting and ending in `"Alte Teichstr."` (1) of less than 20 minutes (2) from the relation and to rewrite the resulting moving labels, in order to have only one unit per walk (3) containing the respective time interval and the label `"short walk"` (4). This is obtained by executing `rewrite` with the pattern from section 2 in SECONDO. For the sake of brevity, only the query parts related to our operators are discussed. Initially, the operator converts the unit pattern sequence into an NFA whose states represent the unit patterns, except for the accepting state which is active only in case of a complete match with the moving label. This automaton performs filter step (1), after which the variable bindings are stored. Subsequently, the *v.attr* parts of the condition (note that 20 minutes are entered as 0 days and 1,200,000 ms) are replaced by the respective values, such that the expression can be evaluated by SECONDO, excluding too long walks (2). The result part contains one new variable, hence each resulting moving label consists of one unit label (3). Since the variable `A` is not attached to a unit pattern, the assignment section has to provide its components with data that are either user-typed (`"short walk"`) or dependent on the trajectory and on the variable bindings (`X.start`, `Z.end`) (4). We present one of the resulting moving labels:

((("2012-04-02-08:10:05" "2012-04-02-08:26:27" TRUE TRUE) "short walk"))

The runtime of this query amounts to 0.047 sec on an AMD Phenom II X6 3.3 GHz running openSUSE 11.4, with 8 GBytes of memory. The walks remaining after the selections (1) and (2) are depicted in Figure 1.

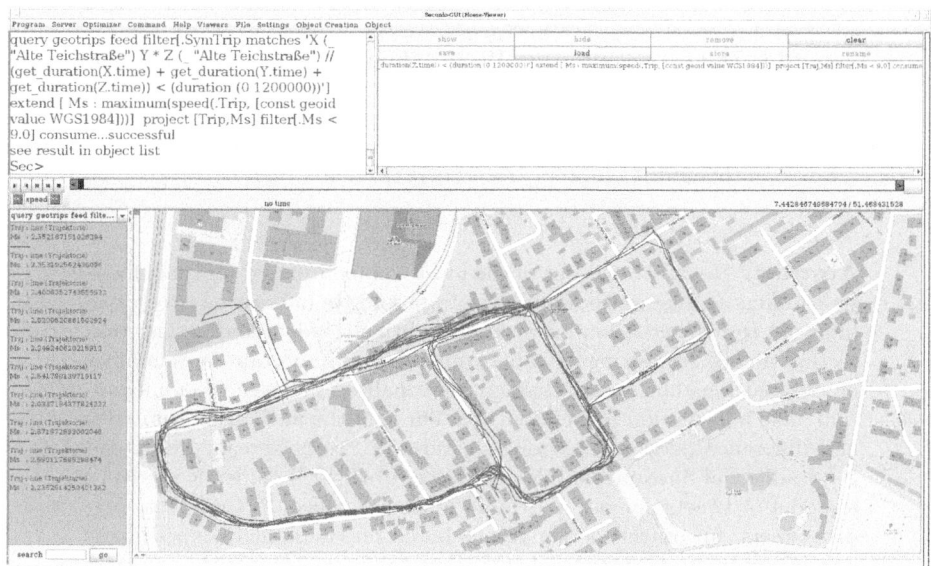

Fig. 1. A screenshot of the filtered walks from the SECONDO GUI

Note that SECONDO is freely available and may be downloaded from [8].

References

1. Hadjieleftheriou, M., Kollios, G., Bakalov, P., Tsotras, V.J.: Complex spatio-temporal pattern queries. In: Proc. VLDB, pp. 877–888 (2005)
2. Mokhtar, H.M.O., Su., J.: A query language for moving object trajectories. In: Proc. SSDBM, pp. 173–184 (2005)
3. Vieira, M.R., Bakalov, P., Tsotras, V.J.: Querying trajectories using flexible patterns. In: Proc. EDBT, pp. 406–417 (2010)
4. Vieira, M.R., Bakalov, P., Tsotras, V.J.: Flextrack: A system for querying flexible patterns in trajectory databases. In: Int. Symp. SSTD, pp. 475–480 (2011)
5. Vieira, M.R., Tsotras, V.J.: Complex motion pattern queries for trajectories. In: ICDE Workshops, pp. 280–283 (2011)
6. du Mouza, C., Rigaux, P.: Mobility patterns. GeoInformatica 9(4), 297–319 (2005)
7. Güting, R.H., Behr, T., Düntgen, C.: Secondo: A platform for moving objects databases research and for publishing and integrating research implementations. IEEE Data Eng. Bull. 33(2), 56–63 (2010)
8. http://dna.fernuni-hagen.de/Secondo.html (2012)
9. http://www.openstreetmap.org (2012)

$ReTweet^p$: Modeling and Predicting Tweets Spread Using an Extended Susceptible-Infected-Susceptible Epidemic Model

Yiping Li, Zhuonan Feng, Hao Wang, Shoubin Kong, and Ling Feng

Dept. of Computer Science & Technology, Tsinghua University, Beijing, China
liyp09@mails.tsinghua.edu.cn, fzn0302@163.com, wanghaomails@gmail.com,
bin8118@126.com, fengling@tsinghua.edu.cn

Abstract. Retweeting is one of the most commonly used tools on Twitter. It offers an easy yet powerful way to propagate interesting tweets one has read to his/her followers without auditing. Understanding and predicting tweets' retweeting extents is valuable and important for a number of tasks such as hot topic detection, personalized message recommendation, fake information prevention, etc. Through the analysis of similarity and difference between epidemic spread and tweets spread, we extend the traditional Susceptible-Infected-Susceptible (SIS) epidemic model as a model of tweets spread, and build a system called $ReTweet^p$ to predict tweets' future retweeting trends based on the model. Experiments on Chinese micro-blog Tencent show that the proposed model is superior compared to the traditional prediction methods.

1 Introduction

A tweet is a post, limited to 140 characters, on the micro-blog service - Twitter. When a micro-blog user composes a tweet, his/her direct followers will see it instantly. If these followers want to share it with respective followers, they can just simply forward the tweet with/without comments by simply pressing the retweeting button. The message will spread extensively after several rounds of retweeting. As each micro-blog user can easily post, comment, and forward tweets without original author's permission, information dissemination among micro-blog users is much faster and wider than that on any traditional medium. Studying the characteristics of such tweets retweeting is thus important for a number of tasks, such as hot topic detection, personalized message recommendation, fake information prevention, etc.

In the literature, [2,4,5] applied the classification approaches to predict the range of propagation of a tweet. The $ReTweet^p$ system to be demonstrated in this paper, however, aims to predict the exact retweeting number rather than a scope. It is more challenging, given the fact that a large span of retweeting amounts different tweets have in Twitter. Interestingly, tweets spread bears some similarity to infectious diseases spread. A classic epidemic model for the spread of infectious diseases is the Susceptible-Infected-Susceptible (SIS) model [1, 3]. It classifies individuals as occupying two categories, namely, *susceptible* meaning they do not have the disease, and *infected* meaning that they do have the disease. The disease can be transmitted to a susceptible person when they come into contact with an infected person. An infected person may recover from the disease to re-enter an susceptible state.

This is similar to tweets spread among micro-blog users. A tweet firstly propagates to the direct followers of the tweet's author. Some of these followers will retweet the tweet to their respective followers by pressing a retweeting button, who may then choose to retweet the same message again or not to their followers. For consistency, the propagation of the tweet from its author to the author's followers can also be viewed as a retweeting action by the author. In the paper, we call a user who posts or a retweets a tweet message a *retweeter* of the message. An analogy between epidemic spread and tweets spread is illustrated in Table 1.

Table 1. Analogy between epidemic spread and tweets spread

Epidemic Spread	Tweets Spread
infectious disease	tweet message
infect	retweet
infected individual	retweeter of a tweet message
susceptible individual	direct follower of a retweeter

We draw inspirations from the dynamic epidemic spread and simulate the dissemination of tweets by the extension of the classic SIS model to tailor to some specific features of tweets spread. A retweeting prediction system called $ReTweet^p$ is built to predict tweets' retweeting extents at a future time point. Our experiments on Chinese micro-blog Tencent demonstrate the effectiveness of the approach compared to other traditional prediction methods.

2 Modeling and Predicting Tweets Spread on Micro-Blog

In the standard SIS model, the total amount of individuals at either state *susceptible* or *infected* is fixed without considering individuals birth. Such a fixed population is not valid when modeling tweets spread due to the openness of micro-blog, where external users may spontaneously read and retweet a message on their own initiative, without following any retweeting user. This is different from the SIS model where the infection must only come from an infected person. Besides, a retweeting user may retweet the same message several times with no following relationship to any retweeting user.

To address the differences, we add a state *external* whereby an external visitor spontaneously retweets a tweet at a rate γ, without following any retweeting (infected) users. Also, an retweeting (infected) user may retweet again the same message at his/her own will. A diagrammatic representation of the extended SIS model, which we call SISe, is shown in Fig. 1. There are four processes by which a user's state may change. 1) A message retweeter transmits a tweet (disease) to his/her direct followers (susceptible users) with rate β, part of whom will press the retweeting button to become retweeters. 2) After retweeting, the retweeter

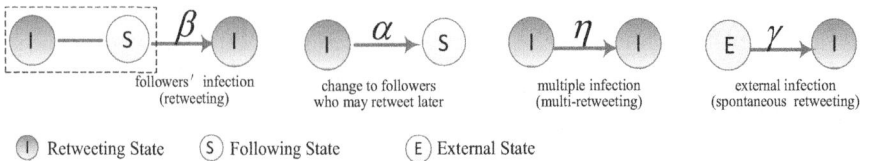

Fig. 1. The tweets retweeting spread model

user returns to the follower state if s/he is a direct follower of a certain current retweeter at rate α, or 3) it remains as a retweeter by retweeting the same tweet multiple times at rate η. 4) An external visitor spontaneously retweets the tweet at rate γ without following any retweeter.

Formally, given a tweet message, at time t, let function $I(t), S(t), E(t)$ return the number of message retweeters, number of direct followers of retweeters at t, and number of external spontaneous visitors, respectively. Upon the birth of a tweet message where $t = t_0$, $I(t_0) = 1$, $E(t_0) = 0$, and $S(t_0)$ is the number of direct followers of the tweet's author.

$$S(t+1) - S(t) = -\beta I(t)S(t) + \alpha I(t)$$
$$I(t+1) - I(t) = \beta I(t)S(t) + \eta I(t) + \gamma E(t) - \alpha I(t) \quad (1)$$
$$E(t+1) - E(t) = \omega I(t)$$

where β is the retweeting transmission rate, α is the retweeter-to-follower change rate, η is the multi-retweeting rate, γ is the externally spontaneous retweeting rate, and ω is the proportion of external spontaneous visitor with respect to the current retweeters. According to Formula 1, we build a system called $ReTweet^p$ which can predict a tweet's retweeting amount at a future time point.

3 System Implementation and Demonstration

The $ReTweet^p$ system consists of three modules. 1) Data Preparation, involving seed data collection and initial rates training, 2) retweeting prediction, and 3) user interaction. We collect 1200 tweets on the *self-cultivation* topic from the Chinese Tencent micro-blog from September 1, 2012 to November 11, 2012, where each tweet was retweeted over 1500 times, with 6871 as the largest retweeting amount. We capture the initial rates of β, α, η, and γ in Formula 1 from the seed tweets, and adjust the values dynamically along with the time.

Fig. 2 is the system screenshot, divided into four parts. *Panel 1* shows the user-requested tweet whose retweeting amount is to be predicted. *Panel 2* allows the user to indicate the prediction time span and frequency. Requirements of retweeters (direct followers or spontaneous visitors) and retweeting with/without comments can also be imposed upon prediction. *Panel 3* shows the tweets hierarchical spread as water waves. A circle represents a direct retweeting follower, and the size of the circle indicates his/her total amount of direct followers. A square denotes an external spontaneous visitor, and the size of the square indicates the total number of direct followers of this visitor. The line between a

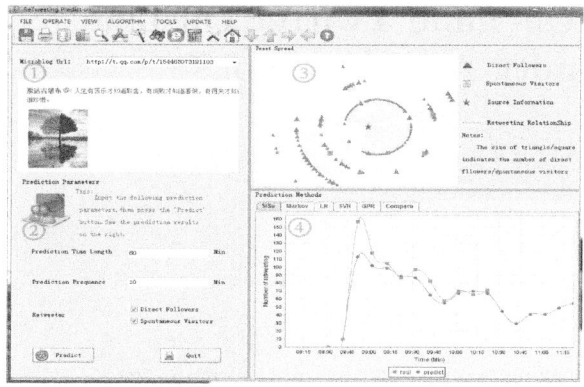
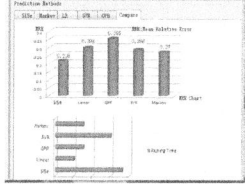

Fig. 2. $ReTweet^p$ screenshot

Fig. 3. Performance comparison

circle/square and a circle is the direct following relationship, while the line between a circle and a square shows the spontaneous tweet reading relationship. *Panel 4* is the retweeting amount prediction result, with the real value in dark color as a reference. A comparison of six different predictive methods (including $ReTweet^p$, Linear Regression LR, Support vector regression SVR, Gaussian processes regression GPR, and Markov) in terms of the mean prediction error and running time is also available when the user clicks the *compare result* button in *Panel 4*, as shown in Fig. 3. For the user-entered prediction request shown in Fig. 2, the $ReTweet^p$ system leads to the least prediction error. This is because it learns and uses the most parameters than the rest and thus takes the most time for high-quality prediction.

Acknowledgement. The work is supported by National Natural Science Foundation of China (60773156, 61073004), Chinese Major State Basic Research Development 973 Program (2011CB302203-2), Important National Science & Technology Specific Program (2011ZX01042-001-002-2), and research fund of Tsinghua-Tencent Joint Laboratory for Internet Innovation Technology.

References

1. Hethcote, H.: A thousand and one epidemic models. Lecture Notes in Biomathematics, vol. 100 (1984)
2. Hong, L., Dan, O., Davison, B.: Predicting popular messages in twitter. In: WWW (2011)
3. Kermack, W., McKendrick, A.: Contributions to the mathematical theory of epidemics. Bulletin of Mathematical Biology 53 (1991)
4. Suh, B., Hong, L., Pirolli, P., Chi, E.: Want to be retweeted: Large scale analytics on factors impacting retweet in twitter network. In: SocialCom (2010)
5. Yang, Z., Guo, J., Tang, J., Zhang, L., Su, Z.: Understanding retweeting behaviors in social networks. In: CIKM (2010)

TwiCube: A Real-Time Twitter Off-Line Community Analysis Tool

Juan Du, Wei Xie, Cheng Li, Feida Zhu, and Ee-Peng Lim

School of Information System,
Singapore Management University
{juandu,weixie,fdzhu,eplim}@smu.edu.sg

Abstract. As a micro-blogging service, Twitter differs from other social network services in two ways: 1) the absence of mutual consent in establishing follow links and 2) being a mixture of news media and social network. A key question to ask in better understanding Twitter user behavior is which part of a user's Twitter network reflects one's real-life social network. TwiCube is an online tool that employs a novel algorithm capable of identifying a user's real-life social community, which we call the user's off-line community, purely from examining the link structure among the user's followers and followees. Based on the identified off-line community, TwiCube provides a summary of the user's interests, tweeting habits and neighborhood popularity analysis. Evaluations from real Twitter users demonstrate that our off-line community detection approach achieves high precision and recall in most cases.

1 Introduction

Twitter distinguishes itself from other SNS like Facebook with two unique characteristics [3]: 1) Twitter functions as a mixture of news media and social network combining features from both; and 2) mutual consent is not required to establish a follow link. And thus, even if two users follow each other, can we conclude that they are friends in person? For example, does the US President Barack Obama personally know all the 670,000 users that he follows? The eluding social characteristics of Twitter network have so far fogged answers to these important questions. Despite the increasingly rich study on Twitter, few have explored the relationship between a user's online and off-line social network, which, on the other hand, has been adequately investigated in standard SNS like Facebook. Indeed some works probing this territory, such as [1, 2]. However, they either used a fairly weak definition of "friend" which refers to anyone to whom the user has directed at least two tweets or they cannot identify whether groups formed by clustering algorithms are actually traces of online or off-line social networks.

TwiCube[1] is an online tool to identify the portion of a user's Twitter follow network that maps to his or her off-line social life, which we call the user's off-line community and the user being studied is then called the target user. The identification of a users off-line community is important in characterizing different

---

[1] http://twitterbud2011.appspot.com/

users and understanding their behavior on Twitter. For example, to build a better model for user interest profiling, on one hand, online friends like medias and celebrities are more informative of the user's interests, hobbies, etc. On the other hand, the close off-line friends are more similar in interests as well. Therefore, what these close friends follow online could complement and reinforce those one follows. For instance, a music-lover may follow three music-related Twitter users, while his or her close friends in the off-line community may altogether follow another twenty music-related ones, giving strong evidence on the users interest in music. By an aggregated analysis on the interests of the off-line community members, we are able to build a more robust and accurate interest profile for the target user. In this paper we give an overview of the architecture of TwiCube and its major modules including a briefing on the algorithms in Section 2. We then demonstrate a case in Section 3.

2 Overview and Architecture

TwiCube is an online tool for off-line community analysis. As Figure 1 shows, an execution loop of TwiCube is triggered by a query for a certain target user and culminates the visualization of the off-line friends network and related statistics. It utilizes two external resources: Twitter and FreeBase. The process of retrieving the network structure from Twitter and computing the off-line community is on-the-fly so that it provides the real-time version and avoid the expenses on storage. The phase of employing FreeBase is off-line since these data are regarded as a dictionary for profiling. Given these raw data, we calculate the closeness score for each pair of nodes and generate the community, along with its corresponding statistics. For efficiency purpose, we cache the query result for further search.

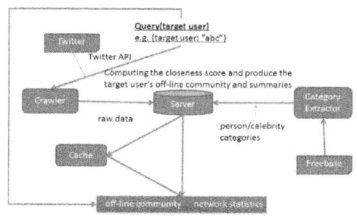

Fig. 1. System overview

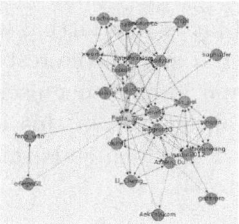

Fig. 2. An example of a user's off-line community

2.1 Data Collection

Category Extractor. Category extractor performs the off-line task of crawling data from knowledge base such as Freebase[2] via FreeBase API. We collected information of two types of entities, persons and organizations, for which a number

---
[2] http://www.freebase.com

of attributes like Twitter account and field are then extracted. The attributes of these persons and organizations are further categorized manually. For example, the profession "Novelist" belongs to the category *Writer*, while the sector "Libertarian Party of San Francisco" is categorized as *Politics*. For those persons and organizations whose Twitter accounts are not available in FreeBase, we used their names as queries in Bing search API to look for their corresponding Twitter home page. Combining our results from the Bing search API and those already obtained in FreeBase, we were able to gather 100 million persons' and organizations' Twitter accounts along with their corresponding categories. Data collected by category extractor are stored in database for generating target user's interest profile.

Twitter Real-Time Follow Network. The online retrieving data source, Twitter, provides a lot of useful APIs to access its user data. We fetch these data by specifying the target user in real-time manner. These data includes the network structure and the target user's recent tweets.

2.2 Data Processing

Off-Line Community Identification. Data obtained from the Twitter real-time follow network is fed to this module to detect the off-line community of a user. In order to accomplish this task, we employ an algorithm in [4]. On a high level, the algorithm works in iterations as follows. Given a target user u, we computed the closeness score between u and all the other users as well as $\hat{v}$, where $\hat{v}$ represents a connection to u almost as weak as any off-line real-life friend should be. A ranking list of all the users together with $\hat{v}$ in decreasing order of the closeness score is thus generated. All the users ranked before $\hat{v}$ are identified as off-line community members, which ends the current iteration. In the next iteration, the key point is that we now treat the whole off-line community identified so far as one virtual user node $\tilde{u}$. Instead of computing the closeness score between u and all the rest users, this time we compute the closeness score between $\tilde{u}$ and every other user. From the ranking list thus generated, if any user jumps ahead of $\hat{v}$ in this iteration, the user will be added to the off-line community of u, which ends this iteration. So on and so forth. For the details, please refer to [4].

User Analysis. By utilizing the off-line community produced for a user, we are able to do some analysis for the target user that is unattainable before.

User Interest Category. The category of user interest is based on the observation on his online community, which usually includes medias, celebrities, etc. Therefore, a target user's interests are built according to one's following celebrities' categories, which are stored in a database by category extractor.

Popularity Score. We analyzed all the members' popularity in both online and off-line communities of a target user. The popularity score is calculated by the number of followers one owns in the target user's off-line community. The leading users in online community indicates the common interests from the off-line

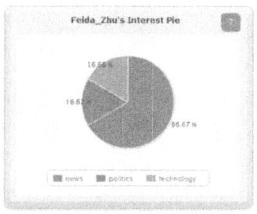  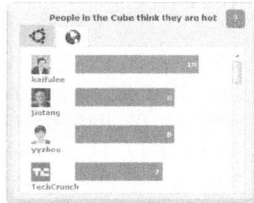

(a) Interest category (b) Tweeting habit (c) Network's popularity

Fig. 3. Case demonstration for a certain user's off-line community analysis

friends' view. And the leading one in off-line community represents the active users.

Tweeting Habit. We also present the user's tweeting time distribution and the retweet number of his tweets, from which to learn the user's tweeting behavior.

3 Demonstration Cases

In this section, we demonstrated a case for analyzing off-line friends network of a query user. We randomly picked up a user from Twitter who has 36 followees and 35 followers. By employing the algorithm to compute the closeness score, we generated the off-line community with 25 users shown as Figure 2. Considering his online community, his interests are mainly about *News*, *Politics* and *Technology* as Figure 3(a) shows. This guy usually publishes tweets during 07:00 to 08:00 in Greenwich Mean time as Figure 3(b) indicates. In addition, among his recent tweets, there are only one which has been highly retweeted as the red one, indicating that he is not active in Twitter. Furthermore, we also present the popularity score for each user on his off-line community's level as Figure 3(c). For more cases, readers could search from our TwiCube search interface.

4 Conclusions

TwiCube is an online tool to detect a Twitter user's off-line social network. Manual evaluations from real Twitter users have demonstrated its high precision and recall. The demo not only finds a target user's off-line community in a real-time manner, but also provides further analysis of the user's off-line community.

Acknowledgments. This project is supported by the Singapore National Research Foundation under its International Research Centre @ Singapore Funding Initiative and administered by the IDM Programme Office.

References

1. Grabowicz, P.A., Ramasco, J.J., Moro, E., Pujol, J., Eguiluz, V.M.: Social features of online networks: the strength of weak ties in online social media. Arxiv preprint arXiv:1107.4009 (2011)
2. Huberman, B., Romero, D., Wu, F.: Social networks that matter: Twitter under the microscope (2008)
3. Kwak, H., Lee, C., Park, H., Moon, S.: What is twitter, a social network or a news media? In: Proceedings of the 19th International Conference on World Wide Web, pp. 591–600. ACM (2010)
4. Xie, W., Li, C., Zhu, F., Lim, E., Gong, X.: When a friend in twitter is a friend in life. In: Proceedings of the 4th International Conference on Web Science, pp. 493–496. ACM (2012)

TaskCardFinder: An Aviation Enterprise Search Engine for Bilingual MRO Task Cards

Qingwei Liu, Hao Wang, Tangjian Deng, and Ling Feng

Dept. of Computer Science & Technology, Tsinghua University, Beijing, China
{liuqw10,wangh08,dtj08}@mails.thu.edu.cn, fengling@tsinghua.edu.cn

Abstract. While conventional search engines demonstrate the power of delivering information to ones' fingertips, the complexity of enterprise information raises a number of challenges to enterprise search engines in the nature of unstructured contents, task-relevance, result presentation and multiple languages. We show how we tackle the challenges in aviation field through the development of an enterprise search engine for English-Chinese MRO (**M**aintenance, **R**epair and **O**verhaul) task cards, called *TaskCardFinder*. It enables technicians and planners to quickly find out bilingual task cards related to a specific service request coming from airlines. Several context-awareness features is demonstrated, including context-aware preference search and recommendation, recall search by context, keywords suggestion, navigational and analysis-oriented search, bilingual search support and dynamic result presentation. A user study is done at an international aviation MRO company. The system is demonstrated in the video, www.youtube.com/watch?v=vj7u_VfRFZw.

1 Introduction

Enterprise search is an essential part of business intelligence [4]. However, the complexity of enterprise information raises many challenges. [3,2] discussed the differences between enterprise search engines and conventional ones in the nature of target content, user behavior and economic motivations, summarized as follows. First, the majority of enterprise information is unstructured and possibly in multi-languages. Second, enterprise search is highly task-relevant, where search context (like user role, activity, company's regulation, etc.) should be considered. Third, domain-specific guided navigation and search refinement are desirable. Fourth, users may have previously seen the wanted information, where searching is recall-based. Fifth, result presentation should include summary, category and aggregate information to enhance the usability. These challenges have led to a formidable problem but also mean enormous potential benefit.

Each year aviation industry spends a whopping amount on MRO services, and second on fuel. In 2007, it reached US $45 billion, and is expected to be US $61 billion by 2017. As per industry projections, the size of the worldwide air transport fleet will expand by nearly 50 percent to 2017, and consequently spur rapid growth of the MRO business [1]. So suppliers of MRO services are under pressure to improve competitive international productivity and cut costs. One particular problem for them is to support quick turn-around time and improve accuracy for cost estimation and airplane maintenance schedule making.

MRO services are performed under the instructions of MRO task cards. For each activity involved, a task card is generated and followed. It is a directive file in MS Word format, including information like aircraft type, airline, main fault, operation procedure, equipment needed, serviceman, etc. One task card may have a few to hundreds of pages in bilingual or multiple languages, containing texts, tables and images. To achieve high-quality cost estimation and service scheduling, leveraging existing task cards is a must. However, managing the large volume of task cards manually is impossible, so *TaskCardFinder*, is built to retrieve bilingual MRO task cards automatically. It resolves the above-mentioned challenges well and aims to provide MRO service planners and technicians with the right information under the right task orders with a high degree of task-relevancy. With several context-awareness features, *TaskCardFinder* has received good feedback and evaluation from a demonstration to an aviation MRO company.

2 System Overview

The framework of *TaskCardFinder* is illustrated in Fig.1.

- Part I (Storage of Task Cards) prepares for searching by parsing, extracting, indexing and storing task cards into a relational DB. Some aggregation analysis results are also computed and stored in the database.
- Part II (Search over Task Cards) is for performing keyword-based or structured search requests with access context. Recall-based search based on users' previous search and viewing history is also provided.
- Part III (User Interaction) facilitates users' easy and simple interaction with the system by means of keywords suggestion, bilingual search support, navigational search and structural view of search result presentation.

Extraction, Segmentation and Indexing. Differently-structured MRO task cards are used by different airline companies and MRO services. To resolve the

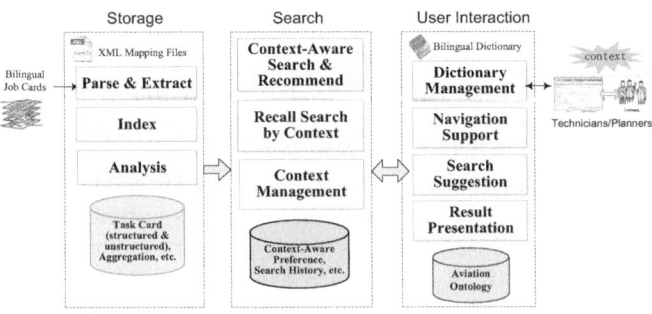

Fig. 1. Framework of *TaskCardFinder*

heterogeneity problem, XML is utilized to extract items from the task card. Every type of task card corresponds to an XML mapping file, characterizing content position, storage property, index strategy (e.g., term vector, analysis and storage method), and other attributes of the task cards. To support keyword-based full-text retrieval, we adopt a Hidden Markov-based Chinese word segmentation open tool ICTCLAS[1] and build fine-grained index for all attributes.

Context-Aware Preference Search and Recommendation. Awareness of user context could enhance the relevancy of search results remarkably and recommend potentially useful task cards. *TaskCardFinder* considers five aspects of aviation context: 1) *search user*-centric; 2) *aviation service company*-centric; 3) *airline customer*-centric; 4) *aviation service*-centric; and 5) *environment*-centric. An aviation service company provides MRO services to different airline customers. Search users work on different services for the company in a specific physical and social environment. To reflect users' different preferences under different contexts, we build a context-aware preference model [5], where a context-aware preference is viewed as a 3-tuple (*Context, Preference, Weight*). Both *context* and *preference* are uniformly described in the form of (*Subject, Property, Object*) via the ontology Protégé OWL tool. *Weight* in [0,1] specifies the holding degree of the context-aware preference.

For example, a context-aware search preference like *"If the search user belongs to the engine team in some MRO services, preferably the work content of the service involved in the returned MRO task cards is related to engine"* can be expressed as follows.

	Subject	**Property**	**Object**
Context:	SearchUser	belongToTeam	"engine team" **AND**
	SearchUser	involvedIn	MROService
Preference:	MROService	hasWorkContent	"engine"
Weight:	1.0		

Recall-Based Search by Context. Recall-based search is a common activity in enterprise search, as the users are normally involved in regular and repetitive duties and need to consult similar task cards frequently. Complementary to keyword-based/structured search, *TaskCardFinder* records user's historic access context for latter re-finding, including search keywords, viewed documents and access date and time.

Keywords Suggestion and Bilingual Search Support. Relevant keywords are suggested based on the current keyword and the aviation MRO ontology. The system first locates the input keyword on the ontology tree, and then decides the association direction such as *upward* to recommend some high-level technical terms, *downward* for a few detailed ones or just *at the same level*. The keyword suggestion could help users refine queries and improve the user experience.

To support bilingual search functionality, TaskCardFinder is equipped with a well-built Chinese-English aviation dictionary containing 153,627 translation

---

[1] http://www.ictclas.org/index.html

pairs, and incrementally populated by more complicated technical terms through dynamically crawling the online translation from the Blue Sky Aviation Lexicon[2], which could be witnessed by the highlighted matched keywords in Fig.3.

Result Presentation and Search Assistance. The search result allows *online preview* and *structural view* (Fig.3) which can help quickly judge if the task card is what is really sought. Navigation according to different MRO topics (e.g., *A/C type, main fault and work content*) is provided (Fig.2). Under each topic, some statistic results (like the number of relevant task cards, total and average man hours)are presented for cost estimation and schedule planning.

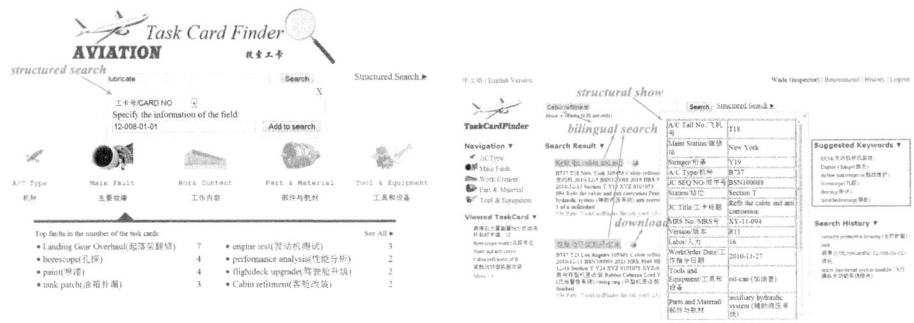

Fig. 2. Main search interface   **Fig. 3.** Search result

3 User Study

We demonstrate $TaskCardFinder$ to the staff from engineering, operation, quality, continuous improvement and R&D departments of an aviation MRO company and receive positive feedback and constructive comments. First, context-aware preference functionality is very desirable, as different airline customers usually have different MRO demands in terms of work content, timeline and cost; e.g., the MRO request of a coastal airline is different from the one on the highland. Besides users also play different roles in MRO services and therefore have different search focuses. Second, version management of task cards is expected. Third, the bilingual issue in bilingual task cards' generation and reviewing remains as a difficulty for the company staff. The system is expected to provide more assistance in these aspects in the future.

Acknowledgement. The work is supported by National Natural Science Foundation of China (60773156, 61073004), Chinese Major State Basic Research Development 973 Program (2011CB302203-2), Important National Science & Technology Specific Program (2011ZX01042-001-002-2), and research fund of Tsinghua-Tencent Joint Laboratory for Internet Innovation Technology.

---

[2] http://air.cidian.cc/

References

1. Gupta, P., Dewangan, S., Gade, S.: Time to enable mobility in aviation MRO software solutions (2009),
http://www.infosys.com/industries/airlines/white-papers/Documents/enable-mobility-aviation.pdf
2. Hawking, D.: Challenges in enterprise search. In: ADC, pp. 15–24 (2004)
3. Mukherjee, R., Mao, J.: Enterprise search: Tough stuff. Queue 2, 36–46 (2004)
4. Schmidt, K., Oberle, D., Deissner, K.: Taking enterprise search to the next level (2009), http://ceur-ws.org/Vol-401/iswc2008pd_submission_22.pdf
5. van Bunningen, A.: Context-Aware Querying: better Answers with Less Effort. PhD thesis, University of Twente, The Netherlands (2008)

EntityManager: An Entity-Based Dirty Data Management System

Hongzhi Wang, Xueli Liu, Jianzhong Li, Xing Tong, Long Yang, and Yakun Li

Harbin Institute of Technology, China,
{wangzh,lijzh,liyakun}@hit.edu.cn, {shally78952286,newyanglong}@163.com,
xiaohuo_2008@yahoo.cn

Abstract. Dirty data exist in many systems. Efficient and effective management of dirty data is in demand. Since data cleaning may result in the the loss of useful data and new dirty data, we attempt to manage dirty data without cleaning and retrieve query result according to the quality requirement of users. Since entity is the unit for understanding objects in the world and many dirty data are led by different descriptions of the same real-world entity, we propose EntityManager, a dirty data management system with entity as the basic unit and keep conflicts in data as uncertain attributes. Even though the query language is SQL , the query in our system has different semantics on dirty data. In the demonstration, we will show a new philosophy for managing dirty data around entities. We will present our prototype allowing load dirty data and query dirty data according to the requirement of users.

1 Introduction

In many systems, dirty data exist because of many reasons. Dirty data will do harm to the applications. Currently, the major method to deal with dirty data is data cleaning. Even though data cleaning could handle dirty data in many cases, such methods have the shortcoming that the repairing in data cleaning may lead to new dirty data and the deletion during data cleaning will result in the loss of data. Without extra information, it is difficult to discover true value of data. Therefore, we attempt to keep dirty data and perform query on dirty data to obtain relative clean results.

Since entity is the basic unit for understanding objects in real world, our idea is to organize tuples according to referred real-world entities. Keeping dirty data, it may occur that an attribute of an entity may have multiple values. Such data could be considered as uncertain data. Even though many uncertain data management systems have been proposed, they are not designed to manage uncertain data generated from entity resolution. They are based on the concept of "possible world", which is difficult to define on dirty data.

Without using the concept of possible world, by managing the data with entity as the basic unit, we develop EntityManager. In our system, entity resolution is performed on the data sets and the tuples corresponding to the same real-world entity are merged as an uncertain tuple which stores different values of an attribute as an uncertain attribute in the relation.

With the consideration of the uncertainty in the attributes and possible errors in the constraint in the query, the queries in our system have different semantics with traditional databases. For the ease of usage, EntityMangager accepts common SQL statements and returns results with possibilities representing the degree that this entity matching the query. With such possibilities, users could judge whether the results should be accepted.

In this demo, the audience has the ability to interact with the system through a graphical interface that allows them to load dirty data, input SQL statements and view results with the uncertainties. Users may input dirty data with various sizes, dirty degrees, attribute numbers and attribute widths. Users are able to change these parameters and observe the impact on performance.

The remaining parts of this paper are organized as follows. Section 2 provides the data model in our system. The query processing methods are discussed briefly in Section 3. In Section 4, the demo scenario is proposed.

2 Data Model

In this section, we introduce the data model in our system as well as the semantics of the queries. In the data model of our system, the definitions of database and relation are the similar with those in traditional database [1], while the definitions of attribute and tuple are different.

With data model different from traditional relational model, the query in our system has different semantics. With the consideration of multiple values of an attribute in the relation, the constraints in the query should take the uncertainty in attributes and constraint into consideration in two aspects. On one hand, the comparison between the value of attributes should consider the uncertainty in the values. On the other hand, since the values in attributes and constraints may contain errors, the comparison between the attributes and the values in the constraints are approximate. The details of the constraints are shown in [8].

As the queries have different semantics, the definitions of some data operators are redefined in our system. The projection operator is the same, but the selection and join operators are different.

3 Query Processing

The framework of query processing in EntityManager is the same as that of traditional relational databases [2]. With different semantics in the query language, we develop following three new techniques for efficiently query processing on the uncertain databases organized according to entities.

1. *Similarity-based Operators*: With the definitions of selection and join operators different from traditional relational database, we develop similarity search algorithm [6,7] and similarity join algorithm [4] for the entities.

2. *Indices*: To process similarity selection and join efficiently, we designed novel index structures. To handle the special operators in our system, the index considers not only more efficient string similarity search on string attributes [6], but also the similarity search of the combination of numerous and string attributes [7], which can be used for the similarity search on weighted strings.
3. *Query Optimization*: With new operators in our system, for the query optimization, even though the query plan selection algorithms [2] in classical relational databases could be applied in our system, the new estimation techniques for the operators should be developed. Thus we design novel result estimation algorithms for the selection and join operators in [9] and [10], respectively. Additionally, since the selectivity of join on multiple dirty relations is difficult to estimate, we propose a random algorithm for selectivity estimation and join order selection algorithm based the selectivity[5].

4 Demonstration

To demonstrate the features of our system, we load some data of books, paper authors crawled from multiple databases on the web. Since data sources may contain errors or inconsistency, the names or prices of the same book may be different. We apply the entity resolution algorithm in [3] to cluster the relation into entities. For the convenience of users, our system provides the same interface as traditional relational databases including query processing and database maintenance. The uncertainties of the values of attributes are computed by the voting. We attempt to demonstrate our system in following steps.

1. *Data Load*: After the relations in the database are created, the data in format of flat text could be loaded in EntityManager. During data loading, entity resolution is performed and the data is stored in the database in form of entities.

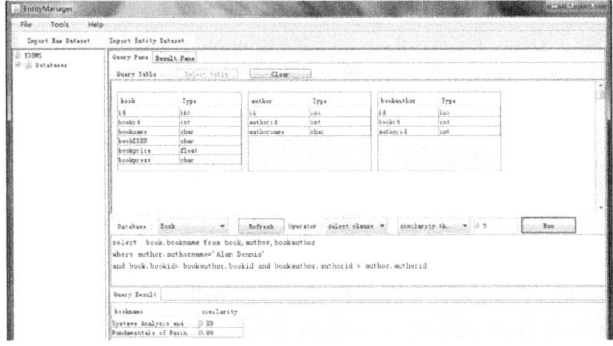

Fig. 1. Query Processing Interfaces of EntityManager

2. *Data Browsing* After the data is loaded into the database, user could browse the results of entity resolution with selecting all data in the relation.
3. *Query Proxcessing* As the basic function of a database system, users input the common SQL query and review query results. As shown in Figure 1, the query processing interface is the same as traditional databases. And the query results can be reviewed in detail. To filter the query results according to the probabilities, users could input the threshold or the number of results with the highest possibilities to be selected.

Acknowledgement. This paper was partially supported by NGFR 973 grant 2012CB316200, NSFC grant 61003046, 61111130189 and NGFR 863 grant 2012AA011004. Doctoral Fund of Ministry of Education of China (No.20102302120054). IBM UR Joint Project (No.MH20110819). Key Laboratory of Data Engineering and Knowledge Engineering (Renmin University of China), Ministry of Education (No.KF2011003).

References

1. Abiteboul, S., Hull, R., Vianu, V.: Foundations of Databases. Addison-Wesley (1995)
2. Garcia-Molina, H., Ullman, J.D., Widom, J.: Database System Implementation. Prentice-Hall (2000)
3. Li, Y., Wang, H., Gao, H.: Efficient entity resolution based on sequence rules. In: Shen, G., Huang, X. (eds.) CSIE 2011, Part I. CCIS, vol. 152, pp. 381–388. Springer, Heidelberg (2011)
4. Liu, X., Wang, H., Li, J., Gao, H.: Es-join: Similarity join algorithm based on entity. Research Report HITDB-12-001, Harbin Institute of Technology (October 2012)
5. Liu, X., Wang, H., Li, J., Gao, H.: Multi-similarity join order selection in entity database. Journal of Frontiers of Computer Science and Technology 6(10), 865 (2012)
6. Tong, X., Wang, H.: Fgram-tree: An index structure based on feature grams for string approximate search. In: Gao, H., Lim, L., Wang, W., Li, C., Chen, L. (eds.) WAIM 2012. LNCS, vol. 7418, pp. 241–253. Springer, Heidelberg (2012)
7. Tong, X., Wang, H., Li, J., Gao, H.: A top-k query algorithm for weighted string based on the tree structure index. In: National Database Conference of China (2012)
8. Wang, H., Li, J., Wang, J., Gao, H.: Dirty data management in cloud database. In: Grid and Cloud Database Management, pp. 133–150 (2011)
9. Zhang, Y., Yang, L., Wang, H.: Range query estimation for dirty data management system. In: Gao, H., Lim, L., Wang, W., Li, C., Chen, L. (eds.) WAIM 2012. LNCS, vol. 7418, pp. 152–164. Springer, Heidelberg (2012)
10. Zhang, Y., Yang, L., Wang, H.: Similarity join size estimation with threshold for dirty data. Journal of Computers 35(10), 2159–2168 (2012)

RelRec: A Graph-Based Triggering Object Relationship Recommender System

Yuwen Dai, Guangyao Li, and Ruoyu Li

Institute of Massive Computing
East China Normal University
{littleday.d,leegyao,isabella.voidmain}@gmail.com

Abstract. Numerous services (email, motion sensor, etc.) emerge and tend to function more comprehensively. What comes with this is the increasing attention to collaboration between them. For example, *IFTTT*[1] (IF This Then That) enables people to set triggering relationships between various services to be automatically implemented in the cloud. *RelRec* is an triggering object relationship recommender system, building a bipartite graph representing the relationships between services. We propose an algorithm to rate relationships by similarity, and diversify the results by a modified classic method from graph theory.

1 Introduction

Recommender system is one of the most popular and profitable applications using state-of-art knowledge over the past decades. Many companies have developed systems to provide meaningful recomendations to users. *Amazon*[2] developed commodity recommender services by learning user activities. In the back-end, a number of approaches are devised. Collaborative Filtering is one of the classical ways that cluster either similarity by their preferences to the same products or items by their features. Control theory is also employed by monitoring feedbacks from recommendation performance and applying control model on input data [1]. In addition, Graph is often used to design and verify recommender systems [2].

Currently, existing recommender systems are mainly focusing on objects themselves. However, some services about relationships between these objects spring out. *IFTTT* is an online service that automatically executes user-defined triggering relationships between different services. It uses *channel* as the representative of service, like email, motion sensor, or light switch. Every channel provides some functionalities and events, called as *actions* and *triggers* respectively. Moreover, the triggering relationship between a trigger and an action is *recipe*. According to the recipes set by the user, *IFTTT* automatically executes corresponding action when a trigger happens. The huge potential of linking services in this way could port enormous automation and intelligence to daily lives from the cloud, especially after connecting physical world to the virtual world using sensor technologies. Nevertheless, *IFTTT* employs a recommender mechanism that merely displays the most popular recipes, which may be either irrelevant

---
[1] http://ifttt.com/
[2] http://www.amazon.com/

or homogeneous because of *Matthew Effect* regarding the usage or the convergence of recipes with similar popular features.

In this paper, we introduce a novel triggering relationship recommender system - *RelRec*, which recommends recipes rather than conventional objects to *IFTTT* users. *RelRec* takes both ratings and diversity into consideration for recommendation. Our primitive experiments show that usage data is sufficient to rate recipes and diversification is feasible on graph.

2 Recommendation Mechanism

2.1 Bipartite Graph Model

We implemented *RelRec* using public data from *IFTTT*. Intuitively, the dichotomous feature of recipes (i.e. triggers and actions) is modeled as a complete bipartite graph, with edge weights indicating usage frequencies of corresponding recipes. In particular, we denote C_i ($i = 1, 2, \ldots, N$) one of N channels, and T_i^j ($i = 1, 2, \ldots, N; j = 1, 2, \ldots, P_i$) and A_i^k ($i = 1, 2, \ldots, N; k = 1, 2, \ldots, Q_i$) as triggers and actions of the i-th channel. A recipe between trigger T_u^j and action A_v^k is denoted as $R_{u,v}^{j,k}$. Moreover, the set of all channels, triggers, actions, or recipes, are denoted as $\mathbb{C}, \mathbb{T}, \mathbb{A}, \mathbb{R}$, respectively. Similarly, $\mathbb{C}^+, \mathbb{T}^+, \mathbb{A}^+, \mathbb{R}^+$ denote the corresponding items that only exist in the dataset. In addition, we define the *usage density* of a recipe $R_{u,v}^{j,k}$ as the usage number over a unit time after the trigger is created, so

$$Den_{u,v}^{j,k} = \frac{\text{Usage Count of } R_{u,v}^{j,k}}{\text{Existing Time of } R_{u,v}^{j,k}}, \text{ if } R_{u,v}^{j,k} \in \mathbb{R}^+, \text{ otherwise } 0 \quad (1)$$

where $Den_{u,v}^{j,k}$ is the *usage density*. "Usage Count of $R_{u,v}^{j,k}$" is stored in our dataset and updated regularly. And "Existing Time of $R_{u,v}^{j,k}$" is calculated using difference between created and current time-stamp, and is updated in timely manner as well.

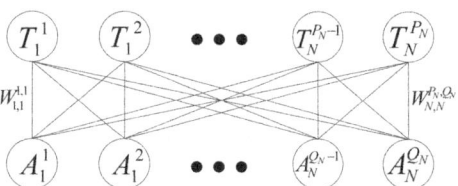

Fig. 1. Complete Bipartite Graph of All Triggers and Actions

Now we are able to build a weighted complete bipartite graph $G = (\mathcal{T} + \mathcal{A}, \mathcal{R})$ (c.f. Figure 1), where $\mathcal{T}$ and $\mathcal{A}$ denote the set of nodes representing all triggers and actions respectively, and $\mathcal{R}$ denotes the set of edges representing all recipes.[3] The weight of every edge is assigned as the *usage density* of the recipe, hence

$$W_{u,v}^{j,k} = Den_{u,v}^{j,k} \tag{2}$$

2.2 Rate Recipes

We assume user firstly choose channel C_r as her interest, so we get a set of triggers $\mathbb{T}_r$ and a set of actions $\mathbb{A}_r$ from C_r. For every trigger in $\mathbb{T}_r$, we rate all actions in $\mathbb{A}$, and choose top-k results from them that are mostly recommended to be connected to this trigger. We also rate all triggers in $\mathbb{T}$ using $\mathbb{A}_r$ symmetrically.

The first step is to compute *similarity* between any one trigger in $\mathbb{T}$ and any one trigger in $\mathbb{T}_r$ based on the *usage densities* calculated in Section 2.1. Specifically, we denote the *usage density* between trigger T and action A as $U_{T,A}$, and the average *usage density* of recipes with action A as $\overline{U_A}$. For one trigger $T_r^k \in \mathbb{T}_r$, and one trigger $T_i^j \in \mathbb{T}$, we use *Adjusted Cosine Similarity* theory [3] to compute their similarity.

$$Sim_{r,i}^{k,j} = \frac{\sum_{Res}[(U_{T_r^k,A} - \overline{U_A}) \cdot (U_{T_i^j,A} - \overline{U_A})]}{\sqrt{\sum_{Res}(U_{T_r^k,A} - \overline{U_A})^2} \cdot \sqrt{\sum_{Res}(U_{T_i^j,A} - \overline{U_A})^2}} \tag{3}$$

where Res is the set of restrictions for each term that is accumulated. Res includes $A \in \mathbb{A}$, $U_{T_r^k,A} \neq 0$, and $U_{T_i^j,A} \neq 0$.

Given any one trigger T_r^k, the *recommendation rate* of action A regarding T_r^k is

$$Rate_{A,T_r^k} = \sum_{T_i^j \in \mathbb{T}} Sim_{r,i}^{k,j} \cdot U_{T_i^j,A} \tag{4}$$

in which $Sim_{r,i}^{k,j}$ is calculate by Equation 3. Thereby, we select top-k actions, and choose corresponding edges for subgraph. After doing these for all P_r triggers in $\mathbb{T}_r$, we will get a bipartite subgraph G^* with $k \times P_r$ edges. The value of $k \times P_r$ is decided by *RelRec* to make sure there are enough recipes to show, unless the maximum value cannot reach the amount of expected recommendations.

2.3 Guarantee Diversity

Diversity is playing a relevant role in recommender systems [2]. Multiple identical triggers or actions are superfluous to inspire users to make the most of *IFTTT* services. By noticing one recommended recipe, user could come up with similar functionalities by herself. For example, if *RelRec* recommends her to update *Twitter*[4] profile picture when *Facebook*[5] profile picture is changed, she can definitely think about updating her *Google+*[6] profile picture without any external hint.

---

[3] For ease of discussion, nodes and corresponding triggers/actions are used interchangeably in this paper, so as edges and corresponding recipes.
[4] http://twitter.com/
[5] http://www.facebook.com/
[6] https://plus.google.com/

According to the subgraph G^* generated in Section 2.2, we can refine the results with respect to diversity in a novel and specific way for relationship recommendations. Diversity is regarded as the large number of kinds of triggers or actions in recommended recipes. So we can formulate our problem as finding most nodes with least edges (i.e. largest $\frac{node\#}{edge\#}$) in bipartite graph, which is almost equivalent to the *Maximum Cardinality Bipartite Matching* problem.

To solve our problem, we employ a modified algorithm based on *Hungarian Method*. The basic idea of *Hungarian Method* is to increase matching size by reversing selections of edges on *Augment Path*. However, after we get the *Maximum Cardinality Bipartite Matching*, if there are not enough recipes to show, we just simply pick one action linked to left-out triggers in the channel. We re-do this method among unchosen edges until we get enough recipes to show or every recipe in G^* is recommended.

3 Demonstration Scenarios

In this demonstration, *RelRec* will be presented in two phases to show its relevance and diversity features comparing to build-in recommendation services of *IFTTT*. We will first set up the back-end components with latest data, and make it update in a timely manner. Figure 2 illustrates the *User Interface*, which will be showed in a browser during demonstration. What shown on the top of the interface are *IFTTT* tabs. A scrollable horizontal list of available channels is under it. After user selects a channel that she is interested in, recommended recipes will show up in the lower pannel as *RelRec* computes. Currently, we display 6 recipes per recommendation. User can choose one recipe button to get into the IFTTT to start these services.

Fig. 2. RelRec User interface and results, including results after rating channels (top), and results after considering diversity (bottom)

References

1. Jambor, T., Wang, J., Lathia, N.: Using control theory for stable and efficient recommender systems. In: Proc. WWW 2012 (2012)
2. Adomavicius, G., Kwon, Y.: Maximizing aggregate recommendation diversity: A graph-theoretic approach. In: Proc. DiveRS 2001 (2001)
3. Sarwar, B., Karypis, G., Konstan, J., Riedl, J.: Item-based collaborative filtering recommendation algorithms. In: Proc. WWW 2001 (2001)

IndoorDB: Extending Oracle to Support Indoor Moving Objects Management

Qianyuan Li, Peiquan Jin, Lei Zhao, Shouhong Wan, and Lihua Yue

School of Computer Science and Technology,
University of Science and Technology of China, 230027, Hefei, China
jpq@ustc.edu.cn

Abstract. Managing moving objects in indoor space has been a research focus in recent years, as most people live and work in indoor space, e.g. working in office, living in apartment, etc. In this paper, we present an extension of Oracle named IndoorDB to support indoor moving objects management in a practical way. The extension is developed as a PL/SQL package and can be integrated into Oracle to offer new data types and operations for indoor location-based queries such as indoor navigation, hot spots detection, KNN, range queries, and so on. After an overview of the general features of IndoorDB, we discuss the architecture and implementation of IndoorDB. And finally, a case study of IndoorDB's demonstration is presented.

1 Introduction

Recently, wireless positioning techniques like RFID, Wifi and Bluetooth offer opportunities for us to track indoor moving objects [1]. Indoor space has some unique features, compared with outdoor space. Firstly, the moving of objects is constrained by rooms and doors. Secondly, the distance measurement is different from that in outdoor space. The latter usually employs the Euclidean distance. However, this is not applicable in indoor space, due to the existence of doors and rooms. Finally, the positioning ways in indoor space usually use sensors like RFID and Bluetooth, which are differing from the GPS receiver in outdoor environment.

Previous research on indoor moving objects management were mainly focused on data models [2], indexes [3], and specific indoor query processing [4], whereas little work has been done in the implementation of real database management systems for indoor moving objects. To our best knowledge, there are no real systems built so far.

Aiming at providing practical support of indoor moving objects management for various indoor LBS applications, we present an extension of Oracle named IndoorDB in this paper. IndoorDB is developed using the cartridge technology provided by Oracle. The unique features of IndoorDB can be summarized as follows:

(1) IndoorDB is SQL-compatible and built on a widely-used commercial RDBMS, i.e. Oracle. Thus it can be easily used in real database applications and provides a practical solution for indoor moving objects management under current database architecture.

(2) IndoorDB supports various moving objects types specially designed for indoor LBS applications, such as *indoor position*, *indoor space*, and *indoor moving object*.

Combined with ten types of extended spatiotemporal operations, various indoor LBS queries can be supported by IndoorDB, such as indoor navigation, indoor KNN, trajectory queries, etc.

2 Implementation of IndoorDB

2.1 Architecture of IndoorDB

IndoorDB is implemented as a data cartridge in Oracle using the object-relational database technologies [5, 6]. The detailed implemental architecture of IndoorDB is shown in Fig.1. The PL/SQL specification provides the signature definition and implementation of all the extended data types and functions in the LayeredModel [7]. The IndoorDB cartridge is the component that actually brings indoor moving objects support into Oracle. Once installed, it becomes an integral part of Oracle, and no external modules are necessary. When IndoorDB is installed into Oracle, users can use standard SQL to store and query indoor moving objects in Oracle. No external work imposes on users.

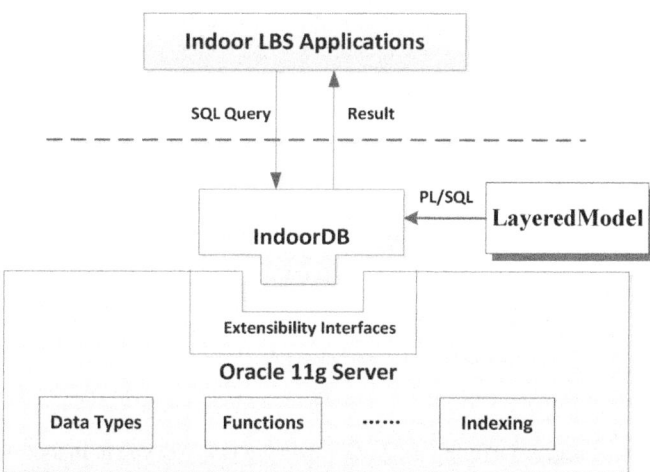

Fig. 1. Architecture of IndoorDB

2.2 Type System of IndoorDB

IndoorDB extends three categories of new data types into Oracle, namely *temporal data types*, *spatial data types*, and *moving objects types* (as shown in Fig.2). The *moving objects types* contain moving base types and an indoor moving object type. The former refers to the numeric, Boolean, or string values changing with time, whereas the latter refers to the indoor moving objects as well as their trajectories. All the new data types are implemented by PL/SQL using the CREATE TYPE statement. Fig.3 shows an example of indoor moving objects and the definition of indoor moving object type in IndoorDB.

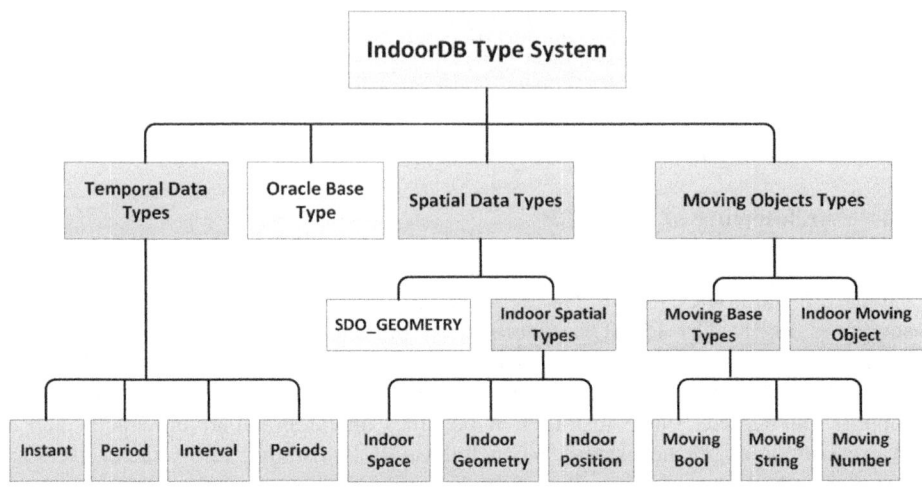

Fig. 2. Type System of IndoorDB

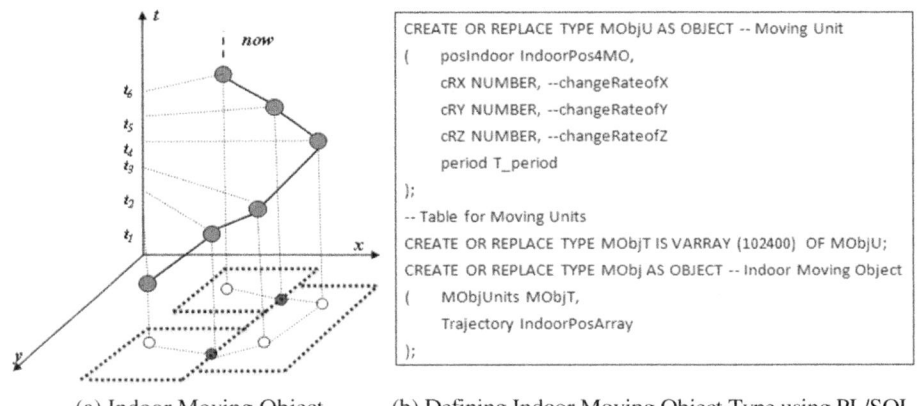

(a) Indoor Moving Object   (b) Defining Indoor Moving Object Type using PL/SQL

Fig. 3. Defining the *Indoor Moving Object* Type in IndoorDB

2.3 Data Operations in IndoorDB

IndoorDB implements ten types of spatiotemporal operations, which are (1) object data management operations, (2) object attribute operations, (3) temporal dimension project operations, (4) value dimension project operations, (5) temporal selection operations, (6) quantification operations, (7) moving Boolean operations, (8) temporal relation operations, (9) object relation operations, and (10) distance operations. All the operations are implemented by PL/SQL and as member functions of extended data types, as shown in Fig.2. For the space limitation, we will not discuss the details about each data operation. However, in the demonstration process, we will show how to use those operations to answer different spatiotemporal queries.

3 Demonstration

We will first use IndoorSTG [8] to generate simulated indoor space and trajectory data, which simulates a 6-floor building consisting of 61 rooms, 7 pass ways, 2 elevators, 74 doors. 95 RFID readers are deployed in the building. We will generate the trajectories of 100 moving objects in one day and then transform them into the database using PL/SQL. After that, we will show how IndoorDB answers different types of indoor LBS queries. All the queries are conducted through a Web-based client interface (see Fig.4).

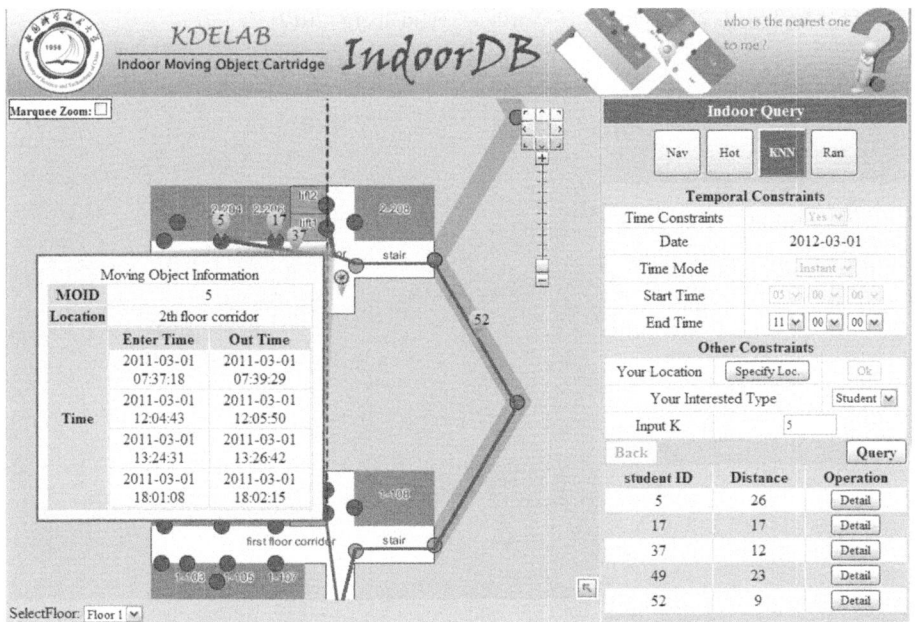

Fig. 4. Client Interface of IndoorDB Demonstration

Acknowledgement. This paper is supported by the National Science Foundation of China (No. 60776801 and No. 71273010), the National Science Foundation of Anhui Province (no. 1208085MG117), and the USTC Youth Innovation Foundation.

References

1. Jensen, C., Hua, L., Bin, Y.: Indoor—A New Data Management Frontier. IEEE Data Engineering Bulletin 33(2), 12–17 (2010)
2. Jensen, C., Hua, L., Bin, Y.: Graph model based indoor tracking. In: Proc. of MDM, Los Alamitos, CA, USA (2009)
3. Jensen, C.S., Lu, H., Yang, B.: Indexing the trajectories of moving objects in symbolic indoor space. In: Mamoulis, N., Seidl, T., Pedersen, T.B., Torp, K., Assent, I. (eds.) SSTD 2009. LNCS, vol. 5644, pp. 208–227. Springer, Heidelberg (2009)

4. Hua, L., Bin, Y., Jensen, C.: Spatio-temporal joins on symbolic indoor tracking data. In: Proc. of ICDE, pp. 816–827 (2011)
5. Zhao, L., Jin, P., Zhang, X., Zhang, L., Wang, H.: STOC: Extending Oracle to Support Spatiotemporal Data Management. In: Du, X., Fan, W., Wang, J., Peng, Z., Sharaf, M.A. (eds.) APWeb 2011. LNCS, vol. 6612, pp. 393–397. Springer, Heidelberg (2011)
6. Jin, P., Sun, P.: OSTM: a Spatiotemporal Extension to Oracle. In: Proc. of NCM. IEEE CS Press, Gyeongju (2008)
7. Jin, P., Zhang, L., Zhao, J., Zhao, L., Yue, L.: Semantics and Modeling of Indoor Moving Objects. International Journal of Multimedia and Ubiquitous Engineering 7(2) (2012)
8. Wang, H., Jin, P., Zhao, L., Zhang, L., Yue, L.: Generating Semantic-Based Trajectories for Indoor Moving Objects. In: Wang, L., Jiang, J., Lu, J., Hong, L., Liu, B. (eds.) WAIM 2011. LNCS, vol. 7142, pp. 13–25. Springer, Heidelberg (2012)

HITCleaner: A Light-Weight Online Data Cleaning System

Hongzhi Wang, Jianzhong Li, Ran Huo, Li Jia, Lian Jin,
Xueying Men, and Hui Xie

Harbin Institute of Technology, China
{wangzh,lijzh}@hit.edu.cn, huoran1988@126.com, 0jialijiali@163.com,
msn19882009@live.cn, ovenwing@yahoo.com.cn, 1587452341@qq.com

Abstract. Data quality is essential in many applications. To reduce the harm of the data in low quality, data cleaning is one of effective solutions. However, existing data clean systems can clean data in some special aspect and require relative complex input. To clean data with complex quality problem for various kinds of users, we develop HITCleaner as a light weight online data cleaning system which could handle various types of data quality problem. HITCleaner provides users an elegant interface to upload dirty data and download cleaned data. It also permits users to clean data with various parameters and components flexibly. In this demonstration, we present a tour of HITCleaner, highlighting a few of its key features. We will demonstrate examples for data cleaning. In particular, we will show the flexibility of HITCleaner for cleaning data.

1 Introduction

Data quality plays an important role in information systems. In many applications, dirty data cause serious problems. Studies in Merrill Lynch [6] report that the 30%-80% time and budget are used in data cleaning instead of system development. Experts estimated that data quality increase 10%-20% costs for each enterprises in average [5].

Because of its importance, data quality draws attentions of the literatures of research and industry. Many systems have been proposed, such as Potter's Wheel [4], AJAX [2]. However, each of current systems focuses on one aspect of data quality problem. Data quality problems have many aspects such as duplication, inaccuracy, inconsistency and incompleteness. Additionally, current systems require users to write statements such as conversion rules or declarative language[4,2]. With the consideration that all users are not experts who even do not know SQL, the data cleaning systems should provide a simple and user-friendly interface.

To make data cleaning effective and easy, we develop HITCleaner at Harbin Institute of Technology, a light-weight online data cleaning system. Comparing with current systems, our system has following benefits.

- *Flexibility.* Since different users may have different requirements for the data cleaning, in this system, we develop data quality detection and cleaning

algorithms for multiple types of data quality problems, including duplication, error, inconsistency and incompleteness.
- *Friendly user interface.* In HITCleaner, users could upload the dirty data file to the system and provide the requirements for data cleaning through web interface. Data cleaning is performed online and user can download clean data file from a web page.
- *High Efficiency.* Since the data in our system require multi-step cleaning, our system applies simple and efficient algorithms for data cleaning. As a result, our system could accomplish the data cleaning task on 1M data within 30 seconds.
- *Light Weight.* Since our system is setup on a server without large disk, the algorithms embedded in our system are all light-weighted. It means that our system does not require a large storage space. It only requires users to upload a file and does not require large support files.

The remaining parts of this paper are organized as followings. Section 2 proposes the architecture of our system. The algorithms used in HITClener are described in Section 3. Section 4 introduces demo scenarios.

2 System Architecture

In this section, we discuss the architecture of our system. To support various types of data quality problems, we design a flexible architecture. The architecture is shown in Figure 1. In our system, each type of data quality problem is handled with one or multiple modules. In the architecture, the interaction modula provides an input interface for the files to clean and the requirements of data cleaning. The result display modula provides a download link for the clean data and a comparison of original data with dirty data.

The modules of entity resolution and truth discovery are used for deduplication, where entity resolution is used to cluster the tuples according to the referred real-world entity and truth discovery determines the true value when conflict occurs. Inconsistency resolution module discovers the part of the data violating dependency rules and correct the data to satisfy the rules. Data imputation module discovers the incomplete parts of data and impute them.

A user could select proper modules meeting the requirements of quality problem of the data. If a user does not know the quality problem of the data. The default processing step is shown as the arrow in the figure. It is because that the complete data will help the entity resolution and inconsistency resolution. And the resolution of conflict data will help the correcting of inconsistency data.

3 Data Cleaning Methods

In this system, different methods are designed for detection and cleaning different types of quality problem. In this section, we will discuss these algorithm briefly.

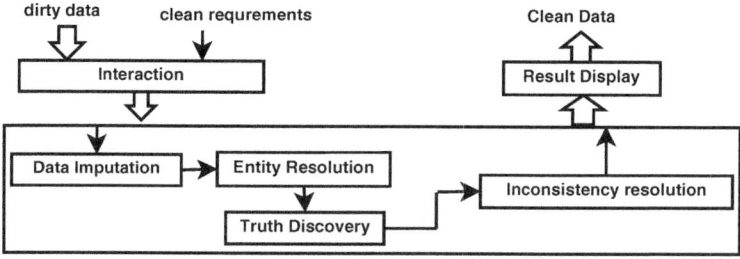

Fig. 1. System Architecture

- *Incompleteness Detection and Data Imputation* According to the type of missing attribute, our system applies two major imputing method, classification and regression. For classification, we use Naive Bayesian classification to deal with the case that both imputed attribute and conditional attributes are classifiable. For regression, we use principle regression to deal with condition where imputed attribute is continuous numeric type.
- *Entity Resolution* We apply a two-phrase entity resolution method. The first focuses on processing the relations and generates similar entity pairs with inverted index on attributes. The second phrase represents the similar pairs generated in the first step as a graph, scan and partition on this graph, each of which represents an entity in the result [3].
- *Truth Discovery* To discover truth for conflict values, we apply an iteration method. We give each value an accuracy and confidence according to the entity resolution results. By weighted voting, based on the accuracy and confidence, the truth could be selected as the value with the max weight. If the confidence and dependency between values are predefined, the context could be used to boost the learning strategy.
- *Inconsistency Discovery and Resolution* Functional dependencies (FD) and Conditional functional dependencies (CFD) are applied to perform the inconsistency discovery and resolution [1]. During inconsistency discovery, a score is assigned to each tuple according to the degree that it violate the rule. During the repairing, we pick the tuple with the smallest score to repair.

4 Demonstration

We plan to demonstrate features of HITCleaner in following 4 parts.

- *Basic Concepts* We demonstrate the basic concept of data cleaning in HITCleaner with a poster, where the system architecture and the flow of algorithms are shown.
- *Standard Data Cleaning* Our system provides non-expert users a default interface for data cleaning. As shown in Figure 2(a), HITCleaner provides a simple interface to upload the file to clean. After cleaning, the user could download the clean results as a flat file.

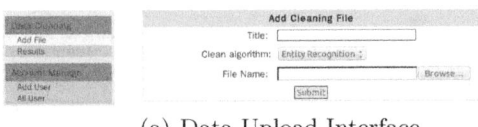

(a) Data Upload Interface

(b) Result Review Interface

Fig. 2. Interfaces of HITCleaner

- *Flexible Data Cleaning* Our system provides friendly interface for users to organize the cleaning components and input the parameters of these components for flexible data cleaning.
- *Data Cleaning Result Review* As the interface shown in Figure 2(b), HIT-Cleaner provides an interface for the comparison between the cleaning results and original data. Such that users can tune the input parameters with the cleaning results.

Acknowledgement. This paper was partially supported by NGFR 973 grant 2012C B316200, NSFC grant 61003046, 61111130189 and NGFR 863 grant 2012AA011004. Doctoral Fund of Ministry of Education of China (No.20102302120054). Key Laboratory of Data Engineering and Knowledge Engineering (Renmin University of China), Ministry of Education (No.KF2011003). the Fundamental Research Funds for the Central Universities(No. HIT. NSRIF. 2013064).

References

1. Fan, W.: Dependencies revisited for improving data quality. In: PODS, pp. 159–170 (2008)
2. Galhardas, H., Florescu, D., Shasha, D., Simon, E., Saita, C.-A.: Declarative data cleaning: Language, model, and algorithms. In: VLDB, pp. 371–380 (2001)
3. Li, L., Wang, H., Gao, H., Li, J.: EIF: A framework of effective entity identification. In: Chen, L., Tang, C., Yang, J., Gao, Y. (eds.) WAIM 2010. LNCS, vol. 6184, pp. 717–728. Springer, Heidelberg (2010)
4. Raman, V., Hellerstein, J.M.: Potter's wheel: An interactive data cleaning system. In: VLDB, pp. 381–390 (2001)
5. Redman, T.C.: Data: An unfolding quality disaster. Information Management Magazine (August 2004)
6. Shilakes, C., Tylman, J.: Enterprise information portals. Merrill Lynch (1998)

Author Index

Abbasi, Rashid I-13
Aberer, Karl II-139
Aksoy, Cem I-299
Allaho, Mohammad Y. II-385
Amaki, Keita I-315
Apers, Peter I-1

Banafaa, Khaled M. I-71
Bao, Zhifeng I-193
Barukh, Moshe Chai II-123
Bellatreche, Ladjel II-64
Benatallah, Boualem II-123
Berkani, Nabila II-64
Bernardino, Jorge II-84
Bi, Yuanjun I-41

Cai, Yi II-179
Cai, Yuanzhe I-25
Chakravarthy, Sharma I-25
Chang, Ya-Hui I-269
Chao, Kun-Mao I-269
Chen, Fangshu I-146
Chen, Feng II-219
Chen, Gang I-424
Chen, Guihai I-176
Chen, Hong II-16
Chen, Hui-zhong II-259
Chen, Jinchuan I-131, I-472, II-108, II-244
Chen, Lei I-361
Chen, Lu I-424
Chen, Luo II-259
Chen, Rui I-392
Chen, Xiaoyun II-324
Chen, Yong-guang II-259
Chen, Yueguo II-108, II-244
Cheng, Jianjun II-324
Chester, Sean I-201
Chin, Alvin II-401
Cui, Jiangtao I-101

Dai, Yuwen II-472
Damiani, Maria Luisa II-450
Daud, Ali I-13

Dayarathna, Miyuru II-164
Deng, Tangjian II-463
Dimitriou, Aggeliki I-299
Ding, Xiaofeng I-346
Du, Juan II-194, II-458
Du, Liang II-401
Du, Xiaoyong I-239, I-472, II-108, II-244

Fang, Qiong II-354
Faust, Martin II-48
Fekete, Alan II-416
Feng, Ling II-454, II-463
Feng, Yansong II-31
Feng, Zhuonan II-454

Gao, Xiaofeng I-176
Gao, Yunjun I-146, I-424
Gu, Xiwu I-71
Gu, Yu II-1, II-301
Guo, Yonggang II-401
Güting, Ralf Hartmut II-450

Han, Hyuck II-416
Hassani, Marwan II-446
Hayduk, Yaroslav I-440
He, Xiaofeng II-370
Higuchi, Ken I-315
Hikida, Satoshi II-99
Huang, Fei I-407
Huang, Liqing I-161
Huang, Zi I-101
Huo, Ran II-481
Huo, Zheng I-377

Jia, Li II-481
Jiang, Di I-209
Jiang, Tao I-424
Jin, Hai I-346
Jin, Lian II-481
Jin, Liang I-3
Jin, Peiquan II-476
Jing, Ning II-259
Jung, Hyungsoo II-416

Khouri, Selma   II-64
Kim, Yunsu   II-446
Kong, Shoubin   II-454
Kou, Yue   II-339
Krueger, Jens   II-48

Le, Hieu Hanh   II-99
Lee, Dik Lun   I-224
Lee, Wang-Chien   II-385
Leng, Mingwei   II-324
Leung, Carson Kai-Sang   I-440
Leung, Kenneth   II-354
Leung, Kenneth Wai-Ting   I-209, I-224
Li, Chen   I-3
Li, Cheng   II-458
Li, Cuiping   II-16
Li, Guangyao   II-472
Li, Hao   I-209
Li, Jiajia   I-456
Li, Jianzhong   II-468, II-481
Li, Jiuyong   I-346
Li, Qianyuan   II-476
Li, Qing   I-424, II-179, II-309
Li, Ruixuan   I-71
Li, Ruoyu   II-472
Li, Xiaodong   II-309
Li, Xuhui   I-193
Li, Yakun   II-468
Li, Yiping   II-454
Li, Yuhua   I-71
Li, Yukun   II-267
Liang, Hongyu   I-331
Lim, Ee-Peng   II-194, II-458
Lin, Huaizhong   I-146
Lin, Qianlu   I-116
Lin, Rung-Ren   I-269
Lin, Xuemin   I-116, II-284
Ling, Tok Wang   I-284
Liu, Fang   I-161
Liu, Jie   I-407
Liu, Jixue   I-346
Liu, Mengchi   I-193
Liu, Qingwei   II-463
Liu, Xueli   II-468
Liu, Yingfan   I-101
Lu, Dongming   I-146
Lu, Hua   I-131
Lu, Xin   I-176
Lu, Zhao   II-431
Lv, Weiming   II-324

Ma, Pengfei   II-210
Mehrotra, Sharad   I-3
Mei, Hong   I-361
Men, Xueying   II-481
Meng, Kangjian   II-401
Meng, Xiaofeng   I-377, I-392
Miao, Xiaoye   I-424
Miklós, Zoltán   II-139
Min, Huaqing   II-179
Muhammad, Faqir   I-13

Ng, Wilfred   I-209, II-354
Nguyen, Quoc Viet Hung   II-139
Nguyen, Thanh Tam   II-139
Nie, Tiezheng   II-339
Nishino, Hiroomi   I-315

Oyama, Satoshi   I-56

Pietsch, Bernhard   II-275
Plattner, Hasso   II-48

Qian, Tieyun   I-193
Qin, Biao   I-239

Rao, Weixiong   I-361
Rasteiro, Deolinda   II-84
Röhm, Uwe   II-416

S Bhowmick, Sourav   II-228
Santos, Ricardo Jorge   II-84
Schwalb, David   II-48
Seidl, Thomas   II-446
Sha, Chaofeng   II-370
Shang, Haichuan   II-284
Shang, Shuo   I-131
Sharaf, Mohamed   I-86
Shen, Derong   II-339
Shen, Xuchuan   II-31
Sheng, Likun   II-16
Shirai, Yasuyuki   I-56
Song, Yuanfeng   II-354
Suzumura, Toyotaro   II-164

Takashima, Hiroyuki   I-56
Tanaka, Katsumi   I-2
Tang, Ruiming   I-284
Tang, Yi   I-161
Tarkoma, Sasu   I-361
Theodoratos, Dimitri   I-299

Thomo, Alex I-201
Tian, Jilei II-31
Tong, Xing II-468
Truong, Ba Quan II-228
Tsuji, Tatsuo I-315
Tsuruma, Koji I-56

Valdés, Fabio II-450
Venkatesh, S. I-201
Vieira, Marco II-84

Wan, Shouhong II-476
Wang, Bin I-254
Wang, Bo I-101
Wang, Botao I-456
Wang, Dong II-31
Wang, Guoren I-456
Wang, Hao II-454, II-463
Wang, Hongzhi II-468, II-481
Wang, Jun II-259
Wang, Li I-41, II-370
Wang, Shan I-239, II-155
Wang, Tao II-179
Wang, Teng II-155
Wang, Wenan II-301
Wang, Xia II-401
Wang, Xiaoyan II-244
Wang, Xiaoye II-267
Wang, Zhigang II-1, II-301
Wei, Jun I-407
Wen, Kunmei I-71
Whitesides, Sue I-201
Wu, Huayu I-284
Wu, Liang II-401
Wu, Weili I-41
Wu, Xiaoying I-299

Xiao, Weidong II-284
Xiao, Yingyuan II-267
Xie, Haoran II-179, II-309
Xie, Hui II-481
Xie, Wei II-458
Xie, Xike I-131, I-472
Xu, Chen I-86, II-219
Xu, Guandong II-401

Yan, Zhixian II-431
Yang, De-Nian II-385
Yang, Long II-468
Yang, Tao II-244
Yang, Xiaochun I-254
Yang, Yongtian I-176
Yao, Yukai II-324
Ye, Dan I-407
Yeom, Heon Y. II-416
Yokota, Haruo II-99
Yu, Ge II-1, II-301, II-339
Yu, Qing I-346
Yuan, Hao I-331
Yuan, Mingxuan I-361
Yue, Lihua II-476

Zhang, Jingjing I-254
Zhang, Min I-472
Zhang, Rong II-370
Zhang, Rui I-377
Zhang, Wenjie I-116, II-284
Zhang, Xiaojian I-392
Zhang, Xiaolu II-108
Zhang, Ying I-116
Zhang, Yinglong II-16
Zhang, Yong II-210
Zhao, Dongyan II-31
Zhao, Lei II-476
Zhao, Pengfei I-224
Zhao, Xiang II-284
Zhao, Xiyan II-267
Zhong, Hua I-407
Zhong, Jiaofei I-176
Zhong, Ming I-193
Zhou, Aoying I-86, II-219, II-370
Zhou, Minqi I-86, II-219
Zhou, Xiaofang I-86
Zhou, Yuanchun II-401
Zhu, Feida II-194, II-458
Zhu, Mingdong II-339
Zhu, Qing II-155
Zhu, Shanfeng II-309
Zimmermann, Roger II-1
Zong, Chuanyu I-254
Zou, Lei II-31, II-108

The manufacturer's authorised representative in the EU is Springer Nature Customer Service Centre GmbH, Europaplatz 3, 69115 Heidelberg, Germany. If you have any concerns regarding our products, please contact ProductSafety@springernature.com

Printed and bound by CPI Group (UK) Ltd, Croydon, CR0 4YY

25/03/2026

02078216-0009